An Introduction to Quantum Theory

An Introduction to Quantum Theory

Jeff Greensite
San Francisco State University, San Francisco, California, USA

IOP Publishing, Bristol, UK

Media content is available from the book information online: https://doi.org/10.1088/978-0-7503-1167-0.

ISBN 978-0-7503-1167-0 (ebook)
ISBN 978-0-7503-1168-7 (print)
ISBN 978-0-7503-1169-4 (mobi)

DOI 10.1088/978-0-7503-1167-0

Version: 20170201

IOP Expanding Physics
ISSN 2053-2563 (online)
ISSN 2054-7315 (print)

British Library Cataloguing-in-Publication Data: A catalogue record for this book is available from the British Library.

Published by IOP Publishing, wholly owned by The Institute of Physics, London

IOP Publishing, Temple Circus, Temple Way, Bristol, BS1 6HG, UK

US Office: IOP Publishing, Inc., 190 North Independence Mall West, Suite 601, Philadelphia, PA 19106, USA

For Phillip, Daniel, and Susanna.

Contents

Part V Advanced topics

Preface

This book is an expanded version of my lectures on undergraduate quantum mechanics, which is a course that I have taught at San Francisco State University for a very long time. While the intended audience is mainly undergraduates, and there is no pretence of rigor, the text does not shy away from abstractions. Linear vector spaces, bra-ket notation, the notion that a physical state is a unit vector in Hilbert space, and the importance of symmetry, are all introduced at an early stage. On the other hand I have tried to avoid presenting quantum mechanics as though it were a branch of mathematics, say by introducing a set of postulates and then working out the consequences. My intention, especially in the early chapters, has been to motivate the rules of quantum mechanics, and its mathematical framework, as a reasonable if not inevitable response to the observed properties of atoms, electrons, and light.

A first semester course based on this text ought to cover the basic principles of quantum mechanics in chapters 1–11, ending with their triumphant application to the spectrum of the hydrogen atom in chapter 12. The second semester covers electron spin, addition of angular momentum, identical particles, perturbation theory and other approximation methods, and scattering theory. There is usually time left in the second semester to cover one or two other topics, such as the interaction of atoms with thermal radiation (chapter 19), quantum entanglement (chapter 24), the Feynman path integral (chapter 26), or how photons emerge from the quantized electromagnetic field (chapter 27). The chapter on the interpretation(s) of quantum mechanics, chapter 25, depends on the preceding chapter 24 on entangled states. Otherwise, chapters 23–27 in Part V (Advanced Topics) can be studied independently. The choice is left to the discretion of the instructor, or the diligent student.

A note on conventions. While SI units are nearly universal, a choice of units is still a matter of taste. In quantum field theory, which is the next step beyond quantum mechanics, Heaviside–Lorentz units are generally preferred. In this book I have used Heaviside–Lorentz units for all cases involving the electromagnetic field. There are simple rules for converting an expression in Heaviside–Lorentz units to or from Gaussian or SI units. These can be found in standard texts on electromagnetism, or online in a Wikipedia article (https://en.wikipedia.org/wiki/Lorentz-Heaviside_units). Regarding operator notation: If Q is an operator, I generally denote the operator in a definite (usually position) representation by $\tilde{Q}$. Thus momentum p becomes operator $\tilde{p} = -i\hbar \mathrm{d}/\mathrm{d}x$. The exception is the position operator $\tilde{x}$ in position representation, where I will usually dispense with the tilde. If Q operates on a bra or ket, which means that the representation is not specified, I will also drop the tilde.

Because this text will be distributed in an e-book format it has been possible to include animations in various places, and I feel these are useful in illustrating topics such as the propagation of wavepackets, both freely and in a box, one-dimensional scattering from a potential well, and resonant scattering in three dimensions. Unfortunately, if the e-book is read on a web browser, there are technical issues which do not arise in a more conventional format. In particular, the animations require Adobe Acrobat. If the e-book is viewed on a browser such as Chrome, Safari, or Firefox, it may be necessary to disable the built-in pdf viewer, and use the Adobe pdf viewer instead. Instructions on how to do this are widely available on the internet (simply google 'how to disable the built-in pdf viewer' followed by the name of your browser). The Mathematica codes which produced those animations can be downloaded from https://dx.doi.org/10.1088/978-0-7503-1167-0 in case the reader would like to experiment with other potentials, or modify the code in some other creative way. The site will also contain a list of corrections to misprints, as they are uncovered. In some cases this textbook, in e-book format, will be freely available to students and faculty through their college or university libraries. I will be pleased if that turns out to be a useful and affordable option for many students.

San Francisco, December 2016

Acknowledgements

I am grateful to the students and colleagues who have provided feedback about the content of this book over the years, and I would also like to thank IoP Publishing for the suggestion that my lecture notes might be worth distributing in the form of an e-book. Thanks are also due to the individuals and organizations who kindly provided some of the illustrations used in this book, and who are noted by name in the corresponding figure captions.

Author biography

Jeff Greensite

Jeff Greensite obtained his doctorate from the University of California at Santa Cruz in 1980, and held research positions at the University of California at Berkeley and the Niels Bohr Institute in Copenhagen before taking up a professorship at San Francisco State University in 1984. His research speciality is in theoretical high-energy physics, especially non-perturbative aspects of quantum chromodynamics and lattice gauge theory, and he is the author of *An Introduction to the Confinement Problem* (Springer, 2011). He lives in San Francisco.

Part I

Fundamentals

IOP Publishing

An Introduction to Quantum Theory

Jeff Greensite

Chapter 1

The classical state

In the first quarter of the 20th century it was discovered that the laws of motion formulated by Galileo, Newton, Lagrange, Hamilton, Maxwell, and many others were inadequate to explain a wide range of phenomena involving electrons, atoms, and light. After a great deal of effort, a new theory (together with a new law of motion) emerged in 1924. That theory is known as 'quantum mechanics', and it is now the basic framework for understanding atomic, nuclear, and subnuclear physics, as well as condensed-matter physics. The laws of motion (due to Galileo, Newton, …) which preceded quantum theory are referred to as 'classical mechanics'.

Although classical mechanics is now regarded as only an approximation to quantum mechanics, it is still true that much of the structure of the quantum theory is inherited from the classical theory that it replaced. So we begin with a lightning review of classical mechanics, whose formulation begins (but does not end!) with Newton's law $F = ma$.

1.1 Baseball, $F = ma$, and the principle of least action

Take a baseball and throw it straight up in the air. After a fraction of a second, or perhaps a few seconds, the baseball will return to your hand. Denote the height of the baseball, as a function of time, as $x(t)$. If we make a plot of x as a function of t, then the curve (we will call it a 'trajectory') in the x–t plane will, in a uniform gravitational field assuming air resistance is negligible, have the form of a parabola. There are an infinite number of possible trajectories. Which one of these the baseball actually follows is determined by the momentum of the baseball at the moment it leaves your hand.

However, if we require that the baseball returns to your hand exactly Δt seconds after leaving it, then there is only one trajectory that the ball can follow. For a baseball moving in a uniform gravitational field it is a simple exercise to determine this trajectory exactly, but we would like to develop a method which can be applied

doi:10.1088/978-0-7503-1167-0ch1

to a particle moving in *any* potential field $V(x)$. So let us begin with Newton's law $F = ma$, which is actually a second-order differential equation

$$m\frac{\mathrm{d}^2x}{\mathrm{d}t^2} = -\frac{\mathrm{d}V}{\mathrm{d}x} \tag{1.1}$$

It is useful to reexpress this second-order equation as a pair of first-order equations

$$\begin{aligned}\frac{\mathrm{d}x}{\mathrm{d}t} &= \frac{p}{m}\\ \frac{\mathrm{d}p}{\mathrm{d}t} &= -\frac{\mathrm{d}V}{\mathrm{d}x}\end{aligned} \tag{1.2}$$

where m is the mass and p is the momentum of the baseball. We want to find the solution of these equations such that $x(t_0) = X_{in}$ and $x(t_0 + \Delta t) = X_f$, where X_{in} and X_f are, respectively, the (initial) height of your hand when the baseball leaves it, and the (final) height of your hand when you catch the ball.[1]

With the advent of the computer, it is often easier to solve equations of motion numerically, rather than struggle to find an analytic solution which may or may not exist (particularly when the equations are non-linear). Although the object of this section is not really to develop numerical methods for solving problems in baseball, we will, for the moment, proceed as though it were. To make the problem suitable for a computer, divide the time interval Δt into N smaller time intervals of duration $\epsilon = \Delta t/N$, and denote, for $n = 0, 1, \ldots, N$,

$$\begin{aligned}&t_n \equiv t_0 + n\epsilon,\\ &x_n = x(t_n), \qquad p_n = p(t_n),\\ &x_0 = X_{in}, \qquad x_N = X_f\end{aligned}$$

An approximation to a continuous trajectory $x(t)$ is given by the set of points $\{x_n\}$ connected by straight lines, as shown in figure 1.1. We can likewise approximate derivatives by finite differences, i.e.

$$\begin{aligned}\left(\frac{\mathrm{d}x}{\mathrm{d}t}\right)_{t=t_n} &\to \frac{x(t_{n+1}) - x(t_n)}{\epsilon} = \frac{x_{n+1} - x_n}{\epsilon}\\ \left(\frac{\mathrm{d}p}{\mathrm{d}t}\right)_{t=t_n} &\to \frac{p(t_{n+1}) - p(t_n)}{\epsilon} = \frac{p_{n+1} - p_n}{\epsilon}\\ \left(\frac{\mathrm{d}^2x}{\mathrm{d}t^2}\right)_{t=t_n} &\to \frac{1}{\epsilon}\left\{\left(\frac{\mathrm{d}x}{\mathrm{d}t}\right)_{t=t_n} - \left(\frac{\mathrm{d}x}{\mathrm{d}t}\right)_{t=t_{n-1}}\right\}\\ &\to \frac{1}{\epsilon}\left\{\frac{(x_{n+1} - x_n)}{\epsilon} - \frac{(x_n - x_{n-1})}{\epsilon}\right\}\end{aligned} \tag{1.3}$$

[1] We will allow these positions to be different, in general, since you might move your hand to another position while the ball is in flight.

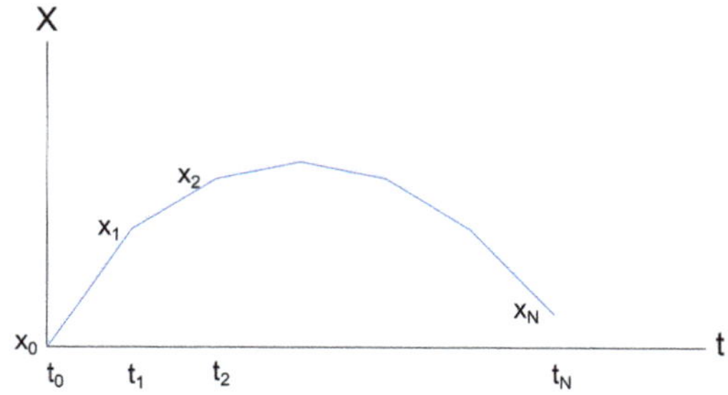

Figure 1.1. A discrete approximation to the continuous path $x(t)$.

and integrals by sums

$$\int_{t_0}^{t_0+\Delta t} \mathrm{d}t\, f(t) \to \sum_{n=0}^{N-1} \epsilon f(t_n) \tag{1.4}$$

where $f(t)$ is any function of time. As we know from elementary calculus, the right-hand side of (1.3) and (1.4) equals the left-hand side in the limit that $\epsilon \to 0$, which is also known as the ***continuum limit***.

We can now approximate the laws of motion, by replacing time derivatives in (1.2) by the corresponding finite differences, and find

$$\begin{aligned} x_{n+1} &= x_n + \left(\frac{p_n}{m}\right)\epsilon \\ p_{n+1} &= p_n - \left(\frac{\mathrm{d}V(x_n)}{\mathrm{d}x_n}\right)\epsilon \end{aligned} \tag{1.5}$$

These are iterative equations. Given position x and momentum p at time $t = t_n$, we can use (1.5) to find the position and momentum at time $t = t_{n+1}$. The finite difference approximation of course introduces a slight error; x_{n+1} and p_{n+1}, computed from x_n and p_n by (1.5) will differ from their exact values by an error of order ϵ^2. This error can be made negligible, in principle, by taking ϵ sufficiently small. It is then possible to use the computer to find an approximation to the trajectory by the 'shooting' method. The method is to make a guess for the initial momentum $p_0 = P_0$, and then use (1.2) to solve for x_1, p_1, x_2, p_2, and so on, until x_N, p_N. If $x_N \approx X_f$, then stop; the set $\{x_n\}$ is the (approximate) trajectory. If not, make a different guess $p_0 = P'_0$, and solve again for $\{x_n, p_n\}$. By trial and error, one can eventually converge on an initial choice for p_0 such that $x_N \approx X_f$. For that choice of initial momentum, the corresponding set of points $\{x_n\}$, connected by straight-line segments, gives the approximate trajectory of the baseball. This process is illustrated in figure 1.2.

Now let us return to the second-order form of Newton's laws, written in equation (1.1). Remarkably, $F = ma$ can be expressed as the condition that a certain function is stationary. Again using (1.3) to replace derivatives by finite differences, the equation $F = ma$ at each time t_n becomes

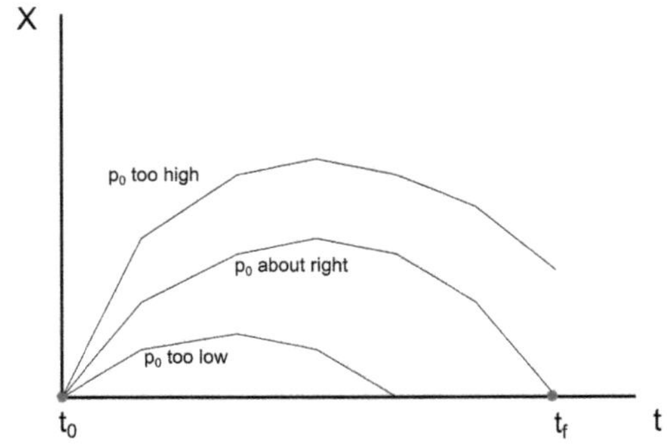

Figure 1.2. The 'shooting method' for finding a trajectory with fixed end points.

$$ma_n \equiv \frac{m}{\epsilon}\left\{\frac{x_{n+1}-x_n}{\epsilon} - \frac{x_n - x_{n-1}}{\epsilon}\right\} = -\frac{\mathrm{d}V(x_n)}{\mathrm{d}x_n} \tag{1.6}$$

The equations have to be solved for $n = 1, 2, \ldots, N-1$, with $x_0 = X_{in}$ and $x_N = X_f$ kept fixed. Now notice that equation (1.6) can be written as a total derivative

$$\frac{\mathrm{d}}{\mathrm{d}x_n}\left\{\frac{1}{2}m\frac{(x_{n+1}-x_n)^2}{\epsilon} + \frac{1}{2}m\frac{(x_n - x_{n-1})^2}{\epsilon} - \epsilon V(x_n)\right\} = 0 \tag{1.7}$$

so that $F = ma$ at some time t_n can be interpreted as a condition that the function in brackets $\{\ldots\}$ should be stationary with respect to variations in the position x_n. But what about all other times? We now introduce a very important expression, crucial in both classical and quantum physics, which is known as the ***action*** of the trajectory. The action is a function which depends on all the points $\{x_n\}$, $n = 0, 1, \ldots, N$ of the trajectory, and in this case it is

$$S[\{x_i\}] \equiv \sum_{n=0}^{N-1}\left[\frac{1}{2}m\frac{(x_{n+1}-x_n)^2}{\epsilon} - \epsilon V(x_n)\right] \tag{1.8}$$

Then Newton's law $F = ma$ can be restated as the condition that the action functional $S[\{x_i\}]$ is stationary with respect to variation of any of the x_i (except for the endpoints x_0 and x_N, which are held fixed). In other words

$$\begin{aligned}
\frac{\mathrm{d}}{\mathrm{d}x_k}S[\{x_i\}] &= \frac{\mathrm{d}}{\mathrm{d}x_k}\sum_{n=0}^{N-1}\left[\frac{1}{2}m\frac{(x_{n+1}-x_n)^2}{\epsilon} - \epsilon V(x_n)\right] \\
&= \frac{\mathrm{d}}{\mathrm{d}x_k}\left\{\frac{1}{2}m\frac{(x_{k+1}-x_k)^2}{\epsilon} + \frac{1}{2}m\frac{(x_k - x_{k-1})^2}{\epsilon} - \epsilon V(x_k)\right\} \\
&= \epsilon\{-ma(t_k) + F(t_k)\} \\
&= 0 \qquad \text{for } k = 1, 2, \ldots, N-1
\end{aligned} \tag{1.9}$$

This set of conditions is known as the ***principle of least action***. It is the principle that the action S is stationary at any trajectory $\{x_n\}$ satisfying the equations of motion $F = ma$, equation (1.6), at every time $\{t_n\}$.

1.2 Euler–Lagrange and Hamilton's equations

In brief, the Euler–Lagrange equations are the second-order form of the equations of motion (1.1), while Hamilton's equations are the first-order form (1.2). In either form, the equations of motion can be regarded as a consequence of the principle of least action. We will now re-write those equations in a very general way, which can be applied to any mechanical system, including those which are much more complicated than a baseball.

We begin by writing

$$S[\{x_i\}] = \sum_{n=0}^{N-1} \epsilon L[x_n, \dot{x}_n] \tag{1.10}$$

where

$$L[x_n, \dot{x}_n] = \frac{1}{2} m\dot{x}_n^2 - V(x_n) \tag{1.11}$$

and where

$$\dot{x}_n \equiv \frac{x_{n+1} - x_n}{\epsilon} \tag{1.12}$$

$L[x_n, \dot{x}_n]$ is known as the ***Lagrangian*** function. Then the principle of least action requires that, for each k, $1 \leqslant k \leqslant N - 1$,

$$\begin{aligned} 0 = \frac{\mathrm{d}}{\mathrm{d}x_k} S[\{x_i\}] &= \sum_{n=0}^{N-1} \epsilon \frac{\mathrm{d}}{\mathrm{d}x_k} L[x_n, \dot{x}_n] \\ &= \epsilon \frac{\partial}{\partial x_k} L[x_k, \dot{x}_k] + \sum_{n=0}^{N-1} \epsilon \frac{\partial L[x_n, \dot{x}_n]}{\partial \dot{x}_n} \frac{d\dot{x}_n}{dx_k} \end{aligned} \tag{1.13}$$

and, since

$$\frac{\mathrm{d}\dot{x}_n}{\mathrm{d}x_k} = \begin{cases} \dfrac{1}{\epsilon} & n = k - 1 \\ -\dfrac{1}{\epsilon} & n = k \\ 0 & \text{otherwise} \end{cases} \tag{1.14}$$

this becomes

$$\frac{\partial}{\partial x_k} L[x_k, \dot{x}_k] - \frac{1}{\epsilon}\left\{ \frac{\partial}{\partial \dot{x}_k} L[x_k, \dot{x}_k] - \frac{\partial}{\partial \dot{x}_{k-1}} L[x_{k-1}, \dot{x}_{k-1}] \right\} = 0 \tag{1.15}$$

Recalling that $x_n = x(t_n)$, this last equation can be written

$$\left(\frac{\partial L[x, \dot{x}]}{\partial x}\right)_{t=t_n} - \frac{1}{\epsilon}\left\{ \left(\frac{\partial L[x, \dot{x}]}{\partial \dot{x}}\right)_{t=t_n} - \left(\frac{\partial L[x, \dot{x}]}{\partial \dot{x}}\right)_{t=t_n-\epsilon} \right\} = 0 \tag{1.16}$$

This is the Euler–Lagrange equation for the baseball. It becomes simpler when we take the $\epsilon \to 0$ limit (the 'continuum' limit). In that limit, we have

$$\begin{gathered} \dot{x}_n = \frac{x_{n+1} - x_n}{\epsilon} \to \dot{x}(t) = \frac{\mathrm{d}x}{\mathrm{d}t} \\ S = \sum_{n=1}^{N-1} \epsilon L[x_n, \dot{x}_n] \to S = \int_{t_0}^{t_0+\Delta t} \mathrm{d}t\ L[x(t), \dot{x}(t)] \end{gathered} \tag{1.17}$$

where the Lagrangian function for the baseball is

$$L[x(t), \dot{x}(t)] = \frac{1}{2} m\dot{x}^2(t) - V[x(t)] \tag{1.18}$$

and the Euler–Lagrange equation, in the continuum limit, becomes

$$\frac{\partial L}{\partial x(t)} - \frac{\mathrm{d}}{\mathrm{d}t}\frac{\partial L}{\partial \dot{x}(t)} = 0 \tag{1.19}$$

In taking partial derivatives of L with respect to x and $\dot{x}$, it is important to understand that we have to pretend that x and $\dot{x}$ are independent variables which have nothing to do with one another, despite the fact that $\dot{x}$ is really the time derivative of x. That is just the meaning of the partial derivative notation.

For the Lagrangian of the baseball, equation (1.18), the relevant partial derivatives are

$$\begin{gathered} \frac{\partial L}{\partial x(t)} = -\frac{\mathrm{d}V[x(t)]}{\mathrm{d}x(t)} \\ \frac{\partial L}{\partial \dot{x}(t)} = m\dot{x}(t) \end{gathered} \tag{1.20}$$

which, when substituted into equation (1.19) give

$$m\frac{\partial^2 x}{\partial t^2} + \frac{\mathrm{d}V}{\mathrm{d}x} = 0 \tag{1.21}$$

This is simply Newton's law $F = ma$, in the second-order form of equation (1.1).

We now want to rewrite the Euler–Lagrange equation in first-order form. Of course, we already know the answer, which is equation (1.2), but let us 'forget' this answer for a moment, in order to introduce a very general method. The reason the Euler–Lagrange equation is second-order in the time derivatives is that $\partial L/\partial \dot{x}$ is first-order in the time derivative. So let us define the ***momentum*** corresponding to the coordinate x to be

$$p \equiv \frac{\partial L}{\partial \dot{x}} \tag{1.22}$$

This gives p as a function of x and $\dot{x}$, but, alternatively, we can solve for $\dot{x}$ as a function of x and p, i.e.

$$\dot{x} = \dot{x}(x, p) \tag{1.23}$$

Next, we introduce the ***Hamiltonian*** function

$$H[p, x] = p\dot{x}(x, p) - L[x, \dot{x}(x, p)] \tag{1.24}$$

Since $\dot{x}$ is a function of x and p, H is also a function of x and p.

The reason for introducing the Hamiltonian is that its first derivatives with respect to x and p have a remarkable property; namely, on a trajectory satisfying the Euler–Lagrange equations, the x and p derivatives of H are proportional to the *time* derivatives of p and x. To see this, first differentiate the Hamiltonian with respect to p,

$$\begin{aligned}\frac{\partial H}{\partial p} &= \dot{x} + p\frac{\partial \dot{x}(x, p)}{\partial p} - \frac{\partial L}{\partial \dot{x}}\frac{\partial \dot{x}(p, x)}{\partial p} \\ &= \dot{x}\end{aligned} \tag{1.25}$$

where we have applied (1.22). Next, differentiating H with respect to x,

$$\begin{aligned}\frac{\partial H}{\partial x} &= p\frac{\partial \dot{x}(x, p)}{\partial x} - \frac{\partial L}{\partial x} - \frac{\partial L}{\partial \dot{x}}\frac{\partial \dot{x}(p, x)}{\partial x} \\ &= -\frac{\partial L}{\partial x}\end{aligned} \tag{1.26}$$

Using the Euler–Lagrange equation (1.19) (and this is where the equations of motion enter), we find

$$\begin{aligned}\frac{\partial H}{\partial x} &= -\frac{\mathrm{d}}{\mathrm{d}t}\frac{\partial L}{\partial \dot{x}} \\ &= -\frac{\mathrm{d}p}{\mathrm{d}t}\end{aligned} \tag{1.27}$$

Thus, with the help of the Hamiltonian function, we have rewritten the single second-order Euler–Lagrange equation (1.19) as a pair of first-order differential equations

$$\begin{aligned}\frac{\mathrm{d}x}{\mathrm{d}t} &= \frac{\partial H}{\partial p} \\ \frac{\mathrm{d}p}{\mathrm{d}t} &= -\frac{\partial H}{\partial x}\end{aligned} \tag{1.28}$$

which are known as ***Hamilton's equations***.

For a baseball, the Lagrangian is given by equation (1.18), and therefore the momentum is

$$p = \frac{\partial L}{\partial \dot{x}} = m\dot{x} \tag{1.29}$$

This is inverted to give

$$\dot{x} = \dot{x}(p, x) = \frac{p}{m} \tag{1.30}$$

and the Hamiltonian is

$$\begin{aligned} H &= p\dot{x}(x, p) - L[x, \dot{x}(x, p)] \\ &= p\frac{p}{m} - \left[\frac{1}{2}m\left(\frac{p}{m}\right)^2 - V(x)\right] \\ &= \frac{p^2}{2m} + V(x) \end{aligned} \tag{1.31}$$

Note that the Hamiltonian for the baseball is simply the kinetic energy plus the potential energy; i.e. the Hamiltonian is an expression for the total energy of the baseball. Substituting H into Hamilton's equations, one finds

$$\begin{aligned} \frac{\mathrm{d}x}{\mathrm{d}t} &= \frac{\partial}{\partial p}\left[\frac{p^2}{2m} + V(x)\right] = \frac{p}{m} \\ \frac{\mathrm{d}p}{\mathrm{d}t} &= -\frac{\partial}{\partial x}\left[\frac{p^2}{2m} + V(x)\right] = -\frac{\mathrm{d}V}{\mathrm{d}x} \end{aligned} \tag{1.32}$$

which is simply the first-order form of Newton's law (1.2).

1.3 Classical mechanics in a nutshell

All the machinery of the least action principle, the Lagrangian function, and Hamilton's equations, is overkill in the case of a baseball. In that case, we knew the equation of motion from the beginning. But for more involved dynamical systems, involving, say, wheels, springs, levers, and pendulums, all coupled together in some complicated way, the equations of motion are often far from obvious, and what is needed is some systematic way to derive them.

For any mechanical system, the ***generalized coordinates*** $\{q^i\}$ are a set of variables needed to describe the configuration of the system at a given time. These could be a set of Cartesian coordinates of a number of different particles, or the angular displacement of a pendulum, or the displacement of a spring from equilibrium, or all of the above. The dynamics of the system, in terms of these coordinates, is given by a Lagrangian function L, which depends on the generalized coordinates $\{q^i\}$ and their first time derivatives $\{\dot{q}^i\}$. Normally, in non-relativistic mechanics, we first specify

1. **The Lagrangian**

$$L\left[\left\{q^i, \dot{q}^i\right\}\right] = \text{Kinetic Energy} - \text{Potential Energy} \tag{1.33}$$

One then defines

2. **The action**

$$S = \int \mathrm{d}t\, L\left[\left\{q^i, \dot{q}^i\right\}\right] \tag{1.34}$$

From the least action principle, following a method similar to the one we used for the baseball, we derive

3. **The Euler–Lagrange equations**

$$\frac{\partial L}{\partial q^i} - \frac{\mathrm{d}}{\mathrm{d}t}\frac{\partial L}{\partial \dot{q}^i} = 0 \tag{1.35}$$

These are the second-order equations of motion. To go to the first-order form, first define

4. **The generalized momenta**

$$p_i \equiv \frac{\partial L}{\partial \dot{q}^i} \tag{1.36}$$

which can be inverted to give the time derivatives $\dot{q}^i$ of the generalized coordinates in terms of the generalized coordinates and momenta

$$\dot{q}^i = \dot{q}^i\left[\left\{q^n, p_n\right\}\right] \tag{1.37}$$

Viewing $\dot{q}$ as a function of p and q, one then defines

5. **The Hamiltonian**

$$H\left[\left\{q^i, p_i\right\}\right] \equiv \sum_n p_n \dot{q}^n - L\left[\left\{q^i, \dot{q}^i\right\}\right] \tag{1.38}$$

Usually the Hamiltonian has the form

$$H[p, q] = \text{Kinetic Energy} + \text{Potential Energy} \tag{1.39}$$

Finally, the equations of motion in first-order form are given by

6. **Hamilton's equations**

$$\begin{aligned} \dot{q}^i &= \frac{\partial H}{\partial p_i} \\ \dot{p}_i &= -\frac{\partial H}{\partial q^i} \end{aligned} \tag{1.40}$$

If the potential energy does not depend explicitly on time, then Hamilton's equations imply that H is time-independent,

$$\begin{aligned} \frac{\mathrm{d}H}{\mathrm{d}t} &= \sum_i \left\{ \frac{\partial H}{\partial q^i}\frac{\partial q^i}{\partial t} + \frac{\partial H}{\partial p_i}\frac{\partial p_i}{\partial t} \right\} \\ &= \sum_i \left\{ -\frac{\partial p_i}{\partial t}\frac{\partial q^i}{\partial t} + \frac{\partial q^i}{\partial t}\frac{\partial p_i}{\partial t} \right\} \\ &= 0 \end{aligned} \tag{1.41}$$

as we would expect if H is the total energy of the system.

Example: the plane pendulum

Our pendulum is a mass m at the end of a weightless rigid rod of length l, which pivots in a plane around the point P. The 'generalized coordinate', which specifies the position of the pendulum at any given time, is the angle θ between the rod and the vertical axis.

1. The Lagrangian

$$L = \frac{1}{2}ml^2\dot{\theta}^2 - (V_0 - mgl\cos(\theta)) \tag{1.42}$$

   where V_0 is the gravitational potential at the height of point P, which the pendulum reaches at $\theta = \pi/2$. Since V_0 is arbitrary, we will just set it to $V_0 = 0$.

2. The action

$$S = \int_{t_0}^{t_1} \mathrm{d}t \left[\frac{1}{2}ml^2\dot{\theta}^2 + mgl\cos(\theta)\right] \tag{1.43}$$

3. The Euler–Lagrange equations

   We have

$$\begin{aligned} \frac{\partial L}{\partial \theta} &= -\, mgl\sin\theta \\ \frac{\partial L}{\partial \dot{\theta}} &= ml^2\dot{\theta} \end{aligned} \tag{1.44}$$

   and therefore

$$ml^2\ddot{\theta} + mgl\sin\theta = 0 \tag{1.45}$$

   is the Euler–Lagrange form of the equations of motion.

4. The generalized momentum

$$p = \frac{\partial L}{\partial \dot{\theta}} = ml^2\dot{\theta} \tag{1.46}$$

5. The Hamiltonian

   Insert

$$\dot{\theta} = \frac{p}{ml^2} \tag{1.47}$$

   into

$$H = p\dot{\theta} - \left[\frac{1}{2}ml^2\dot{\theta}^2 + mgl\cos(\theta)\right] \tag{1.48}$$

to obtain

$$H = \frac{1}{2}\frac{p^2}{ml^2} - mgl\cos(\theta) \tag{1.49}$$

6. Hamilton's equations

$$\begin{aligned}\dot{\theta} &= \frac{\partial H}{\partial p} = \frac{p}{ml^2} \\ \dot{p} &= -\frac{\partial H}{\partial \theta} = -mgl\sin\theta\end{aligned} \tag{1.50}$$

which are easily seen to be equivalent to the Euler–Lagrange equations.

1.4 The classical state

Prediction is rather important in physics, since the only reliable test of a scientific theory is the ability, given the state of affairs at present, to predict the future.

Stated rather abstractly, the process of prediction works as follows. By a slight disturbance known as a ***measurement***, an object is assigned a mathematical representation which we will call its ***physical state***. The laws of motion are mathematical rules by which, given a physical state at a particular time, one can deduce the physical state of the object at some later time. The later physical state is the prediction, which can be checked by a subsequent measurement of the object.

From the discussion so far, it is easy to see that what is meant in classical physics by the 'physical state' of a system is simply its set of generalized coordinates *and* the generalized momenta $\{q^a, p_a\}$. These are supposed to be obtained, at some time t_0, by the measurement process. Given the physical state at some time t, the state at $t + \epsilon$ is obtained by the rule

$$\begin{aligned}q^a(t+\epsilon) &= q^a(t) + \epsilon\left(\frac{\partial H}{\partial p_a}\right)_t \\ p_a(t+\epsilon) &= p_a(t) - \epsilon\left(\frac{\partial H}{\partial q^a}\right)_t\end{aligned} \tag{1.51}$$

In this way, the physical state at any later time can be obtained (in principle) to an arbitrary degree of accuracy, by making the time-step ϵ sufficiently small (or else, if possible, by solving the equations of motion exactly). Note that the coordinates $\{q^a\}$ alone are *not* enough to specify the physical state, because they are not sufficient to predict the future. Information about the momenta $\{p_a\}$ is also required.

The space of all possible $\{q^a, p_a\}$ is known as ***phase space***. For a single particle moving in three dimensions, there are three components of position and three components of momentum, so the 'physical state' is specified by six numbers (x, y, z, p_x, p_y, p_z), which can be viewed as a point in six-dimensional phase space. Likewise, the physical state of a system of N particles consists of three coordinates for each particle ($3N$ coordinates in all), and three components of momentum for each particle ($3N$ momentum components in all), so the state is given by a set of $6N$ numbers, which can be viewed as a single point in $6N$-dimensional space.

As we will see in the next chapters, classical mechanics fails to predict correctly the behavior of both light and matter at the atomic level, and is replaced by quantum mechanics. But classical and quantum mechanics have a lot in common: they both assign physical states to objects, and these physical states evolve according to differential equations which are first order in the time derivatives. The difference lies mainly in the contrast between a physical state as understood by classical mechanics, the 'classical state', and its quantum counterpart, the 'quantum state'. This difference will be explored in the next few chapters.

Problems

1. Two point-like particles moving in three dimensions have masses m_1 and m_2 respectively, and interact via a potential $V(\mathbf{x}_1 - \mathbf{x}_2)$. Find Hamilton's equations of motion for the particles.
2. Suppose, instead of a rigid rod, the mass of the plane pendulum is connected to point P by a weightless spring. The potential energy of the spring is $\frac{1}{2}k(L - L_0)^2$, where L is the length of the spring, and L_0 is its length when not displaced by an external force. Choosing L and θ as the generalized coordinates, find Hamilton's equations.

IOP Publishing

An Introduction to Quantum Theory

Jeff Greensite

Chapter 2

Origins of quantum mechanics

Where do correct ideas come from? Do they drop from the sky? No! Are they innate in the mind? No! They come from social practice, and from it alone.
Mao Tse-Tung

The suggestion that all matter is composed of atoms, and the name 'atom' itself, are due to the Greek thinkers Democritus and Leucippus, who lived about five centuries before Christ. Not until the 19th century, however, did evidence for this hypothesis begin to accumulate, both from chemistry and from thermodynamics. The atomic hypothesis was first stated concretely by John Dalton in 1803, who based it on the fact that the weights of elements which combine to form chemical compounds are always in definite proportions to one another. The assumption that gases are composed of atoms or molecules was the starting point of the kinetic theory of gases, and was essential to the development of thermodynamics and statistical mechanics by luminaries such as Clausius, Maxwell, and Boltzmann. By the early 20th century, improvements in technology and the discovery of radio-activity had enabled physicists to study in some detail the internal structure of atoms, the mass and charge of the electron, and the interaction of atoms with light.

Certain aspects of atomic physics which emerged from these early investigations were puzzling and even paradoxical, in the sense that the observed behavior of electrons, atoms, and light seemed in contradiction with the known laws of mechanics and electromagnetism. These aspects fell roughly into three categories:

1. **The particle-like behavior of light waves**
   Black-body radiation, the photoelectric effect, the Compton effect.
2. **The puzzling stability of the atom**
   Why does the electron not fall into the nucleus? What is the origin of atomic spectra?

doi:10.1088/978-0-7503-1167-0ch2

3. **The wave-like behavior of particles**
   Electron diffraction.

Quantum mechanics emerged as an attempt to explain these phenomena and, as in the Bible, the story begins with light.

2.1 Black-body radiation

Light, when it encounters an obstacle, appears to leave a shadow, with sharp edges, behind the obstacle. Likewise, if an obstacle is placed in front of a stream of tiny particles all traveling in the same direction, that obstacle will also cast a shadow with sharp edges, in the sense that no particles enter the shadow region. Waves, on the other hand, tend to bend around objects. These commonplace observations suggest that light is a stream of particles.

The question of whether light is a stream of particles, or a form of wave motion, was hotly debated in the 17th century. Philosophers from Aristotle and Lucretius up to Hobbes and Descartes had weighed in on the issue, but the first scientific indication that light has wave properties was due to the Italian physicist Francesco Grimaldi, who tried to carefully investigate whether light rays really travel in straight lines. In the process, in the year 1660, Grimaldi discovered the phenomenon of diffraction, and in particular the dark fringes characteristic of diffraction. Isaac Newton was not only aware of the existence of diffraction, but made further experimental contributions to the subject. Nevertheless, Newton advocated on several grounds the theory that light is a stream of particles. For one thing, he was able to explain both the reflection and refraction of light in terms of the particle theory, and for another, from experiments with multiple prisms (particularly his 'trial 44'), he convinced himself that white light is a stream of colored particles. Still, Newton struggled with diffraction, and was unable to provide a convincing explanation.

One of Newton's contemporaries, Christian Huygens, was an advocate of the wave theory of light. Huygens argued that if light were a stream of particles, then one would expect that when two beams of light crossed one another, those particles would collide and scatter, which is not observed.[1] He then showed that the refraction of light could be explained if light moved at different velocities in different media, and that sharp shadows cast by objects, and the (apparent) lack of bending of light around objects, can be understood if the wavelength of light is very much smaller than the size of those objects.

Huygens turned out to be right. More sensitive optical experiments by Young (1801) and Fresnel clearly demonstrated that light has precisely the interference properties expected of wave motion, and this was followed by the measurements of Foucault (1850), showing that the speed of light in water was different from the speed of light in air, just as Huygens had suggested. Then Maxwell, in 1860, by

[1] In fact Huygens was wrong on this point; light *does* scatter light, but the effect is incredibly weak. It can be calculated in the relativistic quantum theory of light and electrons known as quantum electrodynamics, whose consequences were worked out in 1949. But this is getting a little ahead of our story.

Figure 2.1. The founders of quantum mechanics, gathered together at the 1927 Solvay conference. Back row: A Piccard, E Henriot, P Ehrenfest, E Herzen, Th de Donder, E Schrödinger, J Verschaffelt, W Pauli, W Heisenberg, R Fowler, L Brillouin. Second row: P Debye, M Knudsen, W Bragg, H Kramers, P Dirac, A Compton, L Broglie, M Born, N Bohr. Front row: I Langmuir, M Planck, M Curie, H Lorentz, A Einstein, P Langevin, Ch Guye C Wilson, O Richardson. Photo courtesy of the Niels Bohr Archive, Copenhagen.

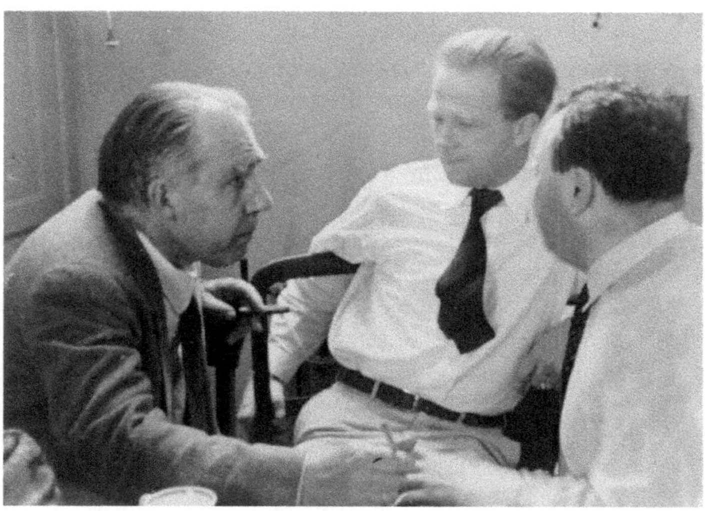

Figure 2.2. Niels Bohr (left), Werner Heisenberg (center), and Wolfgang Pauli (right) in a serious discussion. Photo courtesy of the Niels Bohr Archive, Copenhagen.

unifying and extending the laws of electricity and magnetism, demonstrated that electric and magnetic fields would be able to propagate through space as waves, traveling with a velocity $v = 1/\sqrt{\mu_0\epsilon_0}$, which turned out to equal, within experimental error, the known velocity of light. Experimental confirmation of the existence of electromagnetic waves followed shortly after, and by the 1880s the view that light is a wave motion of the electromagnetic field was universally accepted.

It is a little ironic that following this great triumph of the wave theory of light, evidence began to accumulate that light is, after all, a stream of particles (or, at least, light has particle properties which somehow coexist with its wave properties). The first hint of this behavior came from a study of black-body radiation undertaken by Max Planck, which marks the historical beginning of quantum theory.

Any object, at any finite temperature, emits electromagnetic radiation at all possible wavelengths. The emission mechanism is simple: atoms are composed of negatively charged electrons and positively charged nuclei, and upon collision with other atoms these charges oscillate in some way. According to Maxwell's theory, oscillating charges emit (and can also absorb) electromagnetic radiation. So it is no mystery that if we have a metallic box whose sides are kept at some constant temperature T, the interior of the box will be filled with electromagnetic radiation, which is constantly being emitted and reabsorbed by the atoms which compose the sides of the box. There was, however, some mystery in the energy distribution of this radiation as a function of frequency.

The energy density of radiation in the box, as a function of frequency, is easily worked out using the equipartition principle of statistical mechanics. The total energy is

$$\begin{aligned} E_{\text{rad}} &= \text{no. of degrees of freedom} \times \frac{1}{2}kT \\ &= 2 \times (\text{no. of standing waves}) \times \frac{1}{2}kT \end{aligned} \tag{2.1}$$

where k is Boltzmann's constant and T is the temperature of the box. An electromagnetic field in a box can be thought of as a superposition of an infinite number of standing waves; the 'degrees of freedom' are the amplitudes of each distinct standing wave. The factor of 2 comes from the fact that each standing wave can be in one of two possible polarizations.

It is possible to count the number of standing waves that can exist in a cubical box of volume V, for frequencies in the interval $[f, f + \Delta f]$, and the result is

$$N(f)\Delta f = V\frac{4\pi}{c^3}f^2\,\Delta f \tag{2.2}$$

Then the energy of radiation in this range of frequencies will be

$$\Delta E_{\text{rad}} = 2N(f)\Delta f \times \frac{1}{2}kT = \frac{4\pi kTf^2}{c^3}V\Delta f \tag{2.3}$$

The energy density per unit frequency $\mathcal{E}(f, T)$ is therefore

$$\mathcal{E}(f, T) \equiv \frac{\Delta E_{\mathrm{rad}}}{V\Delta f} = \frac{4\pi k T f^2}{c^3} \tag{2.4}$$

which leads to the Rayleigh–Jeans law for intensity of black-body radiation.

The Rayleigh–Jeans law can be tested by making a hole in the box, and measuring the intensity of radiation emitted from the box as a function of frequency; this intensity is directly proportional to $\mathcal{E}(f, T)$. Radiation from a small hole in a cavity (figure 2.3) is known as 'black-body radiation', because any radiation falling *into* the hole is not reflected out the hole, but is ultimately absorbed by the walls. The experimental result, compared to the prediction, is shown in figure 2.4. Theory disagrees with experiment at high frequencies. In fact, it is clear that there *had* to be something wrong with theory, because the total energy is predicted to be

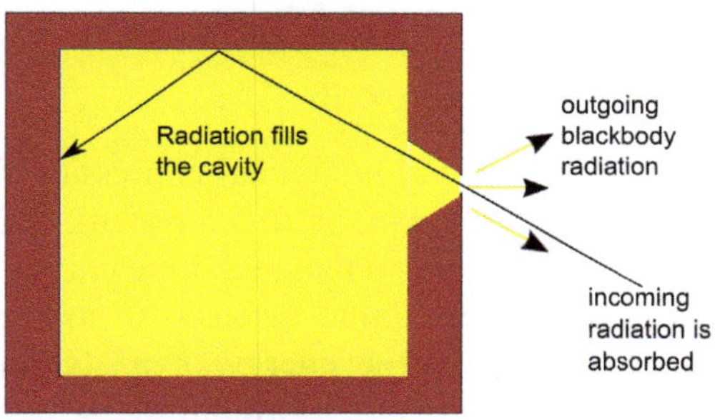

Figure 2.3. By definition a 'black body' absorbs all incident radiation. A cavity with a small opening almost has this property. Thermal radiation escapes through the aperture.

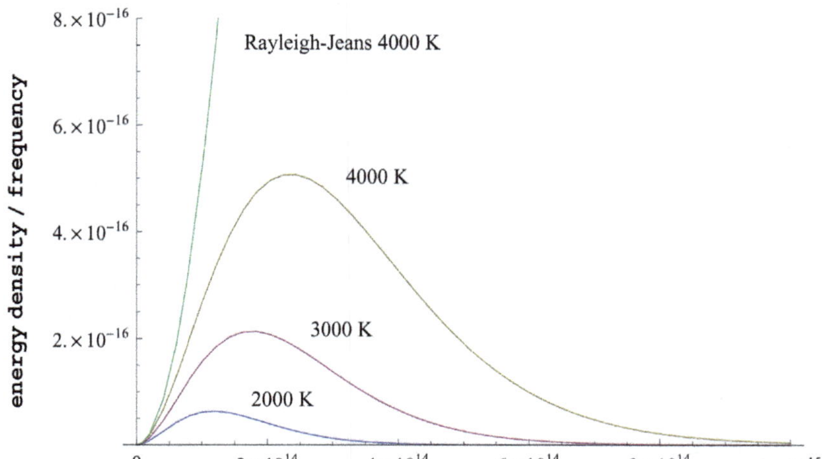

Figure 2.4. Energy density per unit frequency $\mathcal{E}(f, T)$ versus frequency of black-body radiation in a cavity. The distribution at temperatures $T = 2000$, 3000, 4000 K are shown. Also shown is the corresponding Rayleigh–Jeans expression at $T = 4000$ K. Units are SI.

$$E_{\rm rad} = 2 \times (\text{no. of standing waves}) \times \frac{1}{2}kT \qquad (2.5)$$
$$= \infty$$

simply because the range of frequencies is infinite, so there is an infinite number of different standing waves that can be set up in the box. The energy of a box is finite (otherwise its mass, according to special relativity, would be infinite), so this result cannot possibly be correct.

Planck's contribution to this problem was a masterpiece of what is known in physics as ***phenomenology***. The first step of phenomenology is to stare at the data, in this case the experimental curve shown in figure 2.4, and try to find some simple analytical expression that fits it. Planck found that

$$\mathcal{E}(f, T) = \frac{8\pi hf^3}{c^3}\frac{1}{e^{hf/kT} - 1} \qquad (2.6)$$

would do nicely, if one chose the constant h to be

$$h = 6.626 \times 10^{-34}\ \text{J s} \qquad (2.7)$$

The second step is to try to derive the analytical expression for the data, starting from some simple physical assumptions about the system. Planck took aim at the equipartition principle. This principle is only valid if the energy associated with each degree of freedom can take on any value between 0 and ∞, depending on the physical state. In electromagnetism, the energy of a standing wave of a given wavelength is proportional to the square of its amplitude, which can certainly be any number in the range $[0, \infty]$. Planck's suggestion was that, for some unknown reason, the oscillating charges in the walls could only emit or absorb energy in multiples of hf, where f is the frequency of the oscillator. This means that the energy of radiation of frequency f in the box could only have the possible values

$$E_n = nhf \qquad (2.8)$$

where n is an integer. This assumption, combined with the rules of statistical mechanics, is enough to deduce the Planck distribution (2.6).

Note the appearance in Planck's formula of the constant h, known as ***Planck's constant***. It is one of the three most fundamental constants in physics, sharing the honor with c, the speed of light, and G, Newton's constant. All theoretical predictions of quantum physics, to the extent that they disagree with classical physics, have Planck's constant h appearing somewhere in the expression.

2.2 The photoelectric effect

The success of Planck's idea immediately raises the question: why is it that oscillators in the walls can only emit and absorb energies in multiples of hf? The reason for this was supplied by Albert Einstein in 1905, in connection with his explanation of the photoelectric effect.

It was found by Lenard, in 1900, that when light shines on certain metals, the metals emit electrons. This phenomenon is known as the photoelectric effect, and what is surprising about it is that the energy of the emitted electrons is independent of the intensity of the incident light.

The experimental setup, used to measure the energy of the *most energetic* electrons, is shown in figure 2.5(a). The electrons are emitted at the photocathode, fall on a collecting plate, and complete a circuit. The resulting current is measured by an ammeter. A battery can be used to put a potential difference V across the gap between the cathode and the plate, which tends to *repel* the electrons back towards the cathode. For an electron to reach the collecting plate, it must have an energy of *at least* eV, where e is the magnitude of the charge of an electron. As the repelling potential is increased, fewer electrons have sufficient energy to reach the plate, and the current measured by the ammeter decreases. Let V_s denote the voltage where the current just vanishes; i.e. V_s is the lowest voltage such that, at any $V < V_s$, the current is zero. This means that the maximum energy of the emitted electrons is just

$$E_{\max} = eV_s \tag{2.9}$$

It is found experimentally that $E_{\max}$, contrary to expectation, is independent of the intensity of the light falling on the photocathode. As the intensity increases, for fixed light frequency, the *number* of emitted electrons increases, but $E_{\max}$ is constant. On the other hand, when $E_{\max}$ is plotted as a function of frequency, the result is as shown in figure 2.5(b). For frequencies f greater than some minimum $f_0 = E_0/h$, the data for $E_{\max}$ versus f fits a straight line, *whose slope equals Planck's constant!* In other words

$$E_{\max} = hf - E_0 \tag{2.10}$$

Einstein's explanation of this formula is quite simple: light is composed of particles called ***photons***. Each photon has an energy

$$E_{\text{photon}} = hf \tag{2.11}$$

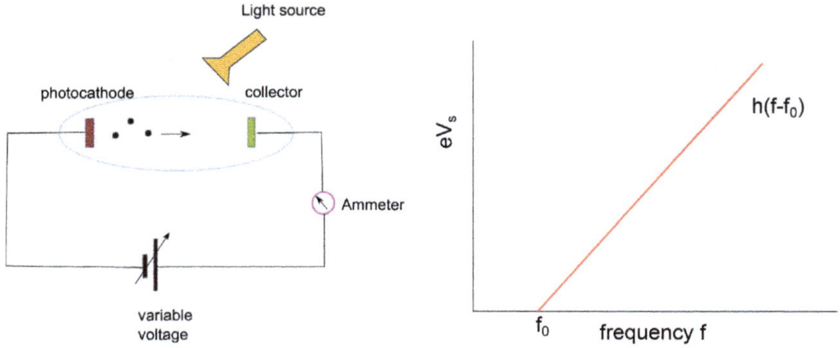

(a) apparatus

(b) maximum electron energy versus frequency

Figure 2.5. The photoelectric effect.

Suppose that the minimum energy required to free an electron from the photocathode is E_0, and an electron absorbs energy from the light by absorbing a photon. Then the maximum possible energy for an emitted electron is the energy of the absorbed photon, less the minimum energy required to free the electron from the metal, i.e.

$$E_{\text{max}} = hf - E_0 \tag{2.10}$$

Raising the intensity of the light simply increases the number of photons hitting the metal per second, which increases the number of emitted electrons per second. But the energy of each electron is dependent only on the energy of the absorbed photon, and this energy depends only on frequency.

Einstein's theory of the photon composition of light immediately explains Planck's condition that the energy of electromagnetic radiation of frequency f, in a box, is restricted to the values

$$E = nhf \tag{2.13}$$

Since there can only be an integer number of photons n at any given frequency, each of energy hf, the energy of the field at that frequency can only be nhf. Planck's restriction on energies is thereby explained in a very natural, appealing way.

Except for one little thing. 'Frequency' is a concept which pertains to waves; yet Einstein's suggestion is that light is composed of particles. The notion that the energy of each 'particle' is proportional to the frequency of the electromagnetic 'wave', which in turn is composed of such 'particles', seems like an inconsistent mix of quite different concepts. However, inconsistent or not, evidence in support of the existence of photons continued to accumulate, as in the case of the Compton effect.

2.3 The Compton effect

Consider an electromagnetic wave incident on an electron at rest. According to classical electromagnetism, the charged electron will begin to oscillate at the frequency of the incident wave, and will therefore radiate further electromagnetic waves at exactly the same frequency as the incident wave. Experiments involving x-rays incident on free electrons show that this is not the case; the x-rays radiated by the electrons are at frequencies lower than that of the incident x-rays. Compton explained this effect in terms of the scattering by electrons of individual photons.

According to special relativity, the relation between energy, momentum, and mass is given by

$$E = \sqrt{p^2c^2 + m^2c^4} \tag{2.14}$$

For particles at rest ($p = 0$), this is just Einstein's celebrated formula $E = mc^2$. For a particle moving at the speed of light, such as a photon, the rest mass $m = 0$; otherwise the momentum would be infinite, since momentum p is related to velocity v via the relativistic expression

$$p = \frac{mv}{\sqrt{1 - \frac{v^2}{c^2}}} \tag{2.15}$$

Then if, for a photon, $m = 0$ and $E = hf$, and given the relation for waves that $v = \lambda f$, we derive a relation between photon momentum and wavelength

$$p = \frac{E}{c} = \frac{hf}{c} = \frac{h}{\lambda} \tag{2.16}$$

where λ is the wavelength of the electromagnetic wave; in this case x-rays.

Now suppose that a photon of the incident x-ray, moving along the z-axis, strikes an electron at rest. The photon is scattered at an angle θ relative to the z-axis, while the electron is scattered at an angle ϕ, as shown in figure 2.6. If $\mathbf{p}_1$ denotes the momentum of the incident photon, $\mathbf{p}_2$ denotes the momentum of the scattered photon, and $\mathbf{p}_e$ is the momentum of the scattered electron, then conservation of momentum tells us that

$$\mathbf{p}_e = \mathbf{p}_1 - \mathbf{p}_2 \tag{2.17}$$

while, from conservation of energy, we have

$$E_1 + mc^2 = E_2 + \sqrt{p_e^2 c^2 + m^2 c^4} \tag{2.18}$$

In this formula E_1 is the energy of the incident photon, and mc^2 is the energy of an electron at rest, while E_2 is the energy of the scattered photon, and $\sqrt{p_e^2 c^2 + m^2 c^4}$ is the energy of the scattered electron. Using the fact that $E = pc$ for photons, we have

$$p_e^2 c^2 + m^2 c^4 = \left[p_1 c - p_2 c + mc^2\right]^2 \tag{2.19}$$

Square the momentum conservation equation to find p_e^2

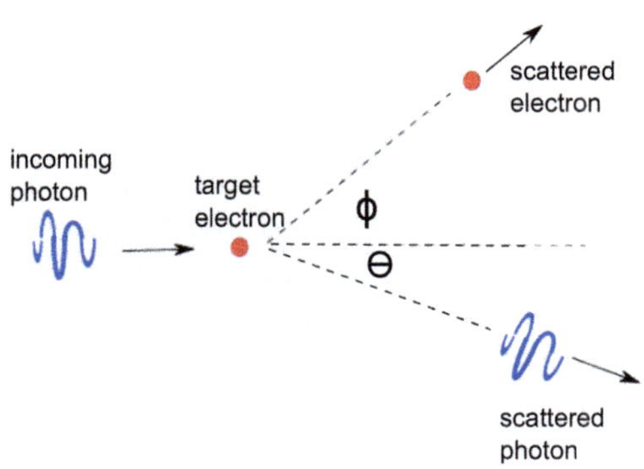

Figure 2.6. The Compton effect: scattering of electrons by photons.

$$\begin{aligned} p_e^2 &= (\mathbf{p}_1 - \mathbf{p}_2) \cdot (\mathbf{p}_1 - \mathbf{p}_2) \\ &= p_1^2 + p_2^2 - 2p_1p_2 \cos\theta \end{aligned} \tag{2.20}$$

and insert the expression for p_e^2 into the energy conservation equation (2.19), to find

$$\frac{1}{p_2} - \frac{1}{p_1} = \frac{1}{mc}(1 - \cos\theta) \tag{2.21}$$

Finally, using the relation (2.16), Compton deduced that

$$\lambda_2 - \lambda_1 = \frac{h}{mc}(1 - \cos\theta) \tag{2.22}$$

which, in fact, agrees with experiment.

2.4 The Heisenberg microscope

The particle-like behavior of light waves has some troubling implications for the concept of a physical state, as understood by classical physics. The classical state of a point-like particle at any moment in time is given by its position *and* its momentum $\mathbf{x}, \mathbf{p}$, which are supposed to be determined by measurement. Of course, this presupposes that one can measure position and momentum simultaneously, to any required degree of precision. But the relation $p = h/\lambda$ suggests that such simultaneous measurements might not be possible, at least, not if such measurements involve the use of light. Roughly, the reason is as follows. In order to determine particle position to an accuracy Δx, it is necessary to use light of wavelength $\lambda < \Delta x$. But that means that the photons composing light of such a wavelength carry momentum $p < h/\Delta x$. In order to observe the particle's position, the particle must scatter light, and in that process the scattered photon can impart much of its momentum to the observed particle, changing the momentum of the particle by some undetermined amount of order $\Delta p \approx h/\Delta x$. The product of the two uncertainties is therefore

$$\Delta x \Delta p \approx \Delta x \frac{h}{\Delta x} = h \tag{2.23}$$

This implies that the more accurately we measure the particle's position (Δx very small), the more uncertainty there is in its momentum (Δp very large), and vice versa.

A concrete, although much oversimplified arrangement for measuring particle position is known as 'Heisenberg's microscope'; the setup is shown in figure 2.7. The idea is to measure the position of a particle moving along the x-axis with a known momentum p_x, to an accuracy Δx. This is done by shining light on the particle. The particle scatters the incident light, and this scattered light is brought to a focal point by a lens of diameter D. However, the ability of a lens to determine position is limited by the resolution of the lens.

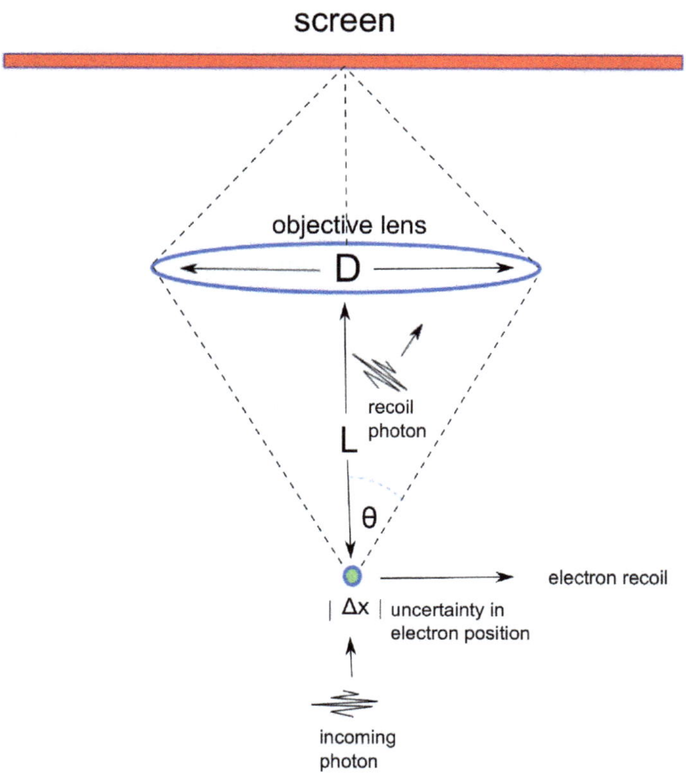

Figure 2.7. A 'thought' experiment: the Heisenberg microscope.

Consider two point sources of light a distance L from the lens, which are separated by an interval Δx. Resolution of these two points requires resolution of an angular separation

$$\phi = \frac{\Delta x}{L} \tag{2.24}$$

On the other hand, according to the Rayleigh criterion, the smallest angular separations which can be resolved by a lens of diameter D is of the order

$$\phi_{\min} \approx \frac{\lambda}{D} \tag{2.25}$$

Equating these two angles, we find that the smallest possible separation between two points which can be resolved by the lens is

$$\Delta x \approx \frac{\lambda}{D/L} \approx \frac{\lambda}{\sin\theta} \tag{2.26}$$

If a lens cannot resolve point sources separated by intervals less than Δx, it is also unable to resolve the position of a single point source to an accuracy better than Δx.

Because a lens focuses all light at a certain image point, we cannot determine at exactly what angle the light was scattered, relative to the y-axis. Scattered light will reach the lens at any scattering angle between 0 and θ. But that means that the x-component of the momentum of the scattered photon is uncertain by approximately

$$\Delta p_x \approx p \sin\theta = \frac{h}{\lambda}\sin\theta \tag{2.27}$$

Multiplying (2.24) by (2.27) gives the uncertainty relation

$$\Delta x \Delta p_x \approx h \tag{2.28}$$

as before.

A physical state is supposed to be determined by measurement, but from the relation (2.28) we see that measurements of position and momentum cannot be made simultaneously, to arbitrarily high precision, if light is used to observe the particle position. This is the first hint that something may be wrong with the classical view of a physical state. Of course, it could be that the relation (2.28) is simply a practical limit on measurement; a particle might have a definite position and momentum despite our inability to measure those quantities simultaneously. But the difficulty could also be much more profound: if a physical state is simply the mathematical representation of the outcome of an accurate measurement process (a view which was advocated in chapter 1) and if accurate numbers (x, p) are *never* an outcome of *any* measurement, then perhaps we are mistaken in thinking that a physical state corresponds to definite values of (x, p). In other words, the origin of the uncertainty could be due to trying to fit a square peg (the true physical state, whatever that may be) into a round hole (the set (x, p)). At the very least, if x and p cannot be measured simultaneously, then there is certainly no experimental *proof* that the classical state is the true physical state. This view is obviously a very radical option; for the moment we only raise it as a possibility, and turn to the mystery of the stability of the atom.

2.5 The Bohr atom

Atoms have radii on the order of 10^{-10} m, and have masses on the order of 10^{-26} kg. In 1911, Ernest Rutherford studied the internal structure of atoms by bombarding gold foil with α-particles from radioactive cesium. By studying the scattering of the α-particles by the gold atoms (a topic we will return to in chapter 22), Rutherford found that almost all the mass of the atom is concentrated in a positively charged nucleus, of radius on the order of 10^{-15} m, i.e. 100 000 times smaller than the radius of the atom itself. The nucleus is surrounded by much lighter, negatively charged electrons, which collectively account for less than 1/2000 of the total mass of the atom. Atomic structure was pictured as analogous to the Solar System, with the nucleus playing the role of the Sun, orbited by much lighter electrons (the 'planets'), bound in their orbits by Coulomb attraction to the nucleus.

However, orbital motion is a form of accelerated motion, and electrons are charged particles. According to electromagnetic theory, an accelerating charged particle radiates electromagnetic waves. As electrons move in their orbits, they should be constantly radiating energy in the form of electromagnetic waves, and as the electrons lose energy, they should spiral into the nucleus; a process which would take about 10^{-11} seconds. By this reasoning, atoms should be about the size of nuclei, but in fact they are roughly 100 000 times larger. So what accounts for the stability of the electron orbits; why don't electrons spiral into the nucleus?

Another mystery of atomic structure was the existence of spectral lines. If a gas is placed in a discharge tube, with a sufficiently large voltage difference maintained at opposite ends of the tube, the gas glows. But, unlike thermal radiation (which occurs due to random collisions among atoms) the light emitted from the discharge tube is not spread diffusely over a broad range of frequencies, but is composed instead of discrete, very definite wavelengths. When such light (in the visible spectrum) is passed through a prism or diffraction grating, it is split into a relatively small number of bright lines, known as 'spectral lines'. It is not hard to understand that an atom, stimulated by an electric current, might emit light; but why should it only emit that light at certain definite frequencies? Empirically, spectroscopists had found that each spectral line of hydrogen could be associated with two integers, m and n, such that the wavelength λ of the spectral line was given by

$$\frac{1}{\lambda} = R_H\left(\frac{1}{n^2} - \frac{1}{m^2}\right) \tag{2.29}$$

where

$$R_H = 109677.576(12)\ \text{cm}^{-1} \tag{2.30}$$

is known as 'Rydberg's constant' for hydrogen. This formula works for hydrogen, modified forms work for certain other gases, but, at the time, nobody could explain its success.

In 1913, Niels Bohr discovered that the formula for the spectral lines of hydrogen could be derived from one crucial assumption about electron orbits: *the angular momentum of an electron can only come in multiples of Planck's constant divided by* 2π. In particular, for circular orbits,

$$L = pr = n\frac{h}{2\pi} \tag{2.31}$$

where p is the electron momentum, and r is the radius of the (assumed) circular orbit. The expression $h/2\pi$ comes up so often in quantum physics that it is given its own symbol

$$\hbar \equiv \frac{h}{2\pi} \tag{2.32}$$

pronounced '*h-bar*'.

The assumption (2.31) can be motivated by the following reasoning. A charged particle rotating in a circular orbit will emit circularly polarized electromagnetic waves which possess both energy and angular momentum. Suppose ΔE is the energy of radiation emitted in a certain time interval Δt. Then according to Maxwell's electrodynamics, the electromagnetic radiation also contains a certain amount of angular momentum ΔL, related to ΔE by the formula

$$\Delta E = 2\pi f \Delta L \tag{2.33}$$

where f is the frequency of the radiation. Now, according to Einstein, the smallest amount of energy emitted is that of one photon, $\Delta E = hf$. Then the smallest amount of angular momentum that could have been emitted would be

$$\Delta L = \frac{h}{2\pi} = \hbar \tag{2.34}$$

As the charged particle radiates photons, it loses angular momentum in units of $\hbar$, and therefore must 'jump' into orbits whose angular momenta are less by integer multiples of $\hbar$. From here one might guess that in the hydrogen atom, the angular momentum of each electron orbit is precisely an integer multiple of $\hbar$. This is the condition (2.31). Let us see how it leads to a prediction for atomic spectra.

The electron is assumed to be moving around the nucleus in a circular orbit. Now, for any circular orbit, the centripetal force must equal the attractive force, i.e.

$$\frac{p^2}{mr} = \frac{e^2}{4\pi r^2} \tag{2.35}$$

in Heaviside–Lorentz units.[2] However, Bohr's quantization condition (2.31) implies

$$p_n = \frac{n\hbar}{r} \tag{2.36}$$

where the subscript indicates that each momentum is associated with a particular integer n. Inserting this expression into (2.35) and solving for r, we find

$$r_n = \frac{4\pi n^2 \hbar^2}{me^2} \tag{2.37}$$

The total energy of the electron, in an orbit of radius r_n, is therefore

$$\begin{aligned} E_n &= \frac{p_n^2}{2m} - \frac{e^2}{4\pi r_n} \\ &= \frac{n^2\hbar^2}{2mr_n^2} - \frac{e^2}{4\pi r_n} \\ &= -\left(\frac{me^4}{2(4\pi)^2\hbar^2}\right)\frac{1}{n^2} \end{aligned} \tag{2.38}$$

The total energy is negative because the electron is in a bound state; energy must be added to bring the electron energy to zero (i.e. to the state of a free electron at rest, far from the nucleus.).

[2] To convert to familiar SI units, simply replace $e^2/4\pi$ by $e^2/4\pi\epsilon_0$ in all expressions where $e^2/4\pi$ appears.

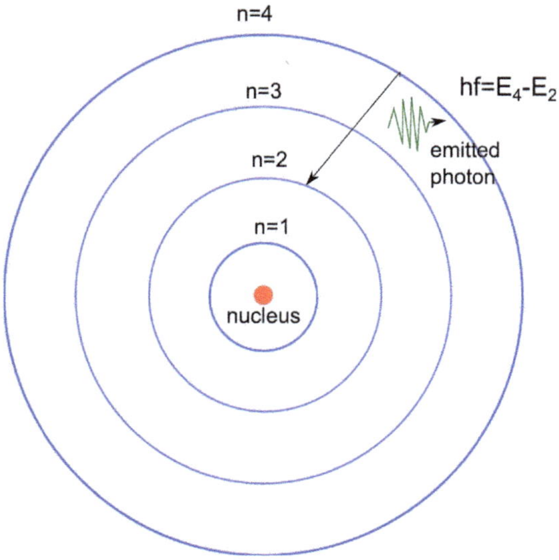

Figure 2.8. Schematic picture of photon emission in the Bohr atom. Unlike the illustration, the radii of orbits are not equally spaced, but increase as n^2.

Bohr's idea was that a hydrogen atom emits a photon when the electron jumps from an energy state E_m, to a lower energy state E_n (figure 2.8). The photon energy is the difference of these two energies

$$\begin{aligned} E_{\text{photon}} &= E_m - E_n \\ hf &= \left(\frac{me^4}{2(4\pi)^2\hbar^2}\right)\left(\frac{1}{n^2} - \frac{1}{m^2}\right) \\ h\frac{c}{\lambda} &= \left(\frac{me^4}{2(4\pi)^2\hbar^2}\right)\left(\frac{1}{n^2} - \frac{1}{m^2}\right) \\ \frac{1}{\lambda} &= \left(\frac{me^4}{2ch(4\pi)^2\hbar^2}\right)\left(\frac{1}{n^2} - \frac{1}{m^2}\right) \end{aligned} \tag{2.39}$$

This final expression not only matches the form of the empirical equation for spectral lines, but an explicit evaluation of the constant factor shows that it is numerically equal to the Rydberg constant

$$R_H = \frac{me^4}{2ch(4\pi)^2\hbar^2} \tag{2.40}$$

In addition to getting the hydrogen spectra right, Bohr's quantization condition (2.31) also implies, if it does not exactly explain, the stability of the atom. For if Bohr's condition is correct, then there is a minimum radius for the electron orbit, namely

$$r_1 = \frac{4\pi\hbar^2}{me^2} = 5.29 \times 10^{-11}\,\text{m} \tag{2.41}$$

which is known as the 'Bohr radius'. The electron can neither spiral nor jump to a radius smaller than this. Therefore, Bohr's theory actually *predicts* the size of the hydrogen atom, which agrees with empirical estimates based on, e.g. the van der Waals equation of state.[3]

After the success of Bohr's model of the atom, attempts were made to generalize his work to elliptical orbits, to include relativistic effects, and to apply it to multi-electron atoms. The outcome of a great deal of effort was this: sometimes the quantization condition worked, and sometimes it did not. Nobody understood the range of validity of the quantization condition, nor why classical mechanics should be subject to such a drastic restriction. A good new idea was needed.

2.6 de Broglie waves

We have already seen that for photons

$$E_{\text{photon}} = hf \qquad \text{and} \qquad p_{\text{photon}} = \frac{h}{\lambda} \tag{2.42}$$

Louis de Broglie's very great contribution to physics, in the year 1924, was the suggestion that these relations also hold true for electrons, i.e.

$$E_{\text{electron}} = hf \qquad \text{and} \qquad p_{\text{electron}} = \frac{h}{\lambda} \tag{2.43}$$

In the case of light, the conceptual difficulty is with the left-hand side of these equations: how can light waves have particle properties? In the case of electrons, it is the right-hand side of the equations which is difficult to understand: what does one mean by the frequency and wavelength of an electron?

de Broglie had the idea that the wave which is somehow associated with an electron would have to be a standing wave along the electron orbits of the hydrogen atom. Now the condition for a standing wave on, e.g. a circular string, is simply that an integral number of wavelengths fit along the circumference, i.e.

$$n\lambda = 2\pi r \tag{2.44}$$

But then, under the assumption that the wavelength of an electron 'wave' is related to its momentum by $p = h/\lambda$, this becomes

$$n\frac{h}{p} = 2\pi r \tag{2.45}$$

or

$$pr = n\hbar \tag{2.46}$$

which is precisely Bohr's quantization condition!

[3] Of course, there is a hidden assumption that $n > 0$, i.e. there are no orbits of radius $r = 0$. In fact, the reader should be aware that despite its success, the Bohr theory is wrong! For example, the orbital angular momentum of the electron in the hydrogen atom ground state is actually zero, which makes no sense at all in Bohr's model. However, we are now roughly following the history, and the historical importance of Bohr's model was immense.

Although deriving the Bohr model is a splendid achievement of de Broglie's assumptions, it is still true that any wave motion should exhibit interference phenomena. By sending light through a pair of slits, Young (in 1801) found the characteristic two-slit interference pattern of wave motion, thereby confirming the wave-like nature of light. What if one does the same experiment with electrons?

Problem

1. Suppose that the condition $L = n\hbar$ applies to the Moon.
   a) What is the quantum number n, taking the Moon's orbit to be exactly circular with radius $R = 385{,}000$ km?
   b) What is the 'Bohr radius' of the Earth–Moon system, taking the Earth and Moon to be point-like objects?

   Helpful constants: Earth mass = 6.0×10^{24} kg, Moon mass = 7.3×10^{22} kg, Newton's constant $G = 6.7 \times 10^{-11}\ \mathrm{m^3\ kg^{-1}\ s^{-2}}$.

IOP Publishing

An Introduction to Quantum Theory

Jeff Greensite

Chapter 3

The wave-like behavior of electrons

Truth is stranger than fiction, but it is because fiction is obliged to stick to possibilities; truth isn't.

Mark Twain

Matter in the solid state consists of atoms in a regular ('crystalline') array of some kind, and the atomic structure of solids is determined by x-ray diffraction. X-rays, being a form of wave motion, reflect off the atoms in the array, and interfere to form a pattern which can be calculated from the principles of physical optics, given a knowledge of the structure of the array, and the wavelength of the x-rays. The inverse problem, finding the structure of the array given the x-ray interference pattern, is the subject of x-ray crystallography.

In 1927 Davisson and Germer, in an effort to check de Broglie's hypothesis that electrons are associated with wave motion, directed a beam of electrons at the surface of crystalline nickel. The electrons were reflected at various angles, and it was found that the intensity of the reflected electrons, as a function of reflection angle, was identical to the interference pattern that would be formed if the beam were instead composed of x-rays. Assuming that the electron beam was indeed some sort of wave, the wavelength could be determined from the intensity pattern of the reflected electrons. The wavelength λ was found to equal, within experimental error, the de Broglie prediction $\lambda = h/p$, where p is the electron momentum, determined by accelerating the incident beam of electrons across a known potential. Apart from brilliantly confirming the existence of 'de Broglie waves', this is an experiment with extraordinary and profound implications. To discuss these implications, it is useful to consider an experimental setup which is simpler to analyze than diffraction from a crystal. In the simplified experiment electrons are directed at an impenetrable barrier containing two, very narrow, slits. (This setup is usually presented as a 'thought' experiment in older textbooks, but in recent times real experiments of this kind have

doi:10.1088/978-0-7503-1167-0ch3

been carried out, see e.g. [1, 2], with results as described below.) First, however, we need an expression for the wave function of de Broglie waves.

3.1 Wave equation for de Broglie waves

Where there is a wave, there is a wave function. The wave function $\psi(x, t)$ of a plane wave of frequency f and wavelength λ may be given, e.g. by a sine wave

$$\sin(kx - \omega t) \tag{3.1}$$

where k is the wavenumber

$$k = \frac{2\pi}{\lambda} = \frac{p}{\hbar} \tag{3.2}$$

and ω is the angular frequency

$$\omega = 2\pi f = \frac{E}{\hbar} \tag{3.3}$$

and where we have used the de Broglie relations to express λ and f in terms of electron momentum p and energy E. Then the electron wave function could have the form of a sine wave

$$\psi(x, t) = \sin(kx - \omega t) = \sin\left(\frac{px - Et}{\hbar}\right) \tag{3.4}$$

or a cosine

$$\psi(x, t) = \cos(kx - \omega t) = \cos\left(\frac{px - Et}{\hbar}\right) \tag{3.5}$$

or any linear combination of sine and cosine, including the complex function

$$\psi(x, t) = \mathrm{e}^{\mathrm{i}(px - Et)/\hbar} \tag{3.6}$$

Normally we would rule out a complex wave function of the form (3.6), on the grounds that, e.g. the displacement of a vibrating string, or the strength of electric and magnetic fields in a radio wave, or the pressure variation in a sound wave, are strictly real quantities.

Given the wave function, what is the wave equation? Waves on strings, sound waves, and light waves, all satisfy wave equations of the form

$$\frac{\partial^2 \psi}{\partial t^2} = \alpha \frac{\partial^2 \psi}{\partial x^2} \tag{3.7}$$

where α is a constant. However, if we plug the sine-wave form (3.4) into this expression we find

$$\frac{E^2}{\hbar^2} \sin \frac{px - Et}{\hbar} = \alpha \frac{p^2}{\hbar^2} \sin \frac{px - Et}{\hbar} \tag{3.8}$$

which implies

$$E^2 = \alpha p^2 \tag{3.9}$$

The same result is obtained, of course, for the cosine. This expression for the energy of a non-relativistic particle in terms of its momentum is simply wrong. For a particle of mass m, the correct expression is

$$E = \frac{p^2}{2m} \tag{3.10}$$

In order to recover this expression, we need a wave equation with one time derivative (which brings down one factor of E), and two space derivatives (which bring down a factor of p^2), i.e.

$$\frac{\partial\psi}{\partial t} = \alpha\frac{\partial^2\psi}{\partial x^2} \tag{3.11}$$

A sine-function will not satisfy this equation, however, since we end up with a cosine on the left-hand side, and a sine on the right-hand side; and vice versa for the cosine wave function. The wave function which *does* work is the complex form (3.6), which, when inserted into (3.11) gives

$$\frac{-\mathrm{i}E}{\hbar}\mathrm{e}^{\mathrm{i}(px-Et)/\hbar} = \alpha\frac{-p^2}{\hbar^2}\mathrm{e}^{\mathrm{i}(px-Et)/\hbar} \tag{3.12}$$

Setting $E = p^2/2m$, we can solve for α to find

$$\frac{\partial\psi}{\partial t} = \frac{\mathrm{i}\hbar}{2m}\frac{\partial^2\psi}{\partial x^2} \tag{3.13}$$

or, equivalently,

$$\mathrm{i}\hbar\frac{\partial\psi}{\partial t} = -\frac{\hbar^2}{2m}\frac{\partial^2\psi}{\partial x^2} \tag{3.14}$$

This is the wave equation for de Broglie waves moving in one dimension. The generalization to three dimensions is a wave function of the form

$$\psi(\mathbf{x}, t) = \mathrm{e}^{\mathrm{i}(\mathbf{k}\cdot\mathbf{x}-\omega t)} = \mathrm{e}^{\mathrm{i}(\mathbf{p}\cdot\mathbf{x}-Et)/\hbar} \tag{3.15}$$

which satisfies the wave equation

$$\mathrm{i}\hbar\frac{\partial\psi}{\partial t} = -\frac{\hbar^2}{2m}\nabla^2\psi \tag{3.16}$$

It is important to realize that in contrast to waves in classical physics, the amplitude of de Broglie waves is *necessarily* complex.

3.2 The double-slit experiment

We consider a beam of electrons which, after acceleration across a potential difference V, acquires a momentum in the x-direction

$$p = \sqrt{2meV} \tag{3.17}$$

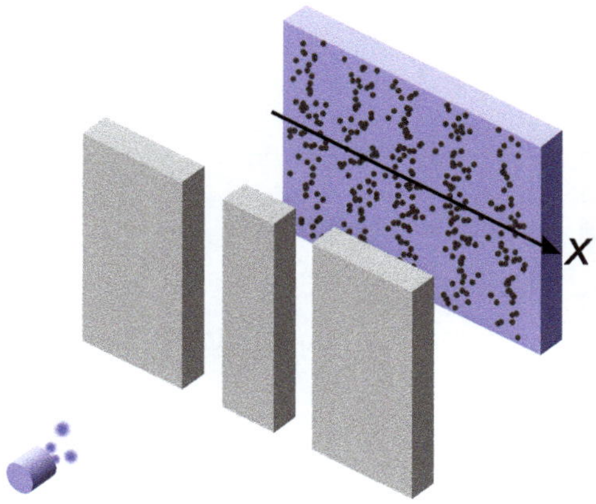

Figure 3.1. A beam of electrons encounters a two-slit barrier. Some of the electrons pass through the slits, and their positions are recorded on a photographic plate.

The beam is directed at a barrier with two horizontal slits of negligible width, separated by a distance d. Those electrons which pass through the slits eventually fall on a screen, which is lined with detectors (Geiger counters or photographic plates), to record the number of electrons falling on the screen, as a function of the distance along the screen (see figure 3.1).

Let us compute the amplitude of the de Broglie wave at the screen, without concerning ourselves, for the moment, with the connection between the de Broglie wave and the actual motion of the electron. Before encountering the barrier, the de Broglie wave is a plane wave moving in the x-direction, as shown in figure 3.2. On the other side of the barrier, the two slits act as coherent sources of cylindrical waves. To compute the amplitude at a point at height y on the screen (see figure 3.3), we simply sum the amplitude of two waves, one originating at slit A, and the other at slit B. The amplitude at the point of height y, of the wave passing through slit A, is

$$\psi_{\mathrm{A}}(y, t) = N \exp\left[\mathrm{i}\left(kL_{\mathrm{A}y} - \omega t\right)\right] = N \exp\left[\mathrm{i}\left(pL_{\mathrm{A}y} - Et\right)/\hbar\right] \tag{3.18}$$

where N is a constant, p is the magnitude of the electron momentum, and $L_{\mathrm{A}y}$ is the distance from slit A to point y on the screen. Similarly, the amplitude of the contribution from slit B is

$$\psi_{\mathrm{B}}(y, t) = N \exp\left[\mathrm{i}\left(kL_{\mathrm{B}y} - \omega t\right)\right] = N \exp\left[\mathrm{i}\left(pL_{\mathrm{B}y} - Et\right)/\hbar\right] \tag{3.19}$$

By the superposition principle of wave motion[1], the amplitude of the de Broglie wave at point y is

[1] Which actually follows from the fact that the wave equation (3.16) is a linear equation.

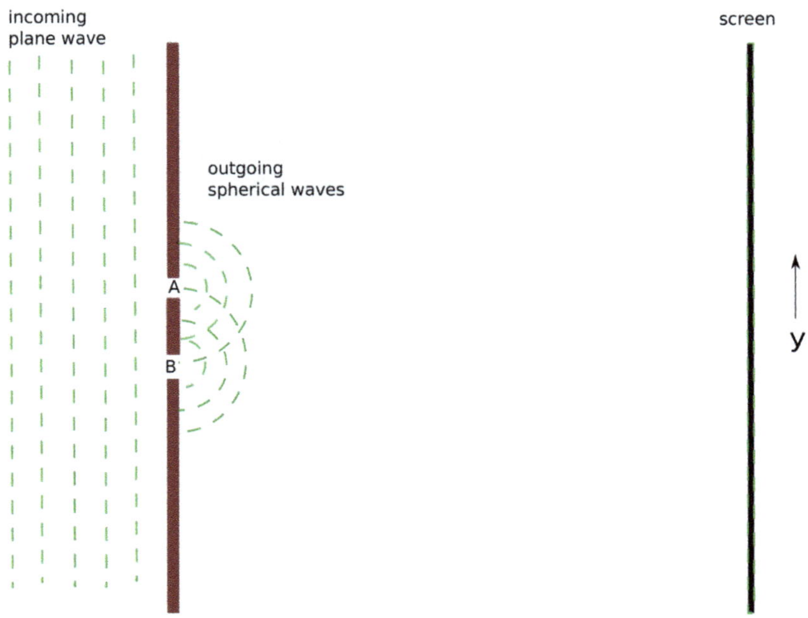

Figure 3.2. Plane waves pass through two slits. Their interference pattern is recorded on a screen.

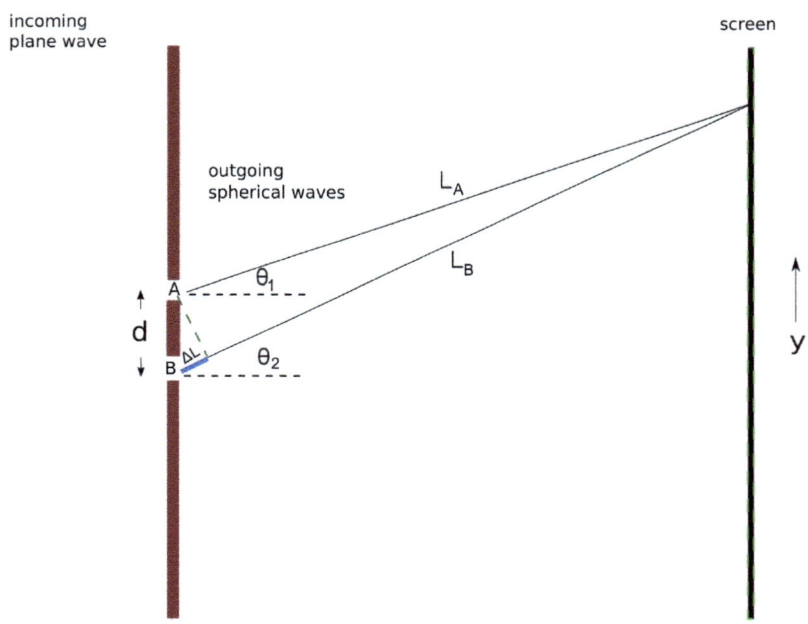

Figure 3.3. Geometry of two-slit interference. L_A is the distance between slit A and a point at height y on the screen. Similarly, L_B is the distance between slit B and that point. Let $\Delta L = L_B - L_A$. When the screen is far from the slits, so that $\theta_1 \approx \theta_2 \approx \theta$, then $\Delta L \approx d \sin\theta$.

$$\begin{aligned}\psi(y, t) &= \psi_A(y, t) + \psi_B(y, t) \\ &= 2N \exp\left[i\left(pL_{av} - Et\right)/\hbar\right]\cos\left(\frac{p\Delta L}{2\hbar}\right)\end{aligned} \tag{3.20}$$

where

$$\begin{aligned}\Delta L &= L_{By} - L_{Ay} \\ L_{av} &= \frac{1}{2}\left(L_{Ay} + L_{By}\right)\end{aligned} \tag{3.21}$$

If the distance to the screen is much greater than the separation d of the slits, then ΔL is approximately

$$\Delta L = d \sin\theta \tag{3.22}$$

where θ is the angle, relative to the x-axis, from the center of the barrier to the point y on the screen.

Now the intensity of a wave is proportional to the square of its amplitude. The generalization to complex amplitudes is that the intensity is proportional to the square of the modulus of the amplitude, i.e.

$$I \propto \psi^*\psi \tag{3.23}$$

The intensity of the de Broglie wave arriving at point y on the screen is therefore

$$I(y) \propto \cos^2\left(\frac{pd \sin\theta}{2\hbar}\right) \tag{3.24}$$

Finally we can make contact with particle motion, because ‘intensity’ is a concept which applies not only to waves, but also to a beam of particles. Intensity is the energy passing through a surface, per unit area, per unit time. Let

$$N(y) = \text{no. of electrons per unit area per unit time} \tag{3.25}$$

which reach the screen in the neighborhood of point y. Since each electron has an energy $E = p^2/2m$, it follows that

$$I(y) = N(y)\frac{p^2}{2m} \tag{3.26}$$

Comparing this expression to (3.23) and (3.24), we have a *prediction* from de Broglie’s theory that

$$\begin{aligned}N(y) &\propto \psi^*\psi \\ &\propto \cos^2\left(\frac{pd \sin\theta}{2\hbar}\right)\end{aligned} \tag{3.27}$$

which can be compared to experiment, simply by counting the number of electrons arriving per second at different regions on the screen. The result of the

Davisson–Germer experiment, and other experiments closely analogous to the double-slit experiment, is that de Broglie's prediction is confirmed. Just as photons have particle-like properties, electrons undoubtedly have wave-like properties.

Suppose that N_{total} electrons get past the barrier per second. Let us ask the question: what is the probability that any one of these electrons will pass through a small area ΔA of the screen, centered at point y, in 1 s? Assuming that the particles move independently, this probability is

$$\text{Prob. to cross } \Delta A/\text{s} = \frac{N(y)\Delta A}{N_{\text{total}}} \tag{3.28}$$

But in order for an electron to pass through a small region of the screen sometime in the interval $t = t_0$ and $t = t_0 + \Delta t$, an electron moving with velocity v must have been located in a small volume

$$\Delta V = \Delta A \times v\Delta t \tag{3.29}$$

in front of the surface ΔA (see figure 3.4). Therefore, the probability that a particle passes the area element ΔA in time Δt, is equal to the probability that the particle, at a given time t_0, is located in a small volume in front of the screen size ΔV, i.e.

$$\begin{aligned} \text{Prob. to be in } \Delta V \text{ at time } t &= \text{prob. to cross } \Delta A/\text{s} \times \Delta t \\ &= \frac{N(y)\Delta A}{N_{\text{total}}}\Delta t \\ &= \frac{1}{vN_{\text{total}}}N(y)\Delta V \\ &\propto \psi^*(y, t)\psi(y, t)\Delta V \end{aligned} \tag{3.30}$$

We see that the probability for an electron to be in a certain volume ΔV is proportional to the square of the modulus of the de Broglie wave, times the volume element. Now if ψ is a solution of the de Broglie wave equation (3.16), so is $\psi' = \text{const.} \times \psi$; this follows from the linearity of the wave equation. Therefore we can always choose a solution of the wave equation such that the proportionality sign in equation (3.30) is replaced by an equals sign. With this choice, we arrive at the interpretation of the de Broglie wave function proposed by Max Born:

The Born interpretation

Denote the wave function associated with an electron by $\psi(x, y, z, t)$. Then the probability to find the electron inside a small volume ΔV centered at point (x, y, z) at time t is

$$\text{Probability} = \psi^*(x, y, z, t)\psi(x, y, z, t)\Delta V$$

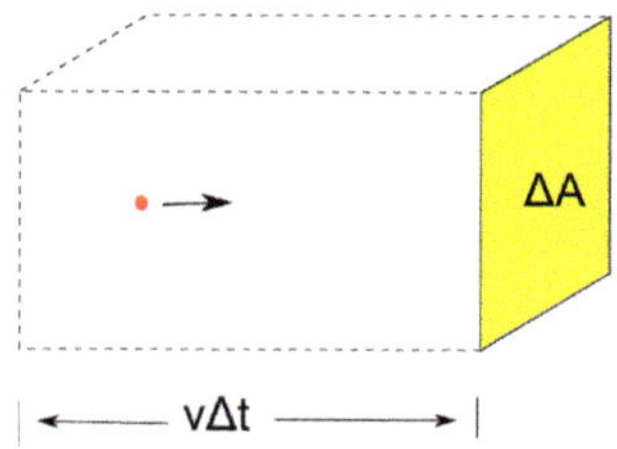

Figure 3.4. A particle with velocity v in the horizontal direction must lie in the volume $v\Delta A\Delta t$ in order to pass through the area ΔA within a time period Δt.

If an electron exists, then the total probability that it is *somewhere* in the Universe is 100%. Therefore, a corollary of the Born interpretation is the ***normalization condition***

$$\int \mathrm{d}x\mathrm{d}y\mathrm{d}z\ \psi^*(x, y, z, t)\psi(x, y, z, t) = 1 \tag{3.31}$$

which must be satisfied by any wave function associated with an electron.

It is useful, when an observable such as position can be any value in a continuous range, to introduce the concept of a ***probability density***. In the case of position, we can write the probability that the electron is found to be in volume ΔV centered at point (x, y, z) as

$$\text{Probability} = P(x, y, z)\Delta V \tag{3.32}$$

where

$$P(x, y, z) = \psi^*(x, y, z, t)\psi(x, y, z, t) \tag{3.33}$$

is the probability density. It can be thought of as probability per unit volume. In general if there is something we can measure, call it Q, which can take on values in some continuous (rather than discrete) range, then the probability that the value of Q lies in a range of width ΔQ centered at the value $Q = Q_0$ is

$$\text{Probability} = P(Q_0)\Delta Q \tag{3.34}$$

where $P(Q)$ is the probability density for the observable Q.

3.3 Do electrons travel along trajectories?

The double-slit experiment is arranged in such a way that equal numbers of electrons, per second, pass through slits A and B. This can be checked by closing slit B, counting the total number of electrons per second reaching the screen, and then closing slit A, and counting again. If the counts/s are equal, then, when both slits are open, the chance that an electron reaching the screen passed through slit A is 50%, with an equal 50% chance that it passed through slit B.

We can determine experimentally the probability $P_A(y)$ that an electron, having passed through slit A, will land on a detector placed at point y on the screen[2]. Close slit B, leaving A open, and count the number of particles n_y that land on the detector, and the total number n_T landing anywhere on the screen. Then

$$P_A(y) = \frac{n_y}{n_T} \tag{3.35}$$

In a similar way, by opening slit B and closing slit A, we can determine the probability $P_B(y)$. Then, with both slits open, the probability that an electron that passed through the barrier will land on the detector at y is

$$\begin{aligned} P(y) &= (\text{prob. of coming from slit A}) \times P_A(y) \\ &\quad + (\text{prob. of coming from slit B}) \times P_B(y) \\ &= \frac{1}{2}\left[P_A(y) + P_B(y)\right] \end{aligned} \tag{3.36}$$

Clearly, if either $P_A(y)$, or $P_B(y)$ (or both) are non-zero, then the probability $P(y)$ is also non-zero.

Now suppose the probabilities $P_A(y)$ and $P_B(y)$ are nearly constant in a sizable region of the screen, which is consistent with the fact that the moduli of the de Broglie waves ψ_A and ψ_B are (nearly) constant. Then $P(y)$ should be nearly constant as well.

In fact, $P(y)$ is not constant at all (see the animation in figure 3.5). Instead, it goes like

$$\begin{aligned} P(y) &\propto \cos^2\left(\frac{pd \sin\theta}{2\hbar}\right) \\ &\propto \psi^*(y, t)\psi(y, t) \end{aligned} \tag{3.37}$$

In particular there are certain points, satisfying

$$pd \sin\theta = 2\hbar\left(n + \frac{1}{2}\right)\pi \tag{3.38}$$

where $P(y) = 0$ *even though* $P_A(y)$ *and* $P_B(y)$ *are non-zero!* In other words, if one closes slit B, leaving A open, there is a finite chance that the electron will reach such a point. If one closes A and leaves B open, there is a finite chance that the electron will reach that point. But if *both* slits A and B are open, the electron will *never* reach such a point, if the angle from the slits to the point satisfies (3.38). Clearly then, in contrast to (3.36),

$$\begin{aligned} P(y) \neq\ &(\text{prob. of coming from slit A}) \times P_A(y) \\ &+ (\text{prob. of coming from slit B}) \times P_B(y) \end{aligned} \tag{3.39}$$

[2] Of course the detector has a finite width Δy, perhaps the width of a pixel. We will neglect this minor complication for now.

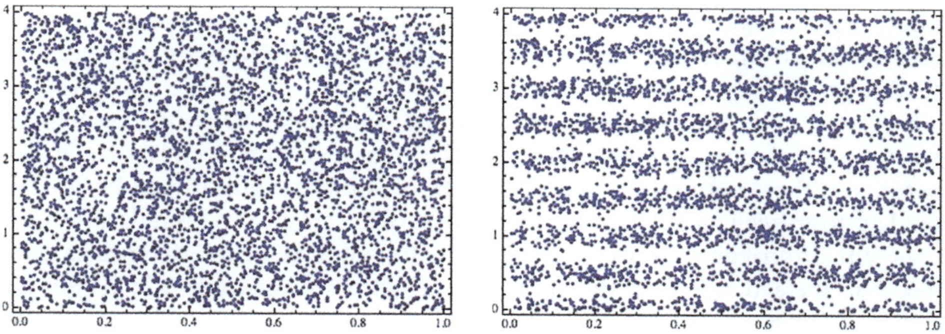

Figure 3.5. A simulation of electrons arriving on a photographic plate, having passed through either a horizontal one-slit (a) or two-slit (b) barrier.

When a conclusion (equation (3.36) in this case) turns out to be false, and the reasoning which led to the conclusion is correct, then there must be something wrong with the premises. The premises in this case are that the electrons in the beam travel independently of one another, and that each electron in the beam passed through *either* slit A *or* slit B.[3] More generally, we have assumed that the electron, as a 'point-like particle', follows a trajectory which takes it through one of the two slits on its way to the screen.

It is easy to check whether the interference effect is due to some collective interaction between electrons in the beam. We can use a beam of such low intensity that the electrons go through the barrier one by one. It takes a little longer to accumulate the data needed to compute $P_A(y)$, $P_B(y)$, and $P(y)$, but the results are ultimately the same. So this is not the explanation.

Then perhaps we should stop thinking of the electron as a point-like object, and start thinking of it as *literally* a de Broglie wave? A wave, after all, can pass through both slits; that is the origin of interference. The problem with this idea is that de Broglie waves expand. At slits A and B, the wave is localized just in the narrow region of the open slits, but as the wave propagates towards the screen it expands, typically spanning (in a real electron diffraction experiment) a region on the order of 10 cm across. Then one would detect 'parts' of an electron. Assuming a de Broglie wave which had expanded to several centimeters across was intense enough to trigger even a single Geiger counter, then if a set of Geiger counters were arranged in a close array, they would all be triggered simultaneously. Likewise, a wave of such dimensions should leave a diffuse glow on a photographic plate, or a thick cylindrical track through a cloud chamber. None of this is ever observed. No matter how big the de Broglie wave of a single electron becomes (as big as a baseball, as big as a house...), only a single 'click' of a Geiger counter is heard, only a sharp 'dot' on

[3] We are also assuming the absence of non-local interactions between the particles and the slits. This means that a particle traveling through slit A cannot be influenced by whether slit B is open or closed, and vice versa. We will touch again on this point much later, in chapter 25.

a photographic plate is recorded. An electron is not, literally, a wave. As far as can be determined from measurements of electron position, electrons are discrete, point-like objects.

If electrons are point-like objects, then one could imagine (with the help of a powerful microscope) actually observing the barrier as the electrons reach it, to determine if they go through slit A, or slit B, or somehow go through both. If one would perform such an experiment, the result is that each electron is indeed observed to pass either through slit A or slit B (never both), but then the interference pattern is wiped out! Instead, one finds the uniform distribution of equation (3.36). Thus if an electron is forced (essentially by observation) to go through one or the other of the two slits, the interference effect is lost. Interference is regained only if the electron is *not* constrained to a trajectory which has passed, with certainty, through one or the other slit.

We are left with the uncomfortable conclusion that electrons are point-like objects which do not follow definite trajectories through space. This sounds like a paradox: how then can electrons get from one point to another? It is actually not a paradox, but it does require a considerable readjustment of our concept of the physical state of electrons (not to mention atoms, molecules, and everything else that makes up the physical Universe). This will be the subject of the next chapter.

Problem

1. Consider a wave function of the form

$$\psi(x, t) = A \sin\left(\frac{px - Et}{\hbar}\right) + B \cos\left(\frac{px - Et}{\hbar}\right) \tag{3.40}$$

where A and B are complex numbers. Assuming that this wave function solves the wave equation (3.11), show that it *must* be proportional to a complex exponential.

References

[1] Bach R, Pope D, S-H Liou and Batelaan H 2013 Controlled double-slit electron diffraction *New J. Phys.* **15** 1–7

[2] Carnal O and Mlynek J 1991 Young's double slit experiment with atoms: a simple atom interferometer *Phys. Rev. Lett.* **66** 2689

IOP Publishing

Jeff Greensite

Chapter 4

The quantum state

We take for granted the idea that every object, at every moment in time, is centered at a definite point in three-dimensional space, and that over a span of time, the location of the object traces a trajectory through that space. We are sure of this because of our long experience, beginning at birth, with the behavior of macroscopic objects, i.e. objects on the scale of a micron or larger, objects which we can *see*, objects which we have the illusion of observing continuously.[1]

It is now time to abandon the view that objects move in trajectories in three-dimensional space, at least in the case of microscopic objects such as electrons, photons, and atoms. This is for two reasons. First, if a particle moves in a trajectory, then it must have a definite position *and* momentum at every moment in time. However, as we have discussed in the case of the Heisenberg microscope, there is no experiment which can verify this. Second, and more importantly, the mere assumption that electrons and photons follow definite trajectories $\mathbf{x}(t)$ leads to conclusions which disagree with experiment, such as the Davisson–Germer experiment discussed in the last chapter, and many other examples to follow later in this book.

How, then, does an electron get from one point to another, without tracing a path through space? The answer is that the electron *does* follow a path; however, it is not a path in the ordinary three-dimensional space we are used to.

4.1 A new representation of motion

Consider a particle moving back and forth, according to the laws of classical physics, in a thin closed pipe of length L. A plot of its trajectory might look like figure 4.1. This is the standard way of representing one-dimensional motion in classical mechanics. To every instant of time t there is a corresponding number $x(t)$,

[1] No observation is truly continuous, because of the discrete, photon composition of light.

doi:10.1088/978-0-7503-1167-0ch4

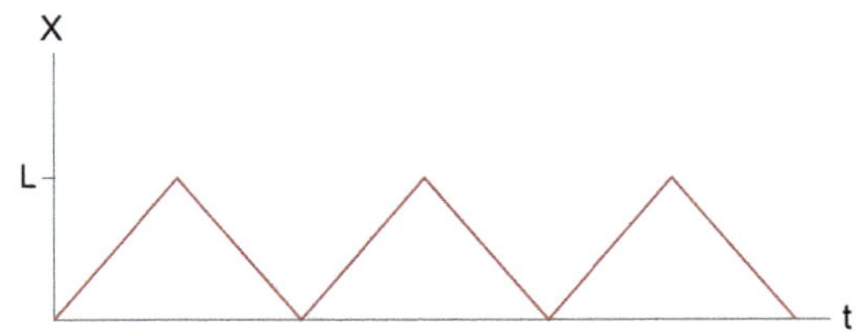

Figure 4.1. Classical motion $x(t)$ of a particle bouncing back and forth in a tube of length L.

which in this case is the distance from one end of the pipe. Is there any other conceivable representation of motion in one dimension?

Let us imagine subdividing the length of the pipe into N equal intervals of length $\epsilon = L/N$, as shown in figure 4.2. If the particle is in the first interval, we choose to represent its state not by a *number*, such as x_1, but rather by an N-dimensional *vector*

$$\mathbf{e}^1 = \begin{bmatrix} 1 \\ 0 \\ 0 \\ \vdots \\ 0 \end{bmatrix} \tag{4.1}$$

If the particle is in the second interval, this will be represented by

$$\mathbf{e}^2 = \begin{bmatrix} 0 \\ 1 \\ 0 \\ \vdots \\ 0 \end{bmatrix} \tag{4.2}$$

and so on, down to the Nth interval which is denoted by

$$\mathbf{e}^N = \begin{bmatrix} 0 \\ 0 \\ 0 \\ \vdots \\ 1 \end{bmatrix} \tag{4.3}$$

The position of the particle is therefore approximated, at any given time, by one of the $\mathbf{e}^n$ unit vectors, and as the particle moves from one interval to another, the unit vector 'jumps' discontinuously from some $\mathbf{e}^k$ to either $\mathbf{e}^{k+1}$ or $\mathbf{e}^{k-1}$, depending on

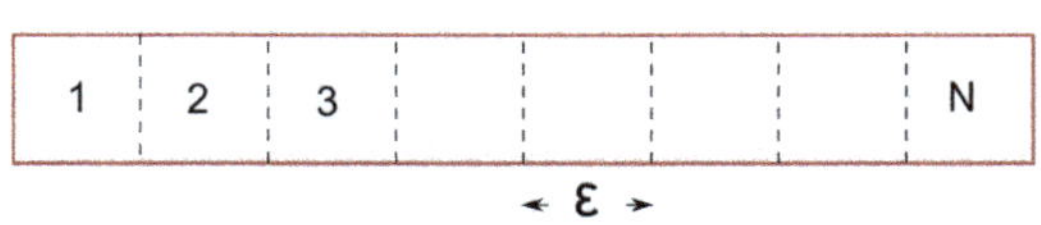

Figure 4.2. Subdivision of a tube of length L into N regions of length ϵ.

which way the particle is moving. If the number of intervals N is large enough, the particle position can be represented, as a function of time, to arbitrary precision.

A reasonable objection to this representation of motion is that, in introducing a set of N orthogonal vectors $\{\mathbf{e}^n\}$, we have also introduced an N-dimensional space which contains linear combinations of those vectors. What, for example, is the physical meaning of a vector such as

$$\mathbf{v} = a\mathbf{e}^1 + b\mathbf{e}^2 = \begin{bmatrix} a \\ b \\ 0 \\ \vdots \\ 0 \end{bmatrix} \tag{4.4}$$

which seems to suggest that the particle could be in intervals 1 and 2 at the same time? In classical mechanics, a particle is definitely in a single interval at any given time, so only the unit vectors $\mathbf{e}^n$ are physically meaningful. The very natural mathematical operation of addition of vectors would therefore have to be regarded, in this representation of classical motion, as physically meaningless.

On the other hand, we have already seen that the de Broglie wave function $\psi(x, t)$ gives *probabilistic* information about the location of the electron. Therefore, let us give the following 'Born interpretation' to the N-dimensional vector, whose components ψ_n are allowed to be complex numbers:

$$\boldsymbol{\psi} = \psi_1\mathbf{e}^1 + \psi_2\mathbf{e}^2 + \cdots + \psi_N\mathbf{e}^N = \begin{bmatrix} \psi_1 \\ \psi_2 \\ \psi_3 \\ \vdots \\ \psi_N \end{bmatrix} \tag{4.5}$$

The probability P_n that an electron, in the state represented by the vector $\boldsymbol{\psi}$, will be found upon measurement to be in the nth interval of the tube, is equal to squared modulus

$$P_n = \psi_n^*\psi_n \tag{4.6}$$

Digression

We must pause for a moment to discuss the inner product of vectors with complex components. The ***norm*** of a vector is defined as square root of the inner product of a vector with itself

$$|v| = \sqrt{\mathbf{v} \cdot \mathbf{v}} \tag{4.7}$$

where the inner product is represented as the product of a row vector times a column vector. If the components of $\mathbf{v}$ are real numbers, this is given by

$$\mathbf{v}\cdot\mathbf{v} = [v_1, v_2, v_3 \ldots, v_N]\begin{bmatrix} v_1 \\ v_2 \\ v_3 \\ \cdot \\ \cdot \\ \cdot \\ v_N \end{bmatrix} = v_1^2 + v_2^2 + \cdots + v_N^2 \tag{4.8}$$

On the other hand, if the components of **v** are complex, this can lead to a negative inner product, and an imaginary norm. In vector algebra the norm of a vector is always real, so the definition of a row vector is modified: its components are the complex conjugate of the corresponding column vector, i.e.

$$\mathbf{v}\cdot\mathbf{v} = [v_1^*, v_2^*, v_3^* \ldots, v_N^*]\begin{bmatrix} v_1 \\ v_2 \\ v_3 \\ \cdot \\ \cdot \\ \cdot \\ v_N \end{bmatrix} = v_1^* v_1 + v_2^* v_2 + \cdots + v_N^* v_N \tag{4.9}$$

In vector algebra a vector comes in two forms, rows and columns. However, the notation **v** does not distinguish between row and column vectors, and sometimes this distinction is useful. We therefore introduce the 'bra-ket' notation in which a 'ket' $|v\rangle$ corresponds to a column vector

$$|v\rangle \Longleftrightarrow \begin{bmatrix} v_1 \\ v_2 \\ v_3 \\ \cdot \\ \cdot \\ \cdot \\ v_N \end{bmatrix} \tag{4.10}$$

and a 'bra' $\langle v|$ to the corresponding row vector

$$\langle v| \Longleftrightarrow [v_1^*, v_2^*, v_3^*, \ldots., v_N^*] \tag{4.11}$$

In this notation, the inner product is

$$\langle v|v\rangle = \mathbf{v}\cdot\mathbf{v} = v_1^* v_1 + v_2^* v_2 + \cdots + v_N^* v_N \tag{4.12}$$

and this is the end of the digression.

Returning to the 'Born interpretation' of the vector $\boldsymbol{\psi}$ (or '$|\psi\rangle$'), we see that the total probability for the electron to be found *somewhere* in the tube is

$$\begin{aligned} P_{\text{total}} &= P_1 + P_2 + \cdots + P_N \\ &= \psi_1^* \psi_1 + \psi_2^* \psi_2 + \cdots + \psi_N^* \psi_N \\ &= \boldsymbol{\psi} \cdot \boldsymbol{\psi} \\ &= \langle \psi | \psi \rangle \end{aligned} \tag{4.13}$$

From the fact that the probability to find the particle somewhere in the tube is 100%, we can conclude that $|\psi\rangle$ is a unit vector, i.e.

$$\langle \psi | \psi \rangle = 1 \tag{4.14}$$

We can now view electron motion in the following way: the physical state of the electron is represented by a unit vector $|\psi\rangle$ in a complex N-dimensional vector space. Since the tip of a unit vector falls on the surface of sphere of radius $R = 1$ in the N-dimensional space, *the motion of the electron corresponds to a trajectory along the surface of a unit sphere in N-dimensional space.* In this way, an electron can travel smoothly and continuously, with no 'jumps', from e.g. the interval 1, represented by vector $\mathbf{e}^1$ to the interval 2, represented by $\mathbf{e}^2$, as shown in figure 4.3. Of course, in going from interval 1 to interval 2, the electron passes through intermediate states such as (4.4), where the electron cannot be viewed as being either in interval 1 *or* in interval 2. But the electron is always located at a definite point on the surface of the unit-sphere, and it is this surface, rather than the line of length L along the tube, which should be regarded as the arena of electron motion.

To complete this new representation of motion we need to take the limit $\epsilon \to 0$, which sends the number of intervals $N \to \infty$. Then the physical states $|\psi\rangle$ are vectors of unit length in an *infinite*-dimensional vector space known as ***Hilbert space***.

4.2 Hilbert space

Any function $f(x)$ with the property of ***square integrability***

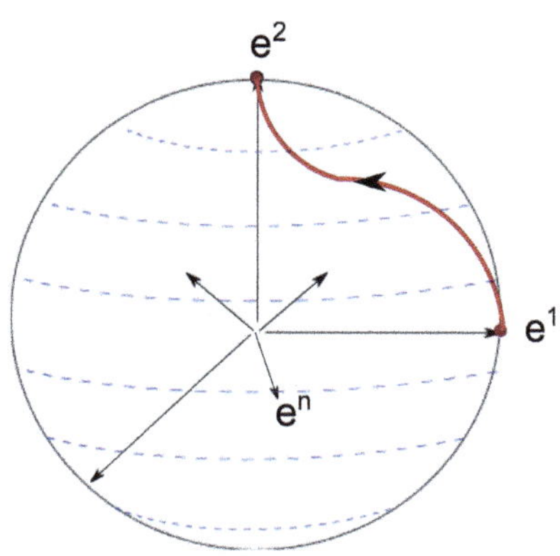

Figure 4.3. In quantum mechanics, the physical state of a particle is represented by a vector of unit length in a multi-dimensional space. A sample trajectory is shown in red. Initially the particle is in region 1 of the tube, and at the end the particle is in region 2. But along the way, the particle is not located in any one region.

$$\int_{-\infty}^{\infty} dx\, f^*(x) f(x) < \infty \tag{4.15}$$

can be regarded as a vector, and a de Broglie wave function $\psi(x, t)$ can, in particular, be regarded as a vector of unit length.

Recall that a vector **v** is represented by a set of numbers (the components of the vector) labeled by an index. There is one number, denoted, v_i for each integer value of the index i in some range $i = 1, 2, \ldots, N$. But a function f is also a set of numbers labeled by an index. There is one number, denoted $f(x)$, for each real-number value of the index x in the range $-\infty < x < \infty$. The argument of a function, x, is completely analogous to the index i of a vector; the value of the function $f(x)$ corresponds to the component v_i of the vector.

The notation of vector algebra has the advantage that when we want to refer to the vector as whole, and not any particular component of the vector, we can write **v**, (or $|v\rangle$ for the column vector in our new 'bra-ket' notation). On the other hand, when we denote a function by $f(x)$, it is sometimes unclear whether we are referring to the function as a whole, or just the particular value at the point x. Let us adopt the notation that $|f\rangle$ refers to the entire function, and $f(x)$ just the value ('component') at the argument ('index') x. Then it is easy to see that functions behave in every way like vectors with a continuous index. For example, one can add vectors

$$\mathbf{v} = a\mathbf{u} + b\mathbf{w} \tag{4.16}$$

which means, in terms of components

$$v_i = au_i + bw_i \tag{4.17}$$

and one can also add functions

$$|f\rangle = a|g\rangle + b|h\rangle \tag{4.18}$$

which means, in terms of 'components'

$$f(x) = ag(x) + bh(x) \tag{4.19}$$

Vectors have inner products with each other

$$\langle u|v\rangle = \mathbf{u} \cdot \mathbf{v} = \sum_{i=1}^{N} u_i^* v_i \tag{4.20}$$

as well as a norm with themselves

$$|v|^2 = \langle v|v\rangle = \sum_{i=1}^{N} v_i^* v_i \tag{4.21}$$

and so do functions:

$$\langle g|f\rangle = \int_{-\infty}^{\infty} dx\, g^*(x)f(x) \tag{4.22}$$

and

$$|f|^2 = \langle f|f\rangle = \int_{-\infty}^{\infty} dx\, f^*(x)f(x) \tag{4.23}$$

If $\{v_i\}$, $i = 1, 2, \ldots, N$ are the components of the column ('ket') vector $|v\rangle$, then $\{v_i^*\}$ are the components of the corresponding row ('bra') vector $\langle v|$. Likewise, $f(x)$, $-\infty < x < \infty$ represent the 'components' of $|f\rangle$, and $f^*(x)$ represents the components of the corresponding bra $\langle f|$.

There is a linear operation called matrix multiplication which turns a vector into another vector

$$\mathbf{v}' = M\mathbf{v} \tag{4.24}$$

or

$$|v'\rangle = M|v\rangle \tag{4.25}$$

in our new notation. In terms of components, matrix multiplication is defined as

$$v_i' = \sum_{j=1}^{N} M_{ij}v_j \tag{4.26}$$

and it has the linearity property that

$$M(a|u\rangle + b|v\rangle) = aM|u\rangle + bM|v\rangle \tag{4.27}$$

where a and b are constants. There is a similar linear operation which turns functions into other functions

$$|f'\rangle = O|f\rangle \tag{4.28}$$

having the linearity property

$$O(a|f\rangle + b|g\rangle) = aO|f\rangle + bO|g\rangle \tag{4.29}$$

In terms of 'components', this operation is written

$$f'(x) = \int_{-\infty}^{\infty} dy\; O(x, y)f(y) \tag{4.30}$$

where $O(x, y)$ is some function of two variables x and y, in complete analogy to the rule for matrix multiplication (4.26). Finally, the expression in linear algebra

$$\begin{aligned}\langle u|M|v\rangle &= \mathbf{u}\cdot M\mathbf{v}\\ &= \sum_{i=1}^{N}\sum_{j=1}^{N} u_i^* M_{ij} v_j\end{aligned} \tag{4.31}$$

corresponds, in the case of functions, to the expression

$$\langle g|O|f\rangle = \int_{-\infty}^{\infty} \mathrm{d}x \int_{-\infty}^{\infty} \mathrm{d}y\ g^*(x)O(x, y)f(y) \tag{4.32}$$

A function, therefore, is just a vector with a continuous index. Since there are an infinite number of 'components' (one component $f(x)$ for each value of x), the vector is a member of an infinite-dimensional space known as 'Hilbert space'. Stated another way, *Hilbert space is the infinite-dimensional vector space of all square-integrable functions,* i.e. functions whose norm (see (4.23)) is finite.

We can think of a plot of $f(x)$ versus x as simply being a simultaneous display of all the components of the Hilbert space vector $|f\rangle$. Such a plot is, in a sense, analogous to the representation of a vector in three-dimensional space by a column of three numbers. From this point of view the animations of wavefunction propagation (e.g. figure 5.1) which are found in later chapters can be thought of as showing how the components of Hilbert space vector $|\psi\rangle$ evolve in time.

4.3 The Dirac delta function

If the linear operation

$$f'(x) = \int_{-\infty}^{\infty} \mathrm{d}y\ O(x, y)f(y) \tag{4.33}$$

is analogous to matrix multiplication in ordinary linear algebra, then what choice of $O(x, y)$ corresponds, in particular, to multiplication by the unit matrix?

In linear algebra, multiplication of any vector by the unit matrix $\mathbb{1}$ is an operation that takes the vector into itself:

$$|v\rangle = \mathbb{1}|v\rangle \tag{4.34}$$

In component form, the unit matrix is known as the ***Kronecker delta*** $\mathbb{1}_{ij} = \delta_{ij}$, which obviously must have the property that

$$v_i = \sum_i \delta_{ij} v_j \tag{4.35}$$

The Kronecker delta satisfying this equation, for any vector $|v\rangle$ whatever, is given by the diagonal matrix

$$\delta_{ij} = \begin{cases} 1 & \text{if} \quad i = j \\ 0 & \text{if} \quad i \neq j \end{cases} \tag{4.36}$$

The corresponding operation for functions

$$|f\rangle = \mathbb{1}|f\rangle \tag{4.37}$$

is written in component form

$$f(x) = \int_{-\infty}^{\infty} \mathrm{d}y\ \delta(x - y)f(y) \tag{4.38}$$

and the function $\delta(x-y)$ which fulfills this equation, for any function $f(x)$ whatever, is known as the ***Dirac delta function***. It is useful not just in quantum mechanics, but throughout mathematical physics.

The Dirac delta function $\delta(x-y)$ is defined as the limit of a sequence of functions $\delta_L(x-y)$, known as a ***delta sequence***, which have the property that

$$f(x) = \lim_{L\to\infty} \int_{-\infty}^{\infty} \mathrm{d}y\ \delta_L(x-y) f(y) \tag{4.39}$$

for any function $f(x)$. Two such sequences which can be shown to have this property are

$$\delta_L(x-y) = \sqrt{\frac{L}{\pi}}\,\mathrm{e}^{-L(x-y)^2} \tag{4.40}$$

and

$$\delta_L(x-y) = \int_{-L}^{L} \frac{\mathrm{d}k}{2\pi}\mathrm{e}^{\mathrm{i}k(x-y)} \tag{4.41}$$

It is reasonable to ask why one should go to the trouble of introducing sequences. Why not take the $L\to\infty$ limit right away, in equation (4.40), and define

$$\delta(x-y) = \begin{cases} \infty & \text{if} \quad x = y \\ 0 & \text{if} \quad x \neq y \end{cases} \tag{4.42}$$

in analogy to the Kronecker delta δ_{ij}? The answer is that the Dirac delta function is not, strictly speaking, a function, any more than ∞ is a number, and the integral of the right-hand side of (4.42) is ill-defined. If one treats ∞ as an ordinary number, it is easy to produce contradictions, e.g.

$$1\times\infty = 2\times\infty \quad\Longrightarrow\quad 1 = 2 \tag{4.43}$$

Instead, one must think of ∞ as the limit of a sequence of numbers, but which is not itself a number. There are many different sequences which can be used, e.g.

$$\begin{aligned} & 1, 2, 3, 4, 5, \ldots. \\ \text{or} \quad & 2, 4, 6, 8, 10, \ldots. \\ \text{or} \quad & 1, 4, 9, 16, 25, \ldots \end{aligned} \tag{4.44}$$

all of which have ∞ as their limit. Similarly, there are many different sequences of functions tending to the same limit, which is not a function in the ordinary sense. Such limits are known as 'generalized functions', of which the Dirac delta function is one very important example.

Let us check that (4.40) and (4.41) satisfy the defining property (4.39) for delta sequences. For the sequence of Gaussians (4.40)

$$\begin{aligned}
&\lim_{L\to\infty}\int_{-\infty}^{\infty} \mathrm{d}y\sqrt{\frac{L}{\pi}}\mathrm{e}^{-L(y-x)^2}f(y)\\
&=\lim_{L\to\infty}\int_{-\infty}^{\infty} \mathrm{d}z\,\sqrt{\frac{L}{\pi}}\mathrm{e}^{-Lz^2}f(x+z)\\
&=\lim_{L\to\infty}\int_{-\infty}^{\infty} \mathrm{d}z\,\sqrt{\frac{L}{\pi}}\mathrm{e}^{-Lz^2}\left[f(x)+\frac{\mathrm{d}f}{\mathrm{d}x}z+\frac{1}{2}\frac{\mathrm{d}^2f}{\mathrm{d}x^2}z^2+\cdots\right]\\
&=\lim_{L\to\infty}\sqrt{\frac{L}{\pi}}\left[\sqrt{\frac{\pi}{L}}f(x)+0+\frac{1}{4L}\sqrt{\frac{\pi}{L}}\frac{\mathrm{d}^2f}{\mathrm{d}x^2}+\cdots\right]\\
&=f(x)
\end{aligned}\tag{4.45}$$

where we have changed variables $z = y - x$, expanded $f(x + z)$ in a Taylor series around the point x, and used the standard formulas of Gaussian integration:

$$\begin{aligned}
&\int_{-\infty}^{\infty} \mathrm{d}z\ \mathrm{e}^{-cz^2}=\sqrt{\frac{\pi}{c}}\\
&\int_{-\infty}^{\infty} \mathrm{d}z\ z\mathrm{e}^{-cz^2}=0\\
&\int_{-\infty}^{\infty} \mathrm{d}z\ z^2\mathrm{e}^{-cz^2}=\frac{1}{2c}\sqrt{\frac{\pi}{c}}
\end{aligned}\tag{4.46}$$

The result (4.45) establishes that the sequence of Gaussians (4.40) is a delta sequence:

$$\delta(x-y)=\lim_{L\to\infty}\sqrt{\frac{L}{\pi}}\mathrm{e}^{-(x-y)^2/L}\tag{4.47}$$

In a similar way, we can prove that (4.41) is a delta sequence

$$\begin{aligned}
&\lim_{L\to\infty}\int_{-\infty}^{\infty} \mathrm{d}y\ f(y)\int_{-L}^{L}\frac{\mathrm{d}k}{2\pi}\mathrm{e}^{ik(x-y)}\\
&=\lim_{L\to\infty}\int_{-\infty}^{\infty} \mathrm{d}y\ f(y)\frac{\sin[L(x-y)]}{\pi(x-y)}
\end{aligned}\tag{4.48}$$

Change variables to $z = L(x - y)$, and this becomes

$$\begin{aligned}
&\lim_{L\to\infty}\int_{-\infty}^{\infty} \mathrm{d}y\ f(y)\int_{-L}^{L}\frac{\mathrm{d}k}{2\pi}\mathrm{e}^{ik(x-y)}\\
&=\lim_{L\to\infty}\int_{-\infty}^{\infty} \mathrm{d}y\ f\left(x-\frac{z}{L}\right)\frac{\sin(z)}{\pi z/L}\\
&=f(x)\int_{-\infty}^{\infty} \mathrm{d}z\frac{\sin(z)}{\pi z}\\
&=f(x)
\end{aligned}\tag{4.49}$$

which establishes that

$$\delta(x-y)=\int_{-\infty}^{\infty}\frac{dk}{2\pi}e^{ik(x-y)} \tag{4.50}$$

A number of useful identities for the Dirac delta function are listed below:

$$f(x)\delta(x-a)=f(a)\delta(x-a) \tag{4.51}$$

$$f(x)\delta[c(x-a)]=f(x)\frac{1}{|c|}\delta(x-a) \tag{4.52}$$

$$f(x)\frac{d}{dx}\delta(x-a)=-\frac{df}{dx}\delta(x-a) \tag{4.53}$$

$$f(x)\delta[g(x)]=f(x)\left|\frac{dg}{dx}\right|^{-1}\delta(x-x_0) \tag{4.54}$$

where, in the last line, $g(x)$ is a function with a zero at $g(x_0)=0$.

Since $\delta(x-y)$ is not a function, these identities should *not* be interpreted as meaning: for a given value of x, the number on the right-hand side equals the number on the left-hand side. Instead, they mean that

$$\int_{-\infty}^{\infty}dx\text{ 'left-hand side'}=\int_{-\infty}^{\infty}dx\text{ 'right-hand side'} \tag{4.55}$$

As an example, we prove the third of these identities using the formula for integration by parts:

$$\begin{aligned}\int_{-\infty}^{\infty}dx\,f(x)\frac{d}{dx}\delta(x-y)&=\lim_{L\to\infty}\int_{-\infty}^{\infty}dx\,f(x)\frac{d}{dx}\delta_L(x-y)\\&=\lim_{L\to\infty}\left[f(x)\delta_L(x-y)|_{x=-\infty}^{x=\infty}-\int_{-\infty}^{\infty}dx\,\frac{df}{dx}\delta_L(x-y)\right]\\&=\lim_{L\to\infty}\int_{-\infty}^{\infty}dx\left[-\frac{df}{dx}\delta_L(x-y)\right]\\&=\int_{-\infty}^{\infty}dx\left[-\frac{df}{dx}\right]\delta(x-y)\end{aligned} \tag{4.56}$$

where the boundary terms are dropped because $\delta_L(\pm\infty)=0$.

4.4 Expectation, uncertainty, and the quantum state

In classical mechanics, the physical state of a system is specified by a set of generalized coordinates and momentum $\{q^i, p_i\}$, which is a point in the phase space of the system. In the course of time, the physical state traces a trajectory through the phase space. In the case of a single particle moving in three dimensions, the physical state is denoted $\{\mathbf{x}, \mathbf{p}\}$, and the phase space is six-dimensional. The projection of the trajectory in the six-dimensional phase space onto the three-dimensional subspace spanned by the x-, y, and z-axes, or in other words, the path $\mathbf{x}(t)$, is the trajectory which we can actually *see* the particle follow.

In quantum mechanics, the physical state of a point-like particle, moving in one dimension, is specified at each moment of time by a wave function $\psi(x, t)$. At any given time t, this wave function is a function only of x, and can be regarded as a vector $|\psi\rangle$ in Hilbert space. Because of the normalization condition imposed by the Born interpretation

$$\langle\psi|\psi\rangle = \int \mathrm{d}x\mathrm{d}y\mathrm{d}z\ \psi^*(x, y, z, t)\psi(x, y, z, t) = 1 \tag{4.57}$$

$|\psi\rangle$ is necessarily a vector of unit length. In the course of time, $|\psi\rangle$ follows a path through Hilbert space. In rough analogy to the motion of unit vectors in finite-dimensional spaces, one could imagine that the tip of the unit vector $|\psi\rangle$ traces a path on the surface of a unit sphere, although in this case the space is infinite-dimensional.

A physical state has the property that, given the state at some time t, we can determine the state at a slightly later time $t + \epsilon$. In classical mechanics, the rule is

$$\begin{aligned} q^a(t + \epsilon) &= q^a(t) + \epsilon\left(\frac{\partial H}{\partial p_a}\right)_t \\ p_a(t + \epsilon) &= p_a(t) - \epsilon\left(\frac{\partial H}{\partial q^a}\right)_t \end{aligned} \tag{4.58}$$

The physical state $|\psi\rangle$ in quantum mechanics also has this property. Given the wave function $\psi(x, t)$ at any particular time t, the wave function at a slightly later time $t + \epsilon$ is determined, for a particle moving freely in one dimension, from the wave equation for de Broglie waves

$$\psi(x, t + \epsilon) = \psi(x, t) + \mathrm{i}\epsilon\frac{\hbar}{2m}\frac{\partial^2}{\partial x^2}\psi(x, t) \tag{4.59}$$

The Born interpretation tells us how to use the wave function $\psi(x, t)$ to make certain experimental predictions. Unlike the predictions of classical mechanics, however, which specify the outcome of any measurement on a single system with certainty, given the physical state $\{q^i, p_i\}$ of the system, the predictions of quantum mechanics are statistical in nature. The Born interpretation tells us that the probability to find the particle in a small interval around the point x, of length ΔL, is

$$P_{\Delta L}(x) = \psi^*(x, t)\psi(x, t)\Delta L \tag{4.60}$$

The meaning of this prediction, in practice, is as follows. Suppose we have a very large number of particles N_p which are in identical physical states described by a certain wave function $\psi(x, t)$. If we measure the position of all of these particles simultaneously, the number n_x which will be observed to be in the interval ΔL around point x is predicted, according to the Born interpretation, to be

$$\frac{n_x}{N_p} = P(x)\Delta L + O\left(N_p^{-1/2}\right)$$
$$= \psi^*(x, t)\psi(x, t)\Delta L + O\left(N_p^{-1/2}\right) \tag{4.61}$$

The term of order $N_p^{-1/2}$ is statistical error, which can be made arbitrarily small by choosing a sufficiently large number of particles N_p.

To understand the origin of the statistical error, consider flipping a coin N times, where N is a very large, even number. Only very rarely does the coin come up heads *exactly* $N/2$ times. Usually the number of heads deviates from $N/2$ by a amount on the order of $\Delta N \sim \sqrt{N}$. The ratio of the deviation ΔN to the total number of coin flips N varies each time one does the full set of N coin flips, but it is generally of order

$$\frac{\Delta N}{N} \sim \frac{1}{\sqrt{N}} \tag{4.62}$$

The deviation of order $N_p^{-1/2}$ in equation (4.61) has the same origin, and is present in any set of measurements which involve random processes. From now on we will suppose that N_p is so large that this deviation is ignorable.

Instead of simultaneously observing a swarm of N_p particles, each of which are in the same state, one could instead perform an experiment in which a single particle is placed in a particular state $\psi(x, t)$, its position is observed, and then the identical experiment is repeated N_p times. If n_x is the number of experiments in which the particle lands in the interval ΔL around point x, then the prediction according to the Born interpretation is again equation (4.61). In fact, this is what is actually happening in the electron two-slit experiment. Electrons can be made to pass one by one through the slits, and the wave function of each electron at the location of the screen is identical. The statistical prediction for the number of electrons reaching a particular small region of the screen is then equation (3.27) of the last chapter.

Let us imagine making a series of experiments on the motion of a particle moving in one dimension, in which the particle is somehow initialized to be in a certain quantum state $|\psi\rangle$, and then the position of the particle is measured. There are two particularly useful numbers which can be predicted by quantum theory, and measured by the series of experiments. One number is the average value $\bar{x}$ of the particle position. Let x_i be the position of the particle in the ith experiment, and N_E be the total number of experiments. Then the average value of the particle position is defined as

$$\bar{x} \equiv \frac{1}{N_E}\sum_{i=1}^{N_E} x_i \tag{4.63}$$

We can also define the average value of x^2 as

$$\bar{x^2} \equiv \frac{1}{N_E}\sum_{i=1}^{N_E} (x_i)^2 \tag{4.64}$$

The other useful number is the average square deviation δx^2 of the observed positions x_i from their average value $\bar{x}$. The average square deviation is defined as

$$\delta x^2 \equiv \frac{1}{N_E}\sum_{i=1}^{N_E}(x_i - \bar{x})^2 \tag{4.65}$$

or, expanding the quantity in parentheses

$$\begin{aligned}\delta x^2 &= \frac{1}{N_E}\sum_{i=1}^{N_E}\left[x_i^2 - 2x_i\bar{x} + \bar{x}^2\right] \\ &= \overline{x^2} - 2\bar{x}\ \bar{x} + \bar{x}^2 \\ &= \overline{x^2} - \bar{x}^2\end{aligned} \tag{4.66}$$

The ‘root-mean-square’ deviation δx is just the square-root of δx^2. These two numbers, $\bar{x}$ and δx, can be reported as result of the series of measurements.

The Born interpretation enables us to predict the values of these two numbers, given the quantum state $|\psi\rangle$. These predicted values come from probability theory. Suppose we make a series of measurements on the quantity Q, which can only take on the possible values $Q_1, Q_2, Q_3, \ldots, Q_{n_Q}$; and suppose the corresponding probabilities of finding these values, in any given measurement, is $P_1, P_2, P_3, \ldots, P_{n_Q}$. Then the expected average value, or ***expectation value*** of Q is defined as

$$\langle Q\rangle = \sum_{n=1}^{n_Q} Q_n P_n \tag{4.67}$$

where n_Q is the number (which could be infinite) of possible values of the observable Q. This theoretical expectation value should equal (up to statistical error) the average value

$$\bar{Q} = \frac{1}{N_M}\sum_{n=1}^{n_Q} N_n Q_n \tag{4.68}$$

where N_n is the number of measurements that found the value $Q = Q_n$, and N_M is the total number of measurements

$$N_M = \sum_{n=1}^{n_Q} N_n \tag{4.69}$$

(One should be careful not to confuse the index i in equation (4.63), which labels the experiment, with the index n in equation (4.67), which labels the possible values the observable Q.) The reason that $\langle Q\rangle$ and $\bar{Q}$ should equal one another is that the fraction of times that the value $Q = Q_n$ is observed should equal the probability of observing Q_n in any given measurement, i.e.

$$P_n = \frac{N_n}{N_M} \tag{4.70}$$

Inserting this expression for the probability into (4.67), one finds $\langle Q \rangle = \bar{Q}$.

In the case of a particle moving in one dimension, one of the possible observables is the position x. Since x can take any value in a continuous range, the sum in (4.67) is replaced by an integral, and the probability is replaced by the probability density

$$P(x) = \psi^*(x, t)\psi(x, t) \tag{4.71}$$

With these modifications, the formula for expectation values predicts

$$\begin{aligned} \langle x \rangle &= \int_{-\infty}^{\infty} \mathrm{d}x \; xP(x) \\ &= \int_{-\infty}^{\infty} x\psi^*(x, t)\psi(x, t)\mathrm{d}x \end{aligned} \tag{4.72}$$

Likewise, the expectation value for the observable x^2 is

$$\langle x^2 \rangle = \int_{-\infty}^{\infty} x^2\psi^*(x, t)\psi(x, t)\mathrm{d}x \tag{4.73}$$

and the expectation value of the square deviation is given by

$$\begin{aligned} \Delta x^2 &= \langle (x - \langle x \rangle)^2 \rangle \\ &= \left\langle \left(x^2 - 2x\langle x \rangle + \langle x \rangle^2\right)\right\rangle \\ &= \int_{-\infty}^{\infty} \mathrm{d}x \left[x^2 - 2x\langle x \rangle + \langle x \rangle^2\right]\psi^*(x, t)\psi(x, t) \end{aligned} \tag{4.74}$$

Using the fact that $\langle x \rangle$ is a constant which can be taken outside the integral, and the normalization condition (3.31), we find

$$\begin{aligned} \Delta x^2 &= \langle x^2 \rangle - 2\langle x \rangle\langle x \rangle + \langle x \rangle^2 \\ &= \langle x^2 \rangle - \langle x \rangle^2 \end{aligned} \tag{4.75}$$

as in equation (4.66). The square root of this quantity, Δx, is referred to as the ***uncertainty*** in the position of the particle in the quantum state $|\psi\rangle$.

We see that from the Born interpretation we can extract two numbers from the wave function, the expectation value $\langle x \rangle$ and the uncertainty Δx, which can be compared to the experimental values of average position $\bar{x}$ and root-mean-square deviation δx, respectively. Experimentally, however, there are other observables of a particle which can be measured, such as its momentum and its energy, and values of $\bar{p}$, δp, $\bar{E}$, δE can be determined from experiment. The Born interpretation does not tell us how to compute these quantities; this will require a further investigation of the dynamics of the quantum state, which is the topic of the next chapter.

Problems

1. Suppose that at time $t = 0$ the state of a particle moving in one dimension is given by the wave function

$$\psi(x) = N e^{-|x-x_0|/2a} \tag{4.76}$$

   a. What is the normalization constant N?
   b. Compute $\langle x \rangle$ and Δx.

2. Prove that:
   a. $\delta(cx) = \frac{1}{|c|}\delta(x)$ where c is any real non-zero constant.
   b. $f(x)\frac{d^2}{dx^2}\delta(x) = \frac{d^2 f}{dx^2}\delta(x)$
   c. $f(x)\delta[g(x)] = f(x)|\frac{dg}{dx}|^{-1}\delta(x - x_0)$ where $g(x) = 0$ *only* at $x = x_0$.

3. Invent your own personalized delta function. That is, come up with a sequence of functions $\delta_L(x - y)$ that depend on a parameter L, which have the property

$$f(x) = \lim_{L\to\infty} \int_{-\infty}^{\infty} dy \;\; \delta_L(x - y) f(y) \tag{4.77}$$

What you are looking for a curve which depends on L, such that: (i) for any L, the area under the curve is 1; and (ii) as $L \to \infty$, the area is peaked around the point $x = y$.

Chapter 5

Dynamics of the quantum state

The purpose of mechanics is to describe how bodies change their position in space with time. I should load my conscience with grave sins against the sacred spirit of lucidity were I to formulate the aims of mechanics in this way, without serious reflection and detailed explanations.

Albert Einstein

The classical motion of a particle is represented by a trajectory in three-dimensional space, while the quantum-mechanical motion of a particle, according to the last chapter, is supposed to be described by a curve on the unit sphere of an infinite-dimensional space. How can classical physics be in any sense an approximation of quantum physics, if the arena of motion is so very different in the two theories?

The answer is this: it is not the physical state of the particle, but rather our observations of its position, that should, in some limit, approximate a classical trajectory. And although the quantum state of a particle does not correspond, in general, to a definite point in 3-space, it is still true that the *expectation value* of particle position at time t

$$\langle x\rangle \equiv \int dx\, x\psi^*(x, t)\psi(x, t) \tag{5.1}$$

traces a trajectory through ordinary space. Given an equation of motion for the quantum state $\psi(x, t)$, it should be possible to derive laws of motion for $\langle x\rangle$, and compare them with the classical laws of motion for $x(t)$. The simplest possibility is that these two sets of laws look the same. This is known as 'Ehrenfest's principle'.

5.1 Ehrenfest's principle

Let $\{q^a, p_a\}$ be the generalized coordinates and momenta of a mechanical system. Ehrenfest's principle is the statement that Hamilton's equations of motion are valid as expectation value equations in quantum mechanics:

doi:10.1088/978-0-7503-1167-0ch5

Ehrenfest's principle

$$\frac{\mathrm{d}}{\mathrm{d}t}\langle q^a\rangle = \left\langle \frac{\partial H}{\partial p_a} \right\rangle$$
$$\frac{\mathrm{d}}{\mathrm{d}t}\langle p_a\rangle = -\left\langle \frac{\partial H}{\partial q^a} \right\rangle \tag{5.2}$$

In particular, for a particle of mass m moving in a potential $V(x)$ in one dimension,

$$\frac{\mathrm{d}}{\mathrm{d}t}\langle x\rangle = \left\langle \frac{p}{m} \right\rangle$$
$$\frac{\mathrm{d}}{\mathrm{d}t}\langle p\rangle = \left\langle -\frac{\partial V}{\partial x} \right\rangle \tag{5.3}$$

Let us check whether these equations are satisfied by the equation for de Broglie waves in one dimension

$$\mathrm{i}\hbar\frac{\partial \psi}{\partial t} = -\frac{\hbar^2}{2m}\frac{\partial^2 \psi}{\partial x^2} \tag{5.4}$$

From the first equation of (5.3), we obtain an expression for the expectation value of momentum

$$\begin{aligned}\langle p\rangle &= m\partial_t\langle x\rangle \\ &= m\partial_t \int \mathrm{d}x\ \psi^*(x,t)x\psi(x,t) \\ &= m\int \mathrm{d}x \left\{ \frac{\partial \psi^*}{\partial t}x\psi + \psi^* x\frac{\partial \psi}{\partial t} \right\}\end{aligned} \tag{5.5}$$

Then, applying the equation (5.4) for de Broglie waves

$$\langle p\rangle = m\left(\frac{\mathrm{i}\hbar}{2m}\right)\int \mathrm{d}x \left\{ -\left(\frac{\partial^2}{\partial x^2}\psi^*\right)x\psi + \psi^* x\frac{\partial^2}{\partial x^2}\psi \right\} \tag{5.6}$$

and using the integration by parts formula[1]

$$\int \mathrm{d}x \left(\frac{\partial^2}{\partial x^2}F\right)G = \int \mathrm{d}x\ F\frac{\partial^2}{\partial x^2}G \tag{5.7}$$

[1] We assume that the wave function ψ and its derivatives vanish at $x = \pm\infty$, so there are no boundary terms.

this becomes

$$\begin{aligned}\langle p\rangle &= m\left(\frac{\mathrm{i}\hbar}{2m}\right)\int \mathrm{d}x\ \psi^*\left\{x\frac{\partial^2}{\partial x^2}\psi - \frac{\partial^2}{\partial x^2}(x\psi)\right\} \\ &= m\left(\frac{\mathrm{i}\hbar}{2m}\right)\int \mathrm{d}x\ \psi^*\left\{x\frac{\partial^2}{\partial x^2}\psi - \frac{\partial}{\partial x}\left(\psi + x\frac{\partial}{\partial x}\psi\right)\right\} \\ &= \int \mathrm{d}x\psi^*\left(-\mathrm{i}\hbar\frac{\partial}{\partial x}\right)\psi\end{aligned} \tag{5.8}$$

Equation (5.8) should be understood as a prediction, given the quantum state $\psi(x, t)$, for the expectation value of momentum. Next, taking the time derivative of $\langle p\rangle$, so as to compare to the second equation of (5.3),

$$\begin{aligned}\partial_t\langle p\rangle &= \partial_t\int \mathrm{d}x\ \psi^*\left(-\mathrm{i}\hbar\frac{\partial}{\partial x}\right)\psi \\ &= \left(\frac{\mathrm{i}\hbar}{2m}\right)(-\mathrm{i}\hbar)\int \mathrm{d}x\left\{-\left(\frac{\partial^2}{\partial x^2}\psi^*\right)\frac{\partial}{\partial x}\psi + \psi^*\frac{\partial}{\partial x}\frac{\partial^2}{\partial x^2}\psi\right\} \\ &= \left(\frac{\mathrm{i}\hbar}{2m}\right)(-\mathrm{i}\hbar)\int \mathrm{d}x\psi^*\left\{-\frac{\partial^2}{\partial x^2}\frac{\partial}{\partial x}\psi + \frac{\partial}{\partial x}\frac{\partial^2}{\partial x^2}\psi\right\} \\ &= 0\end{aligned} \tag{5.9}$$

we conclude that $\partial_t\langle p\rangle = 0$, which agrees with Ehrenfest's principle for the case that the potential $V = 0$.

In the case that $V \neq 0$, the equation for de Broglie waves does not satisfy the second of equations (5.3). So the de Broglie wave equation must be generalized so as to agree with Ehrenfest's principle for *any* potential $V(x)$. This generalized wave equation was found by Schrödinger; it is as central to quantum mechanics as $F = ma$ is to classical mechanics.

5.2 The Schrödinger wave equation

The quantum-mechanical law of motion found by Schrödinger, describing a particle of mass m moving in one dimension in an arbitrary potential $V(x)$, is known as

The Schrödinger wave equation

$$\mathrm{i}\hbar\frac{\partial\psi}{\partial t} = -\frac{\hbar^2}{2m}\frac{\partial^2\psi}{\partial x^2} + V(x)\psi \tag{5.10}$$

It is easy to verify that this equation does, in fact, satisfy Ehrenfest's principle. Once again, starting with the first of the Ehrenfest equations (5.3),

$$\begin{aligned}\langle p\rangle &= m\partial_t\langle x\rangle \\ &= m\int \mathrm{d}x\left\{\frac{\partial\psi^*}{\partial t}x\psi + \psi^* x\frac{\partial\psi}{\partial t}\right\}\end{aligned} \tag{5.11}$$

Then rewriting the Schrödinger equation (and its complex conjugate) as

$$\begin{aligned}\frac{\partial\psi}{\partial t} &= \frac{\mathrm{i}\hbar}{2m}\frac{\partial^2\psi}{\partial x^2} - \frac{\mathrm{i}}{\hbar}V\psi \\ \frac{\partial\psi^*}{\partial t} &= -\frac{\mathrm{i}\hbar}{2m}\frac{\partial^2\psi^*}{\partial x^2} + \frac{\mathrm{i}}{\hbar}V\psi^*\end{aligned} \tag{5.12}$$

and inserting into (5.11), we have

$$\begin{aligned}\langle p\rangle = m\int \mathrm{d}x&\left\{\left(-\frac{\mathrm{i}\hbar}{2m}\frac{\partial^2\psi^*}{\partial x^2} + \frac{\mathrm{i}}{\hbar}V\psi^*\right)x\psi\right. \\ &\left.+\,\psi^* x\left(\frac{\mathrm{i}\hbar}{2m}\frac{\partial^2\psi}{\partial x^2} - \frac{\mathrm{i}}{\hbar}V\psi\right)\right\} \\ = m\left(\frac{\mathrm{i}\hbar}{2m}\right)&\int \mathrm{d}x\left\{-\left(\frac{\partial^2}{\partial x^2}\psi^*\right)x\psi + \psi^* x\frac{\partial^2}{\partial x^2}\psi\right\} \\ = \int \mathrm{d}x\,\psi^*&\left(-\mathrm{i}\hbar\frac{\partial}{\partial x}\right)\psi\end{aligned} \tag{5.13}$$

which is the same rule for computing $\langle p\rangle$ as in the previous $V = 0$ case. Then

$$\begin{aligned}\partial_t\langle p\rangle &= -\mathrm{i}\hbar\int \mathrm{d}x\left[\frac{\partial\psi^*}{\partial t}\frac{\partial\psi}{\partial x} + \psi^*\frac{\partial^2\psi}{\partial t\partial x}\right] \\ &= -\mathrm{i}\hbar\int \mathrm{d}x\left[\frac{\partial\psi^*}{\partial t}\frac{\partial\psi}{\partial x} - \frac{\partial\psi^*}{\partial x}\frac{\partial\psi}{\partial t}\right] \\ &= -\mathrm{i}\hbar\int \mathrm{d}x\left\{\left(-\frac{\mathrm{i}\hbar}{2m}\frac{\partial^2\psi^*}{\partial x^2} + \frac{\mathrm{i}}{\hbar}V\psi\right)\frac{\partial\psi}{\partial x}\right. \\ &\qquad\left.-\frac{\partial\psi^*}{\partial x}\left(\frac{\mathrm{i}\hbar}{2m}\frac{\partial^2\psi}{\partial x^2} - \frac{\mathrm{i}}{\hbar}V\psi\right)\right\} \\ &= -\frac{\hbar^2}{2m}\int \mathrm{d}x\left[\frac{\partial^2\psi^*}{\partial x^2}\frac{\partial\psi}{\partial x} + \frac{\partial\psi^*}{\partial x}\frac{\partial^2\psi}{\partial x^2}\right] \\ &\quad + \int \mathrm{d}x\left[\psi^* V\frac{\partial\psi}{\partial x} + \frac{\partial\psi^*}{\partial x}V\psi\right]\end{aligned} \tag{5.14}$$

Again applying integration by parts to the first term of the first integral, and the second term of the second integral, we find

$$
\begin{aligned}
\partial_t\langle p\rangle &= -\frac{\hbar^2}{2m}\int \mathrm{d}x\left[-\frac{\partial\psi^*}{\partial x}\frac{\partial^2\psi}{\partial x^2}+\frac{\partial\psi^*}{\partial x}\frac{\partial^2\psi}{\partial x^2}\right] \\
&\quad + \int \mathrm{d}x\left[\psi^* V\frac{\partial\psi}{\partial x} - \psi^*\frac{\partial}{\partial x}(V\psi)\right] \\
&= \int \mathrm{d}x\left[\psi^* V\frac{\partial\psi}{\partial x} - \psi^* V\frac{\partial\psi}{\partial x} - \psi^*\frac{\partial V}{\partial x}\psi\right] \\
&= \int \mathrm{d}x\ \psi^*\left(-\frac{\partial V}{\partial x}\right)\psi \\
&= \left\langle -\frac{\partial V}{\partial x}\right\rangle
\end{aligned}
\tag{5.15}
$$

exactly as required by Ehrenfest's principle.

We see that Ehrenfest's principle and the Schrödinger equation, taken together, make the following prediction for the expectation value of particle momentum at time t, regardless of the potential $V(x)$

$$
\langle p\rangle = \int \mathrm{d}x\ \psi^*(x, t)\ \tilde{p}\ \psi(x, t) \tag{5.16}
$$

where $\tilde{p}$ is the differential operator known as the ***momentum operator***

$$
\tilde{p} \equiv -\mathrm{i}\hbar\frac{\partial}{\partial x} \tag{5.17}
$$

This supplies an important piece of information. The Born interpretation of the wave function tells us how to compute the expectation value of position $\langle x\rangle$ of a particle, given its wave function $\psi(x, t)$. But position is not the only observable of a particle that can be measured; for example, one can also measure a particle's momentum, and the Born interpretation does not tell us how to compute the corresponding expectation value $\langle p\rangle$. Now we have the rule for finding $\langle p\rangle$; it is contained in equations (5.16) and (5.17) above.

Equation (5.17) can be used to express any x derivative in terms of the momentum operator, and in particular

$$
\frac{\partial^2}{\partial x^2} = -\frac{1}{\hbar^2}\tilde{p}^2 \tag{5.18}
$$

Then the Schrödinger equation can be written in the compact form

$$
\begin{aligned}
\mathrm{i}\hbar\partial_t\psi &= \left(\frac{\tilde{p}^2}{2m} + V\right)\psi \\
&= \tilde{H}\psi
\end{aligned}
\tag{5.19}
$$

where

$$H[p, x] = \frac{p^2}{2m} + V \tag{5.20}$$

is just the Hamiltonian for a particle of mass m, moving in a potential field V, and $\tilde{H}$ is the ***Hamiltonian operator***

$$\tilde{H} = H[\tilde{p}, x] \tag{5.21}$$

obtained by replacing momentum p by the momentum operator $\tilde{p}$ in the Hamiltonian function.

Finally, we need to check the consistency of the Schrödinger equation with the Born interpretation. According to the Born interpretation, at any instant of time t,

$$\langle\psi|\psi\rangle = \int \mathrm{d}x\ \psi^*(x, t)\psi(x, t) = 1 \tag{5.22}$$

for a particle moving in one dimension. Now suppose this normalization condition is true at some time t_0, and the wave function evolves according to the Schrödinger equation. Will this condition also be valid at later times $t < t_0$? To check this, we differentiate $\langle\psi|\psi\rangle$ with respect to time

$$\begin{aligned}\partial_t\langle\psi|\psi\rangle &= \int \mathrm{d}x\{(\partial_t\psi^*)\psi + \psi^*(\partial_t\psi)\}\\ &= \int \mathrm{d}x\left\{\left(-\frac{\mathrm{i}\hbar}{2m}\partial_x^2\psi^* + \frac{\mathrm{i}}{\hbar}V\psi^*\right)\psi + \psi^*\left(\frac{\mathrm{i}\hbar}{2m}\partial_x^2\psi - \frac{\mathrm{i}}{\hbar}V\psi\right)\right\}\\ &= \frac{\mathrm{i}\hbar}{2m}\int \mathrm{d}x\{-(\partial_x^2\psi^*)\psi + \psi^*\partial_x^2\psi\}\\ &= 0\end{aligned} \tag{5.23}$$

where we have once again used the integration by parts formula (5.7). This result means that the norm of the wave function is constant in time, which is known as ***conservation of probability***. The inner product $\langle\psi|\psi\rangle$ is the total probability to find the particle anywhere in space. If the norm is constant, and $\langle\psi|\psi\rangle = 1$ at some time $t = t_0$, then $\langle\psi|\psi\rangle = 1$ at any later time.

The Schrödinger equation for motion in three dimensions is a straightforward generalization of the one-dimensional case:

$$\mathrm{i}\hbar\frac{\partial\psi}{\partial t} = -\frac{\hbar^2}{2m}\nabla^2\psi + V(x, y, z)\psi \tag{5.24}$$

In this case, the generalized coordinates are $q^1 = x,\ q^2 = y,\ q^3 = z$, and the first of the Ehrenfest equations (5.2) tells us

$$
\begin{aligned}
\langle p_x \rangle &= \int d^3x\, \psi^*(x, y, z, t)\, \tilde{p}_x\, \psi(x, y, z, t) \\
\langle p_y \rangle &= \int d^3x\, \psi^*(x, y, z, t)\, \tilde{p}_y\, \psi(x, y, z, t) \\
\langle p_z \rangle &= \int d^3x\, \psi^*(x, y, z, t)\, \tilde{p}_z\, \psi(x, y, z, t)
\end{aligned}
\tag{5.25}
$$

where

$$
\begin{aligned}
\tilde{p}_x &\equiv -i\hbar \frac{\partial}{\partial x} \\
\tilde{p}_y &\equiv -i\hbar \frac{\partial}{\partial y} \\
\tilde{p}_z &\equiv -i\hbar \frac{\partial}{\partial z}
\end{aligned}
\tag{5.26}
$$

and one can then readily verify the second Ehrenfest equation

$$\partial_t \langle \mathbf{p} \rangle = \langle -\nabla V \rangle \tag{5.27}$$

In terms of the $\tilde{p}$-operators, the Schrödinger equation in three dimensions can be written

$$
\begin{aligned}
i\hbar \frac{\partial \psi}{\partial t} &= \left[\frac{1}{2m} \left(\tilde{p}_x^2 + \tilde{p}_y^2 + \tilde{p}_z^2 \right) + V(x, y, z) \right] \psi \\
&= \tilde{H} \psi
\end{aligned}
\tag{5.28}
$$

5.3 The time-independent Schrödinger equation

When the potential $V(x)$ is time-independent we can simplify the Schrödinger equation by the method of separation of variables. Write

$$\psi(x, t) = \phi(x) T(t) \tag{5.29}$$

and substitute into the (one-dimensional) Schrödinger equation

$$i\hbar \phi(x) \frac{\partial T}{\partial t} = T \tilde{H} \phi \tag{5.30}$$

divide both sides by ϕT

$$i\hbar \frac{1}{T(t)} \frac{\partial}{\partial t} T(t) = \frac{1}{\phi(x)} \tilde{H} \phi(x) \tag{5.31}$$

Since the left-hand side depends only on t, and the right-hand side only on x, the only way this equation can be true is if both sides equal a constant, call it E:

$$
\begin{aligned}
i\hbar\frac{1}{T(t)}\frac{\partial}{\partial t}T(t) &= E \\
\frac{1}{\phi(x)}\tilde{H}\phi(x) &= E
\end{aligned}
\tag{5.32}
$$

The first of these two differential equations can be solved immediately:

$$T(t) = e^{-iEt/\hbar} \tag{5.33}$$

while the second equation

$$\tilde{H}\phi = E\phi \tag{5.34}$$

or, more explicitly,

$$\left[-\frac{\hbar^2}{2m}\frac{d^2}{dx^2} + V(x)\right]\phi(x) = E\phi(x) \tag{5.35}$$

is known as the ***time-independent Schrödinger equation***.

Equation (5.34) is an example of an ***eigenvalue equation***. This is a linear differential equation in which one has to solve simultaneously for a set of constants, known as the ***eigenvalues***, and a set of corresponding functions known as the ***eigenfunctions***. In the case of the Schrödinger equation, the constant E is called the 'energy eigenvalue', and the function $\phi(x)$ is called the 'energy eigenfunction' or 'energy eigenstate'.

To each energy eigenvalue E there is at least one (and sometimes more than one) energy eigenstate, and to each eigenstate there corresponds a solution

$$\psi(x, t) = \phi(x)e^{-iEt/\hbar} \tag{5.36}$$

of the time-dependent Schrödinger equation. Such solutions are also known as ***stationary states*** because the time dependence is contained entirely in an overall phase factor $e^{-iEt/\hbar}$. This means that the probability density to find a particle in the neighborhood of point x,

$$P(x) = \psi^*(x, t)\psi(x, t) = \phi^*(x)\phi(x) \tag{5.37}$$

is constant in time.

Let $\{\phi_\alpha(x), E_\alpha\}$ denote a complete set of eigenstates and corresponding eigenvalues, in the sense that any other eigenstate which is not in the set can be expressed as a linear combination of those that are. Then the general solution to the time-*dependent* Schrödinger equation is a sum

$$\psi(x, t) = \sum_n a_n\phi_n(x)e^{-iE_nt/\hbar} \tag{5.38}$$

if the energy eigenvalues are a discrete set, or an integral

$$\psi(x, t) = \int \mathrm{d}\alpha \; c_\alpha \phi_\alpha(x) \mathrm{e}^{-\mathrm{i}E_\alpha t/\hbar} \tag{5.39}$$

if the energy eigenvalues span a continuous range, or a combination of sum and integral

$$\psi(x, t) = \sum_n a_n \phi_n(x) \mathrm{e}^{-\mathrm{i}E_n t/\hbar} + \int \mathrm{d}\alpha \; c_\alpha \phi_\alpha(x) \mathrm{e}^{-\mathrm{i}E_\alpha t/\hbar} \tag{5.40}$$

if some eigenvalues are discrete and others are found in a continuous range.

Most of the effort in quantum mechanics goes into solving the time-independent Schrödinger equation (5.35). Once a complete set of solutions to that equation is found, the general solution (equation (5.40)) to the time-dependent problem follows immediately.

5.4 The free particle

When the potential $V(x)$ vanishes everywhere, the Schrödinger equation in one dimension reduces to the equation for de Broglie waves

$$\mathrm{i}\hbar \frac{\partial \psi}{\partial t} = -\frac{\hbar^2}{2m} \frac{\partial^2 \psi}{\partial x^2} \tag{5.41}$$

and this equation was deduced from the wave function for plane waves, corresponding to a particle with *definite* momentum p and energy $E_p = p^2/2m$

$$\begin{aligned} \psi_p(x, t) &= N \exp\left[\mathrm{i}(px - E_p t)/\hbar\right] \\ &= N \exp\left[\mathrm{i}\left(px - \frac{p^2}{2m} t/\hbar\right]\right. \end{aligned} \tag{5.42}$$

The Schrödinger equation (whether $V = 0$ or not) is a linear equation, and any linear equation has the property that, if $\psi_{p_1}, \psi_{p_1}, \psi_{p_3}, \ldots$ are all solutions of the equation, then so is any linear combination

$$\psi = c_{p_1}\psi_{p_1} + c_{p_2}\psi_{p_2} + c_{p_3}\psi_{p_3} + \cdots \tag{5.43}$$

The general solution is

$$\psi(x, t) = \int_{-\infty}^{\infty} \frac{\mathrm{d}p}{2\pi\hbar} f(p) \exp\left[\mathrm{i}\left(px - \frac{p^2}{2m} t\right)/\hbar\right] \tag{5.44}$$

which is a linear combination of plane wave solutions (5.42). This means that (5.44) is a solution of the free particle Schrödinger equation (5.41), for any choice of function $f(p)$ whatsoever.

Equation (5.44) is an example of the general solution (5.39) of a time-dependent Schrödinger equation. The time-independent Schrödinger equation in this case is

$$-\frac{\hbar^2}{2m}\frac{\partial^2\phi}{\partial x^2} = E\phi \tag{5.45}$$

For each positive eigenvalue $E \in [0, \infty]$ there are two linearly independent solutions

$$\phi_p(x) = \mathrm{e}^{\mathrm{i}px/\hbar} \qquad p = \pm\sqrt{2mE} \tag{5.46}$$

such that any other solution for a given E can be expressed as a linear combination

$$\phi(x) = c_1\mathrm{e}^{\mathrm{i}\sqrt{2mE}\,x/\hbar} + c_2\mathrm{e}^{-\mathrm{i}\sqrt{2mE}\,x/\hbar} \tag{5.47}$$

The set of eigenfunctions and corresponding eigenvalues

$$\left\{\phi_p(x) = \mathrm{e}^{\mathrm{i}px/\hbar},\ E_p = \frac{p^2}{2m}\right\} \qquad -\infty < p < \infty \tag{5.48}$$

is therefore a complete set of solutions of the time-independent Schrödinger equation. Inserting these solutions into equation (5.39), we arrive at the general solution (5.44).

It is easy to show that equation (5.44) is, in fact, the general solution of the free particle Schrödinger wave equation, in the sense that *any* solution of the free particle equation can be put into this form. First of all, the wave function $\psi(x, t)$ at all times is uniquely determined by the wave function at one initial time, say $t = 0$, denoted

$$\phi(x) \equiv \psi(x, 0) \tag{5.49}$$

This is because the Schrödinger equation is first-order in the time derivative, so that given $\psi(x, t)$ at some time t, the wave function is determined an infinitesimal time $t + \epsilon$ later by the rule given in equation (4.59). Next, for any given $\phi(x)$ we can always find a function $f(p)$ such that

$$\phi(x) = \int_{-\infty}^{\infty}\frac{\mathrm{d}p}{2\pi\hbar}\,f(p)\mathrm{e}^{\mathrm{i}px/\hbar} \tag{5.50}$$

To determine $f(p)$, multiply both sides of the equation by $\exp[-\mathrm{i}p'/\hbar]$, and integrate with respect to x:

$$\begin{aligned}\int_{-\infty}^{\infty}\mathrm{d}x\ \phi(x)\mathrm{e}^{-\mathrm{i}p'x/\hbar} &= \int_{-\infty}^{\infty}\mathrm{d}x\int_{-\infty}^{\infty}\frac{\mathrm{d}p}{2\pi\hbar}\,f(p)\mathrm{e}^{\mathrm{i}(p-p')x/\hbar}\\ &= \int_{-\infty}^{\infty}\frac{\mathrm{d}p}{2\pi\hbar}\,f(p)\int_{-\infty}^{\infty}\mathrm{d}x\ \exp\left[\mathrm{i}\frac{p-p'}{\hbar}x\right]\\ &= \int_{-\infty}^{\infty}\frac{\mathrm{d}p}{2\pi\hbar}\,f(p)2\pi\delta\left(\frac{p-p'}{\hbar}\right)\end{aligned} \tag{5.51}$$

where we have used the integral representation of the Dirac delta function (4.50). Using also the identity (4.52):

$$\delta\left(\frac{p-p'}{\hbar}\right) = \hbar\delta(p-p') \tag{5.52}$$

we get

$$\int_{-\infty}^{\infty} \mathrm{d}x\ \phi(x)\mathrm{e}^{-\mathrm{i}p'x/\hbar} = \int_{-\infty}^{\infty} \mathrm{d}p\ f(p)\delta(p-p') = f(p') \tag{5.53}$$

The conclusion is that *any* $\phi(x)$ can be expressed in the form (5.50), for $f(p)$ chosen to equal

$$f(p) = \int_{-\infty}^{\infty} \mathrm{d}x\ \phi(x)\mathrm{e}^{-\mathrm{i}px/\hbar} \tag{5.54}$$

The terminology is that $f(p)$ is the ***Fourier transform*** of $\phi(x)$ (equation (5.54)), and $\phi(x)$ is the ***inverse Fourier transform*** of $f(p)$ (equation (5.50)).[2]

We now have a prescription, given the wave function at any initial time $t = 0$, for finding the wave function at any later time: from $\phi(x) = \psi(x, 0)$, compute the Fourier transform $f(p)$ using (5.54), then insert $f(p)$ into the right-hand side of equation (5.44). The integral of equation (5.44) will then give the wave function $\psi(x, t)$ at any later time.

5.5 Gaussian wavepackets

A wave function of the form

$$\phi(x) = N\delta_L(x) \tag{5.55}$$

goes to zero, as $L \to \infty$, everywhere except at the point $x = 0$. At the other extreme, a plane wave

$$\phi(x) = N\mathrm{e}^{\mathrm{i}p_0x/\hbar} \tag{5.56}$$

has a constant modulus $|\phi| = N$ at every point along the entire line $-\infty < x < \infty$. A wave function which interpolates between these two extreme cases is the 'Gaussian wavepacket'

$$\phi(x) = N\mathrm{e}^{-x^2/2a^2}\mathrm{e}^{\mathrm{i}p_0x/\hbar} \tag{5.57}$$

in which a Gaussian damping factor is used to modulate a plane wave. As $a \to 0$, at fixed p_0, this wave function is proportional to $\delta_L(x)$ (where $L = 1/2a^2$), while as $a \to \infty$, $\phi(x)$ approaches a plane wave.

Gaussian wavepackets provide a very useful example of how the wave function of a free particle evolves in time, and for this purpose we make use of the Gaussian integration formulas (4.46). Assume that the initial wave function $\psi(x, 0)$ has the form of the Gaussian wavepacket (5.57), with some fixed value of the constant a. The first task is to ***normalize*** the wave function, which means: choose the constant N in (5.57) so that the normalization condition

[2] In standard terminology, however, there is no factor of $1/\hbar$ in the exponent.

$$\int \mathrm{d}x\ \psi^*(x, t)\psi(x, t) = 1 \tag{5.58}$$

is fulfilled at $t = 0$ (which implies that it will be fulfilled for all t, by conservation of probability). The condition is

$$\begin{aligned} 1 &= \int \mathrm{d}x\ \psi^*(x, 0)\psi(x, 0) \\ &= N^2 \int \mathrm{d}x \mathrm{e}^{-x^2/a^2} \\ &= N^2\sqrt{\pi a^2} \end{aligned} \tag{5.59}$$

Therefore

$$N = \left(\frac{1}{\pi a^2}\right)^{1/4} \tag{5.60}$$

We begin by computing the initial expectation values $\langle x\rangle_0$, $\langle p\rangle_0$ and the initial uncertainty Δx_0 in position, at time $t = 0$. First

$$\begin{aligned} \langle x\rangle_0 &= \int \mathrm{d}x\ x\phi^*(x)\phi(x) \\ &= \left(\frac{1}{\pi a^2}\right)^{1/2} \int_{-\infty}^{\infty} \mathrm{d}x\ x\mathrm{e}^{-x^2/a^2} \\ &= 0 \end{aligned} \tag{5.61}$$

and

$$\begin{aligned} \langle p\rangle_0 &= \int \mathrm{d}x\ \phi^*(x)\left(-\mathrm{i}\hbar\frac{\partial}{\partial x}\right)\phi(x) \\ &= -\frac{\mathrm{i}\hbar}{\sqrt{\pi a^2}} \int \mathrm{d}x\ \exp\left(-\mathrm{i}\frac{p_0 x}{\hbar} - \frac{x^2}{2a^2}\right)\frac{\partial}{\partial x}\exp\left(\mathrm{i}\frac{p_0 x}{\hbar} - \frac{x^2}{2a^2}\right) \\ &= -\mathrm{i}\frac{\mathrm{i}\hbar}{\sqrt{\pi a^2}} \int \mathrm{d}x\left(\mathrm{i}\frac{p_0}{\hbar} - \frac{x}{a^2}\right)\mathrm{e}^{-x^2/a^2} \\ &= p_0 \end{aligned} \tag{5.62}$$

These are the relevant expectation values at $t = 0$ of position and momentum. Next, the squared uncertainty is

$$\begin{aligned} \Delta x_0^2 &= \langle x^2\rangle_0 - \langle x\rangle_0^2 = \langle x^2\rangle_0 \\ &= \int \mathrm{d}x\ x^2\phi^*(x)\phi(x) \\ &= \left(\frac{1}{\pi a^2}\right)^{1/2} \int_{-\infty}^{\infty} \mathrm{d}x\ x^2\mathrm{e}^{-x^2/a^2} \\ &= \frac{1}{2}a^2 \end{aligned} \tag{5.63}$$

The initial uncertainty in position at time $t = 0$ is therefore

$$\Delta x_0 = \frac{a}{\sqrt{2}} \tag{5.64}$$

To find the time evolution of the wave function, we first compute the Fourier transform

$$\begin{aligned} f(p) &= \int \mathrm{d}x \; \phi(x) \mathrm{e}^{-\mathrm{i}px/\hbar} \\ &= \left(\frac{1}{\pi a^2}\right)^{1/4} \int \mathrm{d}x \; \exp\left[-\frac{x^2}{2a^2} - \mathrm{i}\frac{(p - p_0)}{\hbar} x\right] \end{aligned} \tag{5.65}$$

Using the formula for Gaussian integration

$$\int_{-\infty}^{\infty} \mathrm{d}z \; \mathrm{e}^{-Az^2 - Bz} = \sqrt{\frac{\pi}{A}} \mathrm{e}^{B^2/4A} \tag{5.66}$$

this becomes

$$f(p) = (4\pi a^2)^{1/4} \mathrm{e}^{-a^2(p-p_0)^2/2\hbar^2} \tag{5.67}$$

Then, substituting $f(p)$ into (5.44), we find the wave function at any later time t

$$\begin{aligned} \psi(x, t) &= (4\pi a^2)^{1/4} \int \frac{\mathrm{d}p}{2\pi\hbar} \mathrm{e}^{-a^2(p-p_0)^2/2\hbar^2} \exp\left[\mathrm{i}\left(px - \frac{p^2}{2m}t\right)/\hbar\right] \\ &= \frac{(4\pi a^2)^{1/4}}{2\pi\hbar} \int \mathrm{d}q \; \exp\left[-\frac{a^2}{2\hbar^2}q^2 + \mathrm{i}\frac{q + p_0}{\hbar}x - \mathrm{i}\frac{(q + p_0)^2}{2\,m\hbar}t\right] \\ &= \frac{(4\pi a^2)^{1/4}}{2\pi\hbar} \exp\left[\mathrm{i}\left(p_0 x - \frac{p_0^2}{2m}t\right)\Big/\hbar\right] \\ &\quad \times \int \mathrm{d}q \; \exp\left[-\left(\frac{a^2}{2\hbar^2} + \frac{\mathrm{i}t}{2m\hbar}\right)q^2 + \mathrm{i}\left(\frac{x}{\hbar} - \frac{p_0 t}{m\hbar}\right)q\right] \\ &= \frac{(4\pi a^2)^{1/4}}{2\pi\hbar} \left[\frac{\pi}{\dfrac{a^2}{2\hbar^2} + \dfrac{\mathrm{i}t}{2m\hbar}}\right]^{1/2} \exp\left\{-\frac{(x - v_0 t)^2}{2(a^2 + \mathrm{i}\hbar t/m)}\right\} \mathrm{e}^{\mathrm{i}\left(p_0 x - \frac{p_0^2}{2m}t\right)\Big/\hbar} \end{aligned} \tag{5.68}$$

where

$$v_0 \equiv \frac{p_0}{m} \tag{5.69}$$

Although (5.68) looks like a complicated expression, its modulus is very simple:

$$\begin{aligned}|\psi|^2 &= \psi^*(x, t)\psi(x, t)\\ &= \frac{1}{\sqrt{\pi a^2(t)}}\exp\left[-\frac{(x - v_0 t)^2}{a^2(t)}\right]\end{aligned} \tag{5.70}$$

where we have defined

$$a(t) \equiv \sqrt{a^2 + \frac{\hbar^2}{m^2 a^2}t^2} \tag{5.71}$$

Note that the modulus of the wave function remains a Gaussian at all times, but the width $a(t)$ of the Gaussian increases, i.e. the wave function 'spreads out' as time passes.

We can then calculate $\langle x\rangle$ and Δx at any time $t > 0$:

$$\begin{aligned}\langle x\rangle &= \frac{1}{\sqrt{\pi a^2(t)}}\int \mathrm{d}x\; x \exp\left[-\frac{(x - v_0 t)^2}{a^2(t)}\right]\\ &= \frac{1}{\sqrt{\pi a^2(t)}}\int \mathrm{d}x'\; (x' + v_0 t)\exp\left[-\frac{x'^2}{a^2(t)}\right]\\ &= v_0 t\end{aligned} \tag{5.72}$$

and also

$$\begin{aligned}\Delta x^2 &= \langle (x - \langle x\rangle)^2\rangle = \langle (x - v_0 t)^2\rangle\\ &= \frac{1}{\sqrt{\pi a^2(t)}}\int \mathrm{d}x (x - v_0 t)^2 \exp\left[-\frac{(x - v_0 t)^2}{a^2(t)}\right]\\ &= \frac{1}{\sqrt{\pi a^2(t)}}\int \mathrm{d}x'\; x'^2 \exp\left[-\frac{x'^2}{a^2(t)}\right]\\ &= \frac{1}{2}a^2(t)\end{aligned} \tag{5.73}$$

so that

$$\Delta x = \frac{a(t)}{\sqrt{2}} = \frac{1}{\sqrt{2}}\sqrt{a^2 + \frac{\hbar^2}{m^2 a^2}t^2} \tag{5.74}$$

To compute $\langle p \rangle$, we make use again of the representation (5.44)

$$\begin{aligned}
\langle p \rangle &= \int \mathrm{d}x\, \psi^*(x, t)\left(-\mathrm{i}\hbar\frac{\partial}{\partial x}\right)\psi(x, t) \\
&= \int \mathrm{d}x \left\{\int \frac{\mathrm{d}p_1}{2\pi\hbar} f(p_1)\mathrm{e}^{\mathrm{i}(p_1 x - E_{p_1}t)/\hbar}\right\}^* \left(-\mathrm{i}\hbar\frac{\partial}{\partial x}\right) \int \frac{\mathrm{d}p_2}{2\pi\hbar} f(p_2)\mathrm{e}^{\mathrm{i}(p_2 x - E_{p_2}t)/\hbar} \\
&= \int \frac{\mathrm{d}p_1}{2\pi\hbar} f^*(p_1)\mathrm{e}^{\mathrm{i}E_{p_1}t/\hbar} \int \frac{\mathrm{d}p_2}{2\pi\hbar}\, p_2 f(p_2)\mathrm{e}^{-\mathrm{i}E_{p_2}t/\hbar} \int \mathrm{d}x\, \mathrm{e}^{\mathrm{i}(p_2 - p_1)x/\hbar} \\
&= \int \frac{\mathrm{d}p_1}{2\pi\hbar} f^*(p_1)\mathrm{e}^{\mathrm{i}E_{p_1}t/\hbar} \int \frac{\mathrm{d}p_2}{2\pi\hbar}\, p_2 f(p_2)\mathrm{e}^{-\mathrm{i}E_{p_2}t/\hbar} 2\pi\hbar\delta(p_1 - p_2) \\
&= \int \frac{\mathrm{d}p}{2\pi\hbar} p f^*(p) f(p)
\end{aligned} \tag{5.75}$$

Note that the time dependence has vanished completely from this equation. It follows that

$$\langle p \rangle = \langle p \rangle_0 = p_0 \tag{5.76}$$

and the expectation values of position and momentum together obey

$$\begin{aligned}
\langle x \rangle_t &= \langle x \rangle_0 + \frac{\langle p \rangle_0}{m} t \\
\langle p \rangle_t &= p_0
\end{aligned} \tag{5.77}$$

With the '⟨⟩' signs removed, these are the equations of a classical trajectory in phase space for a free particle of mass m.

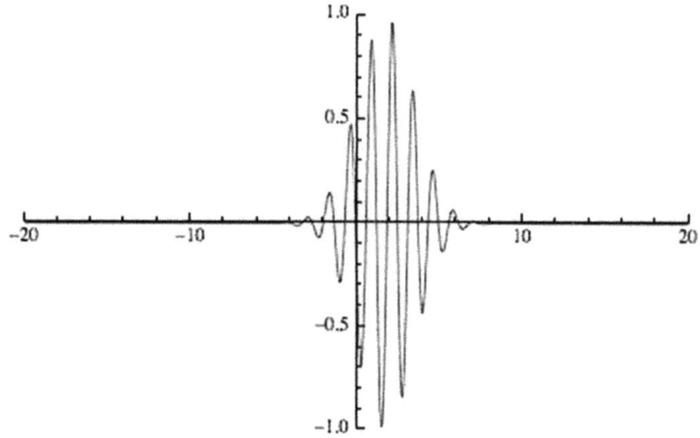

Figure 5.1. Time evolution of a Gaussian wavepacket for a free particle. Only the real part of the complex wave function is shown. To illustrate the motion we employ 'periodic boundary conditions', meaning that a classical particle moving past the right boundary reappears at the left boundary, and vice versa.

On the other hand, the positional uncertainty Δx of the quantum state, which is proportional to the width $a(t)$ of the Gaussian, increases in time; this is known as the 'spreading of the wavepacket' (see the animation in figure 5.1, which illustrates the time evolution of a Gaussian wavepacket). As noted in chapter 3, one of the reasons that the electron cannot be regarded as being *literally* a wave is that its wave function expands to macroscopic size quite rapidly, and it is quite certain that electrons are not macroscopic objects. Let us consider the time t_{2a} that it takes a free particle wave function to double in size ($a(t) = 2a(0)$):

$$t_{2a} = \sqrt{3}\frac{ma^2}{\hbar} \tag{5.78}$$

and the time it takes to expand to some very much larger size $a(t) = A \gg a$

$$t_A = \frac{maA}{\hbar} \tag{5.79}$$

As an example, choose the initial spread of the wavepacket to be on the order of the diameter of atomic hydrogen

$$a = 10^{-10}\text{ m} \tag{5.80}$$

and A to be a macroscopic distance, e.g.

$$A = 10\text{ cm} \tag{5.81}$$

Begin by considering an electron, which has a mass $m_e = 9.11 \times 10^{-31}$ kg; and $\hbar = 1.05 \times 10^{-34}$ J-s. Then the time required for the wave function to expand from $a = 10^{-10}$ m to 2×10^{-10} m is

$$t_{2a} = \sqrt{3}\frac{m_e a^2}{\hbar} = 1.5 \times 10^{-16}\text{ s} \tag{5.82}$$

Likewise, the time required for the wave function to spread to $A = 10$ cm is

$$t_A = \frac{m_e aA}{\hbar} = 8.7 \times 10^{-6}\text{ s} \tag{5.83}$$

If the energy of the electron is $E = 10$ electron volts (eV), then the velocity is

$$v = \sqrt{2E/m} = 1.87 \times 10^6\text{ m s}^{-1} \tag{5.84}$$

Therefore, for the wave function of a 10 eV electron to expand from an atomic diameter to a width of 10 cm, the electron need only travel

$$d = vt_A = 16.3\text{ m} \tag{5.85}$$

Compare these figures for an electron with the corresponding results for an object the mass of a baseball, say $m_B = 0.25$ kg. Supposing the width of the wave function is also initially $a = 10^{-10}$ m, the corresponding numbers are

$$\begin{aligned} t_{2a} &= 4.1 \times 10^{14}\ \text{s} = 1.3\ \text{million years} \\ t_A &= 2.4 \times 10^{23}\ \text{s} = 1.8 \times 10^{17}\ \text{years} \end{aligned} \tag{5.86}$$

In other words, for all practical purposes, the wave function of an object of macroscopic mass doesn't expand at all.

From estimates like these, we can begin to see how quantum dynamics goes over to classical physics when applied to macroscopic objects. Because if Δx is very small, and stays very small, then $\langle V(x)\rangle = V(\langle x\rangle)$ to a very good approximation, and likewise

$$\left\langle -\frac{\mathrm{d}}{\mathrm{d}x}V(x)\right\rangle \approx -\frac{\mathrm{d}}{\mathrm{d}\langle x\rangle}V(\langle x\rangle)$$

In that case, the Ehrenfest equations really become Hamilton's equations, just by replacing x with $\langle x\rangle$, and $\langle x\rangle_t$ will trace out a trajectory that satisfies Newton's laws. But this is only true if the assumption that Δx is small and stays small is satisfied, which is not the case for atoms and electrons. For such microscopic objects, the expectation values $\langle x\rangle_t$, $\langle p\rangle_t$, do not necessarily obey (for $V(x) \neq 0$) the laws of motion introduced by Isaac Newton, and the use of the underlying Schrödinger equation to describe time evolution is unavoidable. We will return to the relationship between wave motion and classical particle motion much later, in chapter 20.

5.6 Group velocity and phase velocity

We have seen that the expectation value of position $\langle x\rangle$ moves at a velocity $v = \langle p\rangle/m$ when $V(x) = 0$. This is a consequence of Ehrenfest's principle for a free particle. On the other hand, the velocity v_{phase} of any plane wave is given by the usual formula

$$v_{\text{phase}} = \lambda f \tag{5.87}$$

so that for a de Broglie wave of a definite momentum p

$$\begin{aligned} v_{\text{phase}} &= \frac{h}{p}\frac{E_p}{h} \\ &= \frac{E_p}{p} = \frac{p^2/2m}{p} \\ &= \frac{p}{2m} \end{aligned} \tag{5.88}$$

which is half the velocity that one might expect, since a classical particle with momentum p moves at a velocity p/m. There is no contradiction, however. The velocity $v_{\text{phase}} = \lambda f$ of a plane wave is known as the ***phase velocity*** of the wave; it refers to the speed at which a point of given phase on the wave (a crest or a trough, say), propagates through space. However, that is not necessarily the same speed at which the energy carried by a group of waves travels through space. That velocity is known as the ***group velocity***.

Consider a wave which is a superposition of two plane waves of frequencies f_1, f_2, and wavelengths λ_1, λ_2, respectively, and which have the same amplitude A. We will assume that the difference between frequencies Δf is much smaller than the average frequency $\bar{f}$, i.e.

$$\Delta f \ll \bar{f} \qquad \text{and} \qquad \Delta\lambda \ll \bar{\lambda} \tag{5.89}$$

By the superposition principle of wave motion

$$\begin{aligned}\psi(x,\,t) &= \psi_1(x,\,t) + \psi_2(x,\,t)\\ &= A\left[\mathrm{e}^{\mathrm{i}(k_1x-\omega_1t)} + \mathrm{e}^{\mathrm{i}(k_2x-\omega_2t)}\right]\\ &= 2A\mathrm{e}^{\mathrm{i}(\bar{k}x-\bar{\omega}t)}\cos\left[\frac{1}{2}(\Delta kx - \Delta\omega t)\right]\end{aligned} \tag{5.90}$$

which is the product of two waves, one of which is moving with the velocity

$$v_{\text{phase}} = \frac{\bar{\omega}}{\bar{k}} \tag{5.91}$$

and the other moving with velocity

$$v_{\text{group}} = \frac{\Delta\omega}{\Delta k} \tag{5.92}$$

The $\mathrm{e}^{(\bar{k}x-\bar{\omega}t)}$ factor can be expanded into a cosine plus i sine, and the product

$$\cos(\bar{k}x - \bar{\omega}t)\cos\left[\frac{1}{2}(\Delta kx - \Delta\omega t)\right] \tag{5.93}$$

is shown in figure 5.2. The 'packets' of waves move collectively with the group velocity v_{group}; the crest of a wave within each packet travels at the phase velocity v_{phase}.

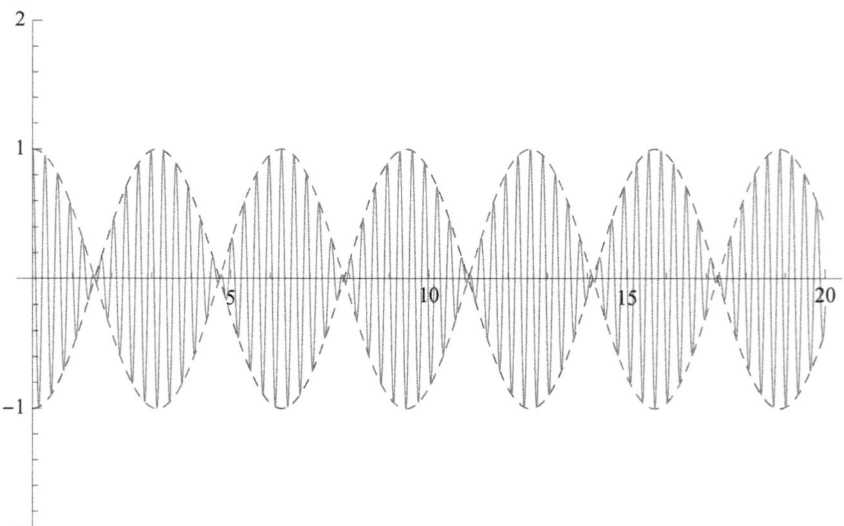

Figure 5.2. Product of a two plane waves with slightly different frequencies and wavelengths.

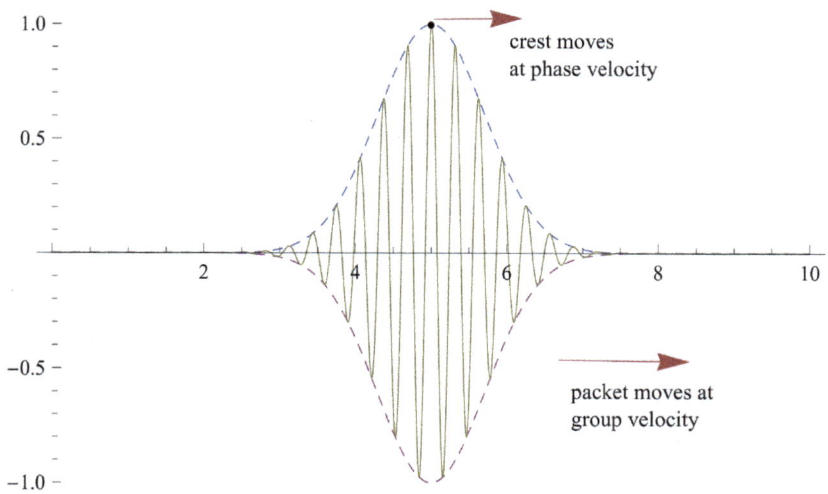

Figure 5.3. An example of a wavepacket. The packet propagates at the group velocity, while crests propagate at the phase velocity.

Now let us consider a general wavepacket of the form

$$\psi(x, t) = \int \mathrm{d}k \ f(k) \mathrm{e}^{\mathrm{i}(kx - \omega(k)t)} \tag{5.94}$$

For a free particle in quantum mechanics, the wavenumber and angular frequency are related to momentum and energy

$$k = \frac{p}{\hbar} \quad \text{and} \quad \omega = \frac{E_p}{\hbar} \tag{5.95}$$

Suppose, as in the case of the Gaussian wavepacket (equation (5.67)), that $f(k)$ is peaked around a particular value k_0. In that case, we can make a Taylor expansion of $\omega(k)$ around $k = k_0$:

$$\omega(k) = \omega_0 + \left(\frac{\mathrm{d}\omega}{\mathrm{d}k}\right)_{k_0} (k - k_0) + O[(k - k_0)^2] \tag{5.96}$$

Inserting this expansion into (5.94), and dropping terms of order $(k - k_0)^2$ and higher, we obtain

$$\begin{aligned}\psi(x, t) &\approx \int \mathrm{d}k \ f(k) \exp\left[\mathrm{i}\left(k_0 x + (k - k_0)x - \omega_0 t - \frac{\mathrm{d}\omega}{\mathrm{d}k}(k - k_0)t\right)\right] \\ &\approx \mathrm{e}^{\mathrm{i}(k_0 x - \omega_0 t)} \int \mathrm{d}k f(k) \exp\left[\mathrm{i}(k - k_0)(x - v_{\mathrm{group}} t)\right] \\ &\approx \mathrm{e}^{\mathrm{i}(k_0 x - \omega_0 t)} F[x - v_{\mathrm{group}} t]\end{aligned} \tag{5.97}$$

where

$$v_{\text{group}} = \left(\frac{\mathrm{d}\omega}{\mathrm{d}k}\right)_{k=k_0} \tag{5.98}$$

This is again a product of two waveforms; a plane wave moving with velocity

$$v_{\text{phase}} = \frac{\omega_0}{k_0} \tag{5.99}$$

and a wavepulse $F[x - v_{\text{group}}t]$ moving with velocity v_{group} of equation (5.98). The product is indicated schematically in figure 5.3. It is clear that the wavepacket propagates collectively at the velocity of the pulse v_{group} rather than the velocity of a crest v_{phase}.[3]

In the case of a free particle, we have

$$\begin{aligned} v_{\text{group}} &= \left(\frac{\mathrm{d}\omega}{\mathrm{d}k}\right)_{k=k_0} \\ &= \left(\frac{\mathrm{d}(E_p/\hbar)}{\mathrm{d}(p/\hbar)}\right)_{p=\langle p\rangle} \\ &= \left(\frac{\mathrm{d}E_p}{\mathrm{d}p}\right)_{p=\langle p\rangle} \\ &= \frac{\langle p\rangle}{m} \end{aligned} \tag{5.100}$$

which is the standard relation between velocity and momentum. The conclusion is that, although the phase velocity of a de Broglie wave is half the classical velocity, a packet of de Broglie waves travels collectively at exactly the classical velocity.

5.7 The particle in a closed tube

Finally we return to the example of chapter 4, in which a particle is moving in a closed tube of length L; collisions between the particle and the ends of the tube are assumed to be elastic. Since the potential barrier against leaving the tube is, for all practical purposes, infinite, the particle is moving in an infinite potential well

$$V(x) = \begin{cases} 0 & 0 \leqslant x \leqslant L \\ \infty & \text{otherwise} \end{cases} \tag{5.101}$$

The problem is to solve the time-independent Schrödinger equation (5.35) in this potential.

[3] According to (5.97) the wavepacket propagates without changing its shape, which is not quite true, as we have seen in the case of the Gaussian wavepacket. The 'spreading' of the wavepacket is due to terms of order $(k - k_0)^2$, which were dropped in equation (5.97).

In the interval $x \in [0, L]$, the time-independent Schrödinger equation is the same as that for a free particle

$$-\frac{\hbar^2}{2m}\frac{\partial^2}{\partial x^2}\phi = E\phi \tag{5.102}$$

and has the same solutions for $E > 0$, namely any linear combination

$$\phi(x) = c_1 \mathrm{e}^{\mathrm{i}px/\hbar} + c_2 \mathrm{e}^{-\mathrm{i}px/\hbar} \qquad \text{where} \qquad p = \sqrt{2mE} \tag{5.103}$$

On the other hand, in the regions $x < 0$ and $x > L$, we have

$$\left[-\frac{\hbar^2}{2m}\frac{\partial^2}{\partial x^2} + \infty\right]\phi = E\phi \tag{5.104}$$

The only possible solution of this equation, for finite E, is

$$\phi(x) = 0 \qquad x < 0 \quad \text{or} \quad x > L \tag{5.105}$$

which simply means that the probability of finding the particle outside the tube is zero.

The solution of a differential equation must be a continuous function. Continuity of the wave function at the points $x = 0$ and $x = L$ gives two boundary conditions

$$\begin{aligned} 0 &= \phi(0) = c_1 + c_2 \\ 0 &= \phi(L) = c_1 \mathrm{e}^{\mathrm{i}pL/\hbar} + c_2 \mathrm{e}^{-\mathrm{i}pL/\hbar} \end{aligned} \tag{5.106}$$

The first condition gives $c_2 = -c_1$, and then the second condition becomes

$$2\mathrm{i}c_1 \sin\left[\frac{pL}{\hbar}\right] = 0 \tag{5.107}$$

The second equation can be recognized as the condition for a standing wave in an interval of length L, i.e.

$$\sin(kL) = 0 \tag{5.108}$$

which is satisfied for wavenumbers

$$k = \frac{n\pi}{L} \tag{5.109}$$

or, in terms of wavelengths $\lambda = 2\pi/k$,

$$L = n\frac{\lambda}{2} \tag{5.110}$$

In the case of a particle in a tube, the wavenumber k is the same as for de Broglie waves

$$k = \frac{2\pi}{\lambda} = \frac{p}{\hbar} \tag{5.111}$$

and the standing wave requirement $\sin(kL) = 0$ implies

$$\begin{aligned} \frac{pL}{\hbar} &= n\pi \\ \Rightarrow \quad p_n &= n\frac{\pi\hbar}{L} \end{aligned} \tag{5.112}$$

where a subscript n has been added to indicate that each $p = \sqrt{2mE}$ is associated with a positive integer $n = 1, 2, 3, \ldots$. The energy eigenstates are therefore

$$\phi_n(x) = \begin{cases} N \sin\left[\dfrac{n\pi}{L}x\right] & 0 \leqslant x \leqslant L \\ 0 & \text{otherwise} \end{cases} \tag{5.113}$$

where $N = 2ic_1$, each with a corresponding eigenvalue

$$E_n = \frac{p_n^2}{2m} = n^2\frac{\pi^2\hbar^2}{2mL^2} \tag{5.114}$$

It is useful to normalize $\phi_n(x)$, so that the solution to the time-dependent Schrödinger equation

$$\psi_n(x, t) = \phi_n(x)e^{-iE_nt/\hbar} \tag{5.115}$$

is a physical state. The normalization condition is

$$\begin{aligned} 1 &= \int_{-\infty}^{\infty} dx\ \psi_n^*(x, t)\psi_n(x, t) \\ &= |N|^2 \int_0^L dx\ \sin^2\left[\frac{n\pi x}{L}\right] \\ &= |N|^2 \frac{L}{2} \end{aligned} \tag{5.116}$$

This determines the normalization constant

$$N = \sqrt{\frac{2}{L}} \tag{5.117}$$

so that

$$\left\{\phi_n(x) = \sqrt{\frac{2}{L}} \sin\left[\frac{n\pi x}{L}\right], \quad E_n = \frac{n^2\pi^2\hbar^2}{2mL^2}\right\}, \quad n = 1, 2, 3, \ldots \tag{5.118}$$

is the complete set of energy eigenvalues and energy eigenfunctions (in the interval $[0, L]$). Then the general solution of the time-dependent Schrödinger equation, according to equation (5.38) is

$$
\begin{aligned}
\psi(x, t) &= \sum_{n=1}^{\infty} a_n \phi_n(x) \mathrm{e}^{-\mathrm{i}E_n t/\hbar} \\
&= \sqrt{\frac{2}{L}} \sum_{n=1}^{\infty} a_n \sin\left[\frac{n\pi x}{L}\right] \mathrm{e}^{-\mathrm{i}E_n t/\hbar}
\end{aligned} \tag{5.119}
$$

Now suppose that the wave function $\psi(x, t)$ is specified at an initial time $t = 0$. How do we find the wave function at any later time $t > 0$, using equation (5.119)? The problem is to find the set of constants $\{a_n\}$ given $\psi(x, 0)$.

The method for finding the a_n, for the particle in a tube, is closely analogous to solving for $f(p)$ for the free particle in equation (5.44), given $\psi(x, 0)$. Begin with

$$
\psi(x, 0) = \sum_{n=1}^{\infty} a_n \phi_n(x) \tag{5.120}
$$

and multiply both sides of the equation by $\phi_k^*(x)$[4]

$$
\phi_k^*(x)\psi(x, 0) = \sum_{n=1}^{\infty} a_n \phi_k^*(x)\phi_n(x) \tag{5.121}
$$

Integrate over x

$$
\int_0^L \mathrm{d}x\ \phi_k^*(x)\psi(x, 0) = \sum_{n=1}^{\infty} a_n \int_0^L \mathrm{d}x\ \phi_k^*(x)\phi_n(x) \tag{5.122}
$$

and use the ***orthogonality relation***

$$
\begin{aligned}
\langle \phi_k | \phi_n \rangle &= \int_{-\infty}^{\infty} \mathrm{d}x\ \phi_k^*(x)\phi_n(x) \\
&= \frac{2}{L} \int_0^L \mathrm{d}x\ \sin\left[k\frac{\pi x}{L}\right] \sin\left[n\frac{\pi x}{L}\right] \\
&= \delta_{kn}
\end{aligned} \tag{5.123}
$$

where δ_{ij} is the Kronecker delta introduced in chapter 4.

Equation (5.123) is referred to as orthogonality because it is the expression for vectors in Hilbert space, which corresponds to the orthogonality relation for unit vectors $\mathbf{e}^i$ in an N-dimensional vector space

$$
\mathbf{e}^i \cdot \mathbf{e}^j = \delta_{ij} \tag{5.124}
$$

In a later chapter we will see that the set of energy eigenstates $\{|\phi_n\rangle\}$ can (and should) be viewed as a set of orthonormal basis vectors in Hilbert space. Applying the orthogonality relation, equation (5.122) becomes

---

[4] Since ϕ_n is a real function, complex conjugation does nothing. It is indicated here because in general, when the energy eigenstates are complex, this operation is necessary.

$$\int_0^L \mathrm{d}x\ \phi_k^*(x)\psi(x,0) = \sum_{n=1}^{\infty} a_n \delta_{kn} \\ = a_k \tag{5.125}$$

This gives us the complete prescription for finding the physical state of the particle at any time $t > 0$, given the state at an initial time $t = 0$. One has to compute the set of coefficients

$$a_n = \langle \phi_n | \psi(t=0) \rangle \\ = \int_0^L \mathrm{d}x\ \phi_n^*(x)\psi(x,0) \tag{5.126}$$

and substitute these into equation (5.119) to get the wave function $\psi(x, t)$.

The expectation values of position and momentum, as a function of time, can also be expressed in terms of the coefficients a_n. For the position

$$\langle x \rangle = \int \mathrm{d}x \psi^*(x,t) x \psi(x,t) \\ = \int \mathrm{d}x \left\{ \sum_{i=1}^{\infty} a_i \phi_i(x) \mathrm{e}^{-\mathrm{i}E_i t/\hbar} \right\}^* x \sum_{j=1}^{\infty} a_j \phi_j(x) \mathrm{e}^{-\mathrm{i}E_j t/\hbar} \\ = \sum_{i=1}^{\infty} \sum_{j=1}^{\infty} a_i^* a_j \mathrm{e}^{\mathrm{i}(E_i - E_j)t/\hbar} X_{ij} \tag{5.127}$$

where we have defined

$$X_{mn} \equiv \int_0^L \phi_m^*(x) x \phi_n(x) \\ = \frac{2}{L} \int_0^L \mathrm{d}x\ x \sin\left[m\frac{\pi x}{L}\right] \sin\left[n\frac{\pi x}{L}\right] \\ = \begin{cases} \frac{1}{2}L & m = n \\ 0 & m - n \text{ even} \\ \frac{2L}{\pi^2}[(m+n)^{-2} - (m-n)^{-2}] & m - n \text{ odd} \end{cases} \tag{5.128}$$

Similarly, for momentum

$$\langle p \rangle = \int \mathrm{d}x \psi^*(x,t) \tilde{p} \psi(x,t) \\ = \int \mathrm{d}x \left\{ \sum_{i=1}^{\infty} a_i \phi_i(x) \mathrm{e}^{-\mathrm{i}E_i t/\hbar} \right\}^* \left(-\mathrm{i}\hbar \frac{\partial}{\partial x} \right) \sum_{j=1}^{\infty} a_j \phi_j(x) \mathrm{e}^{-\mathrm{i}E_j t/\hbar} \\ = \sum_{i=1}^{\infty} \sum_{j=1}^{\infty} a_i^* a_j \mathrm{e}^{\mathrm{i}(E_i - E_j)t/\hbar} P_{ij} \tag{5.129}$$

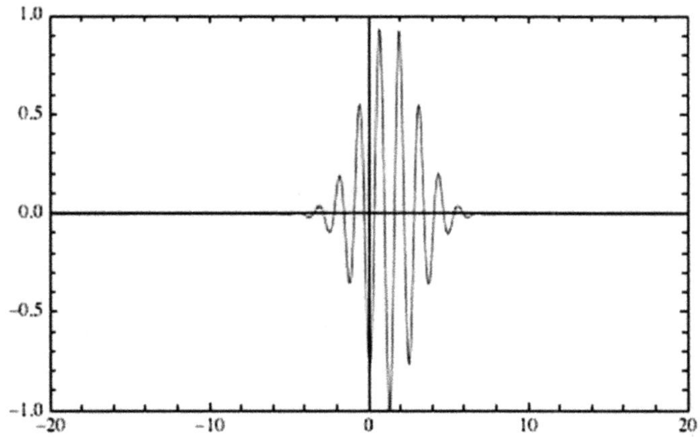

Figure 5.4. Evolution of a Gaussian wavepacket in an infinite square well.

where

$$\begin{aligned} P_{mn} &\equiv \int_0^L \phi_m^*(x)\left(-i\hbar\frac{\partial}{\partial x}\right)\phi_n(x) \\ &= -i\hbar\frac{2\pi n}{L^2}\int_0^L \mathrm{d}x \sin\left[m\frac{\pi x}{L}\right]\cos\left[n\frac{\pi x}{L}\right] \\ &= \begin{cases} -\dfrac{i\hbar}{L}\dfrac{4mn}{m^2-n^2} & m-n \text{ odd} \\ 0 & m-n \text{ even} \end{cases} \end{aligned} \tag{5.130}$$

The animation in figure 5.4 shows the evolution of a Gaussian wavepacket in a finite square well. Note how the particle wave function 'bounces' from one end of the wall to the other while slowly spreading out.

Problems

1. A ball of mass m is moving on a V-shaped track whose height above the floor is given by $y = \tan(\theta)|x|$, where x is the horizontal coordinate measured along the floor. Let the 'generalized coordinates' of this system be simply x (note that y and $\partial y/\partial t$ are determined from $x(t)$), and then:
   a) Find the Lagrangian and the Euler–Lagrange equations of motion of this system.
   b) Work out the Hamiltonian, and the Hamilton equations of motion.
   c) From the Hamiltonian, and expressing p as a differential operator, write down the Schrödinger equation for this system.
2. Verify conservation of probability and the Ehrenfest equations in the three-dimensional case.
3. Show that the wave function in equation (5.44) is a solution of the free particle Schrödinger equation by substituting (5.44) directly into (5.41).
4. Verify that (5.70)–(5.71) follow from (5.68).

5. Let

$$P_T(t) = \langle\psi|\psi\rangle = \int \mathrm{d}x\psi^*(x, t)\psi(x, t) \tag{5.131}$$

be the total probability of finding a particle *somewhere* in space. We have seen that $\partial_t P_T(t) = 0$. But suppose that the potential had a constant imaginary part, i.e.

$$V(x) = V_R(x) - \mathrm{i}V_I \tag{5.132}$$

If $P = 1$ at time $t = 0$, calculate $P(t)$ at any later time. (Complex potentials are sometimes used to describe situations where particles can be absorbed by a 'sink', and disappear.)

6. Total electric charge is conserved, which means that if charge disappears from some region, it must appear somewhere else, and this fact is expressed in the continuity equation

$$\frac{\mathrm{d}\rho}{\mathrm{d}t} = -\nabla \cdot \mathbf{j} \tag{5.133}$$

where ρ is the electric charge density, and $\mathbf{j}$ is the electric current density. But total probability is also conserved, so there must be a similar continuity equation for probability density. If $P(x, y, z, t) = \psi^*(x, y, z, t)\psi(x, y, z, t)$ is the probability density, find the probability current density.

IOP Publishing

An Introduction to Quantum Theory

Jeff Greensite

Chapter 6

Energy and uncertainty

We have seen that for a Gaussian wavepacket

$$\phi(x) = \frac{1}{(\pi a^2)^{1/4}} e^{-(x-x_0)^2/2a^2} e^{ip_0 x/\hbar} \tag{6.1}$$

the uncertainty in position is $\Delta x = a/\sqrt{2}$, which means that this uncertainty vanishes as $a \to 0$. In this limit, $\phi^*\phi$ approaches a Dirac delta function, and the probability density $P(x)$ peaks sharply at the point $x = x_0$. If a particle is in such a quantum state when its position is measured, it is *certain* to be found in the immediate vicinity of the point $x = x_0$; this quantum state is known as an ***eigenstate of position***.

On the other hand, as $a \to \infty$, the Gaussian wavepacket (6.1) approaches a plane wave. According to de Broglie, the plane wave represents a particle moving with a *definite* momentum p_0, therefore the uncertainty in momentum should be $\Delta p = 0$ in that limit. If the particle is in a plane wave state when its momentum is measured, it is *certain* to be found to have momentum $p = p_0$; this quantum state is known as an ***eigenstate of momentum***.

Because the eigenstates of position and momentum are very different, it is impossible for a particle to be in a physical state in which *both* its position *and* its momentum are certain; i.e. $\Delta x = \Delta p = 0$. A simple consequence is that no measurement can determine, precisely and simultaneously, the position and momentum of a particle. If such measurements were possible, they would leave the particle in a physical state having $\Delta x = \Delta p = 0$, and no such quantum state exists. A more quantitative expression of this limitation on measurements is known as the ***Heisenberg uncertainty principle***. To derive it, we need to be able to compute Δp for an arbitrary physical state.

6.1 The expectation value of p^n

According to the Born interpretation, the expectation value of any function $F(x)$ of position is given by

doi:10.1088/978-0-7503-1167-0ch6

$$\begin{aligned}\langle F(x)\rangle &= \int F(x)P(x)\mathrm{d}x \\ &= \int \mathrm{d}x\, F(x)\psi^*(x, t)\psi(x, t)\end{aligned} \tag{6.2}$$

while from Ehrenfest's principle, one finds that the expectation value of momentum is

$$\langle p\rangle = \int \mathrm{d}x\, \psi^*(x)\left(-\mathrm{i}\hbar\frac{\partial}{\partial x}\right)\psi(x, t) \tag{6.3}$$

But we would also like to know how to compute the expectation value of energy

$$\begin{aligned}\langle E\rangle = \langle H\rangle &= \left\langle \frac{p^2}{2m} + V(x)\right\rangle \\ &= \frac{1}{2m}\langle p^2\rangle + \langle V(x)\rangle\end{aligned} \tag{6.4}$$

as well as the uncertainty in momentum

$$\Delta p^2 = \langle p^2\rangle - \langle p\rangle^2 \tag{6.5}$$

and to evaluate either of these quantities, it is necessary to evaluate $\langle p^2\rangle$. What is needed is a probability density for momentum, $P(p)$, analogous to $P(x)$, giving the probability $P(p)\Delta p$ that the particle momentum will be found within a small range of momenta, of width Δp centered at p. This probability density can be deduced from the rule for computing $\langle p\rangle$, and the inverse Fourier transform representation (5.50).

At a given time t_0, the wave function is a function of x only, which we denote, again, by

$$\phi(x) \equiv \psi(x, t_0) \tag{6.6}$$

and the expectation value of momentum at that instant is given by

$$\langle p\rangle = \int \mathrm{d}x\phi^*(x)\left(-\mathrm{i}\hbar\frac{\partial}{\partial x}\right)\phi(x) \tag{6.7}$$

Expressing $\phi(x)$ in terms of its inverse Fourier transform, equation (5.50), and proceeding as in equation (5.75),

$$\begin{aligned}\langle p\rangle &= \int \mathrm{d}x\, \phi^*(x)\left(-\mathrm{i}\hbar\frac{\partial}{\partial x}\right)\phi(x) \\ &= \int \mathrm{d}x\left\{\int \frac{\mathrm{d}p_1}{2\pi\hbar} f(p_1)\mathrm{e}^{\mathrm{i}p_1 x/\hbar}\right\}^*\left(-\mathrm{i}\hbar\frac{\partial}{\partial x}\right)\int \frac{\mathrm{d}p_2}{2\pi\hbar} f(p_2)\mathrm{e}^{\mathrm{i}p_2 x/\hbar} \\ &= \int \frac{\mathrm{d}p_1}{2\pi\hbar} f^*(p_1)\int \frac{\mathrm{d}p_2}{2\pi\hbar} p_2 f(p_2)\int \mathrm{d}x\mathrm{e}^{\mathrm{i}(p_2-p_1)x/\hbar} \\ &= \int \frac{\mathrm{d}p_1}{2\pi\hbar} f^*(p_1)\int \frac{\mathrm{d}p_2}{2\pi\hbar} p_2 f(p_2)2\pi\hbar\delta(p_1 - p_2)\end{aligned} \tag{6.8}$$

Then, using the delta function to eliminate one of the p-integrations, we find

$$\langle p\rangle = \int \mathrm{d}p\, p\frac{f^*(p)f(p)}{2\pi\hbar} \tag{6.9}$$

Comparison of this expression to the corresponding expression for position

$$\langle x\rangle = \int \mathrm{d}x\, xP(x) = \int \mathrm{d}x\, x\psi^*(x)\psi(x) \tag{6.10}$$

shows that the probability density $P(p)$ for momentum must be

$$P(p) = \frac{f^*(p)f(p)}{2\pi\hbar} \tag{6.11}$$

Then the expectation value $\langle G(p)\rangle$ of any function of momentum is given by

$$\langle G(p)\rangle = \int \frac{\mathrm{d}p}{2\pi\hbar}\, G(p)f^*(p)f(p) \tag{6.12}$$

and in particular

$$\langle p^n\rangle = \int \frac{\mathrm{d}p}{2\pi\hbar}\, p^n f^*(p)f(p) \tag{6.13}$$

It is useful to express $\langle p^n\rangle$ directly in terms of the wave function $\phi(x)$, rather than its Fourier transform. Using equation (5.54) to reexpress $f(p)$ in terms of $\phi(x)$, and inserting this expression into (6.13), we have

$$\begin{aligned}
\langle p^n\rangle &= \int \frac{\mathrm{d}p}{2\pi\hbar}p^n \left\{\int \mathrm{d}x_1\, \phi(x_1)\mathrm{e}^{-\mathrm{i}px_1/\hbar}\right\}^* \int \mathrm{d}x_2\, \phi(x_2)\mathrm{e}^{-\mathrm{i}px_2/\hbar} \\
&= \int \mathrm{d}x_1\, \phi^*(x_1) \int \mathrm{d}x_2\, \phi(x_2) \int \frac{\mathrm{d}p}{2\pi\hbar}p^n \mathrm{e}^{\mathrm{i}(x_1-x_2)p/\hbar} \\
&= \int \mathrm{d}x_1\, \phi^*(x_1) \int \mathrm{d}x_2\, \phi(x_2)\left(\mathrm{i}\hbar\frac{\partial}{\partial x_2}\right)^n \int \frac{\mathrm{d}p}{2\pi\hbar}\mathrm{e}^{\mathrm{i}(x_1-x_2)p/\hbar} \\
&= \int \mathrm{d}x_1\, \phi^*(x_1) \int \mathrm{d}x_2\, \phi(x_2)\left(\mathrm{i}\hbar\frac{\partial}{\partial x_2}\right)^n \delta(x_1 - x_2)
\end{aligned} \tag{6.14}$$

Integrating by parts n times with respect to x_2, this becomes

$$\begin{aligned}
\langle p^n\rangle &= \int \mathrm{d}x_1 \int \mathrm{d}x_2 \phi^*(x_1)\delta(x_1 - x_2)\left(-\mathrm{i}\hbar\frac{\partial}{\partial x_2}\right)^n \phi(x_2) \\
&= \int \mathrm{d}x \phi^*(x)\left(-\mathrm{i}\hbar\frac{\partial}{\partial x}\right)^n \phi(x)
\end{aligned} \tag{6.15}$$

The conclusion is that the expectation value $\langle p^n \rangle$ at any time t is just

$$\langle p^n \rangle = \int \mathrm{d}x\, \psi^*(x, t)\, \tilde{p}^n\, \psi(x, t) \tag{6.16}$$

where $\tilde{p}$ is the momentum operator

$$\tilde{p} = -\mathrm{i}\hbar \frac{\partial}{\partial x} \tag{6.17}$$

introduced in the last chapter.

6.2 The Heisenberg uncertainty principle

Let us use (6.16) to compute the momentum uncertainty Δp for the Gaussian wavepacket

$$\phi(x) = \frac{1}{(\pi a^2)^{1/4}} \mathrm{e}^{-x^2/2a^2} \mathrm{e}^{\mathrm{i}p_0 x/\hbar} \tag{6.18}$$

which we already know has a momentum expectation value

$$\langle p \rangle = p_0 \tag{6.19}$$

and a position uncertainty

$$\Delta x = \frac{a}{\sqrt{2}} \tag{6.20}$$

The squared momentum uncertainty is

$$\begin{aligned}
\Delta p^2 &= \langle p^2 \rangle - \langle p \rangle^2 = \langle p^2 \rangle - p_0^2 \\
&= -\hbar^2 \int \mathrm{d}x\, \phi^*(x) \frac{\partial^2}{\partial x^2} \phi(x) - p_0^2 \\
&= -\frac{\hbar^2}{\sqrt{\pi a^2}} \int \mathrm{d}x\, \exp\left(-\mathrm{i}\frac{p_0 x}{\hbar} - \frac{x^2}{2a^2}\right) \\
&\quad \times \left[\left(\mathrm{i}\frac{p_0}{\hbar} - \frac{x}{a^2}\right)^2 - \frac{1}{a^2}\right] \exp\left(\mathrm{i}\frac{p_0 x}{\hbar} - \frac{x^2}{2a^2}\right) - p_0^2 \\
&= \frac{\hbar^2}{2a^2}
\end{aligned} \tag{6.21}$$

or

$$\Delta p = \frac{\hbar}{\sqrt{2}a} \tag{6.22}$$

Multiplying Δx and Δp, the product of uncertainties is simply

$$\Delta x \Delta p = \frac{\hbar}{2} \tag{6.23}$$

We will show in the next chapter (the section on the generalized uncertainty principle) that $\hbar/2$ is the smallest value for the product of $\Delta x \Delta p$ that can be obtained for *any* wave function; wave functions which differ from the Gaussian wavepacket have products greater than $\hbar/2$. Therefore, *for any physical state, the product of position and momentum uncertainties obeys the inequality*

$$\Delta x \Delta p \geqslant \frac{\hbar}{2} \tag{6.24}$$

A corollary of this fact is

The Heisenberg uncertainty principle

It is impossible, by any measurement process, to simultaneously determine the position and momentum of a particle to an accuracy greater than

$$\Delta x \Delta p = \frac{\hbar}{2} \tag{6.23}$$

If in fact we could determine the position and momentum to an accuracy better than (6.23), then the particle would be left in a physical state with $\Delta x \Delta p < \hbar/2$. But according to equation (6.24) there are no such physical states. Therefore, measurements of that kind are impossible.

The uncertainty principle is an unavoidable consequence of quantum theory, and if one could design a measurement process which would be more accurate than the bound (6.23), then quantum theory would be wrong. We have already discussed one attempt to measure x and p simultaneously, in the example of the Heisenberg microscope. In that case it was found that the photon composition of light, combined with the Rayleigh criterion of resolution, results in $\Delta x \Delta p \approx \hbar$, which is in agreement with the uncertainty principle. There have been other ingenious proposals for measurements which would violate the uncertainty principle, especially due to Albert Einstein in his discussions with Niels Bohr.[1] A careful study of these proposals always reveals a flaw. In much the same way that the existence of perpetual motion machines is ruled out by the first and/or second laws of thermodynamics, the existence of an apparatus which would give a precise determination of position and momentum is in conflict with the nature of physical states in quantum mechanics.

However, the example of Heisenberg's microscope often leads to a misunderstanding that the uncertainty principle is simply saying that 'the observation disturbs what is being observed'. It is true that an observation usually changes the physical state of the observed system. But it is not true that this is the full content of the uncertainty principle. If position–momentum uncertainty were only a matter of light disturbing the observed particle, then we would be free to imagine that a particle

[1] c.f. Niels Bohr, *Discussions with Einstein,* in [6].

really has a definite position and definite momentum at every moment in time, but that the physical properties of light prevent their accurate simultaneous determination. This interpretation is wrong, because if a particle had a definite position and momentum at every moment of time, then the particle would follow a definite trajectory. We have already seen that the assumption that particles follow trajectories is inconsistent with electron interference. The Heisenberg principle is best understood, not as a slogan 'the observation disturbs what is observed', but rather as a consequence of the nature of physical states in quantum mechanics, which cannot be simultaneously eigenstates of position and momentum. The fact that uncertainties in position and momentum exist independent of any 'disturbance' of the system by observation, and that these uncertainties have important physical consequences, is very well illustrated by the stability of the hydrogen atom.

Why the hydrogen atom is stable

Suppose the wave function of an electron is concentrated, more or less uniformly, in a sphere of radius R around the nucleus, and falls rapidly to zero outside this sphere. The precise form of the wave function is not so important, because we only want to make a very rough estimate of the electron energy, whose expectation value is given by

$$\begin{aligned}\langle H\rangle &= \langle\text{Kinetic Energy}\rangle + \langle\text{Potential Energy}\rangle \\ &= \left\langle \frac{p^2}{2m} \right\rangle + \left\langle -\frac{e^2}{4\pi r} \right\rangle \\ &= \int \mathrm{d}^3x\, \psi^*(x, y, z, t)\left(-\frac{\hbar^2}{2m}\nabla^2\right)\psi(x, y, z, t) \\ &\quad + \int \mathrm{d}^3x\, \psi^*(x, y, z, t)\left(-\frac{e^2}{4\pi r}\right)\psi(x, y, z, t)\end{aligned} \tag{6.25}$$

First of all we can estimate the expectation value of the potential energy, which is roughly

$$\langle V\rangle \sim -\frac{e^2}{4\pi R} \tag{6.26}$$

Next, the uncertainty in the particle position is

$$\Delta x \approx \Delta y \approx \Delta z \sim R \tag{6.27}$$

which implies, by the uncertainty principle,

$$\Delta p_x \approx \Delta p_y \approx \Delta p_z \sim \frac{\hbar}{2R} \tag{6.28}$$

and the expectation value of kinetic energy is therefore on the order

$$\begin{aligned}\langle KE\rangle &= \frac{1}{2m}\left\langle p_x^2 + p_y^2 + p_z^2\right\rangle \\ &= \frac{1}{2m}\left(\Delta p_x^2 + \Delta p_y^2 + \Delta p_z^2\right) \\ &\sim \frac{3\hbar^2}{8\,mR^2}\end{aligned} \tag{6.29}$$

The expectation value of the total energy is then

$$\langle H\rangle = \frac{3\hbar^2}{8\,mR^2} - \frac{e^2}{4\pi R} \tag{6.30}$$

The total energy is minimized for $\mathrm{d}\langle H\rangle/\mathrm{d}R = 0$, and this minimum is obtained at

$$R = 4\pi\frac{3\hbar^2}{4me^2} \tag{6.31}$$

which is not very far off the Bohr radius

$$r_1 = \frac{4\pi\hbar^2}{me^2} \tag{6.32}$$

We can now understand why the electron does not fall into the nucleus. The potential energy is minimized by an electron localized at $r = 0$. However, the more localized an electron wave function is, the smaller Δx is, and the smaller Δx is, the greater is the uncertainty in Δp. But the greater the value of Δp, the greater is the expectation value of kinetic energy

$$\left\langle \frac{p^2}{2m}\right\rangle \sim \frac{\hbar^2}{2\,m\Delta x^2} \tag{6.33}$$

and at some point this (positive) kinetic energy overwhelms the (negative) Coulomb potential, which goes like $-e^2/4\pi\Delta x$. This is why the minimum electron energy in the hydrogen atom is obtained by a wave function of some finite extent, on the order of the Bohr radius.

The application of the uncertainty principle to the hydrogen atom shows that there is much more to this principle than simply the fact that an observation disturbs an observed object. There are very many hydrogen atoms in the Universe; very few of them are under observation. If Δp were due to a disturbance by observation, then there would be nothing to prevent an unobserved electron from falling into the nucleus. Nevertheless, all hydrogen atoms, observed or not, have a stable ground state, and this is due to the fact that there is no physical state in which an electron is both localized, and at rest.

6.3 The energy of energy eigenstates

We have seen that the expectation value of energy is given by

$$\langle E\rangle = \int \mathrm{d}x\, \psi^*(x, t)\tilde{H}\psi(x, t) \tag{6.34}$$

but of course a series of experiments can give much more information about the energy distribution than just $\langle E\rangle$. According to Bohr's model of the hydrogen atom, an orbiting electron can be found to have only certain definite energies

$$E_n = -\left(\frac{me^4}{2(4\pi)^2\hbar^2}\right)\frac{1}{n^2} \tag{6.35}$$

and there must be some truth to this idea, because it explains atomic spectra so well. We would like to use the Schrödinger equation to compute which energies are possible for an orbiting electron or for any other system, and the probabilities of finding these energies in a measurement process.

A first sign that the energies of bound states may be discrete comes from the particle in a tube, where it was found that the energy eigenvalues are

$$E_n = n^2\frac{\pi^2\hbar^2}{2mL^2} \tag{6.36}$$

The question is: in what way are the energy eigenvalues of the time-independent Schrödinger equation related to the energies that might actually be found in a particular measurement? We now show, using equation (6.34) and the orthogonality property (see below) of the energy eigenstates

$$\langle\phi_n|\phi_m\rangle = 0 \quad \text{if} \quad E_n \neq E_m \tag{6.37}$$

that the set of energy eigenvalues coincides with the set of energies that can be found by measurement, and also show how to compute the probability associated with each energy.

For simplicity, assume that the eigenvalues are discrete and non-degenerate, which means that no two eigenvalues are equal. To prove the orthogonality property (6.37), begin by considering the quantity

$$H_{mn} = \int \mathrm{d}x\, \phi_m^*(x)\, \tilde{H}\, \phi_n(x) \tag{6.38}$$

From the time-independent Schrödinger equation

$$\tilde{H}\phi_k(x) = E_k\phi_k(x) \tag{6.39}$$

this becomes

$$\begin{aligned} H_{mn} &= E_n \int \mathrm{d}x\, \phi_m^*(x)\phi_n(x) \\ &= E_n\langle\phi_m|\phi_n\rangle \end{aligned} \tag{6.40}$$

On the other hand, using the integration by parts formula (5.7),

$$\begin{aligned} H_{mn} &= \int \mathrm{d}x \left\{ -\frac{\hbar^2}{2m}\phi_m^* \frac{\partial^2 \phi_n}{\partial x^2} + V(x)\phi_m^*\phi_n \right\} \\ &= \int \mathrm{d}x \left\{ -\frac{\hbar^2}{2m}\frac{\partial^2 \phi_m^*}{\partial x^2}\phi_n + V(x)\phi_m^*\phi_n \right\} \\ &= \int \mathrm{d}x \left(\tilde{H}\phi_m\right)^* \phi_n \\ &= E_m^* \langle \phi_m | \phi_n \rangle \end{aligned} \tag{6.41}$$

Comparing (6.40) and (6.41),

$$E_n \langle \phi_m | \phi_n \rangle = E_m^* \langle \phi_m | \phi_n \rangle \tag{6.42}$$

For the case $n = m$, this equation implies

$$E_n = E_n^* \tag{6.43}$$

i.e. the energy eigenvalues are real, while for $n \neq m$

$$(E_n - E_m)\langle \phi_m | \phi_n \rangle = 0 \tag{6.44}$$

When the eigenvalues are non-degenerate, it means that

$$n \neq m \Longrightarrow E_n \neq E_m \tag{6.45}$$

and therefore

$$\langle \phi_m | \phi_n \rangle = 0 \qquad (n \neq m) \tag{6.46}$$

Choosing the eigenstates ϕ_n to satisfy the normalization condition

$$\langle \phi_n | \phi_n \rangle = 1 \tag{6.47}$$

establishes the relation

$$\langle \phi_m | \phi_n \rangle = \delta_{mn} \tag{6.48}$$

which we have already seen to be true (equation (5.123)) in the case of the particle in a tube.

According to equation (5.38), the general solution to the time-dependent Schrödinger equation can be expressed in terms of the energy eigenstates as

$$\psi(x, t) = \sum_n a_n \phi_n \mathrm{e}^{-\mathrm{i}E_n t/\hbar} \tag{6.49}$$

Then the energy expectation value is

$$\begin{aligned}\langle E\rangle &= \int \mathrm{d}x \left[\sum_n a_m\phi_m \mathrm{e}^{-\mathrm{i}E_m t/\hbar}\right]^* \tilde{H} \sum_n a_n\phi_n \mathrm{e}^{-\mathrm{i}E_n t/\hbar} \\ &= \sum_m \sum_n a_m^* a_n \mathrm{e}^{-\mathrm{i}(E_n - E_m)t/\hbar} E_n \langle \phi_m | \phi_n \rangle \\ &= \sum_n E_n a_n^* a_n\end{aligned} \tag{6.50}$$

Now recall the expression (4.67), from probability theory, for the expectation value of any quantity Q

$$\langle Q\rangle = \sum_n Q_n P(Q_n) \tag{6.51}$$

where $\{Q_n\}$ are the set of possible values of the quantity, and $P(Q_n)$ is the probability that the value Q_n will be found by measurement. Comparing (6.50) to (6.51), we find that:

I. The set of energies that can be found by measurement is the set of energy eigenvalues $\{E_n\}$ of the time-independent Schrödinger equation; and
II. The probability that a state $\psi(x, t)$ will be found to have an energy E_n is given by the squared modulus of coefficients

$$P(E_n) = |a_n|^2 \tag{6.52}$$

To compute a_n in terms of the wave function $\psi(x, t)$, we multiply both sides of (6.49) by $\phi_m^*(x)$, integrate over x, and use the orthogonality property (6.48)

$$\begin{aligned}\int \mathrm{d}x\ \phi_m^*(x)\psi(x, t) &= \int \mathrm{d}x\ \phi_m^*(x) \sum_n a_n\phi_n \mathrm{e}^{-\mathrm{i}E_n t/\hbar} \\ \langle \phi_m | \psi \rangle &= \sum_n a_n \mathrm{e}^{-\mathrm{i}E_n t/\hbar} \langle \phi_m | \phi_n \rangle \\ &= \sum_n a_n \mathrm{e}^{-\mathrm{i}E_n t/\hbar} \delta_{mn} \\ &= a_m \mathrm{e}^{-\mathrm{i}E_m t/\hbar}\end{aligned} \tag{6.53}$$

So we see that

$$a_n(t) \equiv a_n \mathrm{e}^{-\mathrm{i}E_n t/\hbar} = \langle \phi_n | \psi \rangle \tag{6.54}$$

and the probability for finding the energy E_n, when the system is in the physical state $|\psi\rangle$, is given by

$$P(E_n) = \left|\langle \phi_n | \psi \rangle\right|^2 \tag{6.55}$$

In particular, for the particle in a tube, the lowest possible energy for the particle is the lowest eigenvalue

$$E_1 = \frac{\pi^2\hbar^2}{2mL^2} \tag{6.56}$$

and the corresponding eigenstate

$$\phi_1(x) = \sqrt{\frac{2}{L}}\,\sin\left(\frac{\pi x}{L}\right) \tag{6.57}$$

is known as the ***ground state***. In general, the ground state is the lowest energy state of a quantum system (in quantum field theory, it is also known as the 'vacuum state'). Energy eigenstates of higher energies are known as ***excited states***. The fact that the ground state energy of the particle in the tube E_1 is greater than zero is another example of the Uncertainty Principle at work. Since the particle is confined in a region of length L, it means that $\Delta x \leqslant L$, and therefore

$$\Delta p > \frac{\hbar}{2L} \tag{6.58}$$

Assuming $\langle p \rangle = 0$, this would give a lower bound to the energy of roughly

$$E \approx \frac{(\Delta p)^2}{2m} > \frac{\hbar^2}{8mL^2} \tag{6.59}$$

and this lower bound is in fact smaller, by some numerical factors, than the true ground state energy E_1 given in (6.56), in agreement with the uncertainty principle. Note that the dependence on $\hbar$, m, L of the lower bound is the same as that of E_1. Again, it is not a question of 'the observation disturbing the observed'. There is simply no physical state of a particle in a tube which has an energy lower than E_1.

Energy eigenstates are stationary states, in the sense that the time dependence of the wave function is entirely contained in an overall phase factor

$$\psi_\alpha(x, t) = \phi_\alpha(x)\mathrm{e}^{-\mathrm{i}E_\alpha t/\hbar} \tag{6.60}$$

and this phase cancels out when computing expectation values of any function of x and p, e.g.

$$\begin{aligned} \langle x \rangle &= \int \mathrm{d}x\, \psi_\alpha^*(x, t)\, x\, \psi_\alpha(x, t) \\ &= \int \mathrm{d}x\, \phi_\alpha^*(x)\, x\, \phi_\alpha(x) \\ \langle p \rangle &= \int \mathrm{d}x\, \psi_\alpha^*(x, t)\, \tilde{p}\, \psi_\alpha(x, t) \\ &= \int \mathrm{d}x\, \phi_\alpha^*(x)\, \tilde{p}\, \phi_\alpha(x) \end{aligned} \tag{6.61}$$

Therefore

$$
\begin{aligned}
\partial_t \langle x \rangle &= 0 \\
\left\langle -\frac{\partial V}{\partial x} \right\rangle &= 0
\end{aligned}
\tag{6.62}
$$

where the second equation follows from $\partial_t \langle p \rangle = 0$, and Ehrenfest's principle. Note that these equations are similar to the corresponding expressions for a stationary state in classical mechanics: the particle is static, so

$$
\partial_t x = 0 \tag{6.63}
$$

and it remains static because the force on the particle vanishes, i.e.

$$
F = -\frac{\partial V}{\partial x} = 0 \tag{6.64}
$$

On the other hand, in classical mechanics the kinetic energy of a stationary state is zero. This is not the case in quantum mechanics. For the example of the particle in a tube, the expectation value of momentum vanishes in an energy eigenstate

$$
\begin{aligned}
\langle p \rangle_n &= \int_0^L \mathrm{d}x \, \phi_n^*(x) \, \tilde{p} \, \phi_n(x) \\
&= \frac{2}{L} \int_0^L \mathrm{d}x \, \sin\!\left(\frac{n\pi x}{L}\right) \left(-\mathrm{i}\hbar \frac{\partial}{\partial x}\right) \sin\!\left(\frac{n\pi x}{L}\right) \\
&= 0
\end{aligned}
\tag{6.65}
$$

However, as required by the uncertainty principle, the uncertainty in momentum is *not* zero:

$$
\begin{aligned}
(\Delta p_n)^2 &= \int_0^L \mathrm{d}x \phi_n^*(x) (\tilde{p} - \langle p \rangle_n)^2 \phi_n(x) \\
&= \frac{2}{L} \int_0^L \mathrm{d}x \, \sin\!\left(\frac{n\pi x}{L}\right) \left(-\mathrm{i}\hbar \frac{\partial}{\partial x}\right)^2 \sin\!\left(\frac{n\pi x}{L}\right) \\
&= \frac{n^2 \hbar^2 \pi^2}{L^2}
\end{aligned}
\tag{6.66}
$$

The energy eigenvalues (in this example) are simply given by the kinetic energy which is due to this uncertainty

$$
E_n = \frac{(\Delta p_n)^2}{2m} = \frac{n^2 \pi^2 \hbar^2}{2mL^2} \tag{6.67}
$$

There is no adequate classical picture, either in terms of a particle following a trajectory (in which case $\langle x \rangle$ changes in time), or of a particle at rest (in which case the kinetic energy vanishes), of the energy eigenstates. These are best pictured as standing waves, which, at a more fundamental level, represent the components of vectors in Hilbert space. Nevertheless, when standing waves of different energies are superimposed, it is possible to produce a wavepacket which, as we have seen in the last chapter, roughly follows a classical trajectory according to Ehrenfest's principle. This is an important point to keep in mind: dynamics, non-stationarity, change of any sort in quantum mechanics, implies an uncertainty in the value of energy. If the value of the energy is certain, the system is in an energy eigenstate, and if the system is in an energy eigenstate, the system is stationary.

Problems

1. Consider the wave function

$$\psi(x) = \frac{1}{\sqrt{2a}} e^{-|x|/2a} \tag{6.68}$$

   a) What is $\langle p \rangle$?
   b) Compute $\langle p^2 \rangle$ by first finding the Fourier transform $f(p)$ from equation (5.54), and then using equation (6.13).
   c) Compute $\langle p^2 \rangle$ again by using equation (6.15). Be careful not to differentiate a discontinuous function! Note that, by integration by parts,

$$\int dx\, \phi^* \frac{d^2\phi}{dx^2} = -\int dx\, \frac{d\phi^*}{dx} \frac{d\phi}{dx} \tag{6.69}$$

   Obviously, your answers to parts (b) and (c) should agree.
   d) Using the answers to parts (a), (b), and (c), compute the momentum uncertainty Δp. Take the position uncertainty Δx from problem 1 of chapter 4, and compute the product $\Delta x \Delta p$. Does your answer agree with the Heisenberg uncertainty principle?

2. One billion particles are in a one-dimensional box, with walls at $x = 0$ and $x = L$. Measurements are carried out at time $t = 0$, and at this instant all billion particles are in the same physical state described by

$$\varphi(x) = Nx(x - L) \tag{6.70}$$

   a) Normalize φ and determine N.
   b) How many particles would you expect to find in the interval $0 < x < L/2$?
   c) If the energy of a particle in this state is measured, the probability of being in the nth energy eigenstate is given by equation (6.55), where ϕ_n is the nth energy eigenstate. How many particles would you expect to find in the third energy level?

3. A particle of mass m is in a tube of length L. The wave function of the particle, at time $t = 0$, has the form

$$\psi(x, 0) = N\left[4\phi_2(x) - 3i\phi_4(x) + (2 + i)\phi_5(x)\right] \tag{6.71}$$

   where ϕ_n are the normalized energy eigenstates.

a) Calculate N.
b) What are the possible energies that may be observed, and what is the probability of each?
c) Using the answers to part (b), work out the energy uncertainty ΔE. No integration is necessary.
d) Find $\psi(x, t)$ at any later time. Do the answers to parts (b) and (c) change? Why or why not?

4. A particle of mass m is moving in an infinite square well potential of width L. At time $t = 0$ the wave function of the particle (inside the well) is known to be

$$\psi(x, 0) = N \cos^2\left[\frac{\pi x}{L}\right] \sin\left[\frac{3\pi x}{L}\right] \quad \text{for} \quad 0 \leqslant x \leqslant L \tag{6.72}$$

a) Write $\psi(x, 0)$ as a sum of a few energy eigenstates. Make sure that your wave function is properly normalized.
b) From your answer to part (a), what are the possible energies that could be found by measurement, and what is the probability of each?
c) From your answer to part (b), just using probabilities (no integrals necessary), find $\langle E\rangle$, $\langle E^2\rangle$, and ΔE.
d) Find the wave function at any later time t.

5. Suppose a particle of mass m, in a tube of length L, is in the ground state. The tube is ***suddenly*** stretched to twice its length, by moving one end of the tube to a new position at $x = 2L$. Immediately after the tube is stretched, the wave function of the particle is the same as before (it takes time for a wave function to change). If we now measure the energy of the particle, what energy is most likely to be found?

6. Compute the expectation value of the kinetic and potential energy of a hydrogen atom, from equation (6.26) of the notes, in a state described by the Gaussian wavepacket in three dimensions

$$\phi(x, y, z) = N \exp[-(x^2 + y^2 + z^2)/2R^2] \tag{6.73}$$

(apply the normalization condition to determine N). In computing the expectation value of the potential energy, the formula for spherical coordinates

$$\int \mathrm{d}x\mathrm{d}y\mathrm{d}z\, f(r) = 4\pi \int \mathrm{d}r\, r^2 f(r) \tag{6.74}$$

may be useful. Find the value of R which minimizes $\langle H\rangle$, and compare this value to the Bohr radius. Also compare numerically the minimum value of $\langle H\rangle$ to the ground state energy E_1 of the Bohr atom.

7. Consider a ping-pong ball of mass m and negligible radius, bouncing elastically up and down on a table (ignore motion in any other direction) under the influence of the gravitational potential $V(x) = mgx$ where x is the

height above the table. Estimate the ground state energy of the ping-pong ball, by calculating $\langle E \rangle$ in the state

$$\psi(x) = Nx\mathrm{e}^{-ax^2/2}\text{at} \quad x \geqslant 0 \tag{6.75}$$

and choosing a to minimize $\langle E \rangle$. ($\psi(x) = 0$ at $x < 0$, N is a normalization constant.)

8. A free particle of mass m is initially in the state

$$\psi(x, 0) = \begin{cases} \sqrt{\dfrac{2}{L}} \cos(2\pi x/L) & -\dfrac{1}{2}L \leqslant x \leqslant \dfrac{1}{2}L \\ 0 & \text{otherwise} \end{cases} \tag{6.76}$$

a) Calculate the momentum probability density

$$P(p) = \frac{1}{2\pi\hbar} f^*(p) f(p) \tag{6.77}$$

for this particle, and make a rough sketch of $P(p)$ versus p. Explain the location of the peaks in your momentum probability distribution; why do they occur where they do?

b) Is there any limit to the accuracy of a measurement of p made on this particle? If the wave function were the same, but the particle were in an infinite square well potential (with $V(x) = \infty$ where the wave function is zero), would your answer change?

Reference

[1] Bohr N 1958 *Atomic Physics and Human Knowledge* (New York: Wiley)

IOP Publishing

An Introduction to Quantum Theory

Jeff Greensite

Chapter 7

Operators and observations

Is this a dagger I see before me, the handle toward my hand? Come, let me clutch thee! I have thee not, and yet I see thee still. Art thou not, fatal vision, as sensible to feeling as to sight? Or art thou but a dagger of the mind, a false creation, proceeding from the heat-oppressed brain?

Shakespeare, *Macbeth*

The mathematical core of quantum mechanics is linear algebra in an infinite number of dimensions. Almost every manipulation in quantum theory can be interpreted as an operation involving inner products and/or matrix multiplication. This is for three reasons. First, as we have seen in chapter 4, physical states in quantum mechanics are represented by wave functions, and any function can be regarded as a vector with a continuous index. Second, the dynamical equation of quantum mechanics, i.e. the Schrödinger equation, is a linear differential equation. As we will see, linear differential operators can be interpreted as infinite-dimensional matrices. Finally, the predictions of quantum mechanics, namely, the probabilities of observing such-and-such an event, are in fact determined by the inner products of vectors in Hilbert space. This chapter will be devoted to these mathematical aspects of quantum theory.

7.1 Probabilities from inner products

We have so far learned how to use the quantum state $|\psi\rangle$ to calculate the expectation value of any function of position, any function of momentum, or any sum of such functions, such as energy. In particular

$$\begin{aligned}\langle x\rangle &= \int \mathrm{d}x\, \psi^*(x, t)\, \tilde{x}\, \psi(x, t)\\ \langle p\rangle &= \int \mathrm{d}x\, \psi^*(x, t)\, \tilde{p}\, \psi(x, t)\\ \langle H\rangle &= \int \mathrm{d}x\, \psi^*(x, t)\, \tilde{H}\, \psi(x, t)\end{aligned} \tag{7.1}$$

doi:10.1088/978-0-7503-1167-0ch7

where

$$\tilde{x}\psi(x, t) \equiv x\psi(x, t)$$
$$\tilde{p}\psi(x, t) \equiv -\mathrm{i}\hbar\frac{\partial}{\partial x}\psi(x, t) \tag{7.2}$$
$$\tilde{H}\psi(x, t) \equiv \left\{-\frac{\hbar^2}{2m}\frac{\partial^2}{\partial x^2} + V(x)\right\}\psi(x, t)$$

The probability densities for finding a particle near a certain position x_0, or momentum p_0, and the probability to find that the particle has a certain energy E_n, are given by

$$P_x(x_0) = |\psi(x_0)|^2$$
$$P_p(p_0) = \left|\frac{1}{\sqrt{2\pi\hbar}}f(p_0)\right|^2 \tag{7.3}$$
$$P_E(E_n) = |a_n|^2$$

where

$$a_n = \left\langle \phi_n^E \middle| \psi \right\rangle$$
$$f(p_0) = \int \mathrm{d}x\ \phi(x)\mathrm{e}^{-\mathrm{i}p_0 x/\hbar} \tag{7.4}$$

and where $\phi_n^E(x)$ is a solution of the time-independent Schrödinger equation $\tilde{H}\phi_n^E(x) = E_n\phi_n^E(x)$. The superscript E has been introduced to emphasize that this is a state corresponding to a definite eigenvalue of energy E_n.

A probability is a number, a quantum state $|\psi\rangle$ is a vector. Numbers can be obtained from vectors by taking inner products. The probability $P(E_n)$ is clearly the square modulus of an inner product, and in fact the other two probabilities can also be expressed in that way. Define

$$\phi_{x_0}^x(x) \equiv \delta(x - x_0)$$
$$\phi_{p_0}^p(x) \equiv \frac{1}{\sqrt{2\pi\hbar}}\mathrm{e}^{\mathrm{i}p_0 x/\hbar} \tag{7.5}$$

Then we may write

$$\psi(x_0) = \int \mathrm{d}x\ \delta(x - x_0)\psi(x)$$
$$= \left\langle \phi_{x_0}^x \middle| \psi \right\rangle \tag{7.6}$$

and also

$$\frac{1}{\sqrt{2\pi\hbar}} f(p) = \int \mathrm{d}x \left[\frac{1}{\sqrt{2\pi\hbar}} \mathrm{e}^{\mathrm{i}p_0 x/\hbar}\right]^* \psi(x, t) \\ = \left\langle \phi^p_{p_0} \middle| \psi \right\rangle \tag{7.7}$$

In terms of $\phi^x_{x_0}$, $\phi^p_{p_0}$, ϕ^E_n, the probability and probability densities can be rewritten as follows:

$$P_x(x_0) = \left|\left\langle \phi^x_{x_0} \middle| \psi \right\rangle\right|^2 \\ P_p(p_0) = \left|\left\langle \phi^p_{p_0} \middle| \psi \right\rangle\right|^2 \\ P_E(E_n) = \left|\left\langle \phi^E_n \middle| \psi \right\rangle\right|^2 \tag{7.8}$$

The (generalized) functions $\phi^x_{x_0}$, $\phi^p_{p_0}$, ϕ^E_n all share two things in common. First, they all satisfy ***eigenvalue equations***:

$$\tilde{x}\phi^x_{x_0}(x) = x_0\phi^x_{x_0}(x) \\ \tilde{p}\phi^p_{p_0}(x) = p_0\phi^p_{p_0}(x) \\ \tilde{H}\phi^E_n(x) = E_n\phi^E_n(x) \tag{7.9}$$

The functions $\phi^x_{x_0}$, $\phi^p_{p_0}$, ϕ^E_n are known as ***eigenfunctions*** or ***eigenstates*** (the terms are synonymous), of the operators $\tilde{x}$, $\tilde{p}$, $\tilde{H}$ respectively, and the corresponding numbers x_0, p_0, E_n are known as ***eigenvalues***. (The superscripts x, p, E are introduced to avoid confusing eigenstates of different operators with one another.) Secondly, the inner products of these eigenstates are either delta-functions or Kronecker deltas, depending on whether the eigenvalues belong to a continuous range ($x_0, p_0 \in [-\infty, \infty]$), or a discrete set ($E_n \in \{E_k\}$):

$$\left\langle \phi^x_{x_1} \middle| \phi^x_{x_2} \right\rangle = \int \mathrm{d}x\ \delta(x - x_1)\delta(x - x_2) \\ = \delta(x_1 - x_2) \\ \left\langle \phi^p_{p_1} \middle| \phi^p_{p_2} \right\rangle = \int \mathrm{d}x\ \frac{1}{2\pi\hbar} \mathrm{e}^{\mathrm{i}(p_2 - p_1)x/\hbar} \\ = \delta(p_1 - p_2) \\ \left\langle \phi^E_n \middle| \phi^E_m \right\rangle = \delta_{nm} \tag{7.10}$$

The relationship that exists between observables x, p, H, and operators $\tilde{x}$, $\tilde{p}$, $\tilde{H}$, and between probabilities and inner products, generalizes to all other observable quantities of a system.

7.2 Operators and observables

An ***observable*** is any property of a system that can be measured; e.g. position, momentum, energy, angular momentum, magnetic moment, mass, and so on. In classical mechanics, all observables are functions of the generalized coordinates and momenta $\{q^a, p_a\}$, so a knowledge of the physical state implies a knowledge of all possible observables of the system.

An ***operator*** is a rule for changing functions into other functions. Given any function $|\psi\rangle$ as input, an operator specifies a unique function $|\psi'\rangle$ as output. Symbolically,

$$|\psi'\rangle = O|\psi\rangle \quad \text{or} \quad |O\psi\rangle \tag{7.11}$$

or, in 'component' notation

$$\psi'(x) = \tilde{O}\psi(x) \tag{7.12}$$

A ***linear operator*** has the property that, for any two functions $|\psi_1\rangle$ and $\psi_2\rangle$, and any two constants a and b,

$$O\Big\{a\big|\phi_1\big\rangle + b|\psi_2\rangle\Big\} = aO|\psi_1\rangle + bO|\psi_2\rangle \tag{7.13}$$

Examples of linear operators are multiplication by a constant

$$\tilde{O}\psi(x) = c\psi(x) \tag{7.14}$$

or multiplication by a fixed function $F(x)$,

$$\tilde{O}\psi(x) = F(x)\psi(x) \tag{7.15}$$

or differentiation

$$\tilde{O}\psi(x) = \frac{\partial}{\partial x}\psi(x) \tag{7.16}$$

A non-linear operation could be, e.g. taking a square root

$$\tilde{O}\psi(x) = \sqrt{\psi(x)} \tag{7.17}$$

An important subset of linear operators are the Hermitian operators. A ***Hermitian operator*** is a linear operator with the property that

$$\begin{aligned}\langle\psi|O|\psi\rangle &= \int dx\, \psi^*(x)\tilde{O}\psi(x) \\ &= \text{a real number, for any } \psi(x) \text{ whatsoever}\end{aligned} \tag{7.18}$$

The relation between operators and observables

In quantum mechanics, for every observable O there exists a corresponding Hermitian operator $\tilde{O}$. Knowledge of the physical state $\psi(x, t)$ at some time t implies a knowledge of the expectation value of every observable, according to the rule

$$\begin{aligned}\langle O\rangle &= \langle\psi|O|\psi\rangle \\ &= \int \mathrm{d}x\, \psi^*(x)\, \tilde{O}\, \psi(x)\end{aligned}$$

This is a principle of quantum mechanics which simply generalizes the examples, for position, momentum, and energy, that we have already seen.

It is useful to define the concept of the ***Hermitian conjugate*** $\tilde{\mathbf{O}}^\dagger$, pronounced 'O-dagger', of the operator $\tilde{O}$, which has the property

$$\int \mathrm{d}x \left(\tilde{O}^\dagger\psi_1(x)\right)^* \psi_2(x) = \int \mathrm{d}x\ \psi_1^*(x)\tilde{O}\psi_2(x) \tag{7.19}$$

or, in bra-ket notation,

$$\langle O^\dagger\psi_1|\psi_2\rangle = \langle\psi_1|O|\psi_2\rangle \tag{7.20}$$

for any states ψ_1, ψ_2 whatsoever. A Hermitian operator has the property that $\tilde{O}^\dagger = \tilde{O}$, and it is easy to see why an operator with this property has only real expectation values, i.e. that the condition (7.18) is true for any state ψ. Beginning with the definition (7.19) with $\psi_1 = \psi_2 = \psi$

$$\int \mathrm{d}x\, \psi^*(x)\tilde{O}\psi(x) = \int \mathrm{d}x \left(\tilde{O}^\dagger\psi(x)\right)^* \psi(x) \tag{7.21}$$

take the complex conjugate of both sides, and rearrange the integrand on the right-hand side slightly,

$$\left(\int \mathrm{d}x\, \psi^*(x)\tilde{O}\psi(x)\right)^* = \int \mathrm{d}x\, \psi(x)^*\tilde{O}^\dagger\psi(x) \tag{7.22}$$

or

$$\langle\psi|O|\psi\rangle^* = \langle\psi|O^\dagger|\psi\rangle \tag{7.23}$$

Then if the operator is Hermitian, it follows that the expectation value of the observable in any state is a real number,

$$\langle\psi|O|\psi\rangle^* = \langle\psi|O|\psi\rangle \tag{7.24}$$

and this property is required on physical grounds. The expectation value of an observable, after all, is just the result one would obtain by making measurements of the observable O in a great many identical experiments, and averaging the results.

Anything that can be measured—position, momentum, angular momentum, spin, charge, energy, mass—can only have real values; there is no experimental apparatus which extends somehow into the complex plane, and no observables take on complex values. Therefore the expectation value of any observable, in any quantum state, must also be a real number. This is the reason that observables must correspond to Hermitian operators.

But how do we compute the Hermitian conjugate of an operator, and verify whether the operator is, or is not, Hermitian? The general idea is to start with the right-hand side of equation (7.19), where the operator $\tilde{O}$ operates on ψ_1, and do whatever manipulations are necessary to make the operator act on ψ_2 instead. This procedure is best illustrated by example. First consider the operation of multiplying a state by some fixed function $F(x)$, which may be complex; i.e. $\tilde{O}\psi(x) = F(x)\psi(x)$. Then

$$\begin{aligned}\int \mathrm{d}x\psi_1^*(x)\tilde{O}\psi_2(x) &= \int \mathrm{d}x\psi_1^*(x)F(x)\psi_2(x)\\ &= \int \mathrm{d}x(F^*(x)\psi_1)^*\psi_2(x)\\ &= \int \mathrm{d}x\left(\tilde{O}^\dagger\psi_1(x)\right)^*\psi_2(x)\end{aligned} \tag{7.25}$$

where the last line follows from the definition of Hermitian conjugation. Comparing the second and third lines, we conclude that $\tilde{O}^\dagger\psi = F^*(x)\psi$, and $\tilde{O}$ is Hermitian if and only if $F(x)$ is a real function. This means, in particular, that the operator $\tilde{x}$ is Hermitian, $\tilde{x}^\dagger = \tilde{x}$.

The next example is the derivative operator, $\tilde{O}\psi = \frac{\mathrm{d}}{\mathrm{d}x}\psi$. We have

$$\begin{aligned}\int \mathrm{d}x\psi_1^*(x)\tilde{O}\psi_2(x) &= \int \mathrm{d}x\psi_1^*(x)\frac{\mathrm{d}}{\mathrm{d}x}\psi_2(x)\\ &= -\int \mathrm{d}x\left(\frac{\mathrm{d}}{\mathrm{d}x}\psi_1^*(x)\right)\psi_2(x)\\ &= \int \mathrm{d}x\left(-\frac{\mathrm{d}}{\mathrm{d}x}\psi_1(x)\right)^*\psi_2(x)\\ &= \int \mathrm{d}x\left(\tilde{O}^\dagger\psi_1(x)\right)^*\psi_2(x)\end{aligned} \tag{7.26}$$

where the second line follows from integration by parts. Comparing the third and fourth lines, the conclusion is that

$$\left(\frac{\mathrm{d}}{\mathrm{d}x}\right)^\dagger = -\frac{\mathrm{d}}{\mathrm{d}x} \tag{7.27}$$

so the derivative operator is *not* Hermitian. From this example it should be clear that Hermitian conjugation is not the same thing as complex conjugation. A very closely related example is the Hermitian conjugate of the momentum operator:

$$\begin{aligned}\int \mathrm{d}x \psi_1^*(x) \tilde{p} \psi_2(x) &= \int \mathrm{d}x \psi_1^*(x)(-\mathrm{i}\hbar)\frac{\mathrm{d}}{\mathrm{d}x}\psi_2(x) \\ &= \mathrm{i}\hbar \int \mathrm{d}x\left(\frac{\mathrm{d}}{\mathrm{d}x}\psi_1^*(x)\right)\psi_2(x) \\ &= \int \mathrm{d}x\left(-\mathrm{i}\hbar\frac{\mathrm{d}}{\mathrm{d}x}\psi_1(x)\right)^* \psi_2(x) \\ &= \int \mathrm{d}x(\tilde{p}^\dagger \psi_1(x))^* \psi_2(x)\end{aligned} \tag{7.28}$$

We see that $\tilde{p}^\dagger = -\mathrm{i}\hbar(\mathrm{d}/\mathrm{d}x) = \tilde{p}$, so the momentum operator is Hermitian, as it must be. It is the combination of complex conjugation of the factor i, combined with $(\mathrm{d}/\mathrm{d}x)^\dagger = -\mathrm{d}/\mathrm{d}x$, that makes the momentum operator Hermitian.

A final example is Hermitian conjugation of the Hamiltonian operator $\tilde{H}$. Since energy is an observable, the Hamiltonian operator had better be Hermitian, but we need to check this:

$$\begin{aligned}\int \mathrm{d}x \psi_1^*(x) \tilde{H} \psi_2(x) &= \int \mathrm{d}x \psi_1^*(x)\left(-\frac{\hbar^2}{2m}\frac{\mathrm{d}^2}{\mathrm{d}x^2} + V(x)\right)\psi_2(x) \\ &= -\frac{\hbar^2}{2m}\int \mathrm{d}x \psi_1^*(x)\frac{\mathrm{d}^2}{\mathrm{d}x^2}\psi_2(x) + \int \mathrm{d}x \psi_1^*(x) V(x) \psi_2(x) \\ &= -\frac{\hbar^2}{2m}\int \mathrm{d}x\left(\frac{\mathrm{d}^2}{\mathrm{d}x^2}\psi_1^*(x)\right)\psi_2(x) + \int \mathrm{d}x(V(x)\psi_1(x))^*\psi_2(x) \\ &= \int \mathrm{d}x\left(\left[-\frac{\hbar^2}{2m}\frac{\mathrm{d}^2}{\mathrm{d}x^2} + V(x)\right]\psi_1(x)\right)^*\psi_2(x) \\ &= \int \mathrm{d}x(\tilde{H}^\dagger \psi_1(x))^* \psi_2(x)\end{aligned} \tag{7.29}$$

where in going from the second to the third line we integrate by parts twice. Evidently $\tilde{H}^\dagger = \tilde{H}$; the Hamiltonian operator is Hermitian as required.

We conclude this section with a note about the $\tilde{x}$ operator. While $\tilde{x}$ is an operator, its effect on any function $f(x)$ is simply to multiply the function by the variable x. Likewise, any function of $\tilde{x}$ is also an operator, and e.g. it is more correct to write the potential $V(x)$ which appears in the Hamiltonian operator $\tilde{H}$ as $\tilde{V} = V(\tilde{x})$. But since the effect of $\tilde{V}$ on a function is just to multiply the function by $V(x)$, we will rarely make this distinction. And unless it is necessary for some reason, we will usually not distinguish between $\tilde{x}$ and x in most equations, even though the operator $\tilde{x}$ and the variable x are, mathematically, quite different objects. In contrast, we cannot normally replace the differential operator $\tilde{p}$ by the variable p; this can lead to serious confusion because operating with $-\mathrm{i}\hbar\mathrm{d}/\mathrm{d}x$, and multiplying by the variable p, have very different effects in most expressions.

7.3 The matrix analogy

In linear algebra, the Hermitian conjugate $M^\dagger$ of a matrix M is the transpose complex conjugate, $M^\dagger = (M^T)^*$, which in terms of components means that

$$[M^\dagger]_{ij} = M^*_{ji} \tag{7.30}$$

It is easy to verify that if $\mathbf{u}$, $\mathbf{v}$ are any two vectors, then

$$(M^\dagger \mathbf{u}) \cdot \mathbf{v} = \mathbf{u} \cdot M\mathbf{v} \tag{7.31}$$

In components, from (7.30) and the rules of matrix multiplication

$$\begin{aligned}(M^\dagger \mathbf{u}) \cdot \mathbf{v} &= \sum_i \left(\sum_j [M^\dagger]_{ij} u_j \right)^* v_i \\ &= \sum_i \left(\sum_j M^*_{ji} u_j \right)^* v_i \\ &= \sum_i \sum_j u^*_j M_{ji} v_i \\ &= \mathbf{u} \cdot M\mathbf{v}\end{aligned} \tag{7.32}$$

A Hermitian matrix has the property that $M^\dagger = M$. In particular, a 2×2 Hermitian matrix has the form

$$M = \begin{bmatrix} a & b + \mathrm{i}c \\ b - \mathrm{i}c & d \end{bmatrix} \tag{7.33}$$

where a, b, c, d are real numbers.

Matrix eigenvalue equations are an important concept in linear algebra, and have the form

$$M\mathbf{v}_n = \lambda_n \mathbf{v}_n \tag{7.34}$$

where $\mathbf{v}_n$ is the nth eigenvector and λ_n the corresponding eigenvalue. An $N \times N$ Hermitian matrix has N orthogonal eigenvectors and N real eigenvalues. Some eigenvalues may be the same, in which case the eigenvalues are said to be degenerate.

We have already seen that a function $\psi(x)$ can be thought of as a vector with a continuous index x, i.e. ψ is a vector in an infinite-dimensional space known as Hilbert space. But what are the components of an operator $\tilde{O}$, corresponding to the matrix components M_{ij}? We would expect that the operator could be represented as a function of two arguments ('indices') $O(x, y)$, and that, by analogy with matrix multiplication

$$\tilde{O}f(x) = \int \mathrm{d}y \;\; O(x, y) f(y) \tag{7.35}$$

To see how this works, note that a component of a matrix can be expressed as

$$M_{nm} = \mathbf{e}^n \cdot M\mathbf{e}^m \tag{7.36}$$

where the $\mathbf{e}^n$ denote unit vectors that span the N-dimensional vector space. For example, in three dimensions,

$$\mathbf{e}^1 = \begin{bmatrix} 1 \\ 0 \\ 0 \end{bmatrix}, \quad \mathbf{e}^2 = \begin{bmatrix} 0 \\ 1 \\ 0 \end{bmatrix}, \quad \mathbf{e}^3 = \begin{bmatrix} 0 \\ 0 \\ 1 \end{bmatrix} \tag{7.37}$$

To get the components of an operator, we need the analog of unit vectors $\mathbf{e}^n$. This is supplied by the eigenstates of the $\tilde{x}$ operator

$$\phi^x_{x_0}(x) = \delta(x - x_0) \tag{7.38}$$

Then

$$\begin{aligned} O(x_1, x_2) &= \left\langle \phi^x_{x_1} \middle| O \middle| \phi^x_{x_2} \right\rangle \\ &= \int \mathrm{d}x \delta(x - x_1) \tilde{O}_x \delta(x - x_2) \end{aligned} \tag{7.39}$$

where we have inserted a subscript x on the operator $\tilde{O}$ to indicate that it acts on functions of the argument x. Carrying out the integration

$$O(x_1, x_2) = \tilde{O}_{x_1} \delta(x_1 - x_2) \tag{7.40}$$

Now we can check that

$$\begin{aligned} \int \mathrm{d}y O(x, y) f(y) &= \int \mathrm{d}y \tilde{O}_x \delta(x - y) f(y) \\ &= \tilde{O}_x \int \mathrm{d}y \delta(x - y) f(y) \\ &= \tilde{O} f(x) \end{aligned} \tag{7.41}$$

as expected. The matrix elements of the Hermitian conjugate are then

$$O^\dagger(x, y) = O^*(y, x) \tag{7.42}$$

We leave it as an exercise to compute $\tilde{O}^\dagger$ for a given operator $\tilde{O}$ from this relationship.

Unitary matrices and the evolution operator

Hermitian matrices have the property that $M^\dagger = M$. ***Unitary matrices*** have the property that $M^\dagger = M^{-1}$. A number which has this property is $\mathrm{e}^{\mathrm{i}\theta}$. It can be checked by matrix multiplication that 2×2 matrices of the form

$$U = \mathrm{e}^{\mathrm{i}\theta} \begin{bmatrix} a_0 + \mathrm{i}a_3 & a_2 + \mathrm{i}a_1 \\ -a_2 + \mathrm{i}a_1 & a_0 - \mathrm{i}a_3 \end{bmatrix} \tag{7.43}$$

where a_{0-3} are real numbers and satisfy the constraint

$$a_0^2 + a_1^2 + a_2^2 + a_3^2 = 1 \tag{7.44}$$

are unitary matrices, satisfying the defining property

$$UU^\dagger = \begin{bmatrix} 1 & 0 \\ 0 & 1 \end{bmatrix} \tag{7.45}$$

Unitary matrices have the special property that they preserve inner products, in the sense that

$$\begin{aligned}(U\mathbf{v}) \cdot (U\mathbf{w}) &= \mathbf{v} \cdot U^\dagger U\mathbf{w} \\ &= \mathbf{v} \cdot U^{-1}U\mathbf{w} \\ &= \mathbf{v} \cdot \mathbf{w}\end{aligned} \tag{7.46}$$

A unitary operator acting on vectors in Hilbert space has the same defining property as unitary matrices, namely $U^\dagger = U^{-1}$. The unitary operator which turns up most frequently in quantum mechanics is the ***evolution operator***

$$\begin{aligned}U(t) &= \exp[-\mathrm{i}\tilde{H}t/\hbar] \\ &= \sum_n \frac{1}{n!}\left(-\frac{\mathrm{i}t}{\hbar}\right)^n H^n\end{aligned} \tag{7.47}$$

because the formal solution to the time-dependent Schrödinger equation, given the wave function at time $t = 0$, is

$$\psi(x, t) = \mathrm{e}^{-\mathrm{i}\tilde{H}t/\hbar}\psi(x, 0) \tag{7.48}$$

as can be checked by direct substitution into the Schrödinger equation, using

$$\frac{\partial}{\partial t}\exp[-\mathrm{i}\tilde{H}t/\hbar] = -\frac{\mathrm{i}}{\hbar}\tilde{H}\exp[-\mathrm{i}\tilde{H}t/\hbar] \tag{7.49}$$

Then, because the Hamiltonian operator is Hermitian,

$$\begin{aligned}U^\dagger(t) &= \exp[\mathrm{i}\tilde{H}t/\hbar] \\ &= U^{-1}\end{aligned} \tag{7.50}$$

As with unitary matrices, a unitary operator such as the evolution operator preserves inner products

$$\begin{aligned}\langle U\phi | U\psi\rangle &= \langle\phi | U^\dagger U\psi\rangle \\ &= \langle\phi|\psi\rangle\end{aligned} \tag{7.51}$$

7.4 Eigenstates as states of zero uncertainty

What does a measurement *do*?

A measurement apparatus is designed to determine the value of some observable O, and by doing so it must leave the system in a physical state in which the value of that observable is a definite number. But that means that the system is left in a state for which the uncertainty ΔO of the observable vanishes, at least at the instant the measurement is performed. If, for example, the position of a particle is measured precisely, then the particle is *known* to be at that position the instant the measurement is performed; the physical state must resemble a Dirac delta function. Likewise, if the momentum of a particle is measured, the particle is *known* to have that momentum, and the physical state must resemble a plane wave. In general, whatever the state of the system may be just before the measurement, the outcome of a measurement is a state of 'zero-uncertainty' $\Delta O = 0$ in the observable O. Exactly *how* a measurement apparatus accomplishes this feat is a question which will not concern us now; we will return to the question in chapter 25.

From the fact that every observable $O = x, p, E, \ldots$ is associated with a Hermitian operator $\tilde{O} = \tilde{x}, \tilde{p}, \tilde{H}, \ldots$, we see that a zero-uncertainty state $\phi(x)$ must have the property

$$(\Delta O)^2 = \langle\phi|(O - \langle O\rangle)^2|\phi\rangle = 0 \tag{7.52}$$

Denote the expectation value $\langle O\rangle$ by λ, and define

$$\tilde{D} \equiv \tilde{O} - \lambda \tag{7.53}$$

Since $\tilde{O}$ is Hermitian, and multiplication by a constant λ is a Hermitian operation, the operator $\tilde{D}$ is also Hermitian. Then, using the Hermiticity of $\tilde{D}$

$$\begin{aligned}(\Delta O)^2 &= \langle\phi|(D)^2|\phi\rangle \\ &= \langle\phi|D|D\phi\rangle] \\ &= \langle\phi|D|\phi'\rangle \\ &= \langle D\phi|\phi'\rangle \\ &= \langle\phi'|\phi'\rangle \\ &= \int \mathrm{d}x\, \phi'^*(x)\phi'(x)\end{aligned} \tag{7.54}$$

where we have defined

$$\phi'(x) \equiv \tilde{D}\phi(x) \tag{7.55}$$

Now putting together

$$\begin{aligned}\Delta O &= 0 \\ (\Delta O)^2 &= \int \mathrm{d}x\, \phi'^*(x)\phi'(x) \\ \phi'^*(x)\phi'(x) &\geqslant 0 \qquad \text{for all } x\end{aligned} \tag{7.56}$$

we must conclude that

$$\begin{aligned} 0 &= \phi'(x) \\ &= \tilde{D}\phi(x) \\ &= [\tilde{O} - \lambda]\phi(x) \end{aligned} \tag{7.57}$$

or, in other words, for a zero-uncertainty state

$$\tilde{O}\phi(x) = \lambda\phi(x) \tag{7.58}$$

This an eigenvalue equation, of the form already seen for position, momentum, and energy in equation (7.9). Therefore:

I **Any state with vanishing uncertainty in the observable O is an eigenstate of the corresponding Hermitian operator $\tilde{O}$.**

A measurement process therefore has the property that, whatever the physical state of the object may be at the time just prior to the measurement, the system is left in one of the eigenstates of $\tilde{O}$ at a time just after the measurement.

We can go further. Given that the expectation value of the observable $\langle O \rangle$ in the eigenstate ϕ is the eigenvalue λ, and given that the uncertainty $\Delta O = 0$ vanishes, it follows that a measurement of O could only result in the value $O = \langle O \rangle = \lambda$. Any measurement that was different from the mean value would imply a non-zero uncertainty. But since the result of a measurement is to leave the system in such a zero-uncertainty state, it follows that:

II **A measurement of the observable O can only result in a value which is equal to one of the eigenvalues of the corresponding operator $\tilde{O}$.**

In general there are many solutions $\phi(x)$, λ to the eigenvalue equation (7.58), so it is useful to denote a set of linearly independent solutions as $\{\phi_\alpha(x), \lambda_\alpha\}$, with different solutions distinguished by the value of the subscript α. The eigenvalue equation is written

$$\tilde{O}\phi_\alpha = \lambda_\alpha \phi_\alpha \tag{7.59}$$

with ϕ_α the eigenstate and λ_α the corresponding eigenvalue. If all the eigenvalues are different, the eigenvalues are said to be ***non-degenerate***. If there are n linearly-independent eigenstates $\phi_{\alpha_1}, \phi_{\alpha_2}, \ldots, \phi_{\alpha_n}$ whose eigenvalues are the same, i.e. $\lambda_{\alpha_1} = \lambda_{\alpha_2} = \ldots = \lambda_{\alpha_n}$, then the eigenvalue is said to be ***n-fold degenerate***. For example, the energy eigenvalues of a free particle are two-fold degenerate, because for each eigenvalue E there are two linearly independent eigenstates, e.g.

$$e^{i\sqrt{2mE}/\hbar} \qquad \text{and} \qquad e^{-i\sqrt{2mE}/\hbar} \tag{7.60}$$

which satisfy

$$\tilde{H}\phi = -\frac{\hbar^2}{2m}\frac{\partial^2}{\partial x^2}\phi = E\phi \tag{7.61}$$

Theorem H1
The eigenvalues of a Hermitian operator are real.

Proof. This follows directly from the definition of a Hermitian operator

$$\langle\psi|O|\psi\rangle \quad \text{is real} \tag{7.62}$$

Choose ψ to be an eigenstate ϕ_α

$$\langle\phi_\alpha|O|\phi_\alpha\rangle = \lambda_\alpha\langle\phi_\alpha|\phi_\alpha\rangle \tag{7.63}$$

Since the inner product of any vector with itself is a real number, it follows that the left-hand side is real if and only if λ_α is real.

Theorem H2
If two eigenstates of a Hermitian operator correspond to different eigenvalues, then the two eigenstates are orthogonal.

Proof. Suppose $\phi_{\alpha'}$ and $\phi_{\alpha''}$ are eigenstates of a Hermitian operator $\tilde{O}$, and $\lambda_{\alpha'} \neq \lambda_{\alpha''}$. Then

$$\begin{aligned}
\langle\phi_{\alpha'}|O|\phi_{\alpha''}\rangle &= \langle O\phi_{\alpha'}|\phi_{\alpha''}\rangle \\
\langle\phi_{\alpha'}|\lambda_{\alpha''}|\phi_{\alpha''}\rangle &= \langle\lambda_{\alpha'}\phi_{\alpha'}|\phi_{\alpha''}\rangle \\
\lambda_{\alpha''}\langle\phi_{\alpha'}|\phi_{\alpha''}\rangle &= \lambda_{\alpha'}^*\langle\phi_{\alpha'}|\phi_{\alpha''}\rangle \\
\lambda_{\alpha''}\langle\phi_{\alpha'}|\phi_{\alpha''}\rangle &= \lambda_{\alpha'}\langle\phi_{\alpha'}|\phi_{\alpha''}\rangle
\end{aligned} \tag{7.64}$$

Given that the eigenvalues are different, the only way this equation can hold true is that

$$\langle\phi_{\alpha'}|\phi_{\alpha''}\rangle = 0 \tag{7.65}$$

Theorem H3
For any square-integrable function $\psi(x)$ and any Hermitian operator $\tilde{O}$, the function ψ can always be expressed as some linear combination of the eigenstates of $\tilde{O}$

$$\psi(x) = \sum_\alpha c_\alpha\phi_\alpha(x) \tag{7.66}$$

where the sum is replaced by an integral if the eigenvalues span a continuous range.

We will not try to prove this here, but a plausibility argument from linear algebra may help. The eigenvalues of an $N \times N$ Hermitian matrix M are determined by solutions λ to the secular equation

$$\det(M - \lambda\mathbb{1}) = 0 \tag{7.67}$$

where $\mathbb{1}$ is the $N \times N$ identity matrix. This equation is an Nth-order polynomial, which in general has N solutions which, for a Hermitian matrix, must be real. Suppose, for simplicity, that each of these N eigenvalues λ_n are different. Then the corresponding, normalized eigenvectors $\mathbf{v}_n$ are all orthogonal. But N orthogonal vectors form a basis for an N-dimensional vector space, which means that any vector $\mathbf{w}$ in that space can be expressed as a linear combination

$$\mathbf{w} = \sum_{n=1}^{N} c_n \mathbf{v}_n \tag{7.68}$$

of the eigenvectors. This is the equivalent statement, in a finite-dimensional space, of theorem H3, which applies to an infinite-dimensional Hilbert space.

7.4.1 The generalized Born interpretation

We so far have rules for probabilities (or probability densities) for position, momentum, and energy. What about other observables? Here is the general rule:

> **The generalized Born interpretation**
>
> Suppose that an observable O corresponds to a Hermitian operator $\tilde{O}$ whose eigenvalues λ_n are discrete and non-degenerate, and whose eigenstates ϕ_n are normalized to
>
> $$\langle \phi_n | \phi_m \rangle = \delta_{nm}$$
>
> Denote the quantum state of the system at the time of measurement by $|\psi\rangle$. Then:
>
> 1. The outcome of any measurement of the observable will be one of the eigenvalues $\{\lambda_n\}$.
> 2. The probability $P(\lambda_n)$ that any given eigenvalue will be the result of the measurement is
>
> $$P(\lambda_n) = \left|\langle \phi_n | \psi \rangle\right|^2$$

The argument for the generalized Born rules is identical to the reasoning of chapter 6 concerning energy eigenvalues. This time we will carry the argument through in bra-ket notation. The expectation value is given by

$$\langle O \rangle = \langle \psi | O | \psi \rangle \tag{7.69}$$

From theorem H3

$$|\psi\rangle = \sum_n c_n |\phi_n\rangle \tag{7.70}$$

To determine the coefficients c_n, multiply both sides of equation (7.70) by $\langle\phi_\beta|$

$$\langle\phi_m|\psi\rangle = \sum_n c_n\langle\phi_m|\phi_n\rangle \tag{7.71}$$

Using theorem H2

$$\langle\phi_m|\psi\rangle = \sum_n c_n\delta_{nm} \tag{7.72}$$

therefore

$$c_m = \langle\phi_m|\psi\rangle \tag{7.73}$$

The bra vector $\langle\psi|$ corresponding to the ket $|\psi\rangle$ is

$$\langle\psi| = \sum_n c_n^*\langle\phi_n| \tag{7.74}$$

Substituting these expressions into the expectation value

$$\begin{aligned}\langle O\rangle &= \left[\sum_n c_n^*\langle\phi_n|\right] O \left[\sum_m c_m|\phi_m\rangle\right]\\ &= \sum_n\sum_m c_n^* c_m\langle\phi_n|O|\phi_m\rangle\\ &= \sum_n\sum_m c_n^* c_m\langle\phi_n|\lambda_m|\phi_m\rangle\\ &= \sum_n\sum_m \lambda_m c_n^* c_m\langle\phi_n|\phi_m\rangle\\ &= \sum_n\sum_m \lambda_m c_n^* c_m\delta_{nm}\\ &= \sum_n \lambda_n c_n^* c_n\\ &= \sum_n \lambda_n|\langle\phi_n|\psi\rangle|^2\end{aligned} \tag{7.75}$$

Finally, compare this expression for $\langle O\rangle$ against the expression for an expectation value from probability theory

$$\langle O\rangle = \sum_n O_n P[O_n] \tag{7.76}$$

where $\{O_n\}$ are the possible values that the observable can have. We have already shown, in the beginning of this section, that the only values which can be found by a measurement are the eigenvalues of $\tilde{O}$, i.e.

$$\{O_n\} = \{\lambda_n\} \tag{7.77}$$

then the probability is uniquely determined to be

$$P(\lambda_n) = \left|\langle\phi_n|\psi\rangle\right|^2 \tag{7.78}$$

In the case that the eigenvalues span a continuous range, sums are replaced by integrals and the argument goes through the same way, except that $P(\lambda)$ becomes a probability density. This establishes the generalized Born interpretation.

Let us record and review the solutions to the eigenvalue equations seen thus far:

Eigenstates of position

For

$$\tilde{x} = x \tag{7.79}$$

the eigenvalue equation is

$$x\phi_{x_0}(x) = x_0\phi_{x_0}(x) \tag{7.80}$$

which has solutions

$$\left\{\text{eigenstates} \quad \phi_{x_0}(x) = \delta(x - x_0), \quad \text{eigenvalues} \quad x_0 \in [-\infty, \infty]\right\} \tag{7.81}$$

with inner products

$$\left\langle\phi_{x_1}\middle|\phi_{x_2}\right\rangle = \delta(x_1 - x_2) \tag{7.82}$$

As a consequence of theorem H3, any arbitrary function $\psi(x)$ can be written as a superposition of these eigenstates:

$$\begin{aligned}\psi(x) &= \int \mathrm{d}x_0\; c_{x_0}\phi_{x_0}(x) \\ &= \int \mathrm{d}x_0\; c_{x_0}\delta(x - x_0)\end{aligned} \tag{7.83}$$

which is seen to be satisfied by choosing

$$c_{x_0} = \psi(x_0) \tag{7.84}$$

Eigenstates of momentum

For

$$\tilde{p} = -\mathrm{i}\hbar\frac{\partial}{\partial x} \tag{7.85}$$

the eigenvalue equation is

$$-\mathrm{i}\hbar\frac{\partial}{\partial x}\phi_{p_0}(x) = p_0\phi_{p_0}(x) \tag{7.86}$$

which has solutions

$$\left\{\text{eigenstates} \quad \phi_{p_0}(x) = \frac{1}{\sqrt{2\pi\hbar}} e^{ip_0 x/\hbar}, \quad \text{eigenvalues} \quad p_0 \in [-\infty, \infty]\right\} \tag{7.87}$$

with inner products

$$\left\langle \phi_{p_1} \middle| \phi_{p_2} \right\rangle = \delta(p_1 - p_2) \tag{7.88}$$

As a consequence of theorem H3, any arbitrary function $\psi(x)$ can be written as a superposition of these eigenstates:

$$\psi(x) = \int \mathrm{d}p_0 \; c_{p_0} \phi_{p_0}(x) \tag{7.89}$$

This is an inverse Fourier transform. The coefficient function c_{p_0} is obtained by multiplying both sides of this equation by $\phi^*_{p_1}(x)$, and integrating over x:

$$\begin{aligned} \int \mathrm{d}x \; \phi^*_{p_1}(x)\psi(x) &= \int \mathrm{d}x \int \mathrm{d}p_0 \; c_{p_0}\phi^*_{p_1}(x)\phi_{p_0}(x) \\ \frac{1}{\sqrt{2\pi\hbar}} \int \mathrm{d}x \; \psi(x) e^{-ip_1 x/\hbar} &= \int \frac{\mathrm{d}p_0}{2\pi\hbar} c_{p_0} \int \mathrm{d}x e^{i(p_0 - p_1)x/\hbar} \\ &= \int \mathrm{d}p_0 \; c_{p_0}\delta(p_0 - p_1) \end{aligned} \tag{7.90}$$

or

$$c_p = \frac{1}{\sqrt{2\pi\hbar}} \int \mathrm{d}x \; \psi(x) e^{-ipx/\hbar} \tag{7.91}$$

Eigenstates of energy: the particle in a tube

The Hamiltonian operator is

$$\tilde{H} = -\frac{\hbar^2}{2m}\frac{\partial^2}{\partial x^2} + V(x) \tag{7.92}$$

where $V(x)$ for the particle in a tube was given in equation (5.101). The eigenvalue equation is the time-independent Schrödinger equation

$$\left[-\frac{\hbar^2}{2m}\frac{\partial^2}{\partial x^2} + V(x)\right]\phi_n(x) = E_n \phi_n(x) \tag{7.93}$$

which has solutions

$$\left\{\begin{aligned} &\text{eigenstates} \quad \phi_n(x) = \sqrt{\frac{2}{L}} \sin\left[\frac{n\pi x}{L}\right], \\ &\text{eigenvalues} \quad E_n = n^2 \frac{\hbar^2 \pi^2}{2mL^2}, \quad n = 1, 2, 3, \ldots \end{aligned}\right\} \tag{7.94}$$

The eigenstates have inner products

$$\langle \phi_n | \phi_m \rangle = \delta_{nm} \tag{7.95}$$

Once again, theorem H3 insists that any arbitrary function $\psi(x)$ can be written as a linear combination[1]

$$\begin{aligned} \psi(x) &= \sum_{n=1}^{\infty} c_n \phi_n(x) \\ &= \sqrt{\frac{2}{L}} \sum_{n=1}^{\infty} c_n \sin\left[\frac{n\pi x}{L}\right] \end{aligned} \tag{7.96}$$

This is a Fourier series, and it is well known that any function in the interval [0, L] can be expressed in this form. The coefficients c_n are obtained in the same way as before, i.e. by multiplying both sides by ϕ_m^* and integrating over x:

$$\begin{aligned} \psi(x) &= \sum_{n=1}^{\infty} c_n \phi_n(x) \\ \langle \phi_m | \psi \rangle &= \sum_{n=1}^{\infty} c_n \langle \phi_m | \phi_n \rangle \\ &= \sum_{n=1}^{\infty} c_n \delta_{nm} \end{aligned} \tag{7.97}$$

and we find

$$\begin{aligned} c_m &= \langle \phi_m | \psi \rangle \\ &= \sqrt{\frac{2}{L}} \int_0^L \mathrm{d}x \sin\left[\frac{m\pi x}{L}\right] \psi(x) \end{aligned} \tag{7.98}$$

7.5 The generalized uncertainty principle

Suppose we want to devise a measurement apparatus that will measure two observables A and B simultaneously. From the discussion of previous sections, it is clear that the result of such a measurement will be to leave the system in a state of zero uncertainty in both observables, i.e. $\Delta A = \Delta B = 0$, and that means that the physical state would have to be an eigenstate of both $\tilde{A}$ and $\tilde{B}$. Now, it may or may not be true that such physical states exist, in general, for all possible eigenvalues of $\tilde{A}$ and $\tilde{B}$. If not, then it is not possible to measure the observables A and B simultaneously. For example, in the case of position x and momentum p, there are no physical states which are eigenstates of both $\tilde{x}$ and $\tilde{p}$; this is why it is impossible to measure both x and p simultaneously and precisely.

[1] We exclude cases for which $\langle E \rangle = \infty$, such as wave functions which are non-zero in the regions $x < 0$ or $x > L$.

There is a simple test for whether or not two observables can be measured simultaneously, which requires the concept of the commutator. The ***commutator*** of two operators is defined to be

$$[\tilde{A}, \tilde{B}] \equiv \tilde{A}\tilde{B} - \tilde{B}\tilde{A} \tag{7.99}$$

If $[\tilde{A}, \tilde{B}] = 0$, the two operators are said to commute. Commutators are computed by letting them act on an arbitrary function. For example:

$$[\tilde{x}, \tilde{x}^2]f(x) = (xx^2 - x^2x)f(x) = 0 \tag{7.100}$$

therefore $\tilde{x}$ and $\tilde{x}^2$ commute, i.e.

$$[\tilde{x}, \tilde{x}^2] = 0 \tag{7.101}$$

Similarly

$$\begin{aligned} \left[\tilde{x}, \tilde{p}_y\right]f(x, y, z) &= \left[x\left(-i\hbar\frac{\partial}{\partial y}\right) - \left(-i\hbar\frac{\partial}{\partial y}\right)x\right]f(x) \\ &= -i\hbar\left[\left(x\frac{\partial f}{\partial y} - x\frac{\partial f}{\partial y}\right)\right] \\ &= 0 \end{aligned} \tag{7.102}$$

which means that $\tilde{x}$ and $\tilde{p}_y$ commute,

$$\left[\tilde{x}, \tilde{p}_y\right] = 0 \tag{7.103}$$

However,

$$\begin{aligned} [\tilde{x}, \tilde{p}]f(x) &= \left[x\left(-i\hbar\frac{\partial}{\partial x}\right) - \left(-i\hbar\frac{\partial}{\partial x}\right)x\right]f(x) \\ &= -i\hbar\left(x\frac{\partial f}{\partial x} - \frac{\partial}{\partial x}(xf)\right) \\ &= -i\hbar\left(x\frac{\partial f}{\partial x} - x\frac{\partial f}{\partial x} - f\right) \\ &= i\hbar f \end{aligned} \tag{7.104}$$

From this we can conclude that the operators corresponding to position along the x-axis, and momentum in the x-direction, do *not* commute, and that

$$[\tilde{x}, \tilde{p}] = i\hbar \tag{7.105}$$

The condition that two observables can be measured simultaneously is then stated by:

> **The commutator theorem**
>
> Two observables A and B are simultaneously measurable, i.e. they share the same set of eigenstates, if and only their corresponding operators commute, i.e. $[\tilde{A}, \tilde{B}] = 0$.

The fact that position and momentum are not simultaneously observable is a special case of this theorem. For simplicity, we will prove the theorem only in the case that the eigenvalues of $\tilde{A}$ and $\tilde{B}$ are discrete and non-degenerate.

Let us begin with the 'if' part of the theorem: $[\tilde{A}, \tilde{B}] = 0$ implies that eigenstates of $\tilde{A}$ are eigenstates of $\tilde{B}$ and vice versa. Now $[\tilde{A}, \tilde{B}] = 0$ means that

$$\tilde{A}\tilde{B}f(x) = \tilde{B}\tilde{A}f(x) \tag{7.106}$$

and let ϕ_a be any eigenstate of $\tilde{A}$, i.e.

$$\tilde{A}\phi_a = a\phi_a \tag{7.107}$$

Then

$$\begin{aligned}\tilde{A}\tilde{B}\phi_a(x) &= \tilde{B}\tilde{A}\phi_a(x)\\ &= a\tilde{B}\phi_a(x)\end{aligned} \tag{7.108}$$

Now if we define

$$\phi_a' = \tilde{B}\phi_a \tag{7.109}$$

then equation (7.108) becomes

$$\tilde{A}\phi_a' = a\phi_a' \tag{7.110}$$

which means that ϕ_a' is an eigenstate of $\tilde{A}$ with eigenvalue a. But since the eigenvalues of $\tilde{A}$ are non-degenerate, it can only be that ϕ_a' is proportional to ϕ_a, i.e.

$$\phi_a'(x) = b\phi_a(x) \tag{7.111}$$

where b is some constant. In that case, from the definition of ϕ',

$$\tilde{B}\phi_a(x) = b\phi_a(x) \tag{7.112}$$

This proves that every eigenstate of $\tilde{A}$ is an eigenstate of $\tilde{B}$. Such states are zero-uncertainty states of both A and B, so simultaneous measurement is possible.

Next we prove the 'only if' part of the commutator theorem. Suppose the outcome of a measurement is that $A = a$ and $B = b$. Since the eigenvalues of $\tilde{A}$ are non-degenerate, the resulting physical state is ϕ_a, where

$$\tilde{A}\phi_a = a\phi_a \tag{7.107}$$

However, this must also be a zero-uncertainty state for B, and therefore

$$\tilde{B}\phi_a = b\phi_a \tag{7.114}$$

It follows that

$$\begin{aligned} AB|\phi_a\rangle &= bA|\phi_a\rangle \\ &= ba|\phi_a\rangle \end{aligned} \tag{7.115}$$

while

$$\begin{aligned} BA|\phi_a\rangle &= aB|\phi_a\rangle \\ &= ab|\phi_a\rangle \end{aligned} \tag{7.116}$$

therefore

$$\begin{aligned} [A, B]|\phi_a\rangle &= (AB - BA)|\phi_a\rangle \\ &= (ba - ab)|\phi_a\rangle \\ &= 0 \end{aligned} \tag{7.117}$$

Now, according to theorem H3, any square-integrable function can be represented as a linear combination of eigenstates of $\tilde{A}$, i.e.

$$|f\rangle = \sum_a c_a|\phi_a\rangle \tag{7.118}$$

and this means that

$$\begin{aligned} [\tilde{A}, \tilde{B}]|f\rangle &= \sum_a c_a[\tilde{A}, \tilde{B}]|\phi_a\rangle \\ &= 0 \end{aligned} \tag{7.119}$$

If the operator $[\tilde{A}, \tilde{B}]$ acting on any function $f(x)$ gives 0, then the operator itself is zero

$$[\tilde{A}, \tilde{B}] = 0 \tag{7.120}$$

which proves the ‘only if’ portion of theorem.

The proof above extends trivially to continuous non-degenerate eigenvalues, simply replacing a discrete index by a continuous index, and sums by integrals. The commutator theorem is also true if the eigenvalues are degenerate, but those complications will be postponed to a later chapter.

There is a generalization of the Heisenberg uncertainty principle to the case of any two observables which cannot be simultaneously measured:

The generalized uncertainty principle

The uncertainties ΔA and ΔB of any two observables in any physical state $|\psi\rangle$ satisfy the inequality

$$\Delta A \Delta B \geqslant \frac{1}{2}|\langle\psi|[A, B]|\psi\rangle| \tag{7.121}$$

Proof. Begin with the expressions for uncertainty

$$\begin{aligned}(\Delta A)^2 &= \langle\psi|(\tilde{A}-\langle A\rangle)^2|\psi\rangle \\ (\Delta B)^2 &= \langle\psi|(\tilde{B}-\langle B\rangle)^2|\psi\rangle\end{aligned} \tag{7.122}$$

and define the Hermitian operators

$$\tilde{D}_A \equiv A - \langle A\rangle \qquad \text{and} \qquad \tilde{D}_B = B - \langle B\rangle \tag{7.123}$$

and also

$$|\psi_1\rangle = D_A|\psi\rangle \qquad \text{and} \qquad |\psi_2\rangle = D_B|\psi\rangle \tag{7.124}$$

Then

$$\begin{aligned}(\Delta A)^2 &= \langle\psi|D_A D_A|\psi\rangle \\ &= \langle\psi|D_A|\psi_1\rangle \\ &= \langle D_A\psi|\psi_1\rangle \qquad \text{(Hermiticity)} \\ &= \langle\psi_1|\psi_1\rangle\end{aligned} \tag{7.125}$$

Similarly,

$$(\Delta B)^2 = \langle\psi_2|\psi_2\rangle \tag{7.126}$$

Now, making use of the Cauchy–Schwarz inequality for vectors

$$|u||v| \geqslant |\langle u|v\rangle| \tag{7.127}$$

and squaring both sides, we have

$$\begin{aligned}|\langle\psi_1|\psi_1\rangle||\langle\psi_2|\psi_2\rangle| &\geqslant |\langle\psi_1|\psi_2\rangle|^2 \\ (\Delta A)^2(\Delta B)^2 &\geqslant \frac{1}{2}\left[|\langle\psi_1|\psi_2\rangle|^2 + |\langle\psi_2|\psi_1\rangle|^2\right] \\ &\geqslant \frac{1}{2}\left[|\langle D_A\psi|D_B\psi\rangle|^2 + |\langle D_B\psi|D_A\psi\rangle|^2\right] \\ &\geqslant \frac{1}{2}\left[|\langle\psi|D_A D_B\psi\rangle|^2 + |\langle\psi|D_B D_A\psi\rangle|^2\right]\end{aligned} \tag{7.128}$$

Next write

$$D_A D_B = F + G$$
$$D_B D_A = F - G \tag{7.129}$$

where

$$F = \frac{1}{2}(D_A D_B + D_B D_A)$$
$$G = \frac{1}{2}[D_A, D_B] = \frac{1}{2}[A, B] \tag{7.130}$$

Substituting these expressions into (7.128) we obtain

$$\begin{aligned}(\Delta A)^2(\Delta B)^2 &\geqslant \frac{1}{2}[(\langle F\rangle + \langle G\rangle)(\langle F\rangle + \langle G\rangle)^* \\ &\quad + (\langle F\rangle - \langle G\rangle)(\langle F\rangle - \langle G\rangle)^*] \\ &\geqslant |\langle F\rangle|^2 + |\langle G\rangle|^2 \\ &\geqslant |\langle G\rangle|^2 \\ &\geqslant \frac{1}{4}\,|\langle\psi|[A, B]|\psi\rangle|^2\end{aligned} \tag{7.131}$$

Taking the square root of both sides establishes the generalized uncertainty principle. The Heisenberg uncertainty principle is just a special case, and follows by choosing $A = x$ and $B = p$, so that

$$\begin{aligned}\Delta x \Delta p &\geqslant \frac{1}{2}|\langle\psi|\mathrm{i}\hbar|\psi\rangle| \\ &\geqslant \frac{\hbar}{2}\end{aligned} \tag{7.132}$$

as discussed in chapter 6.

Of course, if A and B commute, this does not mean that $\Delta A \Delta B$ is necessarily zero. That is only true if the system is in an eigenstate of both A and B. Otherwise, $|\langle F\rangle|^2$ will be greater than zero, even if $|\langle G\rangle|^2 = 0$.

7.6 The time–energy uncertainty relation

It was noted, at the end of the last chapter, that change of any kind in a quantum-mechanical system implies an uncertainty in the energy of that system. We would like to quantify this remark: how long does it take for a physical state to change substantially, and how is that length of time related to the energy uncertainty?

Of course, it is necessary to define what is meant by a 'substantial' change in the state. Consider position, momentum, or any other observable Q. At a given moment, Q has a certain expectation value $\langle Q \rangle$, and an uncertainty ΔQ. As time passes, $\langle Q \rangle$ usually changes in some way. We will say that $\langle Q \rangle$ has changed 'substantially' when it has increased or decreased by an amount equal to the uncertainty ΔQ of the observable. The relation between the time Δt required for such a change, and the energy uncertainty ΔE, is expressed in the form of an inequality known as

The time-energy uncertainty relation

Let Q be any observable, and let Δt be the time required for the expectation value $\langle Q \rangle$ to change by an amount equal to its uncertainty ΔQ. Then

$$\Delta t \Delta E \geqslant \frac{\hbar}{2} \tag{7.133}$$

where ΔE is the uncertainty in the energy of the system.

To prove this statement, let us begin with an expression for the rate of change of the observable

$$\begin{aligned}
\frac{\mathrm{d}}{\mathrm{d}t}\langle Q \rangle &= \frac{\mathrm{d}}{\mathrm{d}t} \int \mathrm{d}x\, \psi^*(x, t)\, \tilde{Q}\, \psi(x, t) \\
&= \int \mathrm{d}x \left[\frac{\mathrm{d}\psi^*}{\mathrm{d}t} \tilde{Q}\psi + \psi^* \tilde{Q} \frac{\mathrm{d}\psi}{\mathrm{d}t} \right] \\
&= \int \mathrm{d}x \left[\left(\frac{1}{\mathrm{i}\hbar} \tilde{H}\psi \right)^* \tilde{Q}\psi + \psi^* \tilde{Q} \left(\frac{1}{\mathrm{i}\hbar} \tilde{H}\psi \right) \right] \\
&= \frac{1}{\mathrm{i}\hbar} [-\langle H\psi | Q | \psi \rangle + \langle \psi | Q | H\psi \rangle] \\
&= \frac{1}{\mathrm{i}\hbar} [-\langle \psi | HQ | \psi \rangle + \langle \psi | QH | \psi \rangle] \\
&= \langle \psi | \frac{1}{\mathrm{i}\hbar} [Q, H] | \psi \rangle
\end{aligned} \tag{7.134}$$

Then the time Δt required for $\langle Q \rangle$ to change by an amount equal to its uncertainty ΔQ satisfies

$$\begin{aligned}
\left| \frac{\mathrm{d}\langle Q \rangle}{\mathrm{d}t} \right| \Delta t &= \Delta Q \\
\frac{1}{\hbar} |\langle \psi | [Q, H] | \psi \rangle| \Delta t &= \Delta Q
\end{aligned} \tag{7.135}$$

We now apply the generalized uncertainty principle, equation (7.121) with $A = Q$ and $B = H$ (note that $\Delta H \equiv \Delta E$)

$$\frac{1}{\hbar}(\Delta Q \Delta E)\Delta t \geqslant \frac{1}{\hbar}\left(\frac{1}{2}|\langle\psi|[Q, H]|\psi\rangle|\right)\Delta t = \frac{1}{2}\Delta Q \tag{7.136}$$

or, dividing by ΔQ on both sides

$$\Delta E \Delta t \geqslant \frac{\hbar}{2} \tag{7.137}$$

which proves the energy-time uncertainty relation, for *any* observable Q.

As an example of this uncertainty relation, consider the time required for a free particle to move a distance equal to its uncertainty in position Δx; i.e. for $\langle x\rangle$ to change by Δx. The wavepacket is moving with a group velocity

$$v_g = \frac{\langle p\rangle}{m} \tag{7.138}$$

so the time it takes for the packet to move a distance equal to the uncertainty in position is

$$\Delta t = \frac{\Delta x}{v_g} = \frac{m\Delta x}{\langle p\rangle} \tag{7.139}$$

On the other hand, the uncertainty in energy is related to the uncertainty in momentum via

$$\Delta E = \Delta\left(\frac{p^2}{2m}\right) = \frac{\langle p\rangle \Delta p}{m} \tag{7.140}$$

Then

$$\begin{aligned} \Delta t \Delta E &= \frac{m\Delta x}{\langle p\rangle}\frac{\langle p\rangle\Delta p}{m} \\ &= \Delta x \Delta p \\ &\geqslant \frac{\hbar}{2} \end{aligned} \tag{7.141}$$

as predicted by the time-energy uncertainty relation.

Problems

1. Find the Hermitian conjugate $\tilde{O}^\dagger$ of the operator

$$\tilde{O} = \tilde{x}\tilde{p} = -\mathrm{i}\hbar x\frac{\partial}{\partial x} \tag{7.142}$$

from the property that $\langle O^\dagger\psi_1|\psi_2\rangle = \langle\psi_1|O\psi_2\rangle$.

2. Using equation (7.20), show that in general, for any two operators A and B,

$$(AB)^{\dagger} = B^{\dagger}A^{\dagger} \tag{7.143}$$

and use this equation to immediately write down the Hermitian conjugate of the operator $\tilde{O}$ of problem 1.

3. Use the matrix form of operators (7.40), the expression for the Hermitian conjugate (7.42), and equation (7.35), as an alternate method to show that

$$\left(\frac{\mathrm{d}}{\mathrm{d}x}\right)^{\dagger} = -\frac{\mathrm{d}}{\mathrm{d}x} \tag{7.27}$$

4. Consider the Hermitian 2 × 2 matrix

$$\begin{bmatrix} 0 & -i \\ i & 0 \end{bmatrix} \tag{7.145}$$

Find the two normalized solutions to the eigenvector equation $M\mathbf{u}_n = \lambda_n \mathbf{u}_n$, and verify that theorems H1–H3 hold. In other words, show that the eigenvalues are real, that the two eigenvectors $\mathbf{u}_1$, $\mathbf{u}_2$ are orthogonal. Show also that any arbitrary vector $\mathbf{v}$ can be expressed as a linear combination

$$\mathbf{v} = a\mathbf{u}_1 + b\mathbf{u}_2 \tag{7.146}$$

by computing a, b in terms of the two components of vector v.

5. Compute the commutators

$$[H, x], \quad [H, p], \quad [p^2, x^2] \tag{7.147}$$

Your answers can be expressed in terms of x, p, and the first derivative of $V(x)$.

6. Which of these operators are linear? Of those, find the Hermitian conjugates.

$$\begin{aligned} Q_1\psi(x) &= \frac{\mathrm{d}\psi}{\mathrm{d}x} + 10, & Q_2\psi(x) &= \psi^*(x) \\ Q_3\psi(x) &= \mathrm{e}^{\mathrm{i}x}\frac{\mathrm{d}}{\mathrm{d}x}\psi(x), & Q_4\psi(x) &= \sin(\psi(x)) \\ Q_5\psi(x) &= \frac{\mathrm{d}}{\mathrm{d}x}\sin(x)\frac{\mathrm{d}^2}{\mathrm{d}x^2}\psi(x), & Q_6\psi(x) &= 5\psi\left(\frac{x+3}{2}\right) \end{aligned} \tag{7.148}$$

7. Calculate the commutator

$$\left[e^x, \frac{\mathrm{d}}{\mathrm{d}x}e^x\frac{\mathrm{d}}{\mathrm{d}x}\right] \tag{7.149}$$

Your answer should contain only one derivative operator.

8. Any state at any given time can be expanded as a linear combination of energy eigenstates, so let

$$\psi(x, 0) = \sum_{\alpha} c_{\alpha}\phi_{\alpha}(x) \tag{7.150}$$

where $\tilde{H}\phi_{\alpha} = E_{\alpha}\phi_{\alpha}$. Show that (7.48) implies equation (5.38)
9. Consider the operators $\tilde{A} = \tilde{x}$ and $\tilde{B} = \tilde{x}\tilde{p}\,\tilde{x}$.
   a. Compute ΔA and ΔB in the state

$$\phi(x) = N\mathrm{e}^{-x^2/2a^2} \tag{7.151}$$

   b. Calculate the commutator $[\tilde{A}, \tilde{B}]$, and verify that the lower bound on the product $\Delta A \Delta B$ set by the generalized uncertainty principle is satisfied.
10. Let Δt be the time required for the peak of a Gaussian wavepacket to move a distance equal to the width Δx of the wavepacket. Compute $\Delta E \Delta t$ for the Gaussian wavepacket of equation (6.1), and verify that it satisfies the time–energy uncertainty relation. (Hint: it is easiest to use equation (5.67) in evaluating ΔE.)
11. Prove the Cauchy–Schwarz inequality (7.127).

Part II

Exact solutions

IOP Publishing

An Introduction to Quantum Theory

Jeff Greensite

Chapter 8

Rectangular potentials

Most of the effort in solving quantum-mechanical problems goes into solving the time-independent Schrödinger equation (5.35). Given a complete set of solutions to the time-independent equation, the general solution (5.38)–(5.40) of the time-dependent equation follows immediately.

We have already seen that, in the case of a free particle, the energy eigenvalues could take on any value in the range $[0, \infty]$, while in the case of a particle trapped in a tube, the energies could take on only certain discrete values. This distinction between the energies of bound and unbound particles is very general, and in fact the solutions to any time-independent Schrödinger equation, with any potential $V(x)$, are always of two kinds: (i) ***bound states***, which have discrete energy eigenvalues, E_1, E_2, E_3, …; and (ii) ***unbound*** or 'scattering' states, which have energies in a continuous range $E \in [E_{\text{low}}, \infty]$. So we begin this chapter with an explanation of why the energies of bound states are always discrete, and also why the unbound states, which are energy eigenstates and therefore stationary, have something to do with the scattering of particles by a potential, which is of course a process which evolves in time.

8.1 A qualitative sketch of energy eigenstates

Consider a particle of total energy E in the potential $V(x) \leqslant V_{\text{max}}$ shown in figure 8.1. In classical physics, if $E < V_{\text{max}}$, the particle could never be found in the regions where $V(x) > E$, because that would imply that the kinetic energy $KE = E - V$ was negative. Such intervals are referred to as the ***classically forbidden*** regions; intervals where $E > V(x)$ are the ***classically allowed*** regions. A particle in the central area, region II, could never leave that region, and would instead bounce back and forth between the boundaries of the classically allowed region. Likewise, a particle with energy E in regions I or II could never penetrate the forbidden region and enter region II.

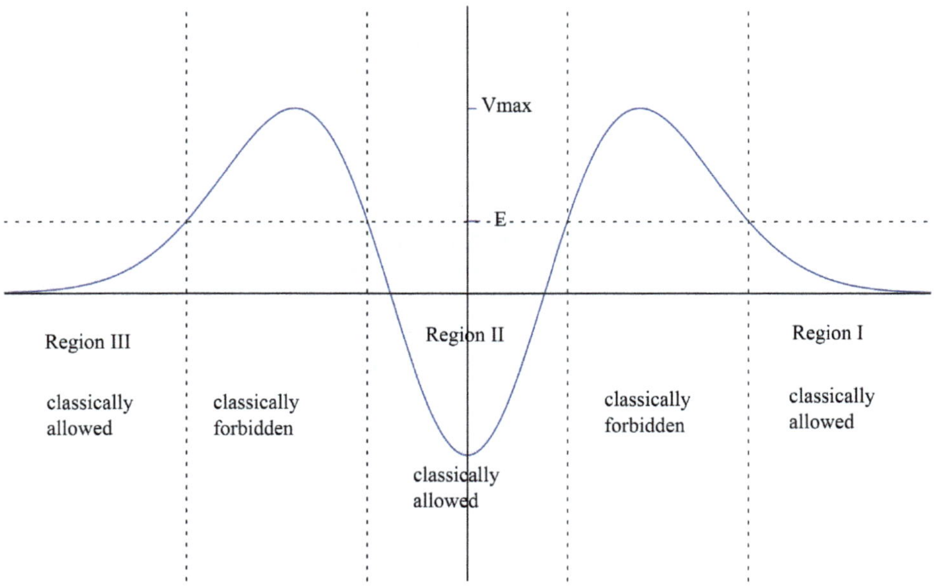

Figure 8.1. A sample potential with $E < V_{max}$, showing the classically allowed ($E > V(x)$) and classically forbidden ($E < V(x)$) regions.

In quantum mechanics the physical state is governed by the Schrödinger equation, and generally there is no reason that the wave function must be exactly zero in regions where $E < V(x)$. This means that there is usually a finite probability to find a particle in the classically forbidden regions, and this fact gives rise to some remarkable phenomena which are inconceivable at the classical level. Nevertheless, the wave function does behave quite differently in the classically allowed and the classically forbidden regions, as one might expect, since classical behavior is an approximation to quantum mechanics, and should become exact in the $\hbar \to 0$ limit.

The time-independent Schrödinger equation is a second-order differential equation

$$\frac{d^2\phi}{dx^2} = -\frac{2m}{\hbar^2}(E - V)\phi \tag{8.1}$$

In general, the first derivative of a function $f' = df/dx$ is the slope of $f(x)$, the second derivative $f'' = d^2f/dx^2$ is the curvature. If $f'' > 0$ at point x, then $f(x)$ is concave upwards at that point; similarly, if $f'' < 0$, then $f(x)$ is concave downwards. From equation (8.1), we can see that:

$E > V$: classically allowed region

$$\begin{array}{lll} \phi > 0 & \phi(x) & \text{concave downwards} \\ \phi < 0 & \phi(x) & \text{concave upwards} \end{array} \tag{8.2}$$

Since the wave function tends to curve downwards when ϕ is positive, and upward when ϕ is negative, the result is typically an oscillating wave function in the classically allowed regions, as shown in figure 8.2(a).

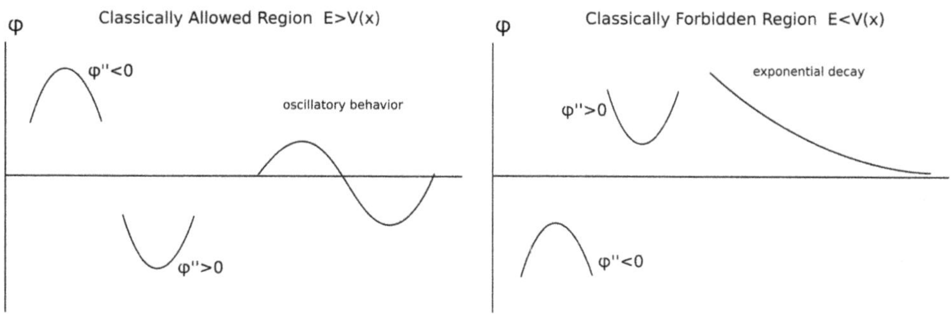

(a) $E > V(x)$ (b) $E < V(x)$

Figure 8.2. Qualitative behavior of energy eigenstates in the classically allowed (a) and classically forbidden (b) regions.

$E < V$: classically forbidden region

$$\begin{array}{ll} \phi > 0 \;\; \phi(x) & \text{concave upwards} \\ \phi < 0 \;\; \phi(x) & \text{concave downwards} \end{array} \tag{8.3}$$

In the classically forbidden regions, oscillatory behavior is impossible, and the wave function tends to either grow or decay exponentially, as shown in figure 8.2(b)

We can use this qualitative information to sketch a rough picture of what an energy eigenstate in the potential shown in figure 8.1 must look like. Let us begin with $E > V_{max}$. In that case there are no classically forbidden regions, and the function oscillates everywhere. The curvature, and therefore the rate of oscillation, depends on the ratio

$$\frac{\phi''}{\phi} = -\frac{\hbar^2}{2m}(E - V) \tag{8.4}$$

so that the greater $E - V$ is, the faster the wave function oscillates. The general behavior is shown in figure 8.3(a); an energy eigenstate will look something like this for *any* energy $E > V_{max}$.

Now suppose the energy is in the range $E \in [0, V_{max}]$. Then there will be classically forbidden regions, and in these regions the wave function will grow or decay exponentially. In the classically allowed regions, the wave function oscillates as shown in figure 8.3(b).

Finally, suppose that $E < 0$. Then, except for a finite interval where $V(x) < 0$ and $E > V(x)$, the entire real line is a classically forbidden region. Typical solutions of the time-independent Schrödinger equation would blow up exponentially as $x \to \pm\infty$, and the integral of $\phi^*(x)\phi(x)$ also diverges. Wave functions of this kind are called ***non-normalizable***. They do not satisfy the condition (4.15), and do not belong to a Hilbert space. Non-normalizable solutions of the time-independent Schrödinger equation are simply thrown away in quantum mechanics. Although such wave functions satisfy the equation of interest, they cannot be used to construct physical states.

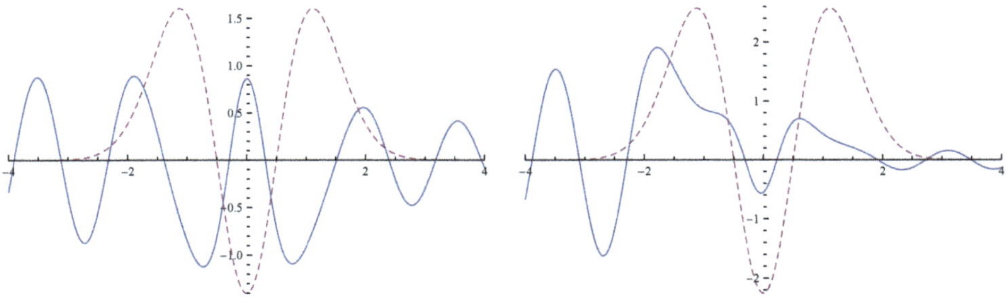

(a) $E > V_{max}$

(b) $0 < E < V_{max}$

Figure 8.3. A sketch of energy eigenstates (blue line) in the example potential (red dashed line), for (a) $E > V_{\max}$, where no region is classically forbidden, and (b) $0 < E < V_{\max}$, where some regions are classically forbidden. These are both eigenstates describing an unbound particle.

Almost all solutions with $E < 0$ are therefore excluded, but not quite all. It is still consistent with the behaviour (8.3) for the wave function to decay exponentially to zero as $x \to \pm\infty$. In wave functions of this sort, the probability of finding the particle outside region II drops exponentially with distance, so it is reasonable to say that the particle is trapped by the potential well, or in other words that the particle is in a 'bound state'. However, it is only possible to have the wave function decay to zero as $x \to \pm\infty$ for certain very special values of the energy eigenvalue E, for reasons which will now be explained.

Suppose the range of the x-axis is taken to be $-L$ to L, rather than $-\infty$ to ∞, where $x = \pm L$ is well inside the classically forbidden region, and we require that $\phi(-L) = \phi(L) = 0$.[1] Let the first derivative $\phi'(-L) = c$, where c is any constant. Given the function and its first derivative at any point, and some value of $E < 0$, the second-order differential equation can be solved for $\phi(x)$ at all x in the range $-L$ to L, numerically if necessary. The function $\phi(x)$ starts to increase exponentially, as x increases, until x reaches the classically allowed region, where the curvature changes, the wave function bends over, and it may even oscillate. But when the wave function leaves the classically allowed region and enters the classically forbidden region at positive x, it begins to again diverge exponentially, and in general $\phi(L) \neq 0$. Changing the initial value of $\phi'(0) = c$ will not help. If this is changed to, say, $\phi'(0) = \alpha c$, where α is some numerical factor, then, because of the linearity of the time-independent Schrödinger equation, all that happens is that the wave function is rescaled everywhere by the same factor, i.e. $\phi(x) \to \alpha\phi(x)$, and the wave function still does not vanish at $x = L$.

However, the energy E can also be varied. Suppose that for some $E = E_a < 0$ the solution diverges towards $+\infty$, as shown in figure 8.4(a), while at $E = E_b < 0$ the solution diverges towards $-\infty$, as seen in figure 8.4(b). Then there is usually a value $E = E_1$ which lies between these two values, where the solution goes smoothly to $\phi(L) = 0$. This case is shown in figure 8.4(c). Changing E even slightly away from E_1 will result in $\phi(L) \neq 0$. On the other hand, upon increasing E to some value E_c,

[1] This is known as a 'two-point boundary value problem'.

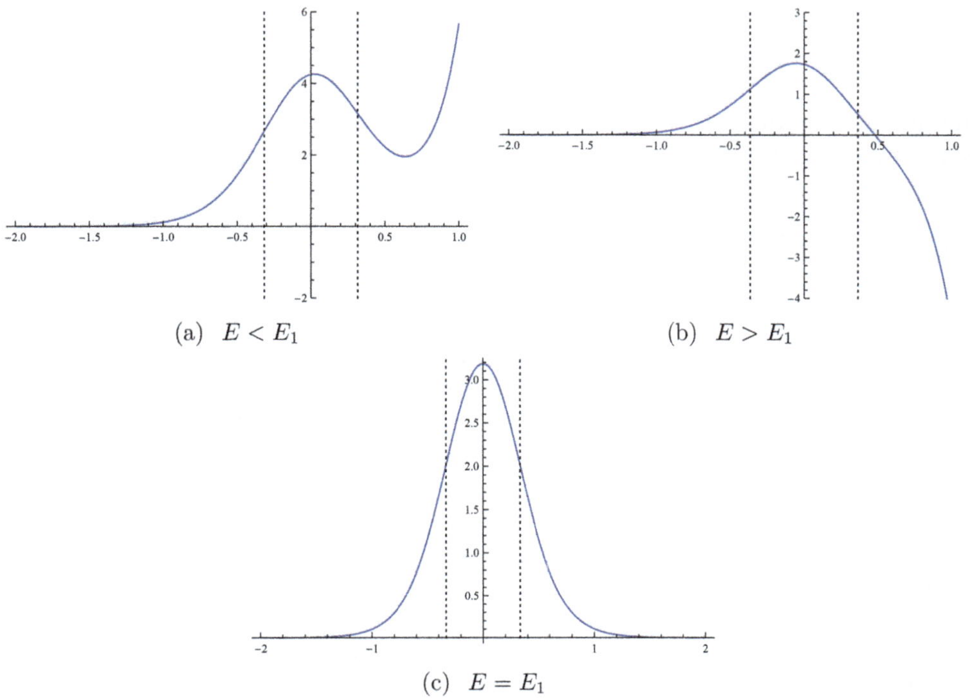

Figure 8.4. For a bound state, it is necessary to impose $\phi(x) \to 0$ as $x \to \pm\infty$. Otherwise the solution diverges exponentially to $\pm$ infinity (subfigures (a) and (b)), and the wave function is not normalizable. But this is possible only at certain special values of E (subfigure (c)), and so the energy eigenvalues of bound states form a discrete set.

the solution may again switch sign at $x = L$. In that case there will be another special value $E = E_2$, where $E_b < E_2 < E_c$, where $\phi(L) = 0$. In general there can be more than one value of E such that the solution satisfies $\phi(-L) = \phi(+L) = 0$. But however many solutions there may be with $E < 0$, they always form a discrete set.

Nothing changes very much as we take the $L \to \infty$ limit, and impose that $\phi(x) \to 0$ as $x \to \pm\infty$. Normalizable solutions, i.e. solutions which do not diverge to $\pm\infty$ in the classically forbidden region, are always a discrete set. This is in contrast to the case where $E > 0$, and only a finite portion of the real line (at most) is classically forbidden. In that case the wave function oscillates, rather than diverges to $\pm\infty$ as $x \to \pm\infty$, consequently it is normalizable (or, more precisely, the overlaps of normalized eigenstates are delta functions), and E can be any real number greater than zero, assuming $V(x) \to 0$ as $x \to \pm\infty$.

The important point to remember is that bound state solutions of the time-independent Schrödinger equation always form a discrete set, while unbound solutions form a continuous set.

8.2 Unbound states and quantum scattering

Consider a particle which is moving in a tube closed at one end. The corresponding potential is

$$V(x) = \begin{cases} 0 & x < 0 \\ \infty & x \geqslant 0 \end{cases} \tag{8.5}$$

Initially, before the particle encounters the end of the tube, its state is represented by some wavepacket indicated in figure 8.5(a), with a momentum expectation value $\langle p \rangle = p_0$. After bouncing off the end of the tube, the 'scattered' wavepacket will be moving in the opposite direction with $\langle p \rangle = -p_0$, as shown in figure 8.5(c). At some intermediate time, the wavepacket is in contact with the end of the tube. Since the wavepacket is finite in extent, there is going to be some delay from the time that the front of the wavepacket reflects off the end of the tube, to the time that the back of the wavepacket reflects off the end. During this time interval, part of the wavepacket which reflected off the end of the tube overlaps with the rest of the incoming wavepacket, which has not yet encountered the end of the tube. This is the situation depicted in figure 8.5(b), where the incoming and reflected wavepackets interfere constructively.

Now if the wavepacket is of some finite extent Δx, then by the uncertainty principle $\Delta p \sim \hbar/\Delta x$. Imagine preparing the initial wavepacket with a very small uncertainty Δp. As $\Delta p \to 0$, then $\Delta x \to \infty$, and the time (and space) interval in which the incoming and reflected waves overlap becomes very large. The limiting case, with $\Delta x = \infty$, is shown in figure 8.6. In this case, the incoming 'wavepacket' at any instant of time is an infinite plane wave

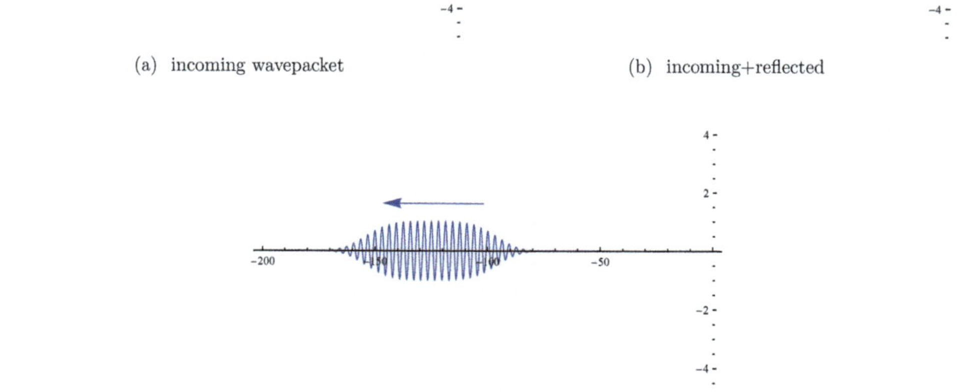

(a) incoming wavepacket

(b) incoming+reflected

(c) reflected wavepacket

Figure 8.5. Reflection of a wavepacket from an impenetrable barrier placed at $x = 0$.

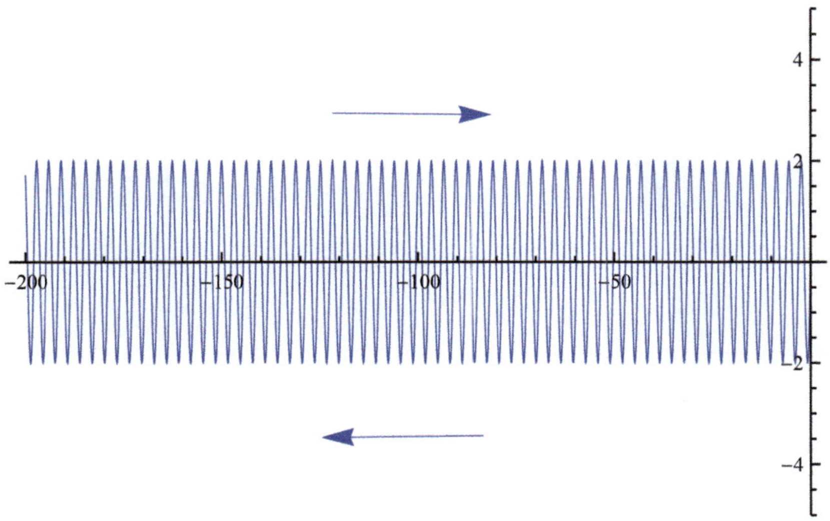

Figure 8.6. An energy eigenstate (a 'stationary' state) as a limit of wavepacket scattering, where the incoming and outgoing waves overlap everywhere, at all times.

$$\phi_{\text{inc}}(x) = A\mathrm{e}^{\mathrm{i}p_0x/\hbar} \tag{8.6}$$

while the reflected wavepacket is also an infinite plane wave of the opposite momentum

$$\phi_{\text{ref}}(x) = B\mathrm{e}^{-\mathrm{i}p_0x/\hbar} \tag{8.7}$$

These plane waves overlap over the entire half-line $x \in [-\infty, 0]$, so the total wave function is the superposition of the incoming and reflected waves

$$\begin{aligned}\phi(x) &= \phi_{\text{inc}}(x) + \phi_{\text{ref}}(x)\\ &= A\mathrm{e}^{\mathrm{i}p_0x/\hbar} + B\mathrm{e}^{-\mathrm{i}p_0x/\hbar}\end{aligned} \tag{8.8}$$

Because $\phi(x) = 0$ for $x > 0$ and because the wave function is continuous at $x = 0$, we must have

$$\phi(x) = 0 \quad \Rightarrow \quad A = -B \tag{8.9}$$

or

$$\phi(x) \propto \sin\left[p_0 x/\hbar\right] \tag{8.10}$$

The state (8.10) is a stationary state. Substituting (8.10) into the time-independent Schrödinger equation, the corresponding energy eigenvalue is found to be

$$E = \frac{p_0^2}{2m} \tag{8.11}$$

Despite the stationarity, this energy eigenstate is obviously just the limit of a non-stationary situation, in which an incoming wavepacket is scattered backwards

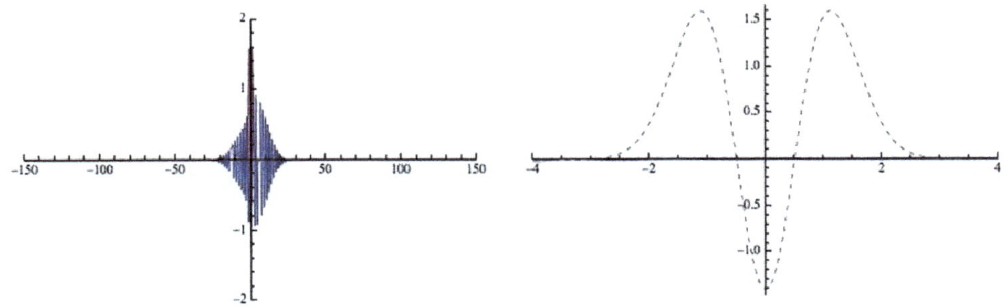

Figure 8.7. Scattering of a wavepacket from a potential, $E > V_{\max}$. (a) Overview; (b) close to the potential well, slow-motion.

by an infinite potential. During the (long) interval in which the incoming wavepacket (with Δp very small) reflects from the end of the tube, the wave function near the end of the tube looks very much like the energy eigenstate (8.10). In fact, in the $\Delta p \to 0$ limit, we can easily identify the part of the eigenstate that corresponds to the incoming wavepacket (ϕ_{inc}) and the part which corresponds to the scattered wavepacket (ϕ_{ref}). This is a general feature of unbound states: an *unbound* stationary state can be viewed as the limit of a dynamical situation, in which an incoming wavepacket of a very precise momentum is scattered by a potential. Part of the unbound state is identified as the incoming wave, other parts represent the scattered waves.

As a second example, consider a particle wavepacket, with $\langle E \rangle > V_{\max}$, incident on the potential well of figure 8.1. Upon encountering the potential, the wavepacket splits into a reflected wavepacket, moving backwards towards $x = -\infty$, and a transmitted wavepacket which has passed through the well and is moving on towards $x = +\infty$, as shown in animations (8.7(a))–(8.7(b)).[2] At some intermediate time, the incoming and reflected wavepackets overlap, and as we take $\Delta p \to 0$ for the incoming wave, the incoming and reflected waves overlap over a very large region. The limiting case resembles figure 8.3(a). This limit is again a stationary state, but the wave to the left of the potential well is a superposition of incoming and reflected waves

$$\begin{aligned} \phi_{\text{left}}(x) &= \phi_{\text{inc}}(x) + \phi_{\text{ref}}(x) \\ &= A e^{ip_0 x/\hbar} + B e^{-ip_0 x/\hbar} \end{aligned} \tag{8.12}$$

while the wave function to the right of the potential well is the outgoing transmitted wave

$$\begin{aligned} \phi_{\text{right}}(x) &= \phi_{\text{trans}}(x) \\ &= C e^{ip_0 x/\hbar} \end{aligned} \tag{8.13}$$

[2] Incidentally, this is another reason why a particle in quantum mechanics cannot be taken *literally* as a wave. Although the wavepacket splits into reflected and transmitted packets, the particle itself does not split: there is some probability that the particle would be found to reflect, and some probability that the particle is transmitted, but the particle itself never splits into a reflected particle and a transmitted particle.

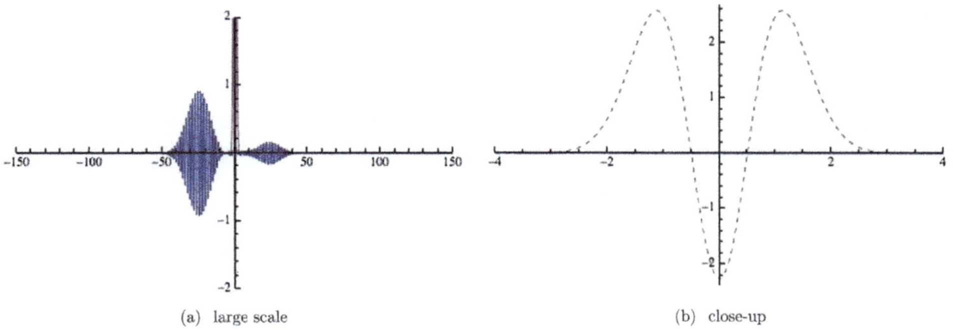

Figure 8.8. Scattering of a wavepacket from a potential, $0 < E < V_{max}$. (a) Overview; (b) close to the potential well, slow-motion. Note, in (a), the sudden increase in amplitude of the wavepacket as it strikes the well. This is due to constructive interference between the incoming and reflected waves.

Scattering of a wavepacket for $\langle E \rangle < V_{max}$ is illustrated in animations (8.8(a))–(8.8(b)). The difference is that for $\langle E \rangle < V_{max}$, most of the wavepacket is reflected, while for $\langle E \rangle > V_{max}$ most of the wavepacket is transmitted.

In any real scattering experiment, of course, there is always some finite uncertainty in the momentum of the incoming particle, the wavepacket is always finite in extent, and that means that eventually the incident wave is completely replaced by the scattered waves. Nevertheless, to compute the quantities of interest in a scattering experiment, it is convenient to consider the $\Delta p = 0$ limit, and work with stationary states in which the incoming and reflected waves always overlap. The 'quantities of interest' in a scattering experiment are the ratio of intensities of the incident and scattered waves, because such ratios can be directly related to observable quantities.

In connection with the two-slit experiment discussed in chapter 3, we learned that if we have a beam of n_p particles of average velocity $v = \langle p \rangle / m$, with the wave function of each particle represented at some time t by $\phi(x)$, then the intensity of the beam at point x at time t is given by

$$\begin{aligned} I &= n_p v \phi^*(x)\phi(x) \\ &= \text{average no. of particles} \\ &\quad \text{passing point } x \text{ per unit time} \end{aligned} \tag{8.14}$$

The goal of scattering theory is to find the intensity of the scattered beam as compared to the intensity of the incoming beam. In the simple case of an incoming particle of momentum p_0 approaching the end of a tube,

$$\begin{aligned} I_{inc} &= \frac{p_0}{m}\phi^*_{inc}(x)\phi_{inc}(x) = \frac{p_0}{m}|A|^2 \\ I_{ref} &= \frac{p_0}{m}\phi^*_{ref}(x)\phi_{ref}(x) = \frac{p_0}{m}|B|^2 \end{aligned} \tag{8.15}$$

and, since $A = -B$,

$$I_{ref} = I_{inc} \tag{8.16}$$

Physically this means that if the flux of incoming particles past some point x_0 is, say, 1000 particles per second, then the flux of reflected particles past that point is also

1000 particles per second; i.e. every incident particle is reflected, no particles are transmitted past the infinite potential barrier at $x = 0$. For a finite potential barrier such as figure 8.1 we also have, in addition to I_{inc} and I_{ref}, an intensity for the transmitted wave, representing particles which have gone through the potential barrier

$$I_{trans} = \frac{p_0}{m}\phi^*_{trans}(x)\phi_{trans}(x) = \frac{p_0}{m}|C|^2 \tag{8.17}$$

The quantities we need to calculate, which can then be compared with experiment, are the ***reflection coefficient***

$$\begin{aligned} R &= \frac{I_{ref}}{I_{inc}} = \frac{|B|^2}{|A|^2} \\ &= \frac{\text{no. of particles/s reflected}}{\text{no. of particles/s incident}} \end{aligned} \tag{8.18}$$

and the ***transmission coefficient***

$$\begin{aligned} T &= \frac{I_{trans}}{I_{inc}} = \frac{|C|^2}{|A|^2} \\ &= \frac{\text{no. of particles/s transmitted}}{\text{no. of particles/s incident}} \end{aligned} \tag{8.19}$$

Of course, the ratios B/A and C/A can only be obtained by actually solving the time-independent Schrödinger equation, which we will do, in this chapter, for simple rectangular potentials.

8.3 The step potential

Consider a particle moving in a potential which rises suddenly from $V(x) = 0$, at $x > 0$, to $V(x) = V$, at $x \geqslant 0$. If $E > V$ then the entire real line is a classically allowed region; an particle moving to the right will pass the potential barrier and continue moving to the right. On the other hand, if $E < V$, the half-line $x > 0$ is a classically forbidden region; an incoming particle will be reflected at $x = 0$ and move away to the left. We will consider these two cases separately:

Energies $E > V$

The Schrödinger equation in region I ($x < 0$) is the equation for a free particle

$$-\frac{\hbar^2}{2m}\frac{\partial^2\phi_1}{\partial x^2} = E\phi_1 \tag{8.20}$$

with the usual free particle solution

$$\phi_1(x) = Ae^{ip_1x/\hbar} + Be^{-ip_1x/\hbar} \quad \text{where} \quad p_1 = \sqrt{2mE} \tag{8.21}$$

In region II ($x > 0$) the equation is

$$-\frac{\hbar^2}{2m}\frac{\partial^2\phi_2}{\partial x^2} = (E - V)\phi_2 \tag{8.22}$$

with the solution

$$\phi_2(x) = Ce^{ip_2x/\hbar} + De^{-ip_2x/\hbar} \quad \text{where} \quad p_2 = \sqrt{2m(E-V)} \tag{8.23}$$

The scattering region is at $x = 0$, this is where $\partial V/\partial x \neq 0$. The part of the wave function that represents a (very long) wavepacket moving towards the scattering region is

$$\phi_{\text{inc}}(x) = \begin{cases} Ae^{ip_1x/\hbar} & x < 0 \\ De^{-ip_2x/\hbar} & x > 0 \end{cases} \tag{8.24}$$

The choice of A and D is a choice of the initial state. If $D = 0$ then the particle is initially approaching the potential step from the left; if $A = 0$ the particle is approaching from the right. If both A and D are non-zero, then the initial wavepacket is split into two pieces, one approaching from the left and one from the right. It is possible to have an initial wavepacket which is discontinuous in this way, but it is uncommon. The standard experimental setup would be to have a particle source located to one side or the other of the scattering region. We will suppose that the particles are coming in from the left, and set $D = 0$. In that case the remaining pieces of the wave function have the interpretation as transmitted and reflected waves

$$\begin{aligned} \phi_{\text{ref}}(x) &= Be^{-ip_1x/\hbar} \qquad (x < 0) \\ \phi_{\text{trans}}(x) &= Ce^{ip_2x/\hbar} \qquad (x > 0) \end{aligned} \tag{8.25}$$

To determine the reflection and transmission coefficients, we need B and C in terms of A.

The solution of the Schrödinger equation must be a continuous function, and its first derivative must also be continuous. If this were not true, the second derivative would be infinite, which violates the Schrödinger equation if the energy and potential are finite.[3] Imposing continuity of the wave function ϕ and its first derivative ϕ' at $x = 0$ give the conditions

$$\begin{aligned} \underline{\phi_1(0) = \phi_2(0)} &\Longrightarrow A + B = C \\ \underline{\phi_1'(0) = \phi_2'(0)} &\Longrightarrow p_1(A - B) = p_2 C \end{aligned} \tag{8.26}$$

These two equations allow us to solve for B and C in terms of A:

$$\begin{aligned} B &= \frac{p_1 - p_2}{p_1 + p_2} A \\ C &= \frac{2p_1}{p_1 + p_2} A \end{aligned} \tag{8.27}$$

The resulting wave function is sketched in figure 8.9(a).

Recall that intensity is proportional to the (group) velocity, $I = v\phi^*\phi$, and in this case the velocity of the transmitted wave $v_2 = p_2/m$ is different from the velocity of the incident and reflected waves $v_1 = p_1/m$. The transmission coefficient is therefore

[3] An exception is if the potential becomes infinite at some point, as in the case of a particle in a tube. In such situations, the slope of the wave function can be discontinuous.

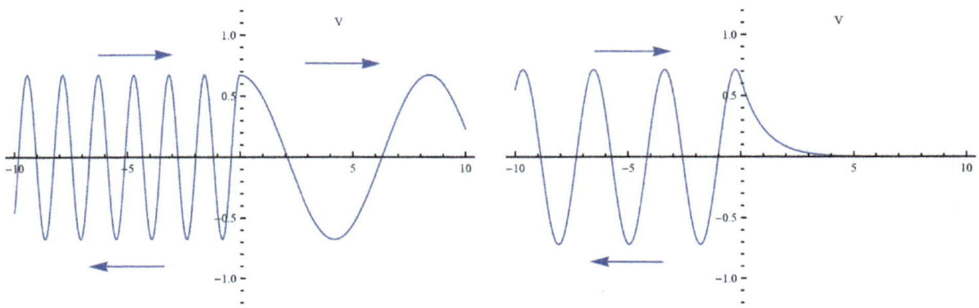

(a) $E > V$ (b) $E < V$

Figure 8.9. Scattering (a) and reflection (b) of a particle encountering a step potential.

$$
\begin{aligned}
T &= \frac{v_2|C|^2}{v_1|A|^2} \\
&= \frac{p_2}{p_1}\frac{4p_1^{\,2}}{(p_2+p_1)^2}
\end{aligned}
\tag{8.28}
$$

while the reflection coefficient is

$$
\begin{aligned}
R &= \frac{v_1|B|^2}{v_1|A|^2} \\
&= \frac{(p_2-p_1)^2}{(p_2+p_1)^2}
\end{aligned}
\tag{8.29}
$$

Notice that

$$R + T = 1 \tag{8.30}$$

which has the interpretation that

$$
\begin{aligned}
&\text{No. of particles reflected/s} + \text{No. of particles transmitted/s} \\
&= \text{No. of particles incident/s}
\end{aligned}
\tag{8.31}
$$

and is a consequence of the conservation of probability by the Schrödinger equation.

It should also be noted that $R \neq 0$, in general. This is in sharp contrast to classical physics, where *no* particles of energy $E > V$ would be reflected.

Energies $E < V$

In this case, the solutions (8.21) in region I and (8.23) in region II are the same as before, however

$$p_2 = \sqrt{2m(E-V)} = \mathrm{i}q_2 \tag{8.32}$$

is an imaginary number. The solution ϕ_2 in the classically forbidden region is then

$$\phi_2(x) = C\mathrm{e}^{-q_2x/\hbar} + D\mathrm{e}^{q_2x/\hbar} \qquad \text{where} \qquad q_2 = \sqrt{2m(V-E)} \tag{8.33}$$

Like any solution in the classically forbidden region, ϕ_2 is, in general, a combination of exponentially growing and exponentially decaying functions. Solutions with $D \neq 0$ are non-normalizable, however, and do not correspond to physical states. So we need only consider solutions with $D = 0$.

Once again applying continuity of the wave function and its first derivative, we have

$$\begin{aligned} \phi_1(0) = \phi_2(0) &\Longrightarrow A + B = C \\ \phi_1'(0) = \phi_1'(0) &\Longrightarrow \mathrm{i}p_1(A - B) = -q_2 C \end{aligned} \tag{8.34}$$

Solving for B and C,

$$\begin{aligned} B &= \frac{p_1 - \mathrm{i}q_2}{p_1 + \mathrm{i}q_2} A \\ C &= \frac{2p_1}{p_1 + \mathrm{i}q_2} A \end{aligned} \tag{8.35}$$

But in this case

$$R = \frac{|B|^2}{|A|^2} = 1 \tag{8.36}$$

which means that all incident particles are reflected; no particles are transmitted towards $x \to +\infty$, as in the classical case. The resulting wave function is shown in figure 8.9(b). *Unlike* the classical case, however, there is a finite probability to find the particle at any finite distance $x > 0$

$$P_{\Delta x}(x > 0) = |C|^2 \mathrm{e}^{-2q_2 x/\hbar} \Delta x \tag{8.37}$$

Once again, there is no way of interpreting this result in classical terms Classically, force is proportional to the derivative of the potential, so there is only a force on the particle at $x = 0$. There is nothing to prevent a particle at $x > 0$ to continue moving to the left. Quantum mechanically, although there is a finite probability to find the particle at $x > 0$, that probability decreases exponentially with x, and any particle of energy $E < V$ is ultimately reflected.

8.4 The finite square well

An attractive potential which is finite at infinity normally has a discrete (finite or infinite) number of bound states, and an infinite number of unbound states with energies in the range $[E_{\min}, \infty]$. The simplest example is the finite square well potential[4]

[4] It is conventional to call it a 'square' well although this does not really make sense, since the x-axis has units of length, while the y-axis has units of energy. The potential for a particle in a tube is known as the 'infinite square well'. Again, the word 'square' should not be taken seriously.

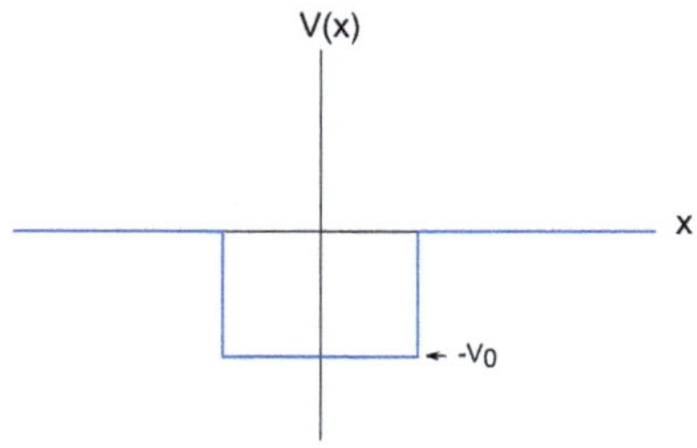

Figure 8.10. The finite attractive square well potential. We are interested in finding the bound states of energy $E < 0$, and the transmission/reflection coefficients for energies $E > 0$.

$$V(x) = \begin{cases} 0 & x < -a \\ -V_0 & -a \leqslant x \leqslant a \\ 0 & x > a \end{cases} \tag{8.38}$$

shown in figure 8.10.

For particle energies $E > 0$ the entire real line is classically allowed; these are the unbound states. For $E < 0$ regions I and II are classically forbidden, and the wave function of a physical state must fall exponentially to zero as $x \to \pm\infty$. These are the bound states.

8.4.1 Bound states $E < 0$

We begin with a discussion of the bound states. In regions I and III the Schrödinger equation is the same as a free particle equation, and the solution is

$$\begin{aligned} \phi_{\mathrm{I}}(x) &= a_1 \mathrm{e}^{\mathrm{i}\sqrt{2mE}x/\hbar} + a_2 \mathrm{e}^{-\mathrm{i}\sqrt{2mE}x/\hbar} \\ \phi_{\mathrm{III}}(x) &= b_1 \mathrm{e}^{\mathrm{i}\sqrt{2mE}x/\hbar} + b_2 \mathrm{e}^{-\mathrm{i}\sqrt{2mE}x/\hbar} \end{aligned} \tag{8.39}$$

However, since $E < 0$, the square-root is imaginary. Writing $E = -\mathcal{E}$, we have

$$\begin{aligned} \phi_{\mathrm{I}}(x) &= a_1 \mathrm{e}^{-\sqrt{2m\mathcal{E}}x/\hbar} + a_2 \mathrm{e}^{\sqrt{2m\mathcal{E}}x/\hbar} \\ \phi_{\mathrm{III}}(x) &= b_1 \mathrm{e}^{-\sqrt{2m\mathcal{E}}x/\hbar} + b_2 \mathrm{e}^{\sqrt{2m\mathcal{E}}x/\hbar} \end{aligned} \tag{8.40}$$

If ψ is a physical state, we must set $a_2 = b_1 = 0$; otherwise the wave function blows up exponentially as $x \to \pm\infty$ and the state is non-normalizable. Then

$$\begin{aligned} \phi_{\mathrm{I}}(x) &= A\mathrm{e}^{-\sqrt{2m\mathcal{E}}x/\hbar} \\ \phi_{\mathrm{III}}(x) &= B\mathrm{e}^{\sqrt{2m\mathcal{E}}x/\hbar} \end{aligned} \tag{8.41}$$

Now suppose we choose A to be a real number. Then $\phi(x)$ must be real everywhere. The reason is that the Schrödinger equation is a linear equation with real coefficients, so writing

$$\phi(x) = \phi^R(x) + \mathrm{i}\phi^{\mathrm{I}}(x) \tag{8.42}$$

it's easy to see that ϕ^R and ϕ^{I} must *both* satisfy the Schrödinger equation

$$\tilde{H}\phi^R = E\phi^R \qquad \tilde{H}\phi^{\mathrm{I}} = E\phi^{\mathrm{I}}. \tag{8.43}$$

Setting A to be a real number means that $\phi^{\mathrm{I}} = 0$ for $x > a$, which means it is zero everywhere, because the Schrödinger equation says that $\phi''(x) \propto \phi(x)$. The conclusion is that $\phi(x)$ is real everywhere; in particular, the constant B is also a real number.

Let $\phi(x)$ be a solution of the time-independent Schrödinger equation with energy E_n. We can show that

$$\varphi(x) = \phi(-x) \tag{8.44}$$

is also an eigenstate of $\tilde{H}$, with the same energy. Start with the time-independent Schrödinger equation

$$-\frac{\hbar^2}{2m}\frac{\mathrm{d}^2}{\mathrm{d}x^2}\phi(x) + V(x)\phi(x) = E\phi(x) \tag{8.45}$$

make the change of variables $x \to -x$

$$\begin{aligned} -\frac{\hbar^2}{2m}\frac{\mathrm{d}^2}{\mathrm{d}x^2}\phi(-x) + V(-x)\phi(-x) &= E\phi(-x) \\ -\frac{\hbar^2}{2m}\frac{\mathrm{d}^2}{\mathrm{d}x^2}\varphi(x) + V(x)\varphi(x) &= E\varphi(x) \end{aligned} \tag{8.46}$$

This proves that $\varphi(x)$ is also a solution. The crucial fact used in (8.46) was the symmetry of $V(x)$ around $x = 0$

$$V(x) = V(-x) \tag{8.47}$$

Without this symmetry, $\varphi(x) = \phi(-x)$ would not be a solution of the Schrödinger equation.

If E_n is degenerate, then $\phi(x)$ and $\varphi(x)$ could be two different states. On the other hand, if E_n is non-degenerate, then the two functions must be proportional to one another:

$$\varphi(x) = c\phi(x) \tag{8.48}$$

where c is a real-valued constant, since ϕ and φ are both real-valued. Finally, the normalization condition requires

$$\begin{aligned} \int \mathrm{d}x\varphi^2(x) &= \int \mathrm{d}x\phi(-x)\phi(-x) \\ &= \int \mathrm{d}x\phi(x)\phi(x) \\ &= 1 \end{aligned} \tag{8.49}$$

On the other hand,

$$\int \mathrm{d}x\varphi^2(x) = c^2\int \mathrm{d}x\phi^2(x) = c^2$$

Comparing the last two equations, we find that $c = \pm 1$. We conclude that if E_n is non-degenerate there are two possibilities:

1. **Even parity**

$$\phi(-x) = +\phi(x) \tag{8.50}$$

2. **Odd parity**

$$\phi(-x) = -\phi(x) \tag{8.51}$$

Assuming that the energy eigenvalues are non-degenerate, the algebra simplifies a lot. In region II, the Schrödinger equation is of the form

$$\frac{\mathrm{d}^2\phi_{\mathrm{II}}}{\mathrm{d}x^2} = -\frac{2m}{\hbar^2}(V_0 - \mathcal{E})\phi_{\mathrm{II}} \tag{8.52}$$

with solutions

$$\phi_{\mathrm{II}} = c_1 \mathrm{e}^{\mathrm{i}\sqrt{2m(V_0-\mathcal{E})}x/\hbar} + c_2 \mathrm{e}^{-\mathrm{i}\sqrt{2m(V_0-\mathcal{E})}x/\hbar} \tag{8.53}$$

However, since ϕ_{II} must be a real function, we expand the exponentials in a combination of sin and cosine

$$\phi_{\mathrm{II}} = C_E \cos\left[\sqrt{2m(V_0 - \mathcal{E})}x/\hbar\right] + C_O \sin\left[\sqrt{2m(V_0 - \mathcal{E})}x/\hbar\right] \tag{8.54}$$

We have seen that the non-degenerate solutions have to be either even or odd parity. For even parity, $C_O = 0$, while for odd parity $C_E = 0$. We deal with these cases separately.

Even parity bound states

The requirement that $\phi(x) = \phi(-x)$ also implies that $A = B$. So we have altogether

$$\begin{aligned} \phi_{\mathrm{I}}(x) &= A\mathrm{e}^{-\sqrt{2m\mathcal{E}}x/\hbar} \\ \phi_{\mathrm{II}}(x) &= C_E \cos\left[\sqrt{2m(V_0 - \mathcal{E})}x/\hbar\right] \\ \phi_{\mathrm{III}}(x) &= A\mathrm{e}^{\sqrt{2m\mathcal{E}}x/\hbar} \end{aligned} \tag{8.55}$$

The continuity of the wave function at $x = \pm a$ requires

$$A\mathrm{e}^{-\sqrt{2m\mathcal{E}}a/\hbar} = C_E \cos\left[\sqrt{2m(V_0 - \mathcal{E})}a/\hbar\right] \tag{8.56}$$

while the continuity of the first derivative at $x = \pm a$ gives

$$\sqrt{2m\mathcal{E}}\,A\mathrm{e}^{-\sqrt{2m\mathcal{E}}a/\hbar} = \sqrt{2m(V_0 - \mathcal{E})}\,C_E \sin\left[\sqrt{2m(V_0 - \mathcal{E})}a/\hbar\right] \tag{8.57}$$

Dividing equation (8.57) by equation (8.56) results in a transcendental equation for $\mathcal{E} = -E$

$$\sqrt{\mathcal{E}} = \sqrt{V_0 - \mathcal{E}}\,\tan\left[\sqrt{2m(V_0 - \mathcal{E})}a/\hbar\right] \tag{8.58}$$

which determines the energy eigenvalues.

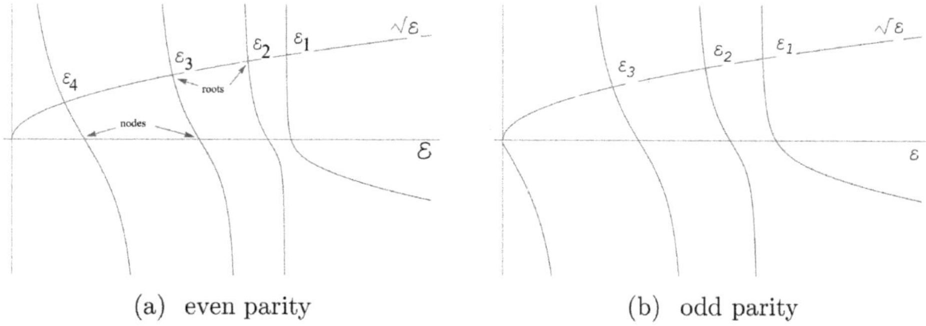

Figure 8.11. Numerical solution of the transcendental equations determining bound states energies; solutions are at the intersections of the lines shown. (a) Even parity bound states; (b) odd parity bound states.

This transcendental equation can be solved graphically, by plotting the left- and right-hand sides of the equation to find the points at which the two curves intersect. The result is shown in figure 8.11(a). We learn from the figure that the solutions, denoted by $\mathcal{E}_n$, are all at $\mathcal{E} < V_0$. This is because for $\mathcal{E} > V_0$ the tangent in equation (8.58) becomes a hyperbolic tangent, so the right-hand side diverges to $-\infty$ as $x \to \infty$, never again intersecting $\sqrt{\mathcal{E}}$. Each solution $\mathcal{E}_n$ is located between points where the tangent equals 0; these 'nodes' occur at

$$\mathcal{E} = V_0 - (k\pi)^2 \frac{\hbar^2}{2ma^2} \qquad (k = 0, 1, 2, \ldots) \tag{8.59}$$

and since $\mathcal{E} > 0$, this means that the number K of even-parity energy eigenvalues is the largest number such that

$$[(K-1)\pi]^2 \frac{\hbar^2}{2ma^2} \leqslant V_0 \tag{8.60}$$

Thus there are a finite number—and no less than one!—of even-parity bound states. The number K of such bound states increases as the well becomes deeper (larger V_0), and wider (larger a).

Odd parity bound states

This time, the requirement that $\phi(-x) = -\phi(x)$ implies that $B = -A$, and $C_E = 0$. The wave function has the form

$$\begin{aligned} \phi_{\rm I}(x) &= A e^{-\sqrt{2m\mathcal{E}}x/\hbar} \\ \phi_{\rm II}(x) &= C_O \sin\left[\sqrt{2m(V_0 - \mathcal{E})}x/\hbar\right] \\ \phi_{\rm III}(x) &= -Ae^{\sqrt{2m\mathcal{E}}x/\hbar} \end{aligned} \tag{8.61}$$

From continuity of the wave function at $x = \pm a$ we have

$$Ae^{-\sqrt{2m\mathcal{E}}a/\hbar} = C_O \sin\left[\sqrt{2m(V_0 - \mathcal{E})}a/\hbar\right] \tag{8.62}$$

and from continuity of the first derivative at $x = \pm a$

$$-\sqrt{2m\mathcal{E}}\,A\mathrm{e}^{-\sqrt{2m\mathcal{E}}\,a/\hbar} = \sqrt{2m(V_0 - \mathcal{E})}\,C_O \cos\left[\sqrt{2m(V_0 - \mathcal{E})}\,a/\hbar\right] \tag{8.63}$$

Dividing equation (8.63) by equation (8.62) gives us the transcendental equation for the energies of odd parity bound states

$$\begin{aligned}\sqrt{\mathcal{E}} &= -\sqrt{V_0 - \mathcal{E}}\,\mathrm{ctn}\left[\sqrt{2m(V_0 - \mathcal{E})}\,a/\hbar\right]\\ &= \sqrt{V_0 - \mathcal{E}}\,\tan\left[\sqrt{2m(V_0 - \mathcal{E})}\frac{a}{\hbar} + \frac{\pi}{2}\right]\end{aligned} \tag{8.64}$$

which can be solved graphically, as shown in figure 8.11(b). Once again, there are as many roots as there are nodes of the tangent; this time the nodes are located at

$$\mathcal{E} = V_0 - \left[\left(k + \frac{1}{2}\right)\pi\right]^2 \frac{\hbar^2}{2ma^2} \qquad (k = 0, 1, 2, \ldots) \tag{8.65}$$

and the number of odd parity nodes is the largest integer K such that

$$\left[\left(K - \frac{1}{2}\right)\pi\right]^2 \frac{\hbar^2}{2ma^2} < V_0 \tag{8.66}$$

Note that for

$$V_0 < \left(\frac{\pi}{2}\right)^2 \frac{\hbar^2}{2ma^2} \tag{8.67}$$

there are no odd-parity bound states.

To sum up, we have found that for a finite square well

1. The number of bound state energies is finite, and there is at least one bound state.
2. The number of bound states increases with the width and depth of the well.
3. The energies are non-degenerate.
4. Each energy eigenstate is either even parity or odd parity, and this is a consequence of the symmetry of the potential $V(x) = V(-x)$, and the fact that the energies are non-degenerate.
5. Although the particle is bound to the well, the wave function of the particle may be non-negligible far outside the well, where (in classical physics) there are no forces at all acting on the particle.

8.4.2 Scattering states $E > 0$

For $E > 0$, the solution of the time-independent Schrödinger equation in all three regions is

$$\begin{aligned}\phi_{\mathrm{III}}(x) &= A\mathrm{e}^{\mathrm{i}\sqrt{2mE}\,x/\hbar} + B\mathrm{e}^{-\mathrm{i}\sqrt{2mE}\,x/\hbar}\\ \phi_{\mathrm{II}}(x) &= C\mathrm{e}^{\mathrm{i}\sqrt{2m(E+V_0)}\,x/\hbar} + D\mathrm{e}^{-\mathrm{i}\sqrt{2m(E+V_0)}\,x/\hbar}\\ \phi_{\mathrm{I}}(x) &= E\mathrm{e}^{\mathrm{i}\sqrt{2mE}\,x/\hbar} + F\mathrm{e}^{-\mathrm{i}\sqrt{2mE}\,x/\hbar}\end{aligned} \tag{8.68}$$

As explained in the introduction, the wave function in regions I and III can be thought of as a superposition of incoming and outgoing wavepackets, in the limit that the uncertainty in momentum of the wavepackets goes to zero. The momentum is either $p = \sqrt{2mE}$, in which case the packet is moving from left to right, or $p = -\sqrt{2mE}$, in which case the packet is moving from right to left. Therefore, we can interpret the components of $\phi_{\rm III}$ and $\phi_{\rm I}$ as representing

$$
\begin{aligned}
&A e^{i\sqrt{2mE}\,x/\hbar} && \text{an incoming wavepacket from the left}\\
&B e^{-i\sqrt{2mE}\,x/\hbar} && \text{an outgoing wavepacket to the left}\\
&E e^{i\sqrt{2mE}\,x/\hbar} && \text{an outgoing wavepacket to the right}\\
&F e^{-i\sqrt{2mE}\,x/\hbar} && \text{an incoming wavepacket from the right}
\end{aligned}
$$

Now suppose we are interested in an experimental situation in which a particle of a fairly precise momentum is sent towards the potential well from the left. In that case there is no incoming wavepacket from the right, so we can set $F = 0$. The incoming wavepacket eventually reaches the potential well. Some of the incoming wavepacket is reflected towards the left; some is transmitted towards the right. So we now have

$$
\begin{aligned}
\phi_{\rm III}(x) &= A e^{i\sqrt{2mE}\,x/\hbar} + B e^{-i\sqrt{2mE}\,x/\hbar}\\
\phi_{\rm II}(x) &= C e^{i\sqrt{2m(E+V_0)}\,x/\hbar} + D e^{-i\sqrt{2m(E+V_0)}\,x/\hbar}\\
\phi_{\rm I}(x) &= E e^{i\sqrt{2mE}\,x/\hbar}
\end{aligned}
\tag{8.69}
$$

where

$$
\begin{aligned}
&A e^{i\sqrt{2mE}\,x/\hbar} && \text{represents the incoming wavepacket}\\
&B e^{-i\sqrt{2mE}\,x/\hbar} && \text{represents the reflected wavepacket}\\
&E e^{i\sqrt{2mE}\,x/\hbar} && \text{represents the transmitted wavepacket}
\end{aligned}
$$

The reflection and transmission coefficients will therefore be

$$
R = \frac{BB^*}{AA^*} \qquad T = \frac{EE^*}{AA^*} \tag{8.70}
$$

so the problem is to compute B and E in terms of A.

We should pause to note that $\phi(x) \neq \pm\phi(-x)$. The reason is that in this case the energy is degenerate; a particle approaching the potential from the left can have exactly the same energy as a particle approaching the well from the right. In this situation, an energy eigenstate does not *necessarily* have the property that $\phi(x) = \pm\phi(-x)$; although $\varphi(x) = \phi(-x)$ is an energy eigenstate, it is not necessarily equivalent to $\phi(x)$ if the energy is degenerate. For the wave function $\phi(x)$ of equation (8.69), which represents a particle approaching the well from the left and then scattering, $\varphi(x)$ represents a particle approaching the potential from the *right*, and then scattering.

Denote

$$
k = \frac{\sqrt{2mE}}{\hbar} \qquad \text{and} \qquad q = \frac{\sqrt{2m(E+V_0)}}{\hbar} \tag{8.71}
$$

so that

$$\begin{aligned}\phi_{\mathrm{III}}(x) &= A\mathrm{e}^{\mathrm{i}kx} + B\mathrm{e}^{-\mathrm{i}kx}\\ \phi_{\mathrm{II}}(x) &= C\mathrm{e}^{\mathrm{i}qx} + D\mathrm{e}^{-\mathrm{i}qx}\\ \phi_{\mathrm{I}}(x) &= E\mathrm{e}^{\mathrm{i}kx}\end{aligned} \tag{8.72}$$

Imposing continuity of the wave function and its first derivative at $x = a$ give us

$$\begin{aligned}&\underline{\phi_{\mathrm{I}}(a) = \phi_{\mathrm{II}}(a)} \Longrightarrow E\mathrm{e}^{\mathrm{i}ka} = C\mathrm{e}^{\mathrm{i}qa} + D\mathrm{e}^{-\mathrm{i}qa}\\ &\underline{\phi_{\mathrm{I}}'(a) = \phi_{\mathrm{II}}'(a)} \Longrightarrow kE\mathrm{e}^{\mathrm{i}ka} = q(C\mathrm{e}^{\mathrm{i}qa} - D\mathrm{e}^{-\mathrm{i}qa})\end{aligned} \tag{8.73}$$

The corresponding continuity conditions at $x = -a$ are

$$\begin{aligned}&\underline{\phi_{\mathrm{III}}(-a) = \phi_{\mathrm{II}}(-a)} \Longrightarrow A\mathrm{e}^{-\mathrm{i}ka} + B\mathrm{e}^{\mathrm{i}ka} = C\mathrm{e}^{-\mathrm{i}qa} + D\mathrm{e}^{\mathrm{i}qa}\\ &\underline{\phi_{\mathrm{III}}'(-a) = \phi_{\mathrm{II}}'(-a)} \Longrightarrow k(A\mathrm{e}^{-\mathrm{i}ka} - B\mathrm{e}^{\mathrm{i}ka}) = q(C\mathrm{e}^{-\mathrm{i}qa} - D\mathrm{e}^{\mathrm{i}qa})\end{aligned} \tag{8.74}$$

We first solve equation (8.73) for C and D in terms of E:

$$\begin{aligned}C &= \frac{1}{2}\left(1 + \frac{k}{q}\right)E\mathrm{e}^{\mathrm{i}(k-q)a}\\ D &= \frac{1}{2}\left(1 - \frac{k}{q}\right)E\mathrm{e}^{\mathrm{i}(k+q)a}\end{aligned} \tag{8.75}$$

substitute into (8.74)

$$\begin{aligned}A\mathrm{e}^{-\mathrm{i}ka} + B\mathrm{e}^{\mathrm{i}ka} &= \left[\frac{1}{2}\left(1 + \frac{k}{q}\right)\mathrm{e}^{\mathrm{i}(k-2q)a} + \frac{1}{2}\left(1 - \frac{k}{q}\right)\mathrm{e}^{\mathrm{i}(k+2q)a}\right]E\\ k(A\mathrm{e}^{-\mathrm{i}ka} - B\mathrm{e}^{\mathrm{i}ka}) &= q\left[\frac{1}{2}\left(1 + \frac{k}{q}\right)\mathrm{e}^{\mathrm{i}(k-2q)a} - \frac{1}{2}\left(1 - \frac{k}{q}\right)\mathrm{e}^{\mathrm{i}(k+2q)a}\right]E\end{aligned} \tag{8.76}$$

Solving for E in terms of A, we find

$$\begin{aligned}2kA\mathrm{e}^{-\mathrm{i}ka} &= \left[(k+q)\frac{1}{2}\left(1 + \frac{k}{q}\right)\mathrm{e}^{\mathrm{i}(k-2q)a} + (k-q)\frac{1}{2}\left(1 - \frac{k}{q}\right)\mathrm{e}^{\mathrm{i}(k+2q)a}\right]E\\ &= \left[\left(k + \frac{k^2}{2q} + \frac{q}{2}\right)\mathrm{e}^{\mathrm{i}(k-2q)a} + \left(k - \frac{k^2}{2q} - \frac{q}{2}\right)\mathrm{e}^{\mathrm{i}(k+2q)a}\right]E\end{aligned} \tag{8.77}$$

so that

$$\frac{E}{A} = \frac{2\mathrm{e}^{-\mathrm{i}ka}}{\left(1 + \frac{k}{2q} + \frac{q}{2k}\right)\mathrm{e}^{\mathrm{i}(k-2q)a} + \left(1 - \frac{k}{2q} - \frac{q}{2k}\right)\mathrm{e}^{\mathrm{i}(k+2q)a}}$$

$$= \frac{\mathrm{e}^{-2\mathrm{i}ka}}{\cos(2qa) + \frac{1}{2}\left(\frac{k}{2q} + \frac{q}{2k}\right)(\mathrm{e}^{-2\mathrm{i}qa} - \mathrm{e}^{2\mathrm{i}qa})} \tag{8.78}$$

$$= \frac{\mathrm{e}^{-2\mathrm{i}ka}}{\cos(2qa) - \frac{\mathrm{i}}{2}\left(\frac{k}{q} + \frac{q}{k}\right) \sin(2qa)}$$

In a similar way, one can solve for B in terms of A

$$\frac{B}{A} = \frac{\frac{1}{2}\mathrm{i}\mathrm{e}^{-2\mathrm{i}ka}\left(\frac{q}{k} - \frac{k}{q}\right) \sin(2qa)}{\cos(2qa) - \frac{1}{2}\mathrm{i}\left(\frac{q}{k} + \frac{k}{q}\right) \sin(2qa)} \tag{8.79}$$

From these ratios, we obtain the transmission and reflection coefficients

$$T = \left|\frac{E}{A}\right|^2 = \frac{1}{\cos^2(2qa) + \frac{1}{4}\left(\frac{k}{q} + \frac{q}{k}\right)^2 \sin^2(2qa)}$$

$$R = \left|\frac{B}{A}\right|^2 = \frac{\frac{1}{4}\left(\frac{q}{k} - \frac{k}{q}\right)^2 \sin^2(2qa)}{\cos^2(2qa) + \frac{1}{4}\left(\frac{k}{q} + \frac{q}{k}\right)^2 \sin^2(2qa)} \tag{8.80}$$

and it is easy to check that

$$R + T = 1 \tag{8.30}$$

as required by conservation of probability.

Now notice that something interesting happens when the energy E of the incoming particle is such that

$$2qa = 2\sqrt{2m(E + V_0)}\frac{a}{\hbar} = n\pi \tag{8.82}$$

In this case

$$T = 1 \qquad \text{and} \qquad R = 0 \tag{8.83}$$

i.e. there is no reflection at all. A particle of this energy is certain to be transmitted past the potential well. The condition (8.82) has a simple interpretation in terms of de Broglie wavelengths. The momentum of the particle inside the well is

$$P = \sqrt{2m(E - V)} = \sqrt{2m(E + V_0)} \tag{8.84}$$

so the condition (8.82) is

$$\begin{aligned} 2a\frac{P}{\hbar} &= n\pi \\ 2a\frac{P}{h}2\pi &= n\pi \\ 2a &= n\frac{\lambda}{2} \end{aligned} \tag{8.85}$$

where we use the de Broglie relation $p = h/\lambda$. Since the width of the potential well is $2a$, it means that there is no reflection whenever the width of the well is exactly an integer number of half-wavelengths across, which is also the condition for a standing wave on an interval of length $2a$.

There is a related phenomenon in atomic physics known as the Ramsauer effect, in which low energy electrons are scattered by a thin film of some material. At certain well-defined energies, the scattering of the electrons is minimized. In this case, an atom can be viewed as an attractive potential well for the electrons. Since an atom is electrically neutral, the potential is zero until the incoming electrons move past the electrons orbiting the nucleus. As the electron approaches closer to the nucleus, there is of course an attractive force, since the nucleus is positive. The analysis of the scattering is more complicated than for the finite square well, the motion is in three dimensions rather than one dimension, and the potential well is not rectangular. Nevertheless, the solid is nearly 'transparent' for incoming electrons of certain definite energies; an effect which is closely analogous to the finite square well case.

8.5 Tunneling

Next we consider a repulsive square well potential

$$V(x) = \begin{cases} 0 & x < -a \\ V_0 & -a \leqslant x \leqslant a \\ 0 & x > a \end{cases} \tag{8.86}$$

Since this potential differs from the attractive square well of the last section only by the sign in front of V_0, there is no need to repeat the algebra of the last section. We can simply take the formulas for the scattering state of the attractive well, replace V_0 everywhere by $-V_0$, and these will be the corresponding formulas for the repulsive well. Then again denote

$$k = \frac{\sqrt{2mE}}{\hbar} \quad \text{and} \quad q = \frac{\sqrt{2m(E - V_0)}}{\hbar} \tag{8.87}$$

so that

$$\begin{aligned}\phi_{\mathrm{III}}(x) &= A\mathrm{e}^{\mathrm{i}kx} + B\mathrm{e}^{-\mathrm{i}kx} \\ \phi_{\mathrm{II}}(x) &= C\mathrm{e}^{\mathrm{i}qx} + D\mathrm{e}^{-\mathrm{i}qx} \\ \phi_{\mathrm{I}}(x) &= E\mathrm{e}^{\mathrm{i}kx}\end{aligned} \tag{8.88}$$

Classically, if $E > V_0$, the incoming particle will travel through the well and continue moving to the right. If $E < V_0$ the incoming particle is reflected at $x = -a$, and travels back to the left.

In quantum mechanics, if $E > V_0$, the situation is qualitatively much like the case for the attractive potential: some of the incoming wave is reflected, and some is transmitted. It is worth noting that no matter how large E is compared to V_0, there is always some finite probability that the particle is reflected. Quantum mechanically, if a bullet is fired at a fixed target of tissue paper, there is always some (exceedingly small) probability that the bullet will bounce off the paper.

If $E < V_0$, then the quantity

$$q = \frac{\sqrt{2m(E - V_0)}}{\hbar} \tag{8.89}$$

is imaginary. Substituting

$$q = \mathrm{i}Q \qquad \text{and} \qquad \sin(2qa) = -\mathrm{i}\sinh(2Qa) \tag{8.90}$$

into equation (8.80) gives us

$$\begin{aligned} T &= \frac{1}{\cosh^2(2Qa) + \frac{1}{4}\left(\frac{k}{Q} + \frac{Q}{k}\right)^2 \sinh^2(2Qa)} \\ R &= \frac{\frac{1}{4}\left(\frac{Q}{k} - \frac{k}{Q}\right)^2 \sinh^2(2Qa)}{\cosh^2(2Qa) + \frac{1}{4}\left(\frac{k}{Q} + \frac{Q}{k}\right)^2 \sinh^2(2Qa)} \end{aligned} \tag{8.91}$$

If the width of the well is large enough, so that

$$2a \gg \frac{1}{Q} = \frac{\hbar}{\sqrt{2m(V_0 - E)}} \tag{8.92}$$

then

$$\sinh^2(2Qa) \approx \cosh^2(2Qa) \approx \frac{1}{4}\mathrm{e}^{2Qa} \tag{8.93}$$

and the transmission coefficient is approximately

$$T \approx \mathrm{e}^{-4Qa} = \exp\left[-4\sqrt{2m(V_0 - E)}\,a/\hbar\right] \tag{8.94}$$

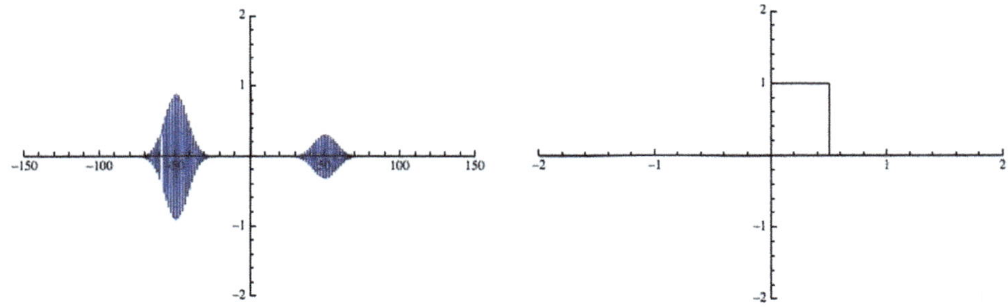

(a) large scale (b) close-up

Figure 8.12. Scattering from a repulsive well with $\langle E \rangle < V_0$, illustrating the phenomenon of tunneling. (a) Overview; (b) close to the potential well, slow-motion.

The situation is illustrated in animations 8.12(a)–8.12(b). Most of the incoming wave is reflected. In the classically forbidden region, the wave function decays exponentially. However, since the width of the classically forbidden region is finite, the wave function at the other end of the barrier is finite, and so there exists a transmitted wave. This means that no matter how high the barrier V_0 is compared to the energy E of the incoming particle, there is always some finite probability for the particles to penetrate through the barrier.

Quantum mechanically, a soap bubble heading for a brick wall has a finite, although exceedingly small, probability of being found on the other side of the brick wall. The quantum-mechanical strategy for breaking out of prison? Sit down and wait!

The fact that a particle can penetrate a potential barrier whose height exceeds the energy of the particle is known as ***tunneling***. It has important applications throughout modern physics, e.g. in the study of semiconductors, and of radioactive decay.

The classical limit

For the case that $E < V_0$, the classical limit is quite straightforward. We should obtain classical behavior as $\hbar \to 0$, and in this case we do. It is easy to see that in the $\hbar \to 0$ limit we find $Q \to \infty$, which means that the transmission coefficient vanishes, the reflection coefficient is $R = 1$, and the wave function vanishes inside the potential barrier. There is no tunneling in the classical limit, as expected.

The case $E > V_0$ is much more subtle; here it is not obvious that $R = 0$ and there is no reflection in the classical limit. However, recall that in arguing that the stationary state solution is relevant to scattering, we had supposed that the size Δx of the incoming wavepacket was far larger than the extension of the potential (in this case $2a$). This is typically not the case for macroscopic objects, and it helps to go back to the Ehrenfest equations. In general the Ehrenfest equations only reduce to Hamiltonian equations when the extension of the particle wavepacket is much smaller than the scale over which the potential varies appreciably, so that $\langle V(x) \rangle \approx V(\langle x \rangle)$, and that this extension remains small as the wavepacket propagates. In other words, classical behavior is only expected when $\Delta x \ll 2a$ for the step potential, and that behavior comes about through interference effects of component waves in the wavepacket, with reflected waves of different wavelengths interfering destructively in the case that $E > V_0$. But this discussion is best left for a later chapter, when we come to the Wentzel–Kramers–Brillouin (WKB) approximation.

8.6 The delta function potential

A main theme in this chapter is that scattering and bound states in a rectangular potential are solved analytically by imposing continuity of the wave function and its first derivative. However, it not *always* the case that the first derivative is continuous; the possible exceptions occur at points where the potential is infinite. This is already clear in the example of the particle in a tube, where the first derivatives of the energy eigenstates are discontinuous at both ends of the tube. Another instructive example in one dimension is the attractive delta function potential

$$V(x) = -g\delta(x) \tag{8.95}$$

The first derivative of the wave function is discontinuous at $x = 0$. To see that, let us integrate both sides of equation (8.1) from $-\epsilon$ to $+\epsilon$, where ϵ is infinitesimally small:

$$\begin{aligned} \int_{-\epsilon}^{\epsilon} \mathrm{d}x \frac{\mathrm{d}^2\phi}{\mathrm{d}x^2} &= -\frac{2m}{\hbar^2} \int_{-\epsilon}^{\epsilon} \mathrm{d}x \ (E + g\delta(x))\phi(x) \\ \left.\frac{\mathrm{d}\phi}{\mathrm{d}x}\right|_{-\epsilon}^{\epsilon} &= -\frac{2m}{\hbar^2} \int_{-\epsilon}^{\epsilon} \mathrm{d}x E\phi(x) - \frac{2m}{\hbar^2} g\phi(0) \end{aligned} \tag{8.96}$$

The integral in the second line can be dropped, as $\epsilon \to 0$, because both E and $\phi(x)$ are finite. So we are left with the discontinuity condition for the slope

$$\left(\frac{\mathrm{d}\phi}{\mathrm{d}x}\right)_{|x=+\epsilon} - \left(\frac{\mathrm{d}\phi}{\mathrm{d}x}\right)_{|x=-\epsilon} = -\frac{2m}{\hbar^2} g\phi(0) \tag{8.97}$$

Now suppose we are looking for a bound state solution with $E = -\mathcal{E} < 0$. Since the wave function must be continuous at $x = 0$, the only possible solution is the even parity state

$$\phi(x) = N \mathrm{e}^{-\sqrt{2m\mathcal{E}}|x|/\hbar} \tag{8.98}$$

Then the discontinuity condition (8.97) becomes

$$-\frac{\sqrt{2m\mathcal{E}}}{\hbar} N - \frac{\sqrt{2m\mathcal{E}}}{\hbar} N = -\frac{2m}{\hbar^2} gN \tag{8.99}$$

which can be solved for $E = -\mathcal{E}$:

$$E = -\frac{mg^2}{2\hbar^2} \tag{8.100}$$

Notice that the probability to find the particle exactly at $x = 0$ is zero, even though the probability density $\phi^*(0)\phi(0)$ is finite, since $x = 0$ is just a single point. So all of the probability distribution is really outside the potential well, where the force on the particle is classically zero. There is nothing to prevent a classical particle, located anywhere but at $x = 0$, from propagating to $\pm\infty$. Yet in quantum mechanics the particle is bound to the well, and, since this is a stationary state, remains bound indefinitely. Once again, there is really no classical picture of this phenomenon.

Problems

1.
   a) Find the bound state energy of a particle in a finite square well potential

$$V(x) = \begin{cases} 0 & x < -a \\ -g/2a & -a < x < a \\ 0 & x > a \end{cases} \tag{8.101}$$

   in the limit $a \to 0$. Compare your answer with the bound state energy in the delta function of equation (8.100).

   b) Calculate the reflection and transmission coefficients in the same limit for an incident particle of momentum p.

2. Compute the reflection/transmission coefficients, for a particle in the delta function potential

$$V(x) = -g\delta(x) \tag{8.102}$$

   and compare your answers with problem 1(b).

3. Find
   a) the transcendental equation which determines the bound state energy of a particle in the potential

$$V(x) = -g\delta(x + a) - g\delta(x - a) \tag{8.103}$$

   b) the transmission and reflection coefficients for a particle of momentum p, propagating in this potential.

4. A particle of mass m moves in the potential

$$V(x) = \begin{cases} -g\delta(x) & x > -a \\ \infty & x < -a \end{cases} \tag{8.104}$$

   a) Without worrying about continuity or boundary conditions, write down the general solution of the Schrödinger equation for a bound state (energy $E < 0$) in regions $-a < x < 0$ and $x > 0$.

   b) In terms of your solution in part (a), write down the boundary condition at $x = -a$ and continuity conditions at $x = 0$ (you will need to use the result proved in the previous set of problems).

   c) From the conditions in part (b), derive a transcendental equation determining the bound-state energy E.

5. For the potential

$$V(x) = \begin{cases} W & x > L \\ 0 & 0 \leqslant x \leqslant L \\ \infty & x < 0 \end{cases} \qquad (8.105)$$

with constant $W > 0$, find the transcendental equation which determines the bound state energies. What are the possible values of L, such that there is a bound state at E equal to W (or, strictly speaking, $E = W - \epsilon$, where ϵ is infinitesimal)?

6. A particle of momentum p is incident from the left on a potential of the form

$$V(x, y) = \begin{cases} 0 & x < -a \\ V_1 & -a < x < a \\ V_2 & x > a \end{cases} \qquad (8.106)$$

The energy $E = p^2/2m$ of the incident particle is greater than V_1 and V_2.

a) Write down the solution of the Schrödinger equation in all three regions ($x < a,\ -a < x < a,\ x > a$), for this situation.

b) Write down the appropriate continuity conditions at $x = \pm a$.

c) Find the transmission coefficient.

IOP Publishing

An Introduction to Quantum Theory

Jeff Greensite

Chapter 9

The harmonic oscillator

There are only a very few potentials for which the Schrödinger equation can be solved analytically. The most important of these is the potential[1]

$$V(x) = \frac{1}{2}k(x - x_0)^2 \tag{9.1}$$

of the harmonic oscillator, because so many of the systems encountered in nuclear physics, condensed matter physics, and elementary particle physics can be treated, to a first approximation, as a set of coupled harmonic oscillators.

There are several reasons why the harmonic oscillator potential shows up so often. First of all, almost *any* smooth potential which has a finite minimum looks like a harmonic oscillator potential in the neighborhood of that minimum. Suppose the minimum of a potential is at the point $x = x_0$. Expand the potential in a Taylor series around x_0

$$V(x) = V(x_0) + \left(\frac{dV}{dx}\right)_{x=x_0}(x - x_0) + \frac{1}{2}\left(\frac{d^2 V}{dx^2}\right)_{x=x_0}(x - x_0)^2 + \cdots \tag{9.2}$$

Because $V(x)$ is a minimum at x_0, the first derivative of the potential vanishes there, so

$$V(x) = V(x_0) + \frac{1}{2}\left(\frac{d^2 V}{dx^2}\right)_{x=x_0}(x - x_0)^2 + \cdots \tag{9.3}$$

For small displacements, this is just a harmonic oscillator.[2]

[1] Most important, that is, for a physicist. For a chemist, who is concerned with how atoms and molecules combine, the Coulomb potential would be the most important case that can be solved analytically, because it gives the spectrum of the simplest atom, namely hydrogen.

[2] Except in very special cases where the quadratic term in the Taylor series is missing, e.g. $V(x) = cx^4$.

doi:10.1088/978-0-7503-1167-0ch9

Secondly, any time the classical equations of motion are linear, it means that we are dealing with a harmonic oscillator, or a set of coupled harmonic oscillators. Newton's law of motion $F = ma$ is generally non-linear, since $F(x)$ is usually a non-linear function of x. However, if F depends linearly on x, it follows that the potential depends quadratically on x, which implies a harmonic oscillator. Now it so happens that any non-dispersive wave equation, e.g.

$$\frac{1}{v^2}\frac{\partial^2 y}{\partial t^2} = \frac{\partial^2 y}{\partial x^2} \tag{9.4}$$

is a linear equation of motion. This equation might describe a transverse wave on a string, or a sound wave in a solid, or an electromagnetic plane wave, propagating in the x-direction. Any system that satisfies a wave equation of this form can be thought of as an infinite set of harmonic oscillators. To see this, we define the Fourier transform $Y(q, t)$ of the wave function $y(x, t)$ at any time t as

$$Y(q, t) = \int_{-\infty}^{\infty} \mathrm{d}x \; y(x, t)\mathrm{e}^{-\mathrm{i}qx} \tag{9.5}$$

with the inverse transform

$$y(x, t) = \int_{-\infty}^{\infty} \frac{\mathrm{d}q}{2\pi} \; Y(q, t)\mathrm{e}^{\mathrm{i}qx} \tag{9.6}$$

The functions $y(x, t)$ and $Y(q, t)$ contain the same amount of information about the wave at time t, since one function can be obtained from the other. Now multiply both sides of (9.4) by $\mathrm{e}^{-\mathrm{i}qx}$, integrate over x, and then integrate by parts:

$$\begin{aligned}
\int \mathrm{d}x \; \frac{1}{v^2}\frac{\partial^2 y}{\partial t^2}\mathrm{e}^{-\mathrm{i}qx} &= \int \mathrm{d}x \; \frac{\partial^2 y}{\partial x^2}\mathrm{e}^{-\mathrm{i}qx} \\
\frac{\partial^2}{\partial t^2}\int \mathrm{d}x \; \frac{1}{v^2}y(x, t)\mathrm{e}^{-\mathrm{i}qx} &= \int \mathrm{d}x \; y(x, t)\frac{\partial^2}{\partial x^2}\mathrm{e}^{-\mathrm{i}qx} \\
\frac{1}{v^2}\frac{\partial^2}{\partial t^2}\int \mathrm{d}x \; y(x, t)\mathrm{e}^{-\mathrm{i}qx} &= -q^2\int \mathrm{d}x \; y(x, t)\mathrm{e}^{-\mathrm{i}qx} \\
\frac{1}{v^2}\frac{\partial^2}{\partial t^2}Y(q, t) &= -q^2 Y(q, t)
\end{aligned} \tag{9.7}$$

But this is simply the equation of motion of a harmonic oscillator

$$\text{Acceleration} \quad \left(\frac{\partial^2 Y}{\partial t^2}\right) \propto -\text{displacement} \quad (Y) \tag{9.8}$$

We conclude that the wave equation (9.4) is secretly an infinite number of harmonic oscillator equations; one equation for each wavenumber q.

In the case of a mass hanging at the end of a spring, we understand that the classical solution of $F = ma$ describes the mass bouncing up and down, with some frequency $\omega = \sqrt{k/m}$. But in the case of wave motion, what is it that is bouncing up

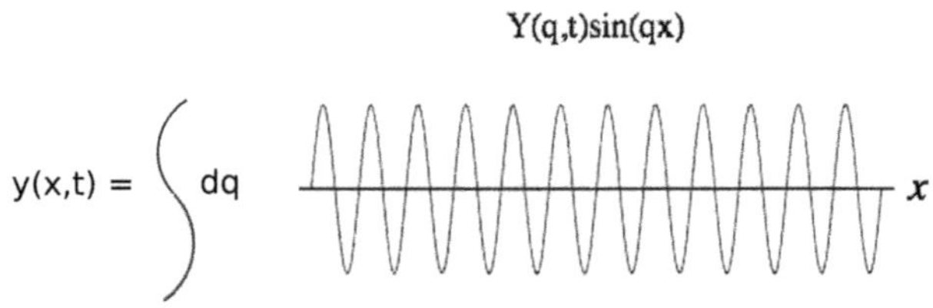

Figure 9.1. Any solution $y(x, t)$ of the wave equation (9.4) is a sum of plane waves e^{iqx} whose amplitude, $Y(q, t)$, oscillates in time as a harmonic oscillator, with angular frequency $\omega_q = qv$. Quantizing a classical system obeying the wave equation amounts to quantizing an infinite number of uncoupled harmonic oscillators.

and down in time? The answer is this: any solution of the wave equation has the form of a sum of infinite plane waves of the form e^{iqx}. That is the meaning of the inverse Fourier transform (9.6). But the amplitude of each wave of wavenumber q, denoted $Y(q, t)$, oscillates in time, and in fact it oscillates exactly like a mass at the end of a spring (see the animation in figure 9.1). It is the amplitude of each component wave of wavenumber q which is 'bouncing up and down' in time, with angular frequency $\omega_q = qv$.

Wave equations show up everywhere in classical physics. Waves propagate in solids, liquids, gases, and plasmas; they propagate in the electromagnetic field, and even (according to Einstein's theory of general relativity) in the gravitational field. All of these systems can therefore be regarded, at some level of approximation, as sets of harmonic oscillators. This is why the harmonic oscillator potential is the most important problem to solve in quantum physics. Fortunately, it is a problem with a simple and elegant solution.

9.1 Raising and lowering operators

A sum of squares can always be interpreted as the squared modulus of a complex number. Let A and B be any two real numbers, and define

$$C = A + iB \tag{9.9}$$

Then there is a simple identity

$$C^*C = A^2 + B^2 \tag{9.10}$$

Now the classical Hamiltonian for a harmonic oscillator, with potential $V(x) = \frac{1}{2}kx^2$, is also a sum of squares:

$$\begin{aligned} H &= \frac{p^2}{2m} + \frac{1}{2}kx^2 \\ &= \left(\frac{p}{\sqrt{2m}}\right)^2 + \left(\sqrt{\frac{k}{2}}x\right)^2 \end{aligned} \tag{9.11}$$

Let us define the complex number

$$a = \sqrt{\frac{1}{\hbar\omega}}\left(\sqrt{\frac{k}{2}}x + \mathrm{i}\frac{p}{\sqrt{2m}}\right) \tag{9.12}$$

where $\omega = \sqrt{k/m}$ is the angular frequency of the oscillator, and the factor $\sqrt{\frac{1}{\hbar\omega}}$ is introduced for later convenience. Then it is easy to see that H is proportional to the squared-modulus of a

$$H = \hbar\omega a^* a \tag{9.13}$$

Next, let us go over to quantum-mechanical operators, with

$$\tilde{H} = \frac{1}{2m}\tilde{p}^2 + \frac{1}{2}k\tilde{x}^2 \tag{9.14}$$

and define the operator $\tilde{a}$ and its Hermitian conjugate $\tilde{a}^\dagger$ as

$$\begin{aligned} \tilde{a} &= \sqrt{\frac{1}{\hbar\omega}}\left(\sqrt{\frac{k}{2}}\tilde{x} + \mathrm{i}\frac{\tilde{p}}{\sqrt{2m}}\right) \\ \tilde{a}^\dagger &= \sqrt{\frac{1}{\hbar\omega}}\left(\sqrt{\frac{k}{2}}\tilde{x} - \mathrm{i}\frac{\tilde{p}}{\sqrt{2m}}\right) \end{aligned} \tag{9.15}$$

where we have used the fact that $\tilde{x}^\dagger = \tilde{x}$, $\tilde{p}^\dagger = \tilde{p}$. We can of course also express $\tilde{x}$ and $\tilde{p}$ in terms of $\tilde{a}$, $\tilde{a}^\dagger$,

$$\begin{aligned} \tilde{x} &= \sqrt{\frac{\hbar}{2m\omega}}(\tilde{a} + \tilde{a}^\dagger) \\ \tilde{p} &= \frac{1}{\mathrm{i}}\sqrt{\frac{m\omega\hbar}{2}}(\tilde{a} - \tilde{a}^\dagger) \end{aligned} \tag{9.16}$$

Now we might guess that, in analogy to the classical case, $\tilde{H} = \hbar\omega\tilde{a}^\dagger\tilde{a}$, but in fact this is not quite right. Substituting (9.16) into (9.14), we find

$$\tilde{H} = \frac{1}{2}\hbar\omega(\tilde{a}^\dagger\tilde{a} + \tilde{a}\tilde{a}^\dagger) \tag{9.17}$$

If operator $\tilde{a}$ commuted with its Hermitian conjugate $\tilde{a}^\dagger$, then the order of operators would not matter. But in fact these operators satisfy a very elegant, and very important, commutation relation

$$\begin{aligned} [\tilde{a}, \tilde{a}^\dagger] &= \frac{1}{\hbar\omega}\left[\sqrt{\frac{k}{2}}\tilde{x} + \frac{\mathrm{i}}{\sqrt{2m}}\tilde{p}, \sqrt{\frac{k}{2}}\tilde{x} - \frac{\mathrm{i}}{\sqrt{2m}}\tilde{p}\right] \\ &= \frac{1}{\hbar\omega}\frac{1}{2}\sqrt{\frac{k}{m}}(-\mathrm{i}[\tilde{x}, \tilde{p}] + \mathrm{i}[\tilde{p}, \tilde{x}]) \\ &= -\frac{\mathrm{i}}{\hbar}[\tilde{x}, \tilde{p}] \\ &= 1 \end{aligned} \tag{9.18}$$

The factor $1/\sqrt{\hbar\omega}$ was introduced in (9.15) for the purpose of having the commutator come out to equal one, instead of some other constant. The commutator relation can also be written as $\tilde{a}\tilde{a}^\dagger = \tilde{a}^\dagger\tilde{a}+1$, and therefore the Hamiltonian (9.17) can be expressed as

$$\tilde{H} = \hbar\omega\left(\tilde{a}^\dagger\tilde{a} + \frac{1}{2}\right) \tag{9.19}$$

Operators $\tilde{a}^\dagger$ and $\tilde{a}$ are called the *raising operator* and *lowering operator* respectively. The reason for calling $\tilde{a}$ the 'lowering' operator is that it has the amazing property of taking an eigenstate of H into another eigenstate, with a lower energy (see figure 9.2). Suppose φ_n is some particular eigenstate, with

$$\tilde{H}\varphi_n = E_n\varphi_n \tag{9.20}$$

Define

$$\phi' = \tilde{a}\varphi_n \tag{9.21}$$

then

$$\begin{aligned} \tilde{H}\phi' &= \hbar\omega\left(\tilde{a}^\dagger\tilde{a} + \frac{1}{2}\right)a\varphi_n \\ &= \hbar\omega\left(\tilde{a}^\dagger\tilde{a}\tilde{a} + \frac{1}{2}\tilde{a}\right)\varphi_n \end{aligned} \tag{9.22}$$

The commutation relation (9.18) implies

$$\tilde{a}^\dagger\tilde{a} = \tilde{a}\tilde{a}^\dagger - 1 \tag{9.23}$$

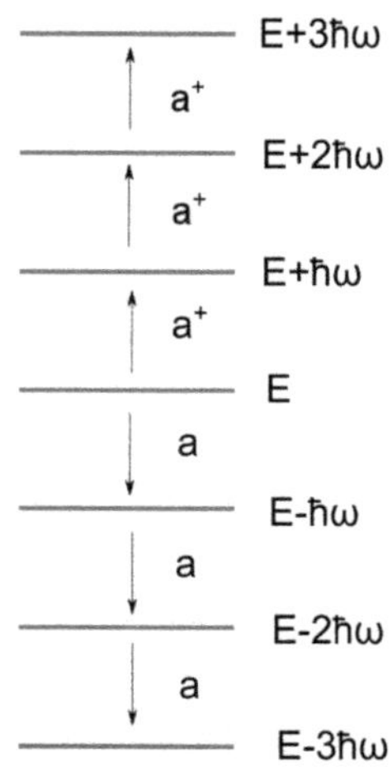

Figure 9.2. Raising and lowering, or 'ladder' operators, applied to energy eigenstates, produce other eigenstates with an energy either higher or lower, respectively, by $\hbar\omega$.

so that

$$\begin{aligned}\tilde{H}\phi' &= \hbar\omega\left[(\tilde{a}\tilde{a}^\dagger - 1)\tilde{a} + \frac{1}{2}\tilde{a}\right]\varphi_n \\ &= \tilde{a}\left[\hbar\omega\left(\tilde{a}^\dagger\tilde{a} + \frac{1}{2}\right) - \hbar\omega\right]\varphi_n \\ &= \tilde{a}[\tilde{H} - \hbar\omega]\varphi_n \\ &= (E_n - \hbar\omega)\tilde{a}\varphi_n \\ &= (E_n - \hbar\omega)\phi'\end{aligned} \tag{9.24}$$

which proves that $\tilde{a}\phi_n$ *is an energy eigenstate with eigenvalue* $E = E_n - \hbar\omega$, although this state is not necessarily normalized to unity. We will reserve the notation φ_n to refer to the normalized eigenstate. The operator $\tilde{a}^\dagger$ is termed the 'raising operator' for similar reasons. Let

$$\phi'' = \tilde{a}^\dagger\varphi_n \tag{9.25}$$

Then

$$\begin{aligned}\tilde{H}\phi'' &= \hbar\omega\left(\tilde{a}^\dagger\tilde{a} + \frac{1}{2}\right)\tilde{a}^\dagger\varphi_n \\ &= \hbar\omega\left(\tilde{a}^\dagger\tilde{a}\tilde{a}^\dagger + \frac{1}{2}\tilde{a}^\dagger\right)\varphi_n\end{aligned} \tag{9.26}$$

The commutation relation (9.18) also says that

$$\tilde{a}\tilde{a}^\dagger = \tilde{a}^\dagger\tilde{a} + 1 \tag{9.27}$$

so

$$\begin{aligned}\tilde{H}\phi'' &= \hbar\omega\left[\tilde{a}\tilde{a}^\dagger(\tilde{a}^\dagger\tilde{a} + 1) + \frac{1}{2}\tilde{a}^\dagger\right]\varphi_n \\ &= \tilde{a}^\dagger\left[\hbar\omega\left(\tilde{a}^\dagger\tilde{a} + \frac{1}{2}\right) + \hbar\omega\right]\varphi_n \\ &= \tilde{a}^\dagger[\tilde{H} + \hbar\omega]\varphi_n \\ &= (E_n + \hbar\omega)\tilde{a}^\dagger\varphi_n \\ &= (E_n + \hbar\omega)\phi''\end{aligned} \tag{9.28}$$

and therefore $\phi'' = \tilde{a}^\dagger\varphi_n$ *is an energy eigenstate with eigenvalue* $E = E_n + \hbar\omega$. Again, it is not necessarily normalized to unity.

Next we use the fact that the harmonic oscillator, like the hydrogen atom, has a lowest energy state (or 'ground' state). That there must be a lowest energy state is clear from the uncertainty principle, but we can also see it from the fact that

$$\begin{aligned}\langle p^2\rangle &= \langle\psi|p^2|\psi\rangle = \langle p\psi|p\psi\rangle \geqslant 0 \\ \langle x^2\rangle &= \langle\psi|x^2|\psi\rangle = \langle x\psi|x\psi\rangle \geqslant 0\end{aligned} \tag{9.29}$$

where we have used the Hermiticity of x and p, and also the fact that the norm of any non-zero vector is greater than zero. Then

$$\langle H\rangle = \frac{1}{2m}\langle p^2\rangle + \frac{1}{2}k\langle x^2\rangle \geqslant 0 \tag{9.30}$$

This proves that a ground state exists, with an energy greater than or equal to zero. However, the existence of a ground state seems to contradict the fact that, by operating on the ground state with the lowering operator a, we get a state with still lower energy. Denote the ground state by φ_0, the ground-state energy by E_0, and let $\varphi' = a\varphi_0$. Then, according to equation (9.24)

$$\tilde{H}(\tilde{a}\varphi_0) = (E_0 - \hbar\omega)(\tilde{a}\varphi_0) \tag{9.31}$$

This equation is satisfied if $\varphi' = \tilde{a}\varphi_0$ is an eigenstate of $\tilde{H}$ with energy $E_0 - \hbar\omega$, which means that φ_0 is not the ground state, *or* if

$$\tilde{a}\varphi_0 = 0 \tag{9.32}$$

Since we have proved that a ground state exists, then the lowering operator must 'annihilate' the ground state; i.e. bring it to 0, as shown in (9.32). Equation (9.32) is a first-order differential equation for the ground state:

$$\left(\sqrt{m\omega}\,x + \hbar\frac{1}{\sqrt{m\omega}}\frac{\partial}{\partial x}\right)\varphi_0 = 0 \tag{9.33}$$

which has the unique solution

$$\varphi_0 = N\mathrm{e}^{-m\omega x^2/2\hbar} \tag{9.34}$$

The constant N is determined from the normalization condition

$$\begin{aligned}1 &= N^2\int_{-\infty}^{\infty} \mathrm{d}x\,\mathrm{e}^{-m\omega x^2/\hbar} \\ &= N^2\sqrt{\frac{\pi\hbar}{m\omega}}\end{aligned} \tag{9.35}$$

so the ground-state eigenfunction is

$$\varphi_0 = \left[\frac{m\omega}{\pi\hbar}\right]^{1/4}\mathrm{e}^{-m\omega x^2/2\hbar} \tag{9.36}$$

The corresponding ground-state energy E_0 is

$$\begin{aligned} \tilde{H}\varphi_0 &= \hbar\omega\left(\tilde{a}^\dagger\tilde{a} + \frac{1}{2}\right)\varphi_0 \\ &= \frac{1}{2}\hbar\omega\varphi_0 \\ \Longrightarrow E_0 &= \frac{1}{2}\hbar\omega \end{aligned} \tag{9.37}$$

The state with the next-to-lowest energy is obtained by operating with the raising operator $\tilde{a}^\dagger$ on the ground state:

$$\varphi_1 = c_1\tilde{a}^\dagger\varphi_0 \tag{9.38}$$

where c_1 is a normalization constant. Since the raising operator raises the energy by $\hbar\omega$, the energy of this state is

$$\begin{aligned} E_1 &= E_0 + \hbar\omega \\ &= \hbar\omega\left(1 + \frac{1}{2}\right) \end{aligned} \tag{9.39}$$

The state with the next higher energy is

$$\begin{aligned} \varphi_2 &\propto \tilde{a}^\dagger\varphi_1 \\ &= c_2(\tilde{a}^\dagger)^2\varphi_0 \end{aligned} \tag{9.40}$$

with energy

$$\begin{aligned} E_2 &= E_1 + \hbar\omega \\ &= \hbar\omega\left(2 + \frac{1}{2}\right) \end{aligned} \tag{9.41}$$

Proceeding in this way, we find an infinite set of eigenstates of the form

$$\begin{aligned} \varphi_n &= c_n(\tilde{a}^\dagger)^n\varphi_0 \qquad n = 0, 1, 2, 3, \ldots \\ E_n &= \hbar\omega\left(n + \frac{1}{2}\right) \end{aligned} \tag{9.42}$$

But how do we know that this is all the eigenstates that there are? Suppose, for example, that there was a state φ' with an energy E' between E_n and E_{n+1}. We could then find another chain of states, with energies less than E' by integer multiples of $\hbar\omega$, by applying the lowering operator multiple times. But eventually the previous argument applies. By operating successively with a, we must reach a lowest state such that $a\varphi = 0$, otherwise the energy eigenvalue would become negative. But this lowest state is unique! We have already solved for it, and it is (9.36). Since φ_0 is reached by applying the lowering operator an integer number of times to φ', it follows that φ' is arrived at by applying the raising operator the same number of times to φ_0 That means that φ' must be proportional to one of the states φ_n, and therefore that there is no state with energy between E_n and E_{n+1}. So the only possible states and energies are those shown in (9.42).

Next, we need to find the normalization constants c_n in (9.42). This is done iteratively. Begin from

$$c_0 = 1 \tag{9.43}$$

and impose the normalization condition

$$\begin{aligned} 1 &= \langle\varphi_n|\varphi_n\rangle \\ &= c_n^2\langle(\tilde{a}^\dagger)^n\varphi_0|(\tilde{a}^\dagger)^n\varphi_0\rangle \\ &= c_n^2\langle\tilde{a}^\dagger(\tilde{a}^\dagger)^{n-1}\varphi_0|(\tilde{a}^\dagger)^n\varphi_0\rangle \\ &= c_n^2\langle(\tilde{a}^\dagger)^{n-1}\varphi_0|\tilde{a}(\tilde{a}^\dagger)^n|\varphi_0\rangle \end{aligned} \tag{9.44}$$

The idea is to commute the $\tilde{a}$ operator to the right, past the $\tilde{a}^\dagger$ operators, until it finally operates on, and annihilates, the ground state φ_0. We have

$$\begin{aligned} \tilde{a}(\tilde{a}^\dagger)^n &= (\tilde{a}\tilde{a}^\dagger)(\tilde{a}^\dagger)^{n-1} \\ &= (\tilde{a}^\dagger\tilde{a} + 1)(\tilde{a}^\dagger)^{n-1} \\ &= \tilde{a}^\dagger\tilde{a}(\tilde{a}^\dagger)^{n-1} + (\tilde{a}^\dagger)^{n-1} \\ &= \tilde{a}^\dagger(\tilde{a}^\dagger\tilde{a} + 1)(\tilde{a}^\dagger)^{n-2} + (\tilde{a}^\dagger)^{n-1} \\ &= (\tilde{a}^\dagger)^2(\tilde{a}^\dagger\tilde{a} + 1)(\tilde{a}^\dagger)^{n-3} + 2(\tilde{a}^\dagger)^{n-1} \\ &= (\tilde{a}^\dagger)^3(\tilde{a}^\dagger\tilde{a} + 1)(\tilde{a}^\dagger)^{n-4} + 3(\tilde{a}^\dagger)^{n-1} \\ &\;\;\vdots \\ &= (\tilde{a}^\dagger)^n\tilde{a} + n(\tilde{a}^\dagger)^{n-1} \end{aligned} \tag{9.45}$$

Equation (9.44) becomes

$$1 = c_n^2\langle(\tilde{a}^\dagger)^{n-1}\varphi_0|[(\tilde{a}^\dagger)^n\tilde{a} + n(\tilde{a}^\dagger)^{n-1}]|\varphi_0\rangle \tag{9.46}$$

Using $a|\varphi_0\rangle = 0$,

$$\begin{aligned} 1 &= nc_n^2\langle(\tilde{a}^\dagger)^{n-1}\varphi_0|(\tilde{a}^\dagger)^{n-1}|\varphi_0\rangle \\ &= n\frac{c_n^2}{c_{n-1}^2}\langle\varphi_{n-1}|\varphi_{n-1}\rangle \\ &= n\frac{c_n^2}{c_{n-1}^2} \end{aligned} \tag{9.47}$$

so that

$$c_n = \frac{c_{n-1}}{\sqrt{n}} \tag{9.48}$$

Using this equation, and $c_0 = 1$, we obtain

$$\begin{aligned} c_1 &= 1 \\ c_2 &= \frac{1}{\sqrt{2 \cdot 1}} \\ c_3 &= \frac{1}{\sqrt{3 \cdot 2 \cdot 1}} \end{aligned} \tag{9.49}$$

or, in general

$$c_n = \frac{1}{\sqrt{n!}} \tag{9.50}$$

We now have the general solution for the energy eigenstates of the harmonic oscillator:

$$\begin{aligned} \varphi_n(x) &= \frac{1}{\sqrt{n!}}(\tilde{a}^\dagger)^n \varphi_0(x) \\ &= \left(\frac{m\omega}{\pi\hbar}\right)^{1/4} \frac{1}{\sqrt{n!}} \left[\frac{1}{\sqrt{2}}\left(\sqrt{\frac{m\omega}{\hbar}}x - \sqrt{\frac{\hbar}{m\omega}}\frac{\partial}{\partial x}\right)\right]^n \mathrm{e}^{-m\omega x^2/2\hbar} \end{aligned} \tag{9.51}$$

Rescaling x

$$y = \sqrt{\frac{m\omega}{\hbar}}x \tag{9.52}$$

The general solution becomes

$$\begin{aligned} \varphi_n(x) &= \frac{1}{\sqrt{n!}}\left(\frac{m\omega}{\pi\hbar}\right)^{1/4}\left(\frac{1}{\sqrt{2}}\right)^n\left(y - \frac{\partial}{\partial y}\right)^n \mathrm{e}^{-y^2/2} \\ &= \frac{1}{\sqrt{n!}}\left(\frac{m\omega}{\pi\hbar}\right)^{1/4}\left(\frac{1}{\sqrt{2}}\right)^n \mathcal{H}_n(y)\mathrm{e}^{-y^2/2} \end{aligned} \tag{9.53}$$

with corresponding eigenvalues

$$E_n = \hbar\omega\left(n + \frac{1}{2}\right) \tag{9.54}$$

The functions $\mathcal{H}_n(y)$ are known as ***Hermite polynomials***. Operating n times on the ground state with the raising operator $a^\dagger$ results in a function which is just the ground state multiplied by an nth-order polynomial in the variable $y = \sqrt{m\omega/\hbar}\,x$. These are the Hermite polynomials. The first several Hermite polynomials are

$$
\begin{aligned}
\mathcal{H}_0(y) &= 1 \\
\mathcal{H}_1(y) &= 2y \\
\mathcal{H}_2(y) &= 4y^2 - 2 \\
\mathcal{H}_3(y) &= 8y^3 - 12y
\end{aligned}
\tag{9.55}
$$

By applying equation (9.53), the Hermite polynomials, and the eigenstates $\varphi_n(x)$, can be determined to any order desired. A selection of harmonic oscillator eigenstates (in units where $\sqrt{m\omega/\hbar} = 1$) is displayed in figure 9.3.

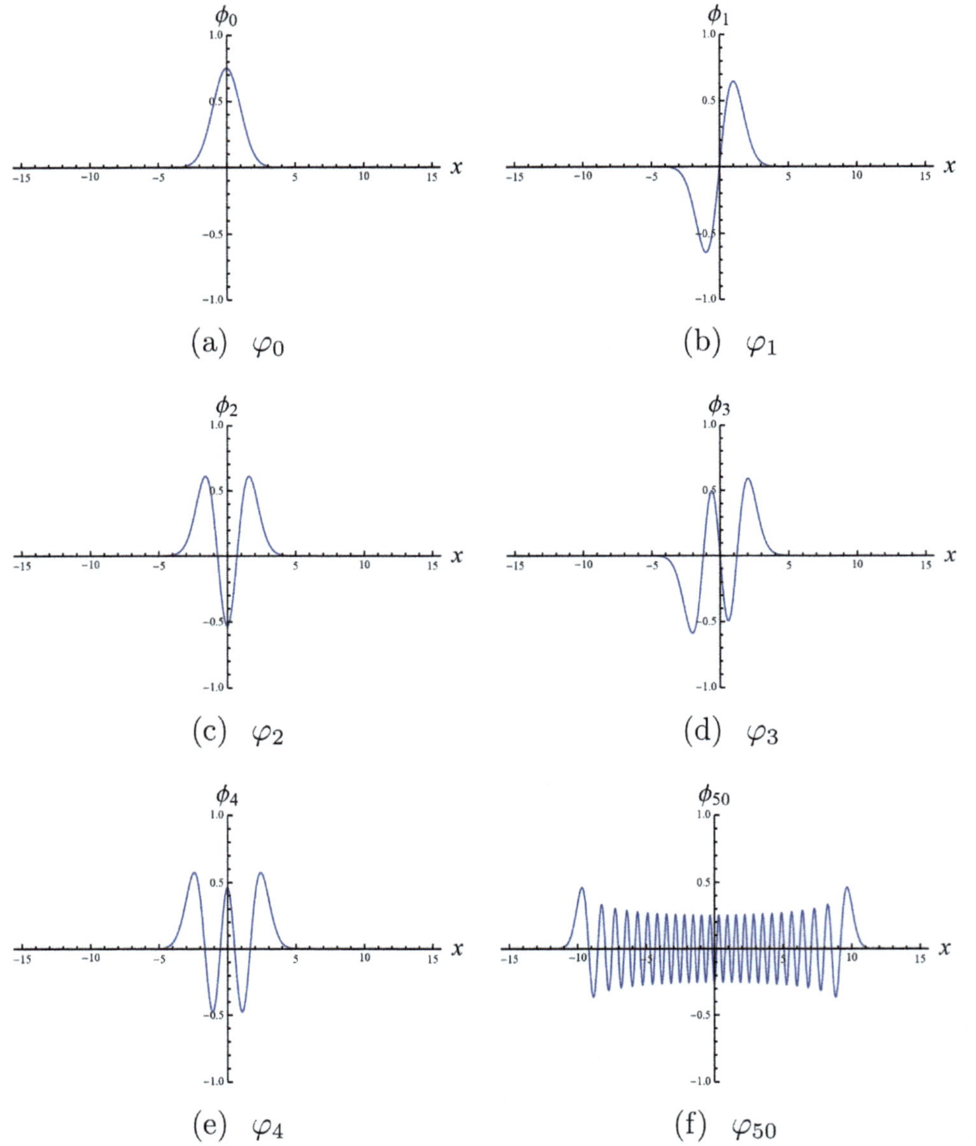

Figure 9.3. A sketch of the first few harmonic oscillator energy eigenstates φ_n, $n = 1$–4, and a highly excited eigenstate φ_{50}. Note that eigenstates with even n are even parity, and odd n are odd parity.

9.2 Algebra and expectation values

One of the most remarkable properties of the quantum harmonic oscillator is that any quantity of this form

$$\langle \varphi_n | F(x, p) | \varphi_m \rangle \tag{9.56}$$

where F is a polynomial in x and p, can be computed algebraically, without ever using the explicit form of the eigenstates $\varphi_n(x)$. The reasons for this are, first, that the x and p operators are linear combinations of the raising and lowering operators (equation (9.16)); and second, that the raising and lowering operators have a simple algebraic action on the eigenstates.

Consider the raising operator acting on the state φ_n. Using equation (9.51) we have

$$\begin{aligned} \tilde{a}^\dagger |\varphi_n\rangle &= \frac{1}{\sqrt{n!}} (\tilde{a}^\dagger)^{n+1} |\varphi_0\rangle \\ &= \sqrt{(n+1)} \frac{1}{\sqrt{(n+1)!}} (\tilde{a}^\dagger)^{n+1} |\varphi_{n+1}\rangle \\ &= \sqrt{n+1} \, |\varphi_{n+1}\rangle \end{aligned} \tag{9.57}$$

and we can also derive a similar equation for $\tilde{a}\varphi_n$:

$$\begin{aligned} \tilde{a} |\varphi_n\rangle &= \frac{1}{\sqrt{n!}} \tilde{a} (\tilde{a}^\dagger)^n |\varphi_0\rangle \\ &= \frac{1}{\sqrt{n!}} \left[(\tilde{a}^\dagger)^n \tilde{a} + n (\tilde{a}^\dagger)^{n-1} \right] |\varphi_0\rangle \\ &= n \frac{1}{\sqrt{n!}} (\tilde{a}^\dagger)^{n-1} |\varphi_0\rangle \\ &= \sqrt{n} \frac{1}{\sqrt{(n-1)!}} (\tilde{a}^\dagger)^{n-1} |\varphi_0\rangle \\ &= \sqrt{n} \, |\varphi_{n-1}\rangle \end{aligned} \tag{9.58}$$

In summary:

$$\begin{aligned} \tilde{a} |\varphi_n\rangle &= \sqrt{n} \, |\varphi_{n-1}\rangle \\ \tilde{a}^\dagger |\varphi_n\rangle &= \sqrt{n+1} \, |\varphi_{n+1}\rangle \\ \tilde{x} &= \sqrt{\frac{\hbar}{2m\omega}} (\tilde{a} + \tilde{a}^\dagger) \\ \tilde{p} &= \frac{1}{\mathrm{i}} \sqrt{\frac{m\omega\hbar}{2}} (\tilde{a} - \tilde{a}^\dagger) \end{aligned} \tag{9.59}$$

As an example of the use of these relations, let us compute the position uncertainty Δx in the nth energy eigenstate. As usual

$$(\Delta x)^2 = \langle x^2 \rangle - \langle x^2 \rangle \tag{9.60}$$

Now

$$\begin{aligned}
\sqrt{\frac{2m\omega}{\hbar}} \langle x \rangle &= \langle \varphi_n | (\tilde{a} + \tilde{a}^\dagger) | \varphi_n \rangle \\
&= \sqrt{n} \left\langle \varphi_n | \varphi_{n-1} \right\rangle + \sqrt{n+1} \left\langle \varphi_n | \varphi_{n+1} \right\rangle \\
&= 0
\end{aligned} \tag{9.61}$$

Next, applying successively the relations (9.59)

$$\begin{aligned}
\frac{2m\omega}{\hbar} \langle x^2 \rangle &= \langle \varphi_n | (\tilde{a} + \tilde{a}^\dagger)(\tilde{a} + \tilde{a}^\dagger) | \varphi_n \rangle \\
&= \langle \varphi_n | (\tilde{a}\tilde{a} + \tilde{a}\tilde{a}^\dagger + \tilde{a}^\dagger\tilde{a} + \tilde{a}^\dagger\tilde{a}^\dagger) | \varphi_n \rangle \\
&= \langle \varphi_n | \tilde{a}\tilde{a} | \varphi_n \rangle + \langle \varphi_n | \tilde{a}\tilde{a}^\dagger | \varphi_n \rangle + \langle \varphi_n | \tilde{a}^\dagger\tilde{a} | \varphi_n \rangle + \langle \varphi_n | \tilde{a}^\dagger\tilde{a}^\dagger | \varphi_n \rangle \\
&= \sqrt{n} \langle \varphi_n | \tilde{a} | \varphi_{n-1} \rangle + \sqrt{n+1} \left\langle \varphi_n | \tilde{a} | \varphi_{n+1} \right\rangle \\
&\quad + \sqrt{n} \langle \varphi_n | \tilde{a}^\dagger | \varphi_{n-1} \rangle + \sqrt{n+1} \left\langle \varphi_n | \tilde{a}^\dagger | \varphi_{n+1} \right\rangle \\
&= \sqrt{n(n-1)} \langle \varphi_n | \varphi_{n-2} \rangle + (n+1) \langle \varphi_n | \varphi_n \rangle \\
&\quad + n \langle \varphi_n | \varphi_n \rangle + \sqrt{(n+1)(n+2)} \left\langle \varphi_n | \varphi_{n+2} \right\rangle \\
&= 0 + (n+1) + n + 0 \\
&= 2n + 1
\end{aligned} \tag{9.62}$$

so the position uncertainty is

$$\Delta x = \sqrt{\frac{\hbar}{2m\omega}(2n+1)} \tag{9.63}$$

Notice that the explicit form of the wave function $\varphi_n(x)$ was not needed to compute Δx; the calculation is completely algebraic, and required only the relations (9.59). It is easy to see that any expression of the form

$$\left\langle \phi_m \middle| F[x, p] \middle| \phi_n \right\rangle \tag{9.64}$$

where $F[x, p]$ is any polynomial in the x and p operators, can also be computed algebraically by the same method.

Problems

1. Find the energy eigenvalues for a particle of mass m moving in a potential

$$V(x) = \begin{cases} \frac{1}{2}kx^2 & x > 0 \\ \infty & x \leqslant 0 \end{cases} \tag{9.65}$$

2. Calculate $\langle\phi_n|xp^2x|\phi_n\rangle$, where ϕ_n is an energy eigenstate of the harmonic oscillator.
3. Let

$$\psi(x, t) = \frac{1}{\sqrt{2}}[\psi_0(x, t) + i\psi_1(x, t)] \tag{9.66}$$

where ψ_0 and ψ_1 are the ground state and first excited state of the harmonic oscillator. Calculate $\langle E\rangle$, $\langle x\rangle$, and $\langle p\rangle$ as a function of time.
4. Let $\phi_m(x)$ and $\phi_n(x)$ be the mth and nth energy eigenstates of the harmonic oscillator. Using the orthogonality relation $\langle\phi_i|\phi_j\rangle = \delta_{ij}$, calculate

$$\langle\phi_m|p^3|\phi_n\rangle \quad \text{and} \quad \langle\phi_m|(a^\dagger)^k|\phi_n\rangle \tag{9.67}$$

where k is any integer.
5. Show that if $\phi = (\tilde{a}^\dagger)^n\varphi_0$, then also $\varphi_0 \propto \tilde{a}^n\phi$.
6. Consider the Hamiltonian

$$H = \frac{p^2}{2m} + \frac{1}{2}kx^2 + gv \tag{9.68}$$

where $v = p/m$ is the velocity, and g is a constant.
   a) Write this Hamiltonian in terms of the usual raising and lowering operators $\tilde{a}^\dagger$, $\tilde{a}$. In this case, when $\tilde{a}$ or $\tilde{a}^\dagger$ operates on an energy eigenstate, the result is *not* an eigenstate of lower or higher energy.
   b) Write H in terms of new operators $\tilde{b} = \tilde{a} - c$, and $\tilde{b}^\dagger = \tilde{a}^\dagger - c^*$, where c is a constant. Choose c so that all terms in the Hamiltonian involving a constant times $\tilde{b}$ or $\tilde{b}^\dagger$ cancel. With this choice of c, prove that $\tilde{b}$ and $\tilde{b}^\dagger$ *do* act as raising and lowering operators when applied to energy eigenstates.
   c) Given the results of part (b), find the energy eigenvalues of H.
   d) Calculate the ground state wave function $\phi_0(x)$ explicitly as a function of x.
7. Define the matrices x_{nm} and p_{nm}, whose matrix elements are given by

$$\begin{aligned} x_{nm} &= \langle\phi_n|x|\phi_m\rangle \\ p_{nm} &= \langle\phi_n|p|\phi_m\rangle \end{aligned} \tag{9.69}$$

where the ϕ_n, ϕ_m are harmonic-oscillator energy eigenstates.

a) Calculate the matrix elements x_{nm} and p_{nm}. Express your answers in terms of Kronecker deltas.
b) Using the results of part (a), calculate the commutator of the two *matrices*

$$[x, p]_{nm} = \sum_{k=0}^{\infty} \{x_{nk}p_{km} - p_{nk}x_{km}\} \tag{9.70}$$

Compare your result to the commutator of *operators* $[\tilde{x}, \tilde{p}] = i\hbar$.

8. Consider a particle of mass m moving in a harmonic oscillator potential $V(x) = \frac{1}{2}kx^2$. Any solution to the time-dependent Schrödinger equation has the form

$$\psi(x, t) = \sum_{n=0}^{\infty} c_n\phi_n(x)e^{-iE_nt/\hbar} \tag{9.71}$$

where the ϕ_n are harmonic oscillator eigenstates.
a) Calculate the expectation value of kinetic energy $p^2/2m$ as a function of time. What is the time-averaged value? Compare to the expectation value of energy.
b) Compute $\langle xp \rangle$. Is your answer necessarily a real number? Should it be?
c) Compute the expectation value of position

$$x(t) = \langle \psi | x | \psi \rangle \tag{9.72}$$

as a function of time. Does $x(t)$ satisfy any simple equation of motion? If so, can you explain why?

9. A particle of mass m is moving in a harmonic oscillator potential $V(x) = \frac{1}{2}kx^2$. At time $t = 0$ the particle is in the state

$$\psi(x, 0) = \frac{1}{\sqrt{3}}\big(\phi_2(x) + \phi_4(x) + \phi_6(x)\big) \tag{9.73}$$

where $\phi_n(x)$ is the nth energy eigenstate. Compute the squared uncertainty Δp^2 of momentum, at any given time t.

10. Prove that

$$\psi = N\exp\left[\lambda\tilde{a}^\dagger\right]\phi_0 \tag{9.74}$$

(known as a 'coherent state') is an eigenstate of the lowering operator $\tilde{a}$, and compute the normalization constant N.

IOP Publishing

An Introduction to Quantum Theory

Jeff Greensite

Chapter 10

Symmetry and degeneracy

We live in a world which is almost, but not quite, symmetric. The Earth is round, nearly, and moves in an orbit around the Sun which is circular, almost. Our Galaxy looks roughly like a spiral. Human beings and most animals have a left–right symmetry; the left-hand side looks the same as the right-hand side, more or less. Starfish have a pentagonal symmetry, table salt has a cubic symmetry, and in general Nature abounds in geometrical shapes of various kinds.

The symmetry which is apparent in large structures (animals, planets, galaxies, …) is even more evident at the very small scales that are the domain of quantum physics. Symmetries appear in the arrangement of atoms in solids, in the arrangement of electrons in atoms, and in the structure of the nucleus. Symmetry principles are especially important for understanding the variety and behavior of the elementary particles, out of which all larger structures are built.

The mathematical expression of a geometrical symmetry is an invariance under some transformation of coordinates. Let us consider some examples which will be important:

- **Reflection symmetry.** A stick-figure sketch is shown in figure 10.1. Suppose we assign to every point on the sketch a set of coordinates (x,y) in the coordinate system shown, and then suppose that every point on the sketch is moved to another point

$$(x, y) \rightarrow (x', y') \tag{10.1}$$

according to the rule

$$x' = x + a \qquad y' = y + b \tag{10.2}$$

where a and b are constants. This is an example of a transformation of coordinates, known as a ***translation***. Under this transformation, the whole figure is moved to somewhere else in the plane, as shown in figure 10.1. This is *not* an invariance of the sketch, since the transformed sketch can be

doi:10.1088/978-0-7503-1167-0ch10

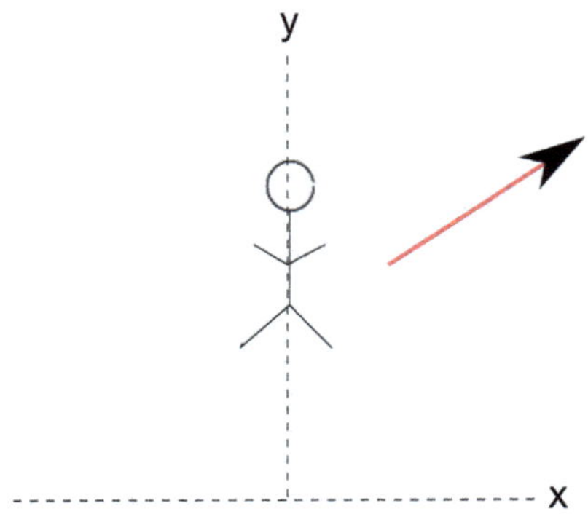

Figure 10.1. The stick figure is unchanged under reflections $x \to -x$, but not under translations $x \to x + a$, $y \to y + b$.

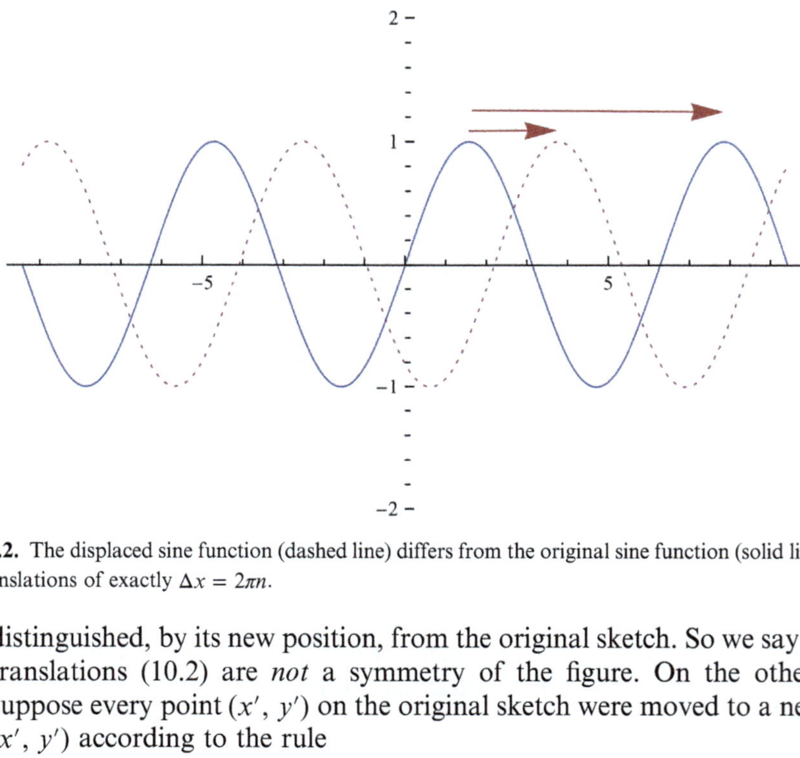

Figure 10.2. The displaced sine function (dashed line) differs from the original sine function (solid line), except under translations of exactly $\Delta x = 2\pi n$.

distinguished, by its new position, from the original sketch. So we say that the translations (10.2) are *not* a symmetry of the figure. On the other hand, suppose every point (x', y') on the original sketch were moved to a new point (x', y') according to the rule

$$x' = -x \qquad y' = y \tag{10.3}$$

This transformation replaces the right-hand side of the figure by the left-hand side, and vice versa. The transformed stick figure is indistinguishable from the original figure, and it is therefore said to be symmetric under ***left–right reflection***.

- **Periodicity (symmetry under finite translations).** Consider next the sine wave shown in figure 10.2. In general, if we transform every point $(x, y) \to (x', y')$ according to the rule (10.2), the figure will end up in a different place, and thus be distinguishable from the untransformed figure. However, if we make the transformation

$$x' = x + 2\pi n \qquad y' = y \tag{10.4}$$

where n is an integer, then the new figure is the same as the old, due to the periodicity of sine waves. These special translations are therefore a symmetry of periodic functions, such as sine and cosine.

- **Rotation symmetry.** As a final example, consider the points on a circle of radius R. Each point on the circle can be assigned polar coordinates (r, θ), where $r = R$ and $0 \leqslant \theta < 2\pi$. A circle is obviously symmetric under arbitrary rotations

$$r' = r \qquad \theta' = \theta + \delta\theta \tag{10.5}$$

where $\delta\theta$ is any constant angle.

Symmetries are important in quantum mechanics whenever the Hamiltonian is invariant under some coordinate transformation:

$$\begin{aligned} x' &= f(x) \\ \frac{\partial}{\partial x'} &= \left(\frac{\partial f}{\partial x}\right)^{-1}\frac{\partial}{\partial x} \end{aligned} \tag{10.6}$$

where the word ‘invariant’ means that

$$\tilde{H}\left[-i\hbar\frac{\partial}{\partial x'}, x'\right] = \tilde{H}\left[-i\hbar\frac{\partial}{\partial x}, x\right] \tag{10.7}$$

Suppose $\phi_\alpha(x)$ is an eigenstate of $\tilde{H}$

$$\tilde{H}\left[-i\hbar\frac{\partial}{\partial x}, x\right]\phi_\alpha(x) = E_\alpha\phi_\alpha(x) \tag{10.8}$$

If we relabel the coordinates $x \to x'$,

$$\tilde{H}\left[-i\hbar\frac{\partial}{\partial x'}, x'\right]\phi_\alpha(x') = E_\alpha\phi_\alpha(x') \tag{10.9}$$

and then use the invariance of the Hamiltonian under $x' = f(x)$, we find

$$\tilde{H}\left[-i\hbar\frac{\partial}{\partial x}, x\right]\phi_\alpha(f(x)) = E_\alpha\phi_\alpha(f(x)) \tag{10.10}$$

which proves that the transformed wave function

$$\phi'(x) = \phi_\alpha(f(x)) \tag{10.11}$$

is also an energy eigenstate, with energy E_α.

The reason that the symmetries of the Hamiltonian are important is that they are invariably associated with conservation laws; i.e. the expectation value of some observable Q (different from the energy) is constant in time

$$\frac{\mathrm{d}}{\mathrm{d}t}\langle Q\rangle = 0 \tag{10.12}$$

For example, symmetry of the Hamiltonian with respect to arbitrary translations (10.2) is associated with conservation of momentum, and symmetry of the Hamiltonian with respect to rotations by an arbitrary angle (10.5) is associated with conservation of angular momentum. It is always possible to choose the set of energy eigenstates $\{\phi_\alpha\}$ to be eigenstates not only of the Hamiltonian, but also of some subset of the conserved observables, and the values of these observables can be used to distinguish between different energy eigenstates which may have the same energy eigenvalue. This chapter will be devoted to some examples of symmetric Hamiltonians, and their corresponding conservation laws.

10.1 The free particle and momentum conservation

The free particle Hamiltonian in one dimension is invariant under arbitrary translations

$$x' = x + a \tag{10.13}$$

because $V(x') = V(x) = 0$ everywhere, and because

$$\frac{\partial}{\partial x'} = \frac{\partial}{\partial x} \tag{10.14}$$

Let us define a linear operator $\tilde{T}$, which acts on functions by transforming the coordinate

$$\tilde{T}F(x) = F(x') = F(x + a) \tag{10.15}$$

Then

$$\begin{aligned}\tilde{T}\tilde{H}\left[\frac{\partial}{\partial x}, x\right]\psi(x) &= \tilde{H}\left[\frac{\partial}{\partial x'}, x'\right]\psi(x') \\ &= \tilde{H}\left[\frac{\partial}{\partial x}, x\right]\psi(x') \\ &= \tilde{H}\left[\frac{\partial}{\partial x}, x\right]\tilde{T}\psi(x)\end{aligned} \tag{10.16}$$

From this equation, we see that the operator $\tilde{T}$ commutes with the Hamiltonian operator

$$[\tilde{T}, \tilde{H}] = 0 \tag{10.17}$$

Now notice that

$$\begin{aligned}\tilde{T}F(x) &= F(x+a)\\ &= F(x) + \frac{\partial F}{\partial x}a + \frac{1}{2}\frac{\partial^2 F}{\partial x^2}a^2 + \cdots\\ &= \sum_{n=0}^{\infty}\frac{a^n}{n!}\frac{\partial^n}{\partial x^n}F(x)\\ &= \exp\left[a\frac{\partial}{\partial x}\right]F(x)\\ &= \exp\left[\mathrm{i}a\tilde{p}/\hbar\right]F(x)\end{aligned} \tag{10.18}$$

which means that $\tilde{T}$ is the exponential of the momentum operator. Since $\tilde{T}$ commutes with $\tilde{H}$ for any displacement a, it follows that the momentum operator p also commutes with H

$$\left[\tilde{p}, \tilde{H}\right] = 0 \tag{10.19}$$

This is easy to check for the free particle Hamiltonian, since $\tilde{H} = \tilde{p}^2/2m$, and $\tilde{p}$ commutes with $\tilde{p}^2$.

In chapter 7, in a discussion of the time-energy uncertainty relation, we found an equation of motion for expectation values

$$\frac{\mathrm{d}}{\mathrm{d}t}\langle Q\rangle = \frac{\mathrm{i}}{\hbar}\langle[Q, H]\rangle \tag{10.20}$$

It follows that *if a Hermitian operator commutes with the Hamiltonian, the corresponding observable is conserved:*

$$\frac{\mathrm{d}}{\mathrm{d}t}\langle Q\rangle = 0 \tag{10.21}$$

In the special case of the free particle, we therefore have conservation of momentum

$$\frac{\mathrm{d}}{\mathrm{d}t}\langle p\rangle = 0 \tag{10.22}$$

We have already noted that the energies of the free particle are two-fold degenerate; i.e. for each energy E there are two linearly independent states with the same energy, and any linear combination of those two states

$$\psi(x) = a\mathrm{e}^{\mathrm{i}\sqrt{2mE}x/\hbar} + b\mathrm{e}^{-\mathrm{i}\sqrt{2mE}x/\hbar} \tag{10.23}$$

is also an eigenstate with energy E. However, since $\tilde{p}$ commutes with $\tilde{H}$, it follows from the commutator theorem that energy E and momentum p are simultaneously observable. This means that we can choose the complete set of energy eigenstates $\{\phi_\alpha\}$ to be eigenstates of $\tilde{H}$ *and* $\tilde{p}$, which is

$$\left\{\phi_p(x) = \frac{1}{\sqrt{2\pi\hbar}}\mathrm{e}^{\mathrm{i}px/\hbar};\;\; E_p = \frac{p^2}{2m};\;\; p \in [-\infty, \infty]\right\} \tag{10.24}$$

Each value of the momentum p singles out one and only one energy eigenstate $\phi_p(x)$, whereas an energy eigenvalue, by itself, is not enough to specify the state uniquely.

10.2 Parity

The free particle potential

$$V(x) = 0 \tag{10.25}$$

the harmonic oscillator potential

$$V(x) = \frac{1}{2}kx^2 \tag{10.26}$$

and the finite square well potential

$$V(x) = \begin{cases} 0 & x < -a \\ -V_0 & -a \leqslant x \leqslant a \\ 0 & x > a \end{cases} \tag{10.27}$$

are all invariant under the ***parity*** transformation

$$x' = -x \tag{10.28}$$

which is a left–right reflection of the x-axis. The kinetic energy term of the Hamiltonian is also invariant under this transformation, since

$$\begin{aligned} -\frac{\hbar^2}{2m}\frac{\partial}{\partial x'}\frac{\partial}{\partial x'} &= -\frac{\hbar^2}{2m}\left(-\frac{\partial}{\partial x}\right)\left(-\frac{\partial}{\partial x}\right) \\ &= -\frac{\hbar^2}{2m}\frac{\partial}{\partial x}\frac{\partial}{\partial x} \end{aligned} \tag{10.29}$$

therefore the Hamiltonians of the free particle, the harmonic oscillator, and the finite square well are invariant under the parity transformation. Define the parity operator as the linear operator which has the property

$$\tilde{P}F(x) = F(-x) \tag{10.30}$$

Once again, we see that

$$\begin{aligned} \tilde{P}\tilde{H}[\partial_x, x]\psi(x) &= \tilde{H}[-\partial_x, -x]\psi(-x) \\ &= \tilde{H}[\partial_x, x]\psi(-x) \\ &= \tilde{H}[\partial_x, x]\tilde{P}\psi(x) \end{aligned} \tag{10.31}$$

so the parity operator commutes with the Hamiltonian

$$\left[\tilde{P}, \tilde{H}\right] = 0 \tag{10.32}$$

and is conserved

$$\frac{d}{dt}\left\langle \tilde{P} \right\rangle = 0 \tag{10.33}$$

It is easy to show that parity is a Hermitian operator.

Suppose ϕ_β is an eigenstate of the parity operator with eigenvalue β

$$\tilde{P}\phi_\beta(x) = \beta\phi_\beta \tag{10.34}$$

Then

$$\tilde{P}\tilde{P}\phi_\beta(x) = \beta\tilde{P}\phi_\beta(x) = \beta^2\phi_\beta(x) \tag{10.35}$$

On the other hand,

$$\tilde{P}\tilde{P}\phi_\beta(x) = \tilde{P}\phi_\beta(-x) = \phi_\beta(x) \tag{10.36}$$

Comparing the last two equations, we see that

$$\beta^2 = 1 \Longrightarrow \beta = \pm 1 \tag{10.37}$$

Eigenstates with $\beta = +1$ are known as 'even-parity' eigenstates, denoted $\phi_+(x)$, and have the property

$$\phi_+(-x) = +\phi_+(x) \tag{10.38}$$

while eigenstates with $\beta = -1$ are known as 'odd-parity' eigenstates, denoted $\phi_-(x)$, with the property

$$\phi_-(-x) = -\phi_-(x) \tag{10.39}$$

By the commutator theorem, the Hamiltonian operator and the parity operator have a common set of eigenstates. We have already seen in chapter 9 that every energy eigenstate of the finite square well was either an even parity state, or an odd parity state. The same is true for the energy eigenstates of the harmonic oscillator. The harmonic oscillator ground state $\phi_0(x)$ is a Gaussian function which is invariant under $x \to -x$; it is therefore an even-parity eigenstate. The raising operator, on the other hand transforms as

$$a^\dagger(-x, -\partial_x) = -a^\dagger(x, \partial_x) \tag{10.40}$$

Using the general form for the harmonic oscillator eigenstates

$$\phi_n(x) = \frac{1}{\sqrt{n!}}(a^\dagger)^n\phi_0(x) \tag{10.41}$$

it is clear that

$$\tilde{P}\phi_n(x) = \phi_n(-x) = (-1)^n\phi_n(x) \tag{10.42}$$

and the energy eigenstates are even-parity for n even, and odd-parity for n odd.

Finally, the free particle Hamiltonian is invariant under parity, so there must be a complete set of energy eigenstates which are also eigenstates of the parity operator. Note that the eigenstates of momentum are *not* eigenstates of parity

$$\tilde{P}e^{ipx/\hbar} = e^{-ipx/\hbar} \neq \pm e^{ipx/\hbar} \tag{10.43}$$

On the other hand, an energy eigenstate does not have to be a momentum eigenstate. As we have seen, an eigenstate of energy E has the general form

$$\phi_E(x) = a e^{ipx/\hbar} + b e^{-ipx/\hbar}, \quad p = \sqrt{2mE} \tag{10.44}$$

If $\phi_E(x)$ is an even-parity eigenstate, it requires that $a = b$, or

$$\phi_{E+}(x) = A \cos\left[\frac{px}{\hbar}\right] \tag{10.45}$$

while if $\phi_E(x)$ is an odd-parity eigenstate, it means that $a = -b$, or

$$\phi_{E-}(x) = B \sin\left[\frac{px}{\hbar}\right] \tag{10.46}$$

So a complete set of energy eigenstates is given by the set

$$\left\{\phi_{E+}(x), \phi_{E-}(x), \quad E \in [0, \infty]\right\} \tag{10.47}$$

This is complete because *any* energy eigenstate of the form (10.44) can be expressed as a linear combination of ϕ_{E+} and ϕ_{E-}. Note that specifying the energy E, and the parity ± 1, determines the energy eigenstate uniquely.

It may seem odd that an eigenstate of momentum is not an eigenstate of parity (and vice versa). After all, both are associated with coordinate transformations which are symmetries of the free particle Hamiltonian. However, the order of these transformations is important. Suppose we have a particle at point x. If we first make a translation $x' = x + a$, and then a transformation $x'' = -x'$, the particle winds up at coordinates $x'' = -x - a$. On the other hand, if we make these transformations in reverse order, first $x' = -x$ and then $x'' = x' + a$, the particle ends up at $x'' = -x + a$, which is a different point. It is not surprising, then, that the momentum operator $\tilde{p}$ and the parity operator $\tilde{P}$ do not commute:

$$\begin{aligned} \tilde{P}\tilde{p}f(x) &= (-\tilde{p})f(-x) \\ &= -\tilde{p}\tilde{P}f(x) \\ &\neq \tilde{p}\tilde{P}f(x) \end{aligned} \tag{10.48}$$

Since

$$\left[\tilde{P}, \tilde{p}\right] \neq 0 \tag{10.49}$$

it follows that $\tilde{P}$ and $\tilde{p}$ cannot have the same set of eigenstates. There are two lessons to be learned from this:

1. When the order of symmetry transformations is important, then the operators associated with those transformations do not commute.
2. When two (or more) operators commute with the Hamiltonian, but not with each other, there are always degenerate energy eigenvalues.

10.3 The particle in a square

As a second example of the two principles stated above, we consider a two-dimensional problem: a particle moving freely inside a square of length L, whose lower left-hand corner is located at the point $x = 0,\ y = 0$. The symmetries of a square include: reflection of the x-axis around the point $x = L/2$:

$$x' = L - x \tag{10.50}$$

reflection of the y-axis around $y = L/2$

$$y' = L - y \tag{10.51}$$

and the interchange of coordinates

$$x' = y \qquad y' = x \tag{10.52}$$

which is equivalent to a reflection around the line $x = y$.

The Hamiltonian for a particle of mass m moving in the square is

$$\tilde{H} = -\frac{\hbar^2}{2m}\left(\frac{\partial^2}{\partial x^2} + \frac{\partial^2}{\partial y^2}\right) + V(x, y) \tag{10.53}$$

where

$$\begin{aligned} V(x, y) &= v(x) + v(y) \\ v(x) &= \begin{cases} 0 & 0 < x < L \\ \infty & \text{otherwise} \end{cases} \end{aligned} \tag{10.54}$$

The variables in the Hamiltonian are separable:

$$\begin{aligned} \tilde{H} &= \tilde{H}_x + \tilde{H}_y \\ \tilde{H}_x &= -\frac{\hbar^2}{2m}\frac{\partial^2}{\partial x^2} + v(x) \\ \tilde{H}_y &= -\frac{\hbar^2}{2m}\frac{\partial^2}{\partial y^2} + v(y) \end{aligned} \tag{10.55}$$

so the method of separation of variables

$$\phi(x, y) = \psi(x)\varphi(y) \tag{10.56}$$

can be applied to solve the time-independent Schrödinger equation

$$\tilde{H}\phi = E\phi \tag{10.57}$$

We obtain

$$\varphi(y)\tilde{H}_x\psi(x) + \psi(x)\tilde{H}_y\varphi(y) = E\psi(x)\varphi(y) \tag{10.58}$$

and dividing by $\phi(x, y)$ on both sides

$$\frac{\tilde{H}_x\psi(x)}{\psi(x)} + \frac{\tilde{H}_y\varphi(y)}{\varphi(y)} = E \tag{10.59}$$

Since the first ratio is only a function of x, and the second is only a function of y, the only way that the sum of the two ratios can be a constant, for all values of x and y, is if each ratio is a constant, i.e.

$$\begin{aligned} \tilde{H}_x\psi_n(x) &= E_n\psi_n(x) \\ \tilde{H}_y\varphi_m(y) &= E_m\varphi_m(y) \end{aligned} \tag{10.60}$$

In this way the two-dimensional equation has been reduced to two one-dimensional equations, each of which is identical to the Schrödinger equation for a particle in a tube, and which have the same solutions:

$$\begin{aligned} \psi_n(x) &= \sqrt{\frac{2}{L}}\sin\left[\frac{n\pi x}{L}\right] \qquad E_n = n^2\frac{\hbar^2\pi^2}{2mL^2} \\ \varphi_m(y) &= \sqrt{\frac{2}{L}}\sin\left[\frac{m\pi y}{L}\right] \qquad E_m = m^2\frac{\hbar^2\pi^2}{2mL^2} \end{aligned} \tag{10.61}$$

A complete set of energy eigenstates is then

$$\left\{\phi_{nm}(x, y) = \frac{2}{L}\sin\left[\frac{n\pi x}{L}\right]\sin\left[\frac{m\pi y}{L}\right],\ E_{nm} = \left(n^2 + m^2\right)\frac{\hbar^2\pi^2}{2mL^2}\right\} \tag{10.62}$$

The energies E_{nm} are two-fold degenerate for $n \neq m$, since

$$E_{nm} = E_{mn} \tag{10.63}$$

Since the Hamiltonian $\tilde{H}$ is invariant under reflections of the x-axis around the point $x = L/2$, and reflections of the y-axis around $y = L/2$, we define the corresponding operators

$$\begin{aligned} \tilde{R}_x f(x, y) &= f(L - x, y) \\ \tilde{R}_y f(x, y) &= f(x, L - y) \end{aligned} \tag{10.64}$$

It is easy to see that $\tilde{R}_x$ and $\tilde{R}_y$ commute,

$$\begin{aligned} \tilde{R}_x\tilde{R}_y f(x, y) &= \tilde{R}_y\tilde{R}_x f(x, y) \\ &= f(L - x, L - y)) \\ &\Longrightarrow \left[\tilde{R}_x, \tilde{R}_y\right] = 0 \end{aligned} \tag{10.65}$$

and that the energy eigenstates of equation (10.62) are eigenstates of both $\tilde{R}_x$ and $\tilde{R}_y$, with eigenvalues ± 1, since

$$\sin\left[\frac{n\pi(L - x)}{L}\right] = \begin{cases} 1 & n \ \text{odd} \\ -1 & n \ \text{even} \end{cases} \times \sin\left[\frac{n\pi x}{L}\right] \tag{10.66}$$

However, the Hamiltonian $\tilde{H}$ is also invariant under an interchange of the x- and y-coordinates

$$\tilde{\mathcal{I}} f(x, y) = f(y, x) \tag{10.67}$$

and this operator does *not* commute with $\tilde{R}_x$ and R_y:

$$\begin{aligned} \tilde{\mathcal{I}}\tilde{R}_x f(x, y) &= f(y, L - x) \\ \tilde{R}_x \tilde{\mathcal{I}} f(x, y) &= f(L - y, x) \end{aligned} \tag{10.68}$$

The eigenvalues of $\tilde{\mathcal{I}}$ are determined by the same reasoning as in the case of parity. Suppose ϕ_β is an eigenstate of $\mathcal{I}$ with eigenvalue β. Then

$$\tilde{\mathcal{I}}\tilde{\mathcal{I}}\phi_\beta(x, y) = \tilde{\mathcal{I}}\phi_\beta(y, x) = \phi_\beta(x, y) \tag{10.69}$$

but also

$$\tilde{\mathcal{I}}\tilde{\mathcal{I}}\phi_\beta(x, y) = \beta\tilde{\mathcal{I}}\phi_\beta(x, y) = \beta^2\phi_\beta(x, y) \tag{10.70}$$

Therefore

$$\beta^2 = 1 \implies \beta = \pm 1 \tag{10.71}$$

Because $\phi_{nm}(x, y)$ and $\phi_{mn}(x, y)$ have the same energy, so does any linear combination

$$\Phi(x, y) = a\phi_{nm}(x, y) + b\phi_{mn}(x, y) \tag{10.72}$$

Requiring Φ to be an eigenstate of $\tilde{\mathcal{I}}$ with eigenvalue $\beta = 1$ means that $\phi(x, y) = \phi(y, x)$, or

$$\begin{aligned} a \sin\left[\frac{n\pi y}{L}\right] \sin\left[\frac{m\pi x}{L}\right] + b \sin\left[\frac{m\pi y}{L}\right] \sin\left[\frac{n\pi x}{L}\right] \\ = a \sin\left[\frac{n\pi x}{L}\right] \sin\left[\frac{m\pi y}{L}\right] + b \sin\left[\frac{m\pi x}{L}\right] \sin\left[\frac{n\pi y}{L}\right] \end{aligned} \tag{10.73}$$

which is satisfied for $a = b$. Likewise, for Φ an eigenstate of $\tilde{\mathcal{I}}$ with eigenvalue $\beta = -1$, which means that $\phi(x, y) = -\phi(y, x)$ we need $a = -b$. Finally, the complete set of energy eigenstates which are also eigenstates of the x–y interchange operator $\tilde{\mathcal{I}}$ are

$$\begin{aligned} \Phi_{nm+} &= \frac{1}{\sqrt{2}}\left[\phi_{nm}(x, y) + \phi_{nm}(y, x)\right], \quad (n \neq m) \\ \Phi_{nm-} &= \frac{1}{\sqrt{2}}\left[\phi_{nm}(x, y) - \phi_{nm}(y, x)\right], \quad (n \neq m) \\ \Phi_{nn+} &= \phi_{nn}(x, y) \end{aligned} \tag{10.74}$$

Once again, we see that there are symmetry operations which commute with the Hamiltonian but do not commute with one another, and that the energy eigenvalues are degenerate. The choice of a complete set of energy eigenstates, (10.62) or (10.74), is determined by requiring that the energy eigenstates, in addition to being eigenstates of the Hamiltonian, are also eigenstates of one of the symmetry operators.

10.4 The quantum corral

The 'quantum corral' refers to a particle moving in the interior of a circle. The potential, in polar coordinates, is given by

$$V(r) = \begin{cases} 0 & r < R \\ \infty & r > R \end{cases} \tag{10.75}$$

which, since it depends only on the radial coordinate, is obviously invariant under arbitrary rotations

$$r' = r \qquad \theta' = \theta + \delta'\theta \tag{10.76}$$

Polar coordinates are related to Cartesian coordinates by

$$x = r\cos\theta \qquad y = r\sin\theta \tag{10.77}$$

and the Laplacian operator ∇^2, in two dimensions, is given in polar coordinates by

$$\begin{aligned} \nabla^2 &\equiv \frac{\partial^2}{\partial x^2} + \frac{\partial^2}{\partial y^2} \\ &= \frac{\partial^2}{\partial r^2} + \frac{1}{r}\frac{\partial}{\partial r} + \frac{1}{r^2}\frac{\partial^2}{\partial \theta^2} \end{aligned} \tag{10.78}$$

Under an arbitrary rotation of the form (10.76), $\partial/\partial\theta' = \partial/\partial\theta$, so the Laplacian is invariant. It follows that the Hamiltonian for a particle moving inside a circle

$$\tilde{H} = -\frac{\hbar^2}{2m}\left(\frac{\partial^2}{\partial r^2} + \frac{1}{r}\frac{\partial}{\partial r} + \frac{1}{r^2}\frac{\partial^2}{\partial \theta^2}\right) + V(r) \tag{10.79}$$

is invariant under rotations.

As in the previous examples, we introduce a rotation operator $\tilde{\mathcal{R}}_{\delta\theta}$ defined by

$$\tilde{\mathcal{R}}_{\delta\theta} f(r, \theta) = f(r, \theta + \delta\theta) \tag{10.80}$$

and, proceeding as in the case of the translation operator

$$\begin{aligned} f(r, \theta + \delta\theta) &= f(r, \theta) + \frac{\partial f}{\partial \theta}\delta\theta + \frac{1}{2}\frac{\partial^2 f}{\partial \theta^2} + \cdots \\ &= \exp\left[\delta\theta\frac{\partial}{\partial\theta}\right] f(r, \theta) \end{aligned} \tag{10.81}$$

so that

$$\tilde{\mathcal{R}}_{\delta\theta} = \exp\left[\delta\theta\frac{\partial}{\partial\theta}\right] \tag{10.82}$$

Since the Hamiltonian is invariant under rotations

$$\left[\tilde{\mathcal{R}}_{\delta\theta}, \tilde{H}\right] = 0 \tag{10.83}$$

it follows from (10.82) that

$$\left[\frac{\partial}{\partial\theta}, \tilde{H}\right] = 0 \tag{10.84}$$

Now, in Cartesian coordinates

$$\begin{aligned}
\frac{\partial}{\partial\theta} &= \frac{\partial x}{\partial\theta}\frac{\partial}{\partial x} + \frac{\partial y}{\partial\theta}\frac{\partial}{\partial y} \\
&= -r\sin\theta\frac{\partial}{\partial x} + r\cos\theta\frac{\partial}{\partial y} \\
&= x\frac{\partial}{\partial y} - y\frac{\partial}{\partial x} \\
&= \frac{1}{-\mathrm{i}\hbar}\left(x\tilde{p}_y - y\tilde{p}_x\right)
\end{aligned} \tag{10.85}$$

In three dimensions, angular momentum around the z-axis is given by

$$L_z = xp_y - yp_x \tag{10.86}$$

In two dimensions, this is the only possible component of angular momentum. The corresponding operator is $\tilde{L}_z = \tilde{x}\tilde{p}_y - \tilde{y}\tilde{p}_x$, and we see that the rotation operator is just the exponentiated angular momentum operator

$$\tilde{\mathcal{R}}_{\delta\theta} = \mathrm{e}^{\mathrm{i}\delta\theta\tilde{L}_z/\hbar} \tag{10.87}$$

and $\tilde{L}_z$ commutes with the Hamiltonian

$$\left[\tilde{L}_z, \tilde{H}\right] = 0 \tag{10.88}$$

It follows that

$$\frac{\mathrm{d}}{\mathrm{d}t}\langle L_z\rangle = 0 \tag{10.89}$$

In other words, symmetry of the potential with respect to rotations around some axis implies conservation of angular momentum around that axis.

By the commutator theorem, the operators $\tilde{L}_z$ and $\tilde{H}$ have a common set of eigenstates. Since

$$\tilde{L}_z = -\mathrm{i}\hbar\frac{\partial}{\partial\theta} \tag{10.90}$$

the Hamiltonian operator contains the angular momentum operator

$$\tilde{H} = -\frac{\hbar^2}{2m}\left(\frac{\partial^2}{\partial r^2} + \frac{1}{r}\frac{\partial}{\partial r} - \frac{1}{\hbar^2 r^2}\tilde{L}_z^2\right) + V(r) \tag{10.91}$$

so we begin by solving for the eigenstates of angular momentum

$$\begin{aligned}\tilde{L}_z\phi(r,\theta) &= \alpha\phi(r,\theta)\\ -\mathrm{i}\hbar\frac{\partial}{\partial\theta}\phi(r,\theta) &= \alpha\phi(r,\theta)\end{aligned} \tag{10.92}$$

which is similar to the momentum eigenvalue equation, and has a similar solution

$$\phi(r,\theta) = \varphi(r)\mathrm{e}^{\mathrm{i}\alpha\theta/\hbar} \tag{10.93}$$

There is one difference, however, between the eigenstates of momentum and angular momentum. Whereas the position $x + a$ is different from position x, for any non-zero constant a, the angle $\theta + 2\pi$ is identical to θ. This imposes an additional constraint

$$\phi(r,\theta + 2\pi) = \phi(r,\theta) \tag{10.94}$$

on the wave function of any physical state. In particular, for the angular momentum eigenstates (10.93), the constraint can only be satisfied if the eigenvalues α of angular momentum take on only discrete values

$$\alpha = n\hbar, \qquad n = 0, \pm 1, \pm 2, \pm 3, \ldots. \tag{10.95}$$

and these are the only values which could result from an accurate measurement. In short, *the values of angular momentum are 'quantized' in units of* $\hbar$. Essentially, it is a consequence of the wave-like properties of matter, combined with the 2π-periodicity of angles.[1] This periodicity also ensures that the L_z operator is Hermitian. Note that if L_z is Hermitian, then its eigenstates must be orthogonal. However, if $\mathrm{e}^{\mathrm{i}\alpha\theta/\hbar}$, $\mathrm{e}^{\mathrm{i}\beta\theta/\hbar}$ are two eigenstates with $\alpha \neq \beta$, it is easy to see that their inner product is zero only if $\alpha - \beta$ is an integer multiple of $2\pi\hbar$.

Denote an eigenstate of angular momentum by

$$\phi_n(r,\theta) = \varphi_n(r)\mathrm{e}^{\mathrm{i}n\theta} \tag{10.96}$$

and substitute into the time-independent Schrödinger equation,

$$\left[-\frac{\hbar^2}{2m}\left(\frac{\partial^2}{\partial r^2} + \frac{1}{r}\frac{\partial}{\partial r} - \frac{1}{\hbar^2 r^2}L_z^2\right) + V(r)\right]\phi_n = E_n\phi_n \tag{10.97}$$

Using the fact that

$$\tilde{L}_z\phi_n = n\hbar\phi_n \tag{10.98}$$

[1] Actually, this 2π periodicity of the wave function is more subtle than it appears. In relativistic quantum mechanics, the wave function of an electron obeys a relativistic version of the Schrödinger equation known as the Dirac equation. This wave function has a 4π, rather than a 2π, angular periodicity. How that is possible is a little beyond the scope of this text.

the Schrödinger equation becomes

$$\left[-\frac{\hbar^2}{2m}\left(\frac{\partial^2}{\partial r^2}+\frac{1}{r}\frac{\partial}{\partial r}-\frac{n^2}{r^2}\right)+V(r)\right]\varphi_n(r)=E_n\varphi_n(r) \tag{10.99}$$

As in the case of the infinite square well, the wave function must be zero where $V(r)=\infty$, i.e.

$$\varphi(r)=0 \text{ when } r\geqslant R \tag{10.100}$$

Inside the circle, at $r\leqslant R$, the Schrödinger equation is

$$\left[\frac{\partial^2}{\partial r^2}+\frac{1}{r}\frac{\partial}{\partial r}+\left(\frac{2mE}{\hbar^2}-\frac{n^2}{r^2}\right)\right]\varphi_n(r)=0 \tag{10.101}$$

Define

$$\begin{aligned} k^2 &= \frac{2mE}{\hbar^2} \\ \rho &= kr \end{aligned} \tag{10.102}$$

In terms of the rescaled radial coordinate ρ, the time-independent equation becomes

$$\left[\frac{\partial^2}{\partial \rho^2}+\frac{1}{\rho}\frac{\partial}{\partial \rho}+\left(1-\frac{n^2}{\rho^2}\right)\right]\varphi_n=0 \tag{10.103}$$

This is one of the standard differential equations of mathematical physics, known as *Bessel's equation*. The solutions are the Bessel functions, denoted $J_n(\rho)$, and Neumann functions, $N_n(\rho)$. The Neumann functions are infinite, and therefore non-differentiable with respect to x and y, at $r=0$. So Neumann functions do not correspond to physical states, and can be discarded. We are therefore left with solutions

$$\varphi_n(r)=J_n(kr) \text{ with energy } E=\frac{\hbar^2k^2}{2m} \tag{10.104}$$

This is a bound state solution, since the particle is bound by the potential inside a circle, and we know that the energies of bound states are discrete. In this case, the possible energies are determined by the condition that

$$\phi_n(R,\theta)=0\Longrightarrow\varphi_n(R)=0\Longrightarrow J_n(kR)=0 \tag{10.105}$$

Bessel functions are oscillating functions (wave functions in classically allowed regions oscillate), so the constant k must be chosen such that kR is a 'zero' of the Bessel function J_n. Denote

$$J_n(x_{nj})=0 \tag{10.106}$$

where $\{x_{n1}, x_{n2}, x_{n3}, \ldots\}$ are the zeros of the nth Bessel function. In particular

$$\begin{aligned} x_{0j} &= 2.405,\ 5.520,\ 8.654,\ \ldots \\ x_{1j} &= 3.832,\ 7.016,\ 10.173,\ \ldots \end{aligned} \tag{10.107}$$

and the possible energies must therefore satisfy

$$kR = x_{nj} \qquad E_{nj} = \frac{\hbar^2 k^2}{2m} \tag{10.108}$$

Finally, the complete set of energy eigenstates and eigenvalues are

$$\left\{\phi_{nj} = J_n\left(\frac{x_{nj}}{R}r\right)e^{in\theta}, \qquad E_{nj} = \frac{\hbar^2 x_{nj}^2}{2mR^2}, \qquad L_z = n\hbar\right\} \tag{10.109}$$

Note that the energy eigenvalues are degenerate, $E_{nj} = E_{(-n)j}$. This means that there must be at least two operators which commute with the Hamiltonian, but not with each other. In this case, an operator which commutes with $\tilde{H}$, but not with $\tilde{L}_z$, is the reflection operator

$$\tilde{P}_\theta f(r, \theta) = f(r, -\theta) \tag{10.110}$$

corresponding to the transformation $\theta' = -\theta$.

Very recent advances in technology, specifically, the invention of the scanning tunneling microscope, and the development of techniques for manipulating matter at the atomic scale, have made it possible to actually trap electrons in a circle, and measure the square modulus of the wave function. The result, displayed on the front cover of this text, is breathtaking. The peaks arranged in a circle show the electron clouds of iron atoms, which are used to form the 'corral'. The circular ripples inside the 'corral' show the wave function (modulus squared) of the trapped electrons.

10.5 Landau levels

Consider a particle of charge q and mass m moving under the influence of an electromagnetic field. The classical equations of motion of a particle of mass m and charge q are given by the Lorentz force law

$$\mathbf{F} = q\mathbf{E} - \frac{q}{c}\mathbf{v} \times \mathbf{B} \tag{10.111}$$

plus Newton's second law $\mathbf{F} = m\mathbf{a}$. We will assume that $\mathbf{E} = 0$, that the particle is constrained to move in the x–y plane, and $\mathbf{B}$ is constant in space and time, and points in the positive z-direction. Since the force on the particle is always orthogonal to the velocity, it serves as a centripetal force which keeps the particle moving in a circle of fixed radius, with an angular frequency—the ***cyclotron frequency***—of qB/mc. This is the classical solution. We now investigate the quantum dynamics.

To begin, we need the Hamiltonian. It is not hard to show that Newton's second law supplemented by the Lorentz force law follows Hamilton's equations applied to the following Hamiltonian:

$$H = \frac{1}{2m}\left(\mathbf{p} - \frac{q}{c}\mathbf{A}\right) \cdot \left(\mathbf{p} - \frac{q}{c}\mathbf{A}\right) + \frac{q}{c}\phi \tag{10.112}$$

$\mathbf{A}$ and ϕ are the vector and electric potentials, related to $\mathbf{E}$, $\mathbf{B}$ via

$$\begin{aligned} \mathbf{E} &= -\frac{1}{c}\frac{\partial \mathbf{A}}{\partial t} - \nabla\phi \\ \mathbf{B} &= \nabla \times \mathbf{A} \end{aligned} \tag{10.113}$$

The vector potential $\mathbf{A}$ is not uniquely determined by $\mathbf{B}$, because transforming $\mathbf{A}$ by a gauge transformation

$$\mathbf{A} \rightarrow \mathbf{A} - \nabla\chi \tag{10.114}$$

where $\chi(x, y, z)$ is an arbitrary function, does not change the magnetic field strength $\mathbf{B}$. In our case $\mathbf{B} = B\hat{\mathbf{e}}_z$ is a position-independent vector in the z-direction, so

$$B = \partial_x A_y - \partial_y A_x \tag{10.115}$$

and for now it is convenient, among all the $\mathbf{A}$ fields which satisfy this condition, to choose the particular solution

$$A_x = 0, \qquad A_y = Bx \tag{10.116}$$

Keep in mind that this choice is a matter of convenience. Nothing essential changes if $\mathbf{A}$ is altered by a gauge transformation. The Hamiltonian operator is then

$$\tilde{H} = \frac{1}{2m}\left(\tilde{p}_x^2 + \tilde{p}_y^2 - 2\frac{q}{c}Bx\tilde{p}_y + \frac{q^2}{c^2}B^2x^2\right) \tag{10.117}$$

This Hamiltonian is obviously invariant under translations $y \rightarrow y + a$ in the y-direction, and for this reason

$$\left[\tilde{H}, \tilde{p}_y\right] = 0 \tag{10.118}$$

That means, by the commutator theorem, that eigenstates of $\tilde{H}$, denoted $\varphi_{kn}(x, y)$, can also be chosen to be eigenstates of $\tilde{p}_y$, i.e.

$$\tilde{H}\varphi_{kn}(x, y) = E_{kn}\varphi_{kn}(x, y), \qquad \tilde{p}_y\varphi_{kn}(x, y) = \hbar k\varphi_{kn}(x, y) \tag{10.119}$$

We can solve the second eigenvalue equation immediately, by separation of variables:

$$\varphi_{kn}(x, y) = \frac{1}{\sqrt{2\pi\hbar}}e^{iky}\phi_{nk}(x), \qquad -\infty < k < \infty \tag{10.120}$$

and substituting this into the time-independent Schrödinger equation we find

$$\frac{1}{2m}\left(\tilde{p}_x^2 + k^2 - 2\frac{q}{c}Bxk + \frac{q^2}{c^2}B^2x^2\right)\phi_{kn}(x) = E_{nk}\phi_{kn}(x) \tag{10.121}$$

or

$$\frac{1}{2m}\left[\tilde{p}_x^2 + \frac{q^2}{c^2}B^2(x - X_k)^2\right]\phi_{kn}(x) = E_{nk}\phi_{kn}(x) \tag{10.122}$$

where $X_k = \frac{kc}{qB}$. This is the equation for a one-dimensional harmonic oscillator centered at $x = X_k$ with a spring constant q^2B^2/mc^2, and the solution is simply

$$\phi_{kn}(x) = \phi_n(x - X_k) \tag{10.123}$$

where the $\phi_n(x)$ are harmonic oscillator eigenstates. Altogether

$$\begin{aligned} \varphi_{kn}(x, y) &= \frac{1}{\sqrt{2\pi\hbar}}\mathrm{e}^{iky}\phi_n(x - X_k), \qquad X_k = \frac{kc}{qB} \\ E_{kn} = E_n &= \hbar\omega\left(n + \frac{1}{2}\right), \qquad \omega = \frac{qB}{mc} \end{aligned} \tag{10.124}$$

The harmonic oscillator energy levels E_n are in this case known as ***Landau levels***, and they are of great importance in solid state physics, particularly in connection with the ***quantum Hall effect***. But the point for now is that each Landau level has an enormous degeneracy, since $E_{kn} = E_n$ depends only on the integer n, and not the continuous eigenvalue $\hbar k$ of p_y. The question is: what is the origin of this degeneracy? To answer that question, let us compute the commutators of $\tilde{H}$ with $\tilde{y}$ and $\tilde{p}_x$:

$$\begin{aligned} \left[\tilde{H}, \tilde{y}\right] &= \frac{i\hbar}{m}\left(\frac{q}{c}B\tilde{x} - \tilde{p}_y\right) \\ \left[\tilde{H}, \tilde{p}_x\right] &= \frac{i\hbar}{m}\left(\frac{q^2}{c^2}B^2x - \frac{q}{c}B\tilde{p}_y\right) = \frac{q}{c}B\left[\tilde{H}, \tilde{y}\right] \end{aligned} \tag{10.125}$$

It follows that the Hermitian operator

$$\tilde{Q} = \tilde{p}_x - \frac{q}{c}B\tilde{y} \tag{10.126}$$

commutes with the Hamilton: $[\tilde{H}, \tilde{Q}] = 0$. But $\tilde{p}_y$ is also a Hermitian operator, it also commutes with the Hamiltonian, and it does *not* commute with $\tilde{Q}$. This is exactly the situation which is covered by a mantra which summarizes the chapter: if $\tilde{H}$ commutes with two or more operators which do not commute with each other, then the eigenvalues of $\tilde{H}$ must be degenerate. And so they are. In this case, while one of the operators is associated with a translation symmetry, the other operator is not obviously associated with a coordinate transformation. But the mantra holds whether or not the two operators correspond to purely spatial symmetries.

Of course, in a real metal of finite area the degeneracy is not continuous because the wave function $\varphi_{kn}(x, y)$ is constrained to vanish at the boundary. To calculate the degeneracy at each Landau level, suppose the metal is a thin rectangular sheet, with length L_x in the x-direction ($0 \leqslant x \leqslant L_x$), and width L_y in the y-direction ($0 \leqslant y \leqslant L_y$). To satisfy the boundary condition that $\varphi_{kn}(x, 0) = \varphi_{kn}(x, L_y)$ we superimpose, as in the infinite square well, two eigenstates with $p_y = \pm\hbar k$, and find

$$\varphi_{kn}(x, y) = \sqrt{\frac{2}{L_y}}\sin(ky)\phi_n(x - X_k), \qquad k = \frac{\pi j}{L_y}, \; j = 1, 2, 3, \ldots, j_{\max} \qquad (10.127)$$

In a finite square well, there is no bound on the integer j. But here there is such a bound, because the center of the harmonic oscillator in the x-direction, namely X_k, must lie between 0 and L_x. Then

$$\begin{aligned} X_k^{\max} &= \frac{k_{\max} c}{qB} = \frac{\pi j_{\max} c}{qBL_y} \\ &= L_x \end{aligned} \qquad (10.128)$$

Solving for $j_{\max}$ gives the number of states at each Landau level, i.e. the energy degeneracy at each level

$$j_{\max} = \frac{1}{\pi}\frac{q}{c}B \times \text{Area} \qquad (10.129)$$

where Area $= L_x \times L_y$ is the area of the metal sheet.

Suppose there are a fixed number of electrons n_e in the sheet. Then for certain discrete values of B (and ignoring electron spin and magnetic moment) the ratio $n_e/j_{\max}$ is an integer, and, because of the Pauli exclusion principle to be discussed in a later chapter, only one electron per quantum state is allowed. This means that $n_e/j_{\max}$ Landau levels are completely filled, and there is a gap of energy $\hbar\omega$ to the first unoccupied level. This is associated with a sudden drop in conductivity known as the ***integer quantum Hall effect***.[2]

10.6 Complete sets of observables

To summarize the lessons of this chapter:

1. If two (or more) Hermitian operators commute with the Hamiltonian, but not with each other, then the energy eigenvalues of the Hamiltonian must be degenerate.

   Reason, If H commutes with A and B, then H and A share a common set of eigenstates, and H and B share a common set of eigenstates. But if A does not commute with B, these two sets of eigenstates must be different. However, if the eigenvalues of H were non-degenerate, then there is only one possible choice of energy eigenstates. Since there are (at least) two

[2] There is also an effect when the Landau levels are only fractionally filled, known as the ***fractional quantum Hall effect*** (see, e.g., Stormer [1]), which is beyond the scope of this text.

different sets of energy eigenstates, the energy eigenvalues must be degenerate.

Example, The free particle Hamiltonian commutes with momentum p and parity P, but $[p, P] \neq 0$. The common set of eigenstates of H and p is

$$\left\{ e^{i\sqrt{2mE}x/\hbar}, \quad e^{-i\sqrt{2mE}x/\hbar}, \qquad E \in [0, \infty] \right\} \tag{10.130}$$

and the common set of eigenvalues of H and P is

$$\left\{ \cos[\sqrt{2mE}x/\hbar], \quad \sin[\sqrt{2mE}x/\hbar], \qquad E \in \left[0, \infty\right] \right\} \tag{10.131}$$

Each of these sets is a ***complete set of states***, in the sense that any energy eigenstate can be written as a linear combination of the eigenstates of either set. Then, by theorem H3, *any* wave function $\phi(x)$ can be written as a sum of eigenstates of either set.

2. Symmetries of the Hamiltonian imply conservation laws.

   Reason, Symmetry under coordinate transformations can be associated with operators which produce those transformations; these operators commute with any Hamiltonian which is invariant under the given transformation. Operators which commute with the Hamiltonian are conserved, according to equation (10.20).

   Examples, Symmetry of the free particle Hamiltonian under translations implies conservation of momentum; symmetry of the Hamiltonian under rotations, e.g. for a particle bound inside a circle, implies conservation of angular momentum.

3. As a general rule, symmetry operations do not commute. Therefore, as a consequence of item 1, the more symmetric the Hamiltonian, the greater is the degeneracy in its energy eigenvalues.

   Example, It will be seen in the next chapter that the energy degeneracy of a spherically symmetric Hamiltonian in three dimensions is greater than the energy degeneracy of a rotationally invariant Hamiltonian in two dimensions.

Items 1 and 3 actually hold true not just for the Hamiltonian, but for all operators. If A commutes with B and C, but B and C do not commute with each other, then the eigenvalues of A are degenerate. The more symmetries an operator has, the greater is the degeneracy in its eigenvalues.

Suppose the operators A, B, C commute, and that the set of eigenstates of A which are also eigenstates of B and C is unique. Then the set $\{A, B, C\}$ is known as a ***complete set of observables***. Measurement of a complete set of observables is sufficient to uniquely determine the quantum state (which must be an eigenstate of all operators in the set). Examples:

a) Momentum p is a complete set of observables. In fact, *any* operator with non-degenerate eigenvalues is a complete set.

b) Energy E is a complete set, for the harmonic oscillator and square well potentials.

c) Energy E and parity P are a complete set for the free particle.
d) Energy E and reflection R_x are a complete set for the particle in a square. So is E and the observable corresponding to x–y interchange $\mathcal{I}$.
e) Energy E and angular momentum L_z are a complete set for the 'quantum corral'.
f) Energy E and p_y are a complete set of observables for a particle moving in a plane, subject to a constant magnetic field in the orthogonal direction.

Problems

1. Prove that the parity operator is Hermitian, using the definition of Hermitian conjugation to determine $\tilde{P}^{\dagger}$.
2. A particle is moving in a rectangular box, in the region $0 < x < L_1$, $0 < y < L_2$, with $L_1 \neq L_2$, and the potential is zero inside and infinite outside the box. What symmetries can you find? Solve for the eigenvalues and eigenstates. Are the energies degenerate? Why or why not?
3. A particle of mass m is moving, in two dimensions, in the potential

$$V(x, y) = \frac{1}{2}k\left(x^2 + y^2\right) \tag{10.133}$$

   a) Using separation of variables in Cartesian (x, y)-coordinates, find all the energy eigenvalues and corresponding eigenstates (you do not have to write the eigenstates explicitly, but may express them in terms of the eigenstates of the one-dimensional harmonic oscillator). What are the energy degeneracies of the lowest four energy eigenvalues?
   b) Define the x-parity and y-parity operators

$$\begin{aligned} \tilde{P}_x f(x, y) &= f(-x, y) \\ \tilde{P}_y f(x, y) &= f(x, -y) \end{aligned} \tag{10.134}$$

   State the eigenvalues of P_x and P_y for each energy eigenstate.
   c) Suppose an impenetrable barrier is erected along the x- and y-axes, so that the wave function must vanish along each axis. What subset of the energy eigenstates in part (a) would satisfy this boundary condition? (Hint: use the results of part (b).)
   d) Degeneracy implies that there must be some operator which commutes with the Hamiltonian, but not with $\tilde{P}_x$ and $\tilde{P}_y$. From the symmetry of the potential, can you think of such an operator?

Reference

[1] Stormer H L 1999 Nobel lecture: the fractional quantum hall effect *Rev. Mod. Phys.* **71** 875–89

IOP Publishing

An Introduction to Quantum Theory

Jeff Greensite

Chapter 11

Angular momentum

When the potential energy in three dimensions depends only on the radial coordinate, the potential is said to be 'spherically symmetric', and the Hamiltonian is invariant under arbitrary rotations; in particular, the Hamiltonian is invariant under rotations around the x-, y-, and z-axes. In the last chapter we found that for rotations around the origin in the x–y plane, the rotation operator is

$$\tilde{\mathcal{R}}_z(\delta\theta) = \mathrm{e}^{\mathrm{i}\tilde{L}_z\delta\theta/\hbar} \tag{11.1}$$

There is nothing special about the z-axis; rotations around the x- and y-axes must have the corresponding forms

$$\begin{aligned} \tilde{\mathcal{R}}_x(\delta\theta) &= \mathrm{e}^{\mathrm{i}\tilde{L}_x\delta\theta/\hbar} \\ \tilde{\mathcal{R}}_y(\delta\theta) &= \mathrm{e}^{\mathrm{i}\tilde{L}_y\delta\theta/\hbar} \end{aligned} \tag{11.2}$$

In the case of rotations around different axes, the order of rotations is important. For example, consider a cylindrical object lying along the z-axis, as shown in figure 11.1, and imagine doing first a rotation by 90° around the z-axis, and then a rotation by 90° along the y-axis. The first rotation, around the z-axis, leaves the object oriented along the z-axis. The second rotation, around the y-axis, brings the object to a final position along the x-axis. Now reverse the order of coordinate transformations: a rotation around the y-axis takes the object to the x-axis, and after the subsequent rotation around the z-axis the object lies along the y-axis. Thus the two rotations, performed in different orders, have different outcomes.

It follows that, although we expect the angular momentum operators to commute with a symmetric Hamiltonian, we do *not* expect the $\tilde{L}_x$, $\tilde{L}_y$, $\tilde{L}_z$ operators to commute with each other. The energy eigenvalues will therefore be degenerate, angular momentum is conserved, and the angular momentum operators can be used to select among different complete sets of energy eigenstates. In this chapter, we will

doi:10.1088/978-0-7503-1167-0ch11

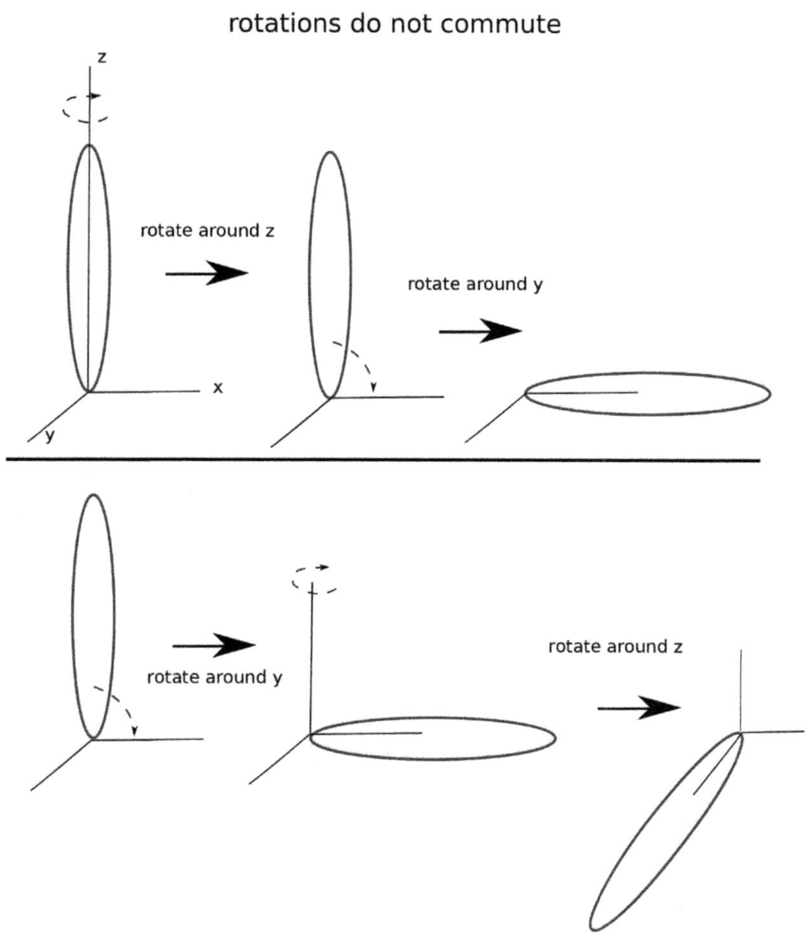

Figure 11.1. Rotations do not commute. In the upper diagrams a cylindrical object is rotated first by 90° around the z-axis, and then by 90° around the y-axis. In the lower diagrams, the order of rotations is reversed.

learn how to construct eigenstates of angular momentum, which is a necessary step towards solving the Schrödinger equation for spherically symmetric potentials.

11.1 The angular momentum commutators

Angular momentum in classical physics is defined as

$$\mathbf{L} = \mathbf{r} \times \mathbf{p} \tag{11.3}$$

or, in components,

$$\begin{aligned} L_x &= yp_z - zp_y \\ L_y &= zp_x - xp_z \\ L_z &= xp_y - yp_x \end{aligned} \tag{11.4}$$

The corresponding quantum-mechanical operators are obtained by replacing p with $\tilde{p}$, i.e.

$$\begin{aligned}
\tilde{L}_x &= y\tilde{p}_z - z\tilde{p}_y = -\mathrm{i}\hbar\left\{y\frac{\partial}{\partial z} - z\frac{\partial}{\partial y}\right\} \\
\tilde{L}_y &= z\tilde{p}_x - x\tilde{p}_z = -\mathrm{i}\hbar\left\{z\frac{\partial}{\partial x} - x\frac{\partial}{\partial z}\right\} \\
\tilde{L}_z &= x\tilde{p}_y - y\tilde{p}_x = -\mathrm{i}\hbar\left\{x\frac{\partial}{\partial y} - y\frac{\partial}{\partial x}\right\}
\end{aligned} \tag{11.5}$$

Spherical coordinates are related to Cartesian coordinates by

$$\begin{aligned}
z &= r\cos\theta \\
x &= r\sin\theta\cos\phi \\
y &= r\sin\theta\sin\phi
\end{aligned} \tag{11.6}$$

and a spherically symmetric potential, also known as a ***central potential***, is a function which is independent of the angular coordinates θ, ϕ, i.e.

$$V(r, \theta, \phi) = V(r) \tag{11.7}$$

We have already deduced that the Hamiltonian should commute with the angular momentum operators, simply because the Hamiltonian of a particle in a central potential is invariant under a rotation of coordinates, and the operators which generate rotations are the exponential of the angular momentum operators. These commutators can of course be checked directly, e.g. in the case of the z-component of angular momentum we have

$$\begin{aligned}
\left[\tilde{L}_z, \tilde{H}\right] &= \left[\tilde{L}_z, \frac{1}{2m}\tilde{p}^2 + V(r)\right] \\
&= \frac{1}{2m}\left[\tilde{L}_z, \tilde{p}^2\right] + \left[\tilde{L}_z, V(r)\right]
\end{aligned} \tag{11.8}$$

where

$$\begin{aligned}
\left[\tilde{L}_z, V(r)\right] &= x\left[\tilde{p}_y, V(r)\right] - y\left[\tilde{p}_x, V(r)\right] \\
&= -\mathrm{i}\hbar\left(x\frac{\partial V}{\partial y} - y\frac{\partial V}{\partial x}\right) \\
&= -\mathrm{i}\hbar\left(x\frac{\partial V}{\partial r}\frac{\partial r}{\partial y} - y\frac{\partial V}{\partial r}\frac{\partial r}{\partial x}\right) \\
&= -\mathrm{i}\hbar\left(x\frac{\partial V}{\partial r}\frac{y}{r} - y\frac{\partial V}{\partial r}\frac{x}{r}\right) \\
&= 0
\end{aligned} \tag{11.9}$$

and

$$\begin{aligned}\left[\tilde{L}_z, \tilde{p}^2\right] &= \left[\left(x\tilde{p}_y - y\tilde{p}_x\right), \tilde{p}_x^2 + \tilde{p}_y^2 + \tilde{p}_z^2\right] \\ &= \left[x\tilde{p}_y, \tilde{p}_x^2\right] - \left[y\tilde{p}_x, \tilde{p}_y^2\right] \\ &= \tilde{p}_y\left[x, \tilde{p}_x^2\right] - \tilde{p}_x\left[y, \tilde{p}_y^2\right] \\ &= 2i\hbar\tilde{p}_y\tilde{p}_x - 2i\hbar\tilde{p}_x\tilde{p}_y \\ &= 0\end{aligned} \tag{11.10}$$

This proves that

$$\left[\tilde{L}_z, \tilde{H}\right] = 0 \tag{11.11}$$

Similar steps show that

$$\left[\tilde{L}_x, \tilde{H}\right] = \left[\tilde{L}_y, \tilde{H}\right] = 0 \tag{11.12}$$

which means that *in a central potential, angular momentum is conserved.*

However, since the order of rotations around the x-, y-, and z-axes are important, the operators $\tilde{L}_x$, $\tilde{L}_y$, $\tilde{L}_z$ cannot commute among themselves. For example,

$$\begin{aligned}\left[\tilde{L}_x, \tilde{L}_y\right] &= \left[\left(y\tilde{p}_z - z\tilde{p}_y\right), \left(z\tilde{p}_x - x\tilde{p}_z\right)\right] \\ &= \left[y\tilde{p}_z, z\tilde{p}_x\right] + \left[z\tilde{p}_y, x\tilde{p}_z\right] \\ &= y\tilde{p}_x\left[\tilde{p}_z, z\right] + x\tilde{p}_y\left[z, \tilde{p}_z\right] \\ &= i\hbar\left(-y\tilde{p}_x + x\tilde{p}_y\right) \\ &= i\hbar\tilde{L}_z\end{aligned} \tag{11.13}$$

Altogether, we have the

Angular momentum commutation relations

$$\begin{aligned}\left[\tilde{L}_x, \tilde{L}_y\right] &= i\hbar\tilde{L}_z \\ \left[\tilde{L}_y, \tilde{L}_z\right] &= i\hbar\tilde{L}_x \\ \left[\tilde{L}_z, \tilde{L}_x\right] &= i\hbar\tilde{L}_y\end{aligned} \tag{11.14}$$

These angular momentum commutators are the fundamental equations that will be applied in this chapter. They will be used to determine both the eigenvalues and the eigenstates of angular momentum, by an algebraic method which closely resembles the treatment of the harmonic oscillator.

The fact that the angular momentum operators commute with the Hamiltonian, but not with each other, means that the energy eigenvalues of the Hamiltonian are degenerate. It also means that all three components of angular momentum cannot be measured simultaneously, which tells us that there does not exist any physical state in which the direction of angular momentum is definite. However, despite the fact that the *direction* of angular momentum is indefinite, there do exist states in which the *magnitude* of angular momentum is definite; these are eigenstates of the squared angular momentum operator

$$\tilde{L}^2 = \tilde{L}_x^2 + \tilde{L}_y^2 + \tilde{L}_z^2 \tag{11.15}$$

Each of the components of angular momentum commutes with $\tilde{L}^2$. For example,

$$\begin{aligned}\left[\tilde{L}_z, \tilde{L}^2\right] &= \left[\tilde{L}_z, \tilde{L}_x^2 + \tilde{L}_y^2 + \tilde{L}_z^2\right] \\ &= \left[\tilde{L}_z, \tilde{L}_x^2\right] + \left[\tilde{L}_z, \tilde{L}_y^2\right]\end{aligned} \tag{11.16}$$

We evaluate the first commutator with the help of the relations in equation (11.14)

$$\begin{aligned}\left[\tilde{L}_z, \tilde{L}_x^2\right] &= \tilde{L}_z\tilde{L}_x\tilde{L}_x - \tilde{L}_x\tilde{L}_x\tilde{L}_z \\ &= \left(\tilde{L}_x\tilde{L}_z + [\tilde{L}_z, \tilde{L}_x]\right)\tilde{L}_x - \tilde{L}_x^2\tilde{L}_z \\ &= \tilde{L}_x\tilde{L}_z\tilde{L}_x + \mathrm{i}\hbar\tilde{L}_y\tilde{L}_x - \tilde{L}_x^2\tilde{L}_z \\ &= \tilde{L}_x\left(\tilde{L}_x\tilde{L}_z + [\tilde{L}_z, \tilde{L}_x]\right) + \mathrm{i}\hbar\tilde{L}_y\tilde{L}_x - \tilde{L}_x^2\tilde{L}_z \\ &= \mathrm{i}\hbar\left(\tilde{L}_x\tilde{L}_y + \tilde{L}_y\tilde{L}_x\right)\end{aligned} \tag{11.17}$$

The second commutator gives a similar result:

$$\begin{aligned}\left[\tilde{L}_z, \tilde{L}_y^2\right] &= \tilde{L}_z\tilde{L}_y\tilde{L}_y - \tilde{L}_y\tilde{L}_y\tilde{L}_z \\ &= \left(\tilde{L}_y\tilde{L}_z + \left[\tilde{L}_z, \tilde{L}_y\right]\right)\tilde{L}_y - \tilde{L}_y^2\tilde{L}_z \\ &= \tilde{L}_y\tilde{L}_z\tilde{L}_y - \mathrm{i}\hbar\tilde{L}_x\tilde{L}_y - \tilde{L}_x^2\tilde{L}_z \\ &= \tilde{L}_y\left(\tilde{L}_y\tilde{L}_z + \left[\tilde{L}_z, \tilde{L}_y\right]\right) - \mathrm{i}\hbar\tilde{L}_x\tilde{L}_y - \tilde{L}_y^2\tilde{L}_z \\ &= -\mathrm{i}\hbar\left(\tilde{L}_y\tilde{L}_x + \tilde{L}_x\tilde{L}_y\right) \\ &= -\left[\tilde{L}_z, \tilde{L}_x^2\right]\end{aligned} \tag{11.18}$$

Adding the two commutators together gives zero, and therefore $\tilde{L}_z$ commutes with $\tilde{L}^2$. The same result is obtained for the other components of angular momentum:

$$\left[\tilde{L}_x, \tilde{L}^2\right] = \left[\tilde{L}_y, \tilde{L}^2\right] = [\tilde{L}_z, \tilde{L}^2] = 0 \tag{11.19}$$

11.2 The eigenvalues of angular momentum

The energy eigenvalues of $\tilde{H}$, for a particle in a central potential, are degenerate, since $\tilde{H}$ commutes with the components of angular momentum, which do not commute with each other. The same can be said for the eigenvalues of $\tilde{L}^2$. These eigenvalues must be degenerate, because $\tilde{L}^2$ also commutes with $\tilde{L}_x$, $\tilde{L}_y$, $\tilde{L}_z$, which do commute with each other. By the commutator theorem $\tilde{L}^2$ and $\tilde{L}_x$ have a common set of eigenstates; so do $\tilde{L}^2$ and $\tilde{L}_y$, and also $\tilde{L}^2$ and $\tilde{L}_z$. But, again by the commutator theorem, these must be *different* sets of eigenstates.

Denote the eigenstates of $\tilde{L}^2$ and, say, $\tilde{L}_z$, as φ_{ab}. Angular momentum has units of action, i.e. energy × time, which are the same units as Planck's constant $\hbar$. So we may as well write the eigenvalues of angular momentum in units of $\hbar$ where

$$\begin{aligned} \tilde{L}^2\varphi_{ab} &= a^2\hbar^2\varphi_{ab} \\ \tilde{L}_z\varphi_{ab} &= b\hbar\varphi_{ab} \end{aligned} \tag{11.20}$$

where a and b are dimensionless numbers. The eigenvalues of $\tilde{L}_z$ may be positive or negative, but the eigenvalues of $\tilde{L}^2$ are necessarily positive, because $\tilde{L}^2$ is a sum of squares of Hermitian operators. Therefore,

$$\begin{aligned} \left\langle \varphi_{ab}\middle|\tilde{L}^2\middle|\varphi_{ab}\right\rangle &= \left\langle \varphi_{ab}\middle|\tilde{L}_x^2\middle|\varphi_{ab}\right\rangle + \left\langle \varphi_{ab}\middle|L_y^2\middle|\varphi_{ab}\right\rangle + \left\langle \varphi_{ab}\middle|L_z^2\middle|\varphi_{ab}\right\rangle \\ a^2\hbar^2\langle\varphi_{ab}|\varphi_{ab}\rangle &= \left\langle \varphi_{ab}\middle|\tilde{L}_x^2\middle|\varphi_{ab}\right\rangle + \left\langle \varphi_{ab}\middle|\tilde{L}_y^2\middle|\varphi_{ab}\right\rangle + \hbar^2b^2\langle\varphi_{ab}|\varphi_{ab}\rangle \\ a^2\hbar^2 &= \langle\tilde{L}_x\varphi_{ab}|\tilde{L}_x\varphi_{ab}\rangle + \left\langle\tilde{L}_y\varphi_{ab}\middle|\tilde{L}_y\varphi_{ab}\right\rangle + \hbar^2b^2 \\ a^2\hbar^2 &= \hbar^2b^2 + \left|\tilde{L}_x\varphi_{ab}\right|^2 + \left|\tilde{L}_y\varphi_{ab}\right|^2 \end{aligned} \tag{11.21}$$

which establishes that the eigenvalues of $\tilde{L}^2$ are positive semi-definite, and also that

$$-a \leqslant b \leqslant a \tag{11.22}$$

The fact that $\tilde{L}^2$ is a sum of squares of Hermitian operators, which commute in a simple way among themselves (equation (11.14)), suggests trying the same trick that we used for the harmonic oscillator; i.e. the trick of introducing raising and lowering operators. So we define, in analogy to $\tilde{a}^\dagger$, $\tilde{a}$

$$\begin{aligned} \tilde{L}_+ &\equiv \tilde{L}_x + \mathrm{i}\tilde{L}_y \\ \tilde{L}_- &\equiv \tilde{L}_x - \mathrm{i}\tilde{L}_y \end{aligned} \tag{11.23}$$

where $\tilde{L}_-$ is the Hermitian conjugate of $\tilde{L}_+$, and express $\tilde{L}^2$ in terms of these operators,

$$\begin{aligned} \tilde{L}^2 &= \left(\tilde{L}_x^2 + \tilde{L}_y^2\right) + \tilde{L}_z^2 \\ &= \left(\tilde{L}_+\tilde{L}_- + \mathrm{i}\left[\tilde{L}_x, \tilde{L}_y\right]\right) + \tilde{L}_z^2 \\ &= \tilde{L}_+\tilde{L}_- - \hbar L_z + \tilde{L}_z^2 \end{aligned} \tag{11.24}$$

Equivalently, since

$$\left[\tilde{L}_+, \tilde{L}_-\right] = -2\mathrm{i}\left[\tilde{L}_x, \tilde{L}_y\right] = 2\hbar\tilde{L}_z \tag{11.25}$$

we have

$$\begin{aligned}\tilde{L}^2 &= \tilde{L}_+\tilde{L}_- - \hbar\tilde{L}_z + \tilde{L}_z^2 \\ &= \tilde{L}_-\tilde{L}_+ + [\tilde{L}_+, \tilde{L}_-] - \hbar\tilde{L}_z + \tilde{L}_z^2 \\ &= \tilde{L}_-\tilde{L}_+ + \hbar\tilde{L}_z + \tilde{L}_z^2\end{aligned} \tag{11.26}$$

We will also need the commutators of $\tilde{L}_z$ with $\tilde{L}_\pm$:

$$\begin{aligned}\left[\tilde{L}_z, \tilde{L}_+\right] &= \left[\tilde{L}_z, \tilde{L}_x\right] + \mathrm{i}\left[\tilde{L}_z, \tilde{L}_y\right] \\ &= \mathrm{i}\hbar\tilde{L}_y + \mathrm{i}(-\mathrm{i}\hbar\tilde{L}_x) \\ &= \hbar\tilde{L}_+\end{aligned} \tag{11.27}$$

and likewise

$$\left[\tilde{L}_z, \tilde{L}_-\right] = -\hbar\tilde{L}_- \tag{11.28}$$

In summary, the commutators of (11.14) in terms of $\tilde{L}_x$, $\tilde{L}_y$, $\tilde{L}_z$ can be replaced by commutators involving $\tilde{L}_\pm$, $\tilde{L}_z$, and $\tilde{L}^2$:

$$\begin{aligned}\left[\tilde{L}_+, \tilde{L}_-\right] &= 2\hbar\tilde{L}_z \\ \left[\tilde{L}_z, \tilde{L}_\pm\right] &= \pm\hbar\tilde{L}_\pm \\ \tilde{L}^2 &= \tilde{L}_+\tilde{L}_- + \tilde{L}_z^2 - \hbar\tilde{L}_z \\ &= \tilde{L}_-\tilde{L}_+ + \tilde{L}_z^2 + \hbar\tilde{L}_z\end{aligned} \tag{11.29}$$

The point of introducing the $\tilde{L}_\pm$ operators is that they act on eigenstates of $\tilde{L}_z$ just like the raising and lowering operators a and $a^\dagger$ act on eigenstates of the harmonic oscillator Hamiltonian. Let

$$\varphi' = \tilde{L}_+\varphi_{ab} \tag{11.30}$$

Then

$$\begin{aligned}\tilde{L}_z\varphi' &= \tilde{L}_z\tilde{L}_+\varphi_{ab} \\ &= \left(\tilde{L}_+\tilde{L}_z + \hbar\tilde{L}_+\right)\varphi_{ab} \\ &= (\hbar b + \hbar)\tilde{L}_+\varphi_{ab} \\ &= \hbar(b + 1)\varphi'\end{aligned} \tag{11.31}$$

and we see that $\varphi' = \tilde{L}_+\varphi_{ab}$ is also an eigenstate of $\tilde{L}_z$, with the eigenvalue $\hbar(b + 1)$. The raising operator has therefore raised the eigenvalue of $\tilde{L}_z$ by the amount $\hbar$.

The 'raised' eigenstate φ is also an eigenstate of L^2, with the same eigenvalue as φ_{ab}:

$$\begin{aligned}\tilde{L}^2\varphi' &= \tilde{L}^2\tilde{L}_+\varphi_{ab} \\ &= \tilde{L}_+\tilde{L}^2\varphi_{ab} \\ &= \hbar^2 a^2 \tilde{L}_+\varphi_{ab} \\ &= \hbar^2 a^2 \varphi'\end{aligned} \tag{11.32}$$

Therefore,

$$\tilde{L}_+\varphi_{ab} = C_{ab}^+\varphi_{a,b+1} \tag{11.33}$$

where C_{ab}^+ is a constant. This establishes the fact that $\tilde{L}_+$ is a raising operator, analogous to the $a^\dagger$ operator used in solving the harmonic oscillator. In a similar way, we can show that $\tilde{L}_-$ is a lowering operator. Define

$$\varphi'' = \tilde{L}_-\varphi_{ab} \tag{11.34}$$

Then

$$\begin{aligned}\tilde{L}_z\varphi'' &= \tilde{L}_z\tilde{L}_-\varphi_{ab} \\ &= \left(\tilde{L}_-\tilde{L}_z - \hbar\tilde{L}_-\right)\varphi_{ab} \\ &= (\hbar b - \hbar)\tilde{L}_-\varphi_{ab} \\ &= \hbar(b-1)\varphi''\end{aligned} \tag{11.35}$$

so that $\varphi'' = \tilde{L}_-\varphi_{ab}$ is also an eigenstate of $\tilde{L}_z$, with the eigenvalue $\hbar(b-1)$. It is also an eigenstate of $\tilde{L}^2$:

$$\begin{aligned}\tilde{L}^2\varphi'' &= \tilde{L}^2\tilde{L}_-\varphi_{ab} \\ &= \tilde{L}_-\tilde{L}^2\varphi_{ab} \\ &= \hbar^2 a^2 \tilde{L}_-\varphi_{ab} \\ &= \hbar^2 a^2 \varphi''\end{aligned} \tag{11.36}$$

Therefore $\tilde{L}_-$ is a lowering operator

$$\tilde{L}_-\varphi_{ab} = C_{ab}^-\varphi_{a,b-1} \tag{11.37}$$

where C_{ab}^- is a constant.

In the case of the harmonic oscillator, we deduced from the fact that $\langle E\rangle \geqslant 0$ that there must be a state of lowest energy φ_0, and therefore a lowest energy eigenvalue E_0. Since the lowering operator a cannot produce a state of lower energy than E_0, it must annihilate the ground state

$$a\varphi_0 = 0 \tag{11.38}$$

The argument for angular momentum is quite similar. We have already found that the eigenvalues of $\tilde{L}_z$ have both a lower *and* an upper bound, for fixed $\hbar^2 a^2$, since

$$-a \leqslant b \leqslant a \tag{11.39}$$

So let us denote the minimum value of b as b_{min} and the maximum value as b_{max}, where $|b_{\text{min}}|$, $|b_{\text{max}}| \leqslant a$. The corresponding eigenstates must have the property that

$$\begin{aligned} \tilde{L}_+ \varphi_{ab_{\text{max}}} &= 0 \\ \tilde{L}_- \varphi_{ab_{\text{min}}} &= 0 \end{aligned} \tag{11.40}$$

Now, we must be able to reach the highest state $\varphi_{ab_{\text{max}}}$ by acting successively on $\varphi_{ab_{\text{min}}}$ with the raising operator

$$\varphi_{ab_{\text{max}}} \propto \left(\tilde{L}_+\right)^n \varphi_{ab_{\text{min}}} \tag{11.41}$$

which implies, since the raising operator raises b to $b + 1$, that

$$b_{\text{max}} = b_{\text{min}} + n \qquad n = \text{a positive integer or } 0 \tag{11.42}$$

or, equivalently,

$$\begin{aligned} b_{\text{max}} &= b_{av} + \frac{n}{2} \\ b_{\text{min}} &= b_{av} - \frac{n}{2} \\ b_{av} &= \frac{1}{2}(b_{\text{max}} + b_{\text{min}}) \end{aligned} \tag{11.43}$$

Next, using the expressions for $\tilde{L}^2$ in (11.29)

$$\begin{aligned} \tilde{L}^2 \varphi_{ab_{\text{max}}} &= \left(\tilde{L}_- \tilde{L}_+ + \tilde{L}_z^2 + \hbar \tilde{L}_z\right) \varphi_{ab_{\text{max}}} \\ \hbar^2 a^2 \varphi_{ab_{\text{max}}} &= \hbar^2 b_{\text{max}}(b_{\text{max}} + 1) \varphi_{ab_{\text{max}}} \end{aligned} \tag{11.44}$$

so that

$$\begin{aligned} a^2 &= b_{\text{max}}(b_{\text{max}} + 1) \\ &= \left(b_{av} + \frac{1}{2}n\right)\left(b_{av} + \frac{1}{2}n + 1\right) \end{aligned} \tag{11.45}$$

Likewise,

$$\begin{aligned} \tilde{L}^2 \varphi_{ab_{\text{min}}} &= \left(\tilde{L}_+ \tilde{L}_- + \tilde{L}_z^2 - \hbar \tilde{L}_z\right) \varphi_{ab_{\text{min}}} \\ \hbar^2 a^2 \varphi_{ab_{\text{min}}} &= \hbar^2 b_{\text{min}}(b_{\text{min}} - 1) \varphi_{ab_{\text{min}}} \end{aligned} \tag{11.46}$$

so that

$$\begin{aligned} a^2 &= b_{\min}(b_{\min} - 1) \\ &= \left(b_{av} - \frac{1}{2}n\right)\left(b_{av} - \frac{1}{2}n - 1\right) \\ &= \left(-b_{av} + \frac{1}{2}n\right)\left(-b_{av} + \frac{1}{2}n + 1\right) \end{aligned} \tag{11.47}$$

Equating the right-hand sides of (11.45) and (11.47)

$$\left(b_{av} + \frac{1}{2}n\right)\left(b_{av} + \frac{1}{2}n + 1\right) = \left(-b_{av} + \frac{1}{2}n\right)\left(-b_{av} + \frac{1}{2}n + 1\right) \tag{11.48}$$

which implies that:

$$\begin{aligned} b_{av} &= 0 \\ b_{\max} &= \frac{n}{2} \\ b_{\min} &= -\frac{n}{2} \\ a^2 &= \frac{n}{2}\left(\frac{n}{2} + 1\right) \end{aligned} \tag{11.49}$$

It is customary to use the notation

$$\begin{aligned} l &\equiv \frac{n}{2} \\ m &\equiv b \end{aligned} \tag{11.50}$$

and to relabel eigenstates φ_{ab} as φ_{lm}. In this notation, we have

The eigenvalues of angular momentum

$$\begin{aligned} \tilde{L}^2\varphi_{lm} &= l(l + 1)\hbar^2\varphi_{lm} \\ \tilde{L}_z\varphi_{lm} &= m\hbar\varphi_{lm} \\ l &= 0, \frac{1}{2}, 1, \frac{3}{2}, 2, \frac{5}{2}, 3, \ldots \\ m &= -l, -l + 1, -l + 2, \ldots, l - 3, l - 2, l - 1, l \end{aligned} \tag{11.51}$$

Note that, as predicted, the eigenvalues of $\tilde{L}^2$ are degenerate: to each eigenvalue $L^2 = l(l + 1)\hbar^2$ there are $2l + 1$ linearly independent eigenstates φ_{lm}, for values of m in the range $-l \leqslant m \leqslant l$, as in (11.51). In other words, each L^2 eigenvalue is $(2l + 1)$-fold degenerate.

11.3 The angular momentum cones

It is time to pause and interpret the remarkable result, equation (11.51), of the last section. What we have found is that angular momentum is 'quantized', in the sense that a measurement of the magnitude of angular momentum will only find one of the discrete set of values

$$|L| = \sqrt{l(l+1)}\,\hbar, \qquad l = 0, \frac{1}{2}, 1, \frac{3}{2}, 2, \frac{5}{2}, 3, \ldots \tag{11.52}$$

and a measurement of the component of angular momentum along a certain axis, e.g. the z-axis, would only find one of the possible values

$$L_z = m\hbar \qquad m \in \{-l, -l+1, -l+2, \ldots, l-2, l-1, l\} \tag{11.53}$$

These eigenvalues have been deduced without solving the Schrödinger equation, or any other differential equation. They were obtained from the commutation relations (11.14) and a bit of clever algebra, nothing more.

Confirmation of these results comes from Nature. All elementary particles, for example, have an intrinsic spin angular momentum. The magnitude of this spin angular momentum depends on the type of elementary particle, but it is *always* one of the values (11.52). The π-meson has spin 0, corresponding to $l = 0$. Electrons, protons, neutrons, and quarks have 'spin $\frac{1}{2}$', i.e. their total spin angular momentum has a magnitude

$$|S| = \sqrt{\frac{3}{4}}\,\hbar \tag{11.54}$$

corresponding to $l = \frac{1}{2}$. The ρ and ω mesons have 'spin 1', i.e.

$$|S| = \sqrt{2}\,\hbar \tag{11.55}$$

corresponding to $l = 1$. The Δ-hyperon has 'spin $\frac{3}{2}$' ($l = \frac{3}{2}$), and so on. The component of spin angular momentum along a particular axis can also be measured experimentally via the particle magnetic moment. The component of magnetic moment along a given axis is proportional to the spin angular momentum along that axis. Experimentally, the values of the component S_z along, e.g. the z-axis, obey the relation (11.53). There are, however, some important differences between intrinsic spin angular momentum (whose magnitude can never change for a given type of particle) and orbital angular momentum, which can change by discrete amounts. We will defer further discussion of spin angular momentum until chapter 13.

Apart from the fact that both the total magnitude $|L|$ and component L_z come in discrete amounts (11.51), angular momentum in quantum mechanics differs from angular momentum in classical mechanics in several ways. Of these, the most striking is the fact that the angular momentum in the z-direction (or in any other direction) is always smaller than the magnitude of total (non-zero) angular momentum. $L_z = l\hbar$ is the largest possible value for L_z, and

$$l\hbar < \sqrt{l(l+1)}\,\hbar \tag{11.56}$$

which means that the angular momentum can never be completely aligned in a particular direction. In fact, if the angular momentum *did* point in a definite direction, then all components L_x, L_y, L_z would be definite. But these components cannot be simultaneously definite, because the corresponding operators do not commute. This means that there is no physical state in which the angular momentum points in a particular direction.

In classical physics, we visualize angular momentum $\vec{L}$ as a vector. In quantum physics, it is better to visualize the angular momentum associated with a given eigenstate φ_{lm} as a cone. In an eigenstate the magnitude of angular momentum $|L|$ and the z-component L_z are fixed,

$$|L| = \sqrt{l(l+1)}\hbar \qquad L_z = m\hbar \tag{11.57}$$

while values of L_x and L_y are indefinite; although their squared expectation values must satisfy

$$\langle L_x^2 \rangle + \langle L_y^2 \rangle + \langle L_z^2 \rangle = \langle L^2 \rangle \tag{11.58}$$

We can picture the angular momentum associated with φ_{lm} as the cone of all vectors $\vec{L}$ satisfying (11.57). The cones for $l = \frac{1}{2}, 1, \frac{3}{2}$ are displayed in figure 11.2.

In classical mechanics it is easy to add two or more angular momenta; it is just a matter of vector addition. In quantum mechanics, as one might imagine, the process is more complicated. For example, suppose one has two electrons which are each in angular momentum eigenstates, and we ask: 'what is the total angular momentum of the system?' (How does one go about adding 'cones' of angular momentum?) The procedure for addition of angular momentum will be explained in chapter 14.

11.4 Eigenfunctions of angular momentum

In the case of the harmonic oscillator, we found the ground state wave function by solving the first-order differential equation

$$a\varphi_0(x) = 0 \tag{11.59}$$

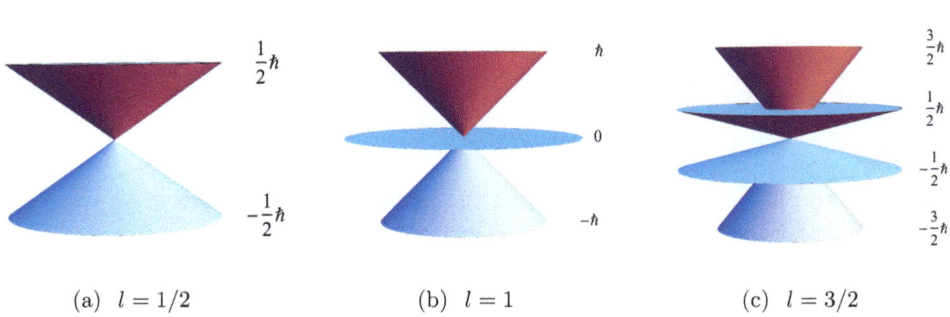

Figure 11.2. Angular momentum cones. A pictorial representation of the fact that, except for $l = 0$, there is no physical state with definite values of all three components L_x, L_y, L_z. Each cone is swept out by vectors **L** with component $L_z = m\hbar$ and length $|L| = \sqrt{l(l+1)}\hbar$.

and then all other eigenstates can be obtained by applying the raising operator $a^\dagger$. In the case of angular momentum, the strategy is very similar. For a given l, we first solve the first-order differential equations

$$\begin{aligned} \tilde{L}_+\varphi_{ll}(x, y, z) &= 0 \\ \tilde{L}_z\varphi_{ll}(x, y, z) &= l\hbar\varphi_{ll} \end{aligned} \tag{11.60}$$

and then obtain all other φ_{lm} wave functions in the multiplet by applying successively the lowering operator $\tilde{L}_-$.

It is much easier to solve these differential equations in spherical coordinates $r,\ \theta,\ \phi$

$$\begin{aligned} z &= r\cos\theta \\ x &= r\sin\theta\cos\phi \\ y &= r\sin\theta\sin\phi \end{aligned} \tag{11.6}$$

because in spherical coordinates the r variable drops out of the angular momentum operators:

$$\begin{aligned} \tilde{L}_x &= \mathrm{i}\hbar\left(\sin\phi\frac{\partial}{\partial\theta} + \cot\theta\cos\phi\frac{\partial}{\partial\phi}\right) \\ \tilde{L}_y &= \mathrm{i}\hbar\left(-\cos\phi\frac{\partial}{\partial\theta} + \cot\theta\sin\phi\frac{\partial}{\partial\phi}\right) \\ \tilde{L}_z &= -\mathrm{i}\hbar\frac{\partial}{\partial\phi} \\ \tilde{L}^2 &= -\hbar^2\left[\frac{1}{\sin\theta}\frac{\partial}{\partial\theta}\sin\theta\frac{\partial}{\partial\theta} + \frac{1}{\sin^2\theta}\frac{\partial^2}{\partial\phi^2}\right] \end{aligned} \tag{11.62}$$

and therefore any eigenstate of angular momentum has the form

$$\varphi_{lm}(x, y, z) = f(r)Y_{lm}(\theta, \phi) \tag{11.63}$$

where

$$\begin{aligned} \tilde{L}^2Y_{lm}(\theta, \phi) &= l(l+1)\hbar^2Y_{lm}(\theta, \phi) \\ \tilde{L}_zY_{lm}(\theta, \phi) &= m\hbar Y_{lm}(\theta, \phi) \end{aligned} \tag{11.64}$$

and where $f(r)$ is any function such that the normalization condition

$$\begin{aligned} 1 &= \int \mathrm{d}x\,\mathrm{d}y\,\mathrm{d}z\ \phi^*(x, y, z)\phi(x, y, z) \\ &= \int_0^\infty \mathrm{d}r\ r^2 \int_0^\pi \mathrm{d}\theta\ \sin\theta \int_0^{2\pi} \mathrm{d}\phi\ f^*(r)f(r)Y^*_{lm}(\theta, \phi)Y_{lm}(\theta, \phi) \end{aligned} \tag{11.65}$$

is satisfied. It is conventional to normalize the Y_{lm} such that the integral over solid angle is also equal to one:

$$\int_0^{\pi} d\theta \, \sin\theta \int_0^{2\pi} d\phi \; Y^*_{lm}(\theta, \phi) Y_{lm}(\theta, \phi) = 1 \tag{11.66}$$

With this normalization, the $Y_{lm}(\theta, \phi)$ are known as ***spherical harmonics***.

In spherical coordinates, the raising and lowering operators are

$$\begin{aligned} \tilde{L}_+ &= \tilde{L}_x + i\tilde{L}_y \\ &= \hbar e^{i\phi}\left(\frac{\partial}{\partial\theta} + i\cot\theta\frac{\partial}{\partial\phi}\right) \\ \tilde{L}_- &= \tilde{L}_x - i\tilde{L}_y \\ &= -\hbar e^{-i\phi}\left(\frac{\partial}{\partial\theta} - i\cot\theta\frac{\partial}{\partial\phi}\right) \end{aligned} \tag{11.67}$$

and we solve the first-order differential equations

$$\tilde{L}_+ Y_{ll} = 0 \qquad \tilde{L}_z Y_{ll} = l\hbar Y_{lm} \tag{11.68}$$

by the method of separation of variables

$$Y_{lm}(\theta, \phi) = A(\theta)B(\phi) \tag{11.69}$$

The $\tilde{L}_z$ eigenvalue equation (for any m) becomes

$$-i\hbar\frac{dB}{d\phi} = m\hbar B \tag{11.70}$$

which has the solution

$$B(\phi) = e^{im\phi} \tag{11.71}$$

As was pointed out in the case of the 'quantum corral', discussed in the last chapter, the angle $\phi = 0$ is the same as the angle $\phi = 2\pi$, so the wave functions must satisfy the periodicity condition

$$Y_{lm}(\theta, \phi + 2\pi) = Y_{lm}(\theta, \phi) \tag{11.72}$$

But this means, since $B(\phi) = e^{im\phi}$, that m must be restricted to the integer values

$$m = 0, \pm1, \pm2, \pm3, \ldots, \pm n, \ldots \tag{11.73}$$

As a result, since $-l \leqslant m \leqslant l$, the possible values for l are

$$l = 0, 1, 2, 3, \ldots, n, \ldots \tag{11.74}$$

So, although we have found algebraically that m could have both integer *and* half-integer values

$$m = 0, \pm\frac{1}{2}, \pm1, \pm\frac{3}{2}, \ldots, \pm n/2, \ldots \tag{11.75}$$

it turns out that the periodicity condition has ruled out the half-integer possibilities for orbital angular momentum. As we will see in chapter 13, the half-integer values are still possible for intrinsic spin angular momentum, where the requirement of angular periodicity of the wave function does not apply.

So far we have

$$Y_{lm}(\theta, \phi) = A(\theta)\mathrm{e}^{\mathrm{i}m\phi} \tag{11.76}$$

and in particular

$$Y_{ll}(\theta, \phi) = A(\theta)\mathrm{e}^{\mathrm{i}l\phi} \tag{11.77}$$

Applying the raising operator to this state, we must have

$$\begin{aligned} 0 &= \tilde{L}_+ Y_{ll} \\ &= \hbar \mathrm{e}^{\mathrm{i}\phi}\left(\mathrm{i}\cot\theta\frac{\partial}{\partial\phi} + \frac{\partial}{\partial\theta}\right)A(\theta)\mathrm{e}^{\mathrm{i}l\phi} \\ &= \hbar \mathrm{e}^{\mathrm{i}(l+1)\phi}\left(-l\cot\theta + \frac{\partial}{\partial\theta}\right)A(\theta) \end{aligned} \tag{11.78}$$

or

$$\frac{\mathrm{d}}{\mathrm{d}\theta}A(\theta) = l\cot\theta A(\theta) \tag{11.79}$$

which is solved by

$$A(\theta) = \text{const.} \times \sin^l\theta \tag{11.80}$$

Then

$$Y_{ll}(\theta, \phi) = N\sin^l\theta\, \mathrm{e}^{\mathrm{i}l\phi} \tag{11.81}$$

where N is a normalization constant, which is determined from the normalization condition

$$\begin{aligned} 1 &= \int_0^\pi \mathrm{d}\theta\sin\theta \int_0^{2\pi} \mathrm{d}\phi\, Y_{ll}^* Y_{ll} \\ &= 2\pi N^2 \int_0^\pi \mathrm{d}\theta\, \sin^{(2l+1)}\theta \\ &= \frac{2\pi^{3/2} l!}{\Gamma\left(l + \frac{3}{2}\right)} N^2 \end{aligned} \tag{11.82}$$

where the Γ-function is a special function with the properties

$$\Gamma(x+1) = x\Gamma(x), \qquad \Gamma\left(\frac{1}{2}\right) = \sqrt{\pi} \tag{11.83}$$

and $\Gamma(n + 1) = n!$ for integer n. Of course, the normalization condition only determines N, even assuming N is real, up to an overall sign. The convention is to choose this sign to be $(-1)^l$, so that finally

$$Y_{ll}(\theta, \phi) = (-1)^l \left[\frac{\Gamma\left(l + \frac{3}{2}\right)}{2\pi^{3/2} l!} \right]^{1/2} \sin^l \theta e^{il\phi} \tag{11.84}$$

From here we can get the other Y_{lm} using the lowering operator

$$\tilde{L}_- Y_{l,m}(\theta, \phi) = C_{lm}^- Y_{l,m-1}(\theta, \phi) \tag{11.85}$$

given the constants C_{lm}^-. Corresponding constants C_{lm}^+ are defined from

$$\tilde{L}_+ Y_{l,m}(\theta, \phi) = C_{lm}^+ Y_{l,m+1}(\theta, \phi) \tag{11.86}$$

and it is consistent to choose the $C_{lm}^\pm$ to be real. To get the C_{lm}^- constants, we again resort to some clever algebra. We have

$$\begin{aligned} \tilde{L}_- \tilde{L}_+ &= \tilde{L}^2 - \tilde{L}_z^2 - \hbar \tilde{L}_z \\ \tilde{L}_+ \tilde{L}_- &= \tilde{L}^2 - \tilde{L}_z^2 + \hbar \tilde{L}_z \end{aligned} \tag{11.87}$$

Then

$$\begin{aligned} \left\langle \varphi_{lm} | \tilde{L}_- \tilde{L}_+ | \varphi_{lm} \right\rangle &= \left\langle \varphi_{lm} \middle| \left(\tilde{L}^2 - \tilde{L}_z^2 - \hbar \tilde{L}_z \right) \middle| \varphi_{lm} \right\rangle \\ \left\langle \tilde{L}_+ \varphi_{lm} | \tilde{L}_+ \varphi_{lm} \right\rangle &= \hbar^2 \left[l(l+1) - m^2 - m \right) \langle \varphi_{lm} | \varphi_{lm} \rangle \\ \left(C_{lm}^+ \right)^* C_{lm}^+ &= \hbar^2 (l - m)(l + m + 1) \end{aligned} \tag{11.88}$$

so that

$$C_{lm}^+ = \hbar \sqrt{(l - m)(l + m + 1)} \tag{11.89}$$

Likewise

$$\begin{aligned} \left\langle \varphi_{lm} | \tilde{L}_+ \tilde{L}_- | \varphi_{lm} \right\rangle &= \left\langle \varphi_{lm} \middle| \left(\tilde{L}^2 - \tilde{L}_z^2 + \hbar \tilde{L}_z \right) \middle| \varphi_{lm} \right\rangle \\ \left\langle \tilde{L}_- \varphi_{lm} | \tilde{L}_- \varphi_{lm} \right\rangle &= \hbar^2 [l(l+1) - m^2 + m) \langle \varphi_{lm} | \varphi_{lm} \rangle \\ \left(C_{lm}^- \right)^* C_{lm}^- &= \hbar^2 (l + m)(l - m + 1) \end{aligned} \tag{11.90}$$

so that

$$C_{lm}^- = \hbar \sqrt{(l + m)(l - m + 1)} \tag{11.91}$$

We can now compute all the Y_{lm} using

$$\begin{aligned} Y_{ll}(\theta, \phi) &= (-1)^l \left[\frac{\Gamma\left(l + \frac{3}{2}\right)}{2\pi^{3/2} l!} \right]^{1/2} \sin^l \theta \mathrm{e}^{\mathrm{i}l\phi} \\ \tilde{L}_- Y_{lm} &= \hbar\sqrt{(l+m)(l-m+1)}\, Y_{l,\, m-1} \\ \tilde{L}_+ Y_{lm} &= \hbar\sqrt{(l-m)(l+m+1)}\, Y_{l,\, m+1} \end{aligned} \tag{11.92}$$

Since Y_{ll} was normalized to 1 in equation (11.82), all of the Y_{lm} obtained by applying the lowering operator will also be normalized to 1. Also, since $\tilde{L}^2$ and $\tilde{L}_z$ are Hermitian operators, spherical harmonics corresponding to different values of l and/or m will be orthogonal:

$$\int_0^\pi \mathrm{d}\theta \sin\theta \int_0^{2\pi} \mathrm{d}\phi \; Y^*_{lm}(\theta, \phi) Y_{l'm'}(\theta, \phi) = \delta_{ll'}\delta_{mm'} \tag{11.93}$$

Example: the $l = 1$ multiplet

As an example of the procedure, let us compute the $l = 1$ multiplet of spherical harmonics, i.e. Y_{11}, Y_{10}, $Y_{1,-1}$. We begin with

$$\begin{aligned} Y_{11} &= (-1) \left[\frac{\Gamma\left(\frac{5}{2}\right)}{2\pi^{3/2}} \right]^{1/2} \sin\theta \mathrm{e}^{\mathrm{i}\phi} \\ &= -\sqrt{\frac{3}{8\pi}} \sin\theta \mathrm{e}^{\mathrm{i}\phi} \end{aligned} \tag{11.94}$$

then

$$\begin{aligned} \tilde{L}_- Y_{11} &= -\sqrt{\frac{3}{8\pi}} \hbar \mathrm{e}^{-\mathrm{i}\phi} \left(i\cot\theta \frac{\partial}{\partial\phi} - \frac{\partial}{\partial\theta} \right) \sin\theta \mathrm{e}^{\mathrm{i}\phi} \\ \hbar\sqrt{2}\, Y_{10} &= \sqrt{\frac{3}{8\pi}} \hbar(\cot\theta \sin\theta + \cos\theta) \\ Y_{10} &= \sqrt{\frac{3}{4\pi}} \cos\theta \end{aligned} \tag{11.95}$$

and applying the lowering operator again

$$\begin{aligned} \tilde{L}_- Y_{10} &= \sqrt{\frac{3}{4\pi}} \hbar \mathrm{e}^{-\mathrm{i}\phi} \left(\mathrm{i}\cot\theta \frac{\partial}{\partial\phi} - \frac{\partial}{\partial\theta} \right) \cos\theta \\ \hbar\sqrt{2}\, Y_{1,-1} &= \sqrt{\frac{3}{4\pi}} \hbar \mathrm{e}^{-\mathrm{i}\phi} \sin\theta \\ Y_{1,-1} &= \sqrt{\frac{3}{8\pi}} \mathrm{e}^{-\mathrm{i}\phi} \sin\theta \end{aligned} \tag{11.96}$$

It is easy to check that applying the lowering operator one more time annihilates the state

$$\tilde{L}_- Y_{1,-1} = 0 \tag{11.97}$$

as it should. In fact, in constructing the spherical harmonics, we could start with the solution of the differential equation

$$\tilde{L}_- Y_{l,-l} = 0 \tag{11.98}$$

and obtain the rest of the Y_{lm} by using the raising operator $\tilde{L}_+$.

Of course, one rarely has to go to the trouble of computing the spherical harmonics, since they are listed in many books. The first few, including the $l = 1$ multiplet we have just computed, are

$$\begin{aligned}
Y_{00} &= \sqrt{\frac{1}{4\pi}} \\
Y_{11} &= -\sqrt{\frac{3}{8\pi}} \sin\theta \mathrm{e}^{\mathrm{i}\phi} \\
Y_{10} &= \sqrt{\frac{3}{4\pi}} \cos\theta \\
Y_{1,-1} &= \sqrt{\frac{3}{8\pi}} \sin\theta \mathrm{e}^{-\mathrm{i}\phi} \\
Y_{22} &= \sqrt{\frac{15}{32\pi}} \sin^2\theta \mathrm{e}^{2\mathrm{i}\phi} \\
Y_{21} &= -\sqrt{\frac{15}{8\pi}} \sin\theta\cos\theta \mathrm{e}^{\mathrm{i}\phi} \\
Y_{20} &= \sqrt{\frac{5}{16\pi}} (3\cos^2\theta - 1) \\
Y_{2,-1} &= \sqrt{\frac{15}{8\pi}} \sin\theta\cos\theta \mathrm{e}^{-\mathrm{i}\phi} \\
Y_{2,-2} &= \sqrt{\frac{15}{32\pi}} \sin^2\theta \mathrm{e}^{-2\mathrm{i}\phi}
\end{aligned} \tag{11.99}$$

There is a nice graphical way of representing the spherical harmonics. The modulus of the spherical harmonics $|Y_{lm}|$ is a function only of θ, and is independent of ϕ (therefore $|Y_{lm}| = |Y_{l,-m}|$). So, in spherical coordinates (r, θ) where θ is the angle away from the z-axis, we plot the function

$$r(\theta) = |Y_{lm}(\theta, \phi)| \tag{11.100}$$

The resulting diagrams are shown in figure 11.3.

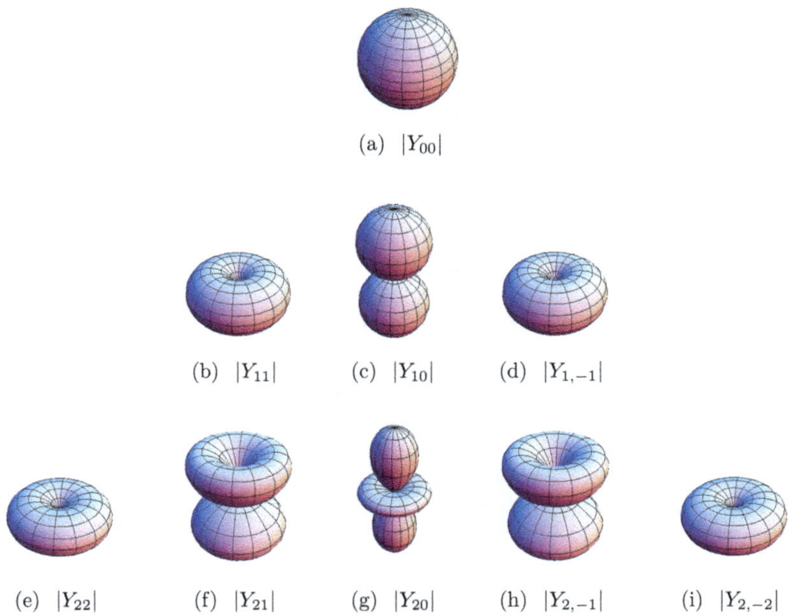

Figure 11.3. Graphical representation of the modulus of the spherical harmonics.

The rigid rotator

As a first application of the spherical harmonics, let us consider a rigid rod with moment of inertia I, which pivots freely around one fixed end at $r = 0$. Since the length of the rod is fixed, the generalized coordinates are the angular position of the rod (θ, ϕ). The classical Hamiltonian describing the dynamics of this simple system is

$$H = \frac{1}{2I}L^2 \tag{11.101}$$

where $\vec{L}$ is the angular momentum. The corresponding Schrödinger equation is then simply

$$\tilde{H}\varphi_{lm}(\theta, \phi) = \frac{1}{2I}\tilde{L}^2\varphi_{lm}(\theta, \phi) = E_{lm}\varphi_{lm}(\theta\phi) \tag{11.102}$$

and the energy eigenstates are angular momentum eigenstates

$$\varphi_{lm}(\theta, \phi) = Y_{lm}(\theta, \phi) \qquad E_{lm} = \frac{1}{2I}l(l+1)\hbar^2 \tag{11.103}$$

We note that the energy eigenvalues are degenerate, as expected

$$E_{lm} = E_{lm'} \qquad -l \leqslant m, m' \leqslant l \tag{11.104}$$

11.5 The radial equation for central potentials

It has already been noted that the angular momentum operators involve only θ and ϕ, so an eigenstate of $\tilde{L}^2$ and $\tilde{L}_z$ has the general form

$$\varphi = f(r)Y_{lm}(\theta, \phi) \tag{11.105}$$

where $f(r)$ is an arbitrary function of r. For central potentials $V(r, \theta, \phi) = V(r)$, the Hamiltonian commutes with the angular momentum operators, so by the commutator theorem $\tilde{H}$, $\tilde{L}^2$, $\tilde{L}_z$ have a common set of eigenstates of the form (11.105). However, if φ is an eigenstate of $\tilde{H}$, then the radial function $f(r)$ is no longer arbitrary, but is determined from an ordinary differential equation in the variable r known as the ***radial equation***.

Let us begin with the classical Hamiltonian

$$H = \frac{\mathbf{p} \cdot \mathbf{p}}{2m} + V(r) \tag{11.106}$$

The momentum vector $\mathbf{p}$, in spherical coordinates, can be expressed as a sum of two vectors, one of which ($\mathbf{p}_r$) is parallel to the radial direction, and the other ($\mathbf{p}_\perp$) perpendicular to the radial direction, i.e.

$$\begin{aligned}
\mathbf{p} &= \mathbf{p}_r + \mathbf{p}_\perp \\
\mathbf{p}_r &\equiv p\cos(\theta_{rp})\hat{\mathbf{e}}_r = \frac{1}{r^2}(\mathbf{r} \cdot \mathbf{p})\mathbf{r} \\
|\mathbf{p}_\perp| &\equiv p\sin(\theta_{rp}) = \frac{|\mathbf{r} \times \mathbf{p}|}{r} \\
&= \frac{|\mathbf{L}|}{r}
\end{aligned} \tag{11.107}$$

so that

$$H = \frac{1}{2m}\left[p_r^2 + \frac{L^2}{r^2}\right] + V(r) \tag{11.108}$$

where

$$p_r \equiv \frac{\mathbf{r} \cdot \mathbf{p}}{r} \tag{11.109}$$

Upon quantization, r, $\mathbf{p}$ and $\mathbf{L}$ become operators,

$$\begin{aligned}
\tilde{r}F(r, \theta, \phi) &= rF(r, \theta, \phi) \\
\tilde{\mathbf{p}}F(r, \theta, \phi) &= -i\hbar\nabla F(r, \theta, \phi) \\
\tilde{\mathbf{L}}F(r, \theta, \phi) &= -i\hbar\mathbf{r} \times \nabla F(r, \theta, \phi)
\end{aligned} \tag{11.110}$$

and we define the operator corresponding to p_r as

$$\tilde{p}_r = \frac{1}{2}\left(\frac{\mathbf{r}}{r} \cdot \tilde{\mathbf{p}} + \tilde{\mathbf{p}} \cdot \frac{\mathbf{r}}{r}\right) \tag{11.111}$$

This ordering of r and $\tilde{p}$ is chosen so that $\tilde{p}_r$ is a Hermitian operator. If we had simply substituted $\mathbf{p} \to \tilde{\mathbf{p}}$ in (11.109), then $\tilde{p}_r$ would not be Hermitian, and neither would the Hamiltonian, which involves $\tilde{p}_r^2$. In terms of differential operators

$$\begin{aligned}
\tilde{p}_r F &= -\frac{1}{2}\mathrm{i}\hbar\left[\frac{1}{r}\mathbf{r}\cdot\nabla + \nabla\cdot\frac{\mathbf{r}}{r}\right]F \\
&= -\frac{1}{2}\mathrm{i}\hbar\left[\frac{\partial}{\partial r} + \frac{\mathbf{r}}{r}\cdot\nabla + \frac{1}{r}(\nabla\cdot\mathbf{r}) + \mathbf{r}\cdot\left(\nabla\frac{1}{r}\right)\right]F \\
&= -\frac{1}{2}\mathrm{i}\hbar\left[2\frac{\partial}{\partial r} + \frac{3}{r} - \frac{1}{r}\right]F \\
&= -\mathrm{i}\hbar\left[\frac{\partial}{\partial r} + \frac{1}{r}\right]F \\
&= -\mathrm{i}\hbar\frac{1}{r}\frac{\partial}{\partial r}rF
\end{aligned} \tag{11.112}$$

so that

$$\tilde{p}_r = -\mathrm{i}\hbar\frac{1}{r}\frac{\partial}{\partial r}r \tag{11.113}$$

In terms of this operator, the Schrödinger equation

$$\tilde{H}\varphi(r, \theta, \phi) = E\varphi(r, \theta, \phi) \tag{11.114}$$

is

$$\left[\frac{1}{2m}\left(\tilde{p}_r^2 + \frac{1}{r^2}\tilde{L}^2\right) + V(r)\right]\varphi(r, \theta, \phi) = E\varphi(r, \theta, \phi) \tag{11.115}$$

or, explicitly

$$\left[-\frac{\hbar^2}{2m}\left(\left(\frac{1}{r}\frac{\partial^2}{\partial r^2}r\right) + \frac{1}{r^2}\left(\frac{1}{\sin\theta}\frac{\partial}{\partial\theta}\sin\theta\frac{\partial}{\partial\theta} + \frac{1}{\sin^2\theta}\frac{\partial^2}{\partial\phi^2}\right)\right) + V(r)\right]\varphi = E\varphi \tag{11.116}$$

This equation could be obtained more quickly by starting from

$$\left[-\frac{\hbar^2}{2m}\nabla^2 + V(r)\right]\varphi = E\varphi \tag{11.117}$$

and then just expressing ∇^2 in spherical coordinates

$$\nabla^2 = \left(\frac{1}{r}\frac{\partial^2}{\partial r^2}r\right) + \frac{1}{r^2}\left(\frac{1}{\sin\theta}\frac{\partial}{\partial\theta}\sin\theta\frac{\partial}{\partial\theta} + \frac{1}{\sin^2\theta}\frac{\partial^2}{\partial\phi^2}\right) \tag{11.118}$$

Comparing ∇^2 to equation (11.62), we see that it contains the operator $\tilde{L}^2$, although the reason for this is not so clear. The appearance of the angular momentum operator, on the other hand, is quite obvious from equation (11.108).

Having found the eigenstates of angular momentum, we write

$$\varphi(r, \theta, \phi) = R_{klm}(r) Y_{lm}(\theta, \phi) \tag{11.119}$$

where the index k is introduced to distinguish between states of different energy, which have the same eigenvalues of $\tilde{L}^2$ and $\tilde{L}_z$. Substitute into the Schrödinger equation

$$\begin{aligned} \left[\frac{1}{2m}\tilde{p}_r^2 + \frac{1}{2mr^2}\tilde{L}^2 + V(r)\right] R_{klm}(r) Y_{lm}(\theta, \phi) &= E_{klm} R_{klm}(r) Y_{lm}(\theta, \phi) \\ Y_{lm}\left[\frac{1}{2m}\tilde{p}_r^2 + \frac{\hbar^2}{2mr^2} l(l+1) + V(r)\right] R_{klm}(r) &= E_{klm} R_{klm}(r) Y_{lm}(\theta, \phi) \end{aligned} \tag{11.120}$$

Cancelling Y_{lm} on both sides of the equation, we note that neither the potential nor the differential operator depend on m, so we can write that $R_{klm} = R_{kl}$ and $E_{klm} = E_{kl}$. We then have an equation which involves only the radial coordinate

$$\frac{1}{2m}\left[\tilde{p}_r^2 + \frac{\hbar^2 l(l+1)}{r^2}\right] R_{kl}(r) + V(r) R_{kl}(r) = E_{kl} R_{kl}(r) \tag{11.121}$$

which, using

$$\begin{aligned} \tilde{p}_r^2 &= -\hbar^2 \frac{1}{r}\frac{\partial^2}{\partial r^2} r \\ &= -\hbar^2 \left[\frac{\partial^2}{\partial r^2} + \frac{2}{r}\frac{\partial}{\partial r}\right] \end{aligned} \tag{11.122}$$

becomes

The radial equation

$$\frac{\mathrm{d}^2 R_{kl}}{\mathrm{d}r^2} + \frac{2}{r}\frac{\mathrm{d}R_{kl}}{\mathrm{d}r} + \left[\frac{2m}{\hbar^2}\{E_{kl} - V(r)\} - \frac{l(l+1)}{r^2}\right] R_{kl}(r) = 0$$

This is the radial equation for energy eigenstates in a spherical potential. An alternative form is obtained by setting $U_{kl}(r) = rR_{kl}(r)$ and applying (11.122):

$$\frac{\mathrm{d}^2 U_{kl}}{\mathrm{d}r^2} + \left[\frac{2m}{\hbar^2}\{E_{kl} - V(r)\} - \frac{l(l+1)}{r^2}\right] U_{kl}(r) = 0 \tag{11.123}$$

11.5.1 The free particle

The simplest possible central potential is $V(r) = 0$. If we write

$$E = \frac{\hbar^2 k^2}{2m} \tag{11.124}$$

Then the radial equation is

$$\frac{d^2 R_{kl}}{dr^2} + \frac{2}{r}\frac{dR_{kl}}{dr} + \left[k^2 - \frac{l(l+1)}{r^2}\right] R_{kl}(r) = 0 \tag{11.125}$$

Rescaling the r-coordinate,

$$u = kr \tag{11.126}$$

this becomes

$$\left[\frac{d^2}{du^2} + \frac{2}{u}\frac{d}{du} + \left(1 - \frac{l(l+1)}{u^2}\right)\right] R_{kl}(r) = 0 \tag{11.127}$$

Writing

$$R_{kl}(r) = Q_l(u) \tag{11.128}$$

we have a second-order differential equation

$$\frac{d^2 Q_l}{du^2} + \frac{2}{u}\frac{dQ_l}{du} + \left(1 - \frac{l(l+1)}{u^2}\right) Q_l = 0 \tag{11.129}$$

which is known as the ***spherical Bessel equation***. Solutions of this equation, which are finite and differentiable at $r = 0$, are the ***spherical Bessel functions*** $Q_l(u) = j_l(u)$, the first few of which are listed below:[1]

$$\begin{aligned} j_0(u) &= \frac{\sin(u)}{u} \\ j_1(u) &= \frac{\sin(u)}{u^2} - \frac{\cos(u)}{u} \\ j_2(u) &= \left(\frac{3}{u^3} - \frac{1}{u}\right)\sin(u) - \frac{3}{u^2}\cos(u) \end{aligned} \tag{11.130}$$

Putting everything together, the eigenstates of energy and angular momentum

$$\left\{\tilde{H},\ \tilde{L}^2, \tilde{L}_z\right\} \tag{11.131}$$

[1] There are also solutions known as spherical Neumann functions $n_l(u)$. These, however, are singular and non-differentiable at $r = 0$ (e.g. $n_0(u) = -\cos(u)/u$ and must be discarded for the case of the free particle. Still, these spherical Neumann functions are solutions of the radial equation for free particles away from $r = 0$, and will become important when we discuss scattering and the method of partial waves in chapter 22.

for $V = 0$ are

$$\varphi_{klm}(r, \theta, \phi) = j_l(kr) Y_{lm}(\theta, \phi) \tag{11.132}$$

with eigenvalues

$$E_k = \frac{\hbar^2 k^2}{2m}, \qquad L^2 = l(l+1)\hbar^2, \qquad L_z = m\hbar \tag{11.133}$$

Since there is one and only one eigenstate corresponding to a given set of eigenvalues of $\{\tilde{H}, \tilde{L}^2, \tilde{L}_z\}$, it follows that E, L^2, L_z is a complete set of observables.

For the free particle in one dimension, we found that E, P was a complete set of observables, where P is parity, and also that p is a complete set of observables, where p is momentum. Similarly, in three dimensions, the three components of momentum $\{p_x, p_y, p_z\}$ are a complete set of observables, because there is one and only one eigenstate of $\{\tilde{p}_x, \tilde{p}_y, \tilde{p}_z\}$ with a given set of eigenvalues

$$\begin{aligned} -\mathrm{i}\hbar\frac{\partial}{\partial x}\psi_p(x, y, z) &= p_x\psi_p(x, y, z) \\ -\mathrm{i}\hbar\frac{\partial}{\partial y}\psi_p(x, y, z) &= p_y\psi_p(x, y, z) \\ -\mathrm{i}\hbar\frac{\partial}{\partial z}\psi_p(x, y, z) &= p_z\psi_p(x, y, z) \end{aligned} \tag{11.134}$$

namely

$$\psi_p(x, y, z) = N \exp\left[\mathrm{i}\left(p_x x + p_y y + p_z z\right)/\hbar\right] \tag{11.135}$$

This wave function is *also* an eigenstate of the Hamiltonian

$$\tilde{H}\psi_p = -\frac{\hbar^2}{2m}\nabla^2\psi_p = E_p\psi_p \tag{11.136}$$

with energy

$$E_p = \frac{1}{2m}\left(p_x^2 + p_y^2 + p_z^2\right) \tag{11.137}$$

The Hamiltonian of a free particle is invariant under translations

$$\vec{r} \to \vec{r} + \vec{a} \tag{11.138}$$

and also under rotations of the coordinates. The energy eigenstates $\{\psi_p\}$ are eigenstates of the Hamiltonian and the translation operator

$$T = \mathrm{e}^{\mathrm{i}\vec{a}\cdot\vec{p}} \tag{11.139}$$

while the energy eigenstates $\{j_l(kr) Y_{lm}(\theta, \phi)\}$ are eigenstates of the Hamiltonian and the rotation operator along the z-axis

$$\tilde{\mathcal{R}}_z(\delta\phi) = \mathrm{e}^{\mathrm{i}\delta\phi\tilde{L}_z} \tag{11.140}$$

as well as the total angular momentum operator $\tilde{L}^2$.

Problems

1. Suppose we make measurements of the component of angular momentum **L** in some specified direction. If the total angular momentum is $\sqrt{30}\,\hbar$, what is the minimum angle there can be between **L** and a line in the given direction?
2. A rigid rotator (mass at the end of a rigid rod) is in the angular momentum state $l = 2$, $m = -1$. What is the probability of finding the mass in the angular region $\theta = \pi/6 \pm 0.01$ radians, and $\phi = \pi/4 \pm 0.02$ radians?
3. Compute $[\tilde{L}_i, \tilde{p}_j]$ for the x-, y-, z-components of momentum and angular momentum. Your result can be expressed compactly using the Levi–Civita symbol ϵ_{ijk}.
4. Show that for a particle in an angular momentum state Y_{lm},
   a. $\langle L_x \rangle = \langle L_y \rangle = 0$
   b. $\langle L_x^2 \rangle = \langle L_y^2 \rangle = \frac{1}{2}\hbar^2[l(l+1) - m^2]$

   From this information, verify the generalized uncertainty principle (equation (7.121)) for $\Delta L_x \Delta L_y$.
5. Obtain the $l = 1$ multiplet by solving (11.98), and then applying the raising operator.
6. We consider a particle of mass m, constrained to move on the surface of a sphere of radius R. The wave function of the particle, at time $t = 0$, is

$$\psi = N\left(\frac{zy}{R} + z\right) \tag{11.141}$$

   a) Write ψ in spherical coordinates (remember that the radial coordinate $r = R$ is fixed). Express ψ as a (properly normalized) superposition of spherical harmonics.
   b) If you measure L^2 and L_z simultaneously for a particle in this state, what are the possible outcomes, and what is the probability of each outcome?
   c) Compute $\langle L^2 \rangle$, $\langle L_z \rangle$, and ΔL_z.
   d) The particle moves freely on the sphere, and the Hamiltonian is $H = L^2/(2I)$, where $I = mR^2$. Find the wave function (in spherical coordinates) at any later time t. Do the answers to parts (b) and (c) change with time? Explain.

   You should not have to perform any integrations to solve this problem.
7. A particle of mass m is constrained to move between two concentric, impermeable spheres of radii $r = a$ and $r = b$. The potential $V(r) = 0$ between the spheres ($a < r < b$), and $V(r) = \infty$ otherwise. Find all of the zero angular momentum ($l = 0$) normalized energy eigenstates, and the energy eigenvalues, by solving the radial equation (11.123).
8. The Hamiltonian of a certain rotator in three dimensions is given by

$$H = \frac{1}{2I_1}\left(\tilde{L}_x^2 + \tilde{L}_y^2\right) + \frac{1}{2I_2}\tilde{L}_z^2$$

   with $I_1 > I_2$. Do any of the operators $\tilde{L}_x$, $\tilde{L}_y$, $\tilde{L}_z$ commute with H? Find the energy eigenvalues of H, and the degeneracy of each eigenvalue.

IOP Publishing

An Introduction to Quantum Theory

Jeff Greensite

Chapter 12

The hydrogen atom

Historically, the first application of the Schrödinger equation (by Schrödinger himself) was to the hydrogen atom, and the result for the frequency of the hydrogen spectral lines was the same as that obtained by Bohr.[1] In this chapter we will retrace Schrödinger's steps in solving the Schrödinger equation for the hydrogen atom. The raising/lowering operator trick will not be used here. Nevertheless, we will see that the energy eigenvalues of the Hamiltonian are still determined, ultimately, by algebra.

The hydrogen atom is a simple system: a light electron bound to a heavy proton. The proton is so much heavier than the electron (about 2000 times heavier) that we may regard its position as fixed, at the center of a system of spherical coordinates. The potential energy between the electron and proton is given, in Heaviside–Lorentz units, by the Coulomb expression

$$V(r) = -\frac{e^2}{4\pi r} \tag{12.1}$$

Since this potential is spherically symmetric, we know from the preceding chapter that the Hamiltonian will commute with the angular momentum operators, and therefore that energy eigenstates may also be chosen to be eigenstates of $\tilde{L}^2$ and $\tilde{L}_z$, i.e.

$$\phi_E(r, \theta, \phi) = R_{kl}(r) Y_{lm}(\theta, \phi) \tag{12.2}$$

[1] The difference is that Bohr's rather ad hoc quantization condition ($mvr = n\hbar$) was successful only for hydrogen and a few other simple atoms. In contrast, the Schrödinger equation is a law of motion, as fundamental as $F = ma$ (which it replaces).

doi:10.1088/978-0-7503-1167-0ch12

The time-independent Schrödinger equation is

$$\left[-\frac{\hbar^2}{2m}\nabla^2 - \frac{e^2}{4\pi r}\right]\phi_E(r, \theta, \phi) = E\phi_E(r, \theta, \phi) \tag{12.3}$$

or

$$\left[\frac{1}{2m}\left(\tilde{p}_r^2 + \frac{1}{r^2}\tilde{L}^2\right) - \frac{e^2}{4\pi r}\right]R_{kl}(r)\,Y_{lm}(\theta, \phi) = ER_{kl}\,Y_{lm} \tag{12.4}$$

which reduces, as shown in the previous chapter, to the 'radial equation' for $R_{kl}(r)$

$$\frac{\mathrm{d}^2R_{kl}}{\mathrm{d}r^2} + \frac{2}{r}\frac{\mathrm{d}R_{kl}}{\mathrm{d}r} + \left[\frac{2m}{\hbar^2}\left\{E_{kl} + \frac{e^2}{4\pi r}\right\} - \frac{l(l+1)}{r^2}\right]R_{kl}(r) = 0 \tag{12.5}$$

or

$$\left[\frac{1}{r}\frac{d^2}{\mathrm{d}r^2}r - \frac{l(l+1)}{r^2} + \frac{2me^2}{4\pi\hbar^2}\frac{1}{r} + \frac{2mE}{\hbar^2}\right]R(r) = 0 \tag{12.6}$$

We are interested in the spectrum of atomic hydrogen, so this means we would like to solve for the bound states, i.e. those states for which $E < 0$. From now on, we take $E = -|E| < 0$. Multiply the above equation by r on the left, and define

$$u(r) = rR(r) \tag{12.7}$$

so that

$$\left[\frac{d^2}{\mathrm{d}r^2} - \frac{l(l+1)}{r^2} + \frac{2me^2}{4\pi\hbar^2}\frac{1}{r} - \frac{2m|E|}{\hbar^2}\right]u(r) = 0 \tag{12.8}$$

which is the radial equation in the form (11.123). This equation involves two constants, namely

$$\frac{2mE}{\hbar^2} \quad \text{and} \quad \frac{2me^2}{4\pi\hbar^2} \tag{12.9}$$

which we can reduce to one constant by rescaling r. Define

$$k^2 = \frac{2m|E|}{\hbar^2} \quad \text{and} \quad r = \frac{\rho}{2k} \tag{12.10}$$

Substitute into (12.8), and we find

$$\frac{\mathrm{d}^2u}{\mathrm{d}\rho^2} - \frac{l(l+1)}{\rho^2}u + \left(\frac{\lambda}{\rho} - \frac{1}{4}\right)u = 0 \tag{12.11}$$

where we have defined

$$\lambda = \frac{me^2}{k4\pi\hbar^2} = \frac{1}{ka_0} \tag{12.12}$$

(a_0 is the Bohr radius $4\pi\hbar^2/me^2$). The problem is reduced to finding values λ and functions $u(\rho)$ which satisfy (12.11).

The strategy for solving (12.11) is to first solve the equation in the asymptotic limits $\rho \to \infty$ and $\rho \to 0$. With the solutions for very large and very small ρ in hand, we then look for a solution which interpolates between these two regions.

Begin with $\rho \to \infty$. Then terms proportional to $1/\rho^2$ and $1/\rho$ can be dropped, and the equation becomes

$$\frac{d^2u}{d\rho^2} - \frac{u}{4} = 0 \tag{12.13}$$

which has the general solution

$$u(\rho) = Ae^{-\rho/2} + Be^{\rho/2} \tag{12.14}$$

The second term, proportional to the constant B, is non-normalizable, therefore $B = 0$ for physical states. Hence, we find that

$$u(\rho) \to Ae^{-\rho/2} \qquad \text{as} \qquad \rho \to \infty \tag{12.15}$$

Next, consider the $\rho \to 0$ regime. Then terms proportional to a constant, or $1/\rho$, are negligible compared to the term proportional to $1/\rho^2$, so the radial equation is approximately

$$\frac{d^2u}{d\rho^2} - \frac{l(l+1)}{\rho^2}u = 0 \tag{12.16}$$

which, one can easily check, is solved by

$$u(\rho) = C\rho^{l+1} + D\rho^{-l} \tag{12.17}$$

For any value $l > 0$ the wave function (proportional to u/r) is non-normalizable for $D \neq 0$, and even at $l = 0$ the ratio u/r is non-differentiable at $r = 0$ if $D \neq 0$. So we will set $D = 0$ also.

Since we now know that $u(\rho) \sim e^{-\rho/2}$ at large ρ, and $u(\rho) \sim \rho^{l+1}$ at small ρ, we can guess that the exact solution might have the form

$$u(\rho) = e^{-\rho/2}\rho^{l+1}F(\rho) \tag{12.18}$$

where $F(\rho)$ is some unknown function, which goes to a constant as $\rho \to 0$, and grows slower than an exponential as $\rho \to \infty$. This kind of educated guess regarding the form of the solution of an equation is called an ***ansatz***. Substituting (12.18) in (12.11) gives an equation for $F(\rho)$

$$\left[\rho\frac{d^2}{d\rho^2} + (2l + 2 - \rho)\frac{d}{d\rho} - (l + 1 - \lambda)\right]F(\rho) = 0 \tag{12.19}$$

The next step is to write $F(\rho)$ as a power series

$$F(\rho) = \sum_{j=0}^{\infty} c_j\rho^j \tag{12.20}$$

and insert into (12.19) to get

$$\begin{gathered}\rho\sum_j c_j j(j-1)\rho^{j-2} + \sum_j(2l + 2 - \rho)c_j j\rho^{j-1} - \sum_j(l + 1 - \lambda)c_j\rho^j = 0\\ \sum_j\{c_j j(j-1) + c_j j(2l+2)\}\rho^{j-1} - \sum_j\{c_j j + c_j(l + 1 - \lambda)\}\rho^j = 0\end{gathered} \tag{12.21}$$

Since this equation has to be true for *every* value of ρ, it follows that the coefficient of each power of ρ must vanish. The coefficient in front of ρ^j is

$$c_{j+1}(j+1)j + c_{j+1}(2l+2)(j+1) - c_j j - c_j(l + 1 - \lambda) = 0 \tag{12.22}$$

or

$$c_{j+1}(j+1)(2l + 2 + j) - c_j(l + 1 + j - \lambda) = 0 \tag{12.23}$$

This equation allows us to compute the coefficients $\{c_j\}$ iteratively:

$$c_{j+1} = \frac{j + l + 1 - \lambda}{(j+1)(j + 2l + 2)}c_j \tag{12.24}$$

However, we have also required that $F(r)$ grow more slowly than an exponential, in order that $u(\rho) \sim \exp[-\rho/2]$ as $\rho \to \infty$. It turns out that this can only be true for very special values of the constant λ. Recall that the asymptotic behavior of a solution of the Schrödinger equation can be *either* $\exp[+\rho/2]$ *or* $\exp[-\rho/2]$. Suppose it turns out that $F(\rho) \sim \exp(\rho)$ as $\rho \to \infty$. Then we would end up with the non-normalizable solution for $u(r)$. In fact, for almost all values of λ, this is exactly what happens. According to (12.24), at large index j

$$c_{j+1} \approx \frac{c_j}{j+1} \tag{12.25}$$

which implies

$$c_j \approx \frac{\text{const.}}{j!} \tag{12.26}$$

This means that at large ρ, where the power series in ρ is dominated by the terms at large index j,

$$F(\rho) = \sum_j c_j \rho^j \approx \text{const.} \times \sum_j \frac{1}{j!}\rho^j \approx \text{const.} \times e^\rho \tag{12.27}$$

in which case

$$u(\rho) \approx \rho^{l+1} e^{\rho/2} \tag{12.28}$$

which is non-normalizable. So, even though we started with an ansatz (12.18) which seemed to incorporate the asymptotic behavior we want, the equation for F ends up generating, for almost any choice of λ, the asymptotic behavior that we do not want. This situation is already familiar from the discussion in chapter 8. Typically, solutions of the Schrödinger equation for $E < 0$ are non-normalizable for almost all values of E. Only at a discrete set of bound-state energies is it possible to have the wave function fall to zero asymptotically.

For the hydrogen atom, the only way to have a normalizable solution is if one of the c_j coefficients vanishes, i.e. for some $j = j_s$, we have $c_{j_s+1} = 0$. Then, from equation (12.24),

$$c_j = 0 \qquad \text{for all} \qquad j > j_s \tag{12.29}$$

and

$$F(\rho) = \sum_{j=0}^{j_s} c_j \rho^j \sim c_{j_s} \rho^{j_s} \qquad \text{as} \qquad \rho \to \infty \tag{12.30}$$

Since $F(\rho)$ is a polynomial, $R(r) = u(\rho)/r$ will be normalizable.

From equation (12.24), we see that the condition for c_{j_s+1} to vanish, given $c_{j_s} \neq 0$, is that

$$\lambda = n = j_s + l + 1 \tag{12.31}$$

where n is an integer. Therefore, the bound-state energies of the hydrogen atom are deduced from the condition that

$$\lambda = \frac{1}{ka_0} = n \tag{12.32}$$

The integer n is known as the ***principal quantum number***. Using

$$k = \frac{\sqrt{2m|E|}}{\hbar} \qquad \text{and} \qquad a_0 = \frac{4\pi\hbar^2}{me^2} \tag{12.33}$$

the bound-state energies, labeled by the integer n, are

$$E_n = -\frac{me^4}{2n^2(4\pi)^2\hbar^2} \tag{12.34}$$

This is the same result (2.38) as that obtained by Niels Bohr; it is in agreement with the experimental result (2.39) of chapter 2 for the hydrogen spectrum.

We can now write

$$\rho = 2kr = \frac{2r}{na_0} \tag{12.35}$$

Putting everything together, the energy eigenstates are

$$\begin{aligned}
\varphi_{nlm}(r, \theta, \phi) &= N_{nl}R_{nl}(r)\,Y_{lm}(\theta, \phi) \\
&= N_{nl}\frac{u_{nl}(2r/na_0)}{r}Y_{lm}(\theta, \phi) \\
\rho &= \frac{2r}{na_0} \\
u_{nl}(\rho) &= \rho^{l+1}\mathrm{e}^{-\rho/2}F_{nl}(\rho) \\
F_{nl}(\rho) &= \sum_{j=0}^{n-l-1} c_j\rho^j \\
c_{j+1} &= \frac{j+l+1-n}{(j+1)(j+2l+2)}c_j \\
c_0 &= 1
\end{aligned} \tag{12.36}$$

with the normalization constants N_{nl} determined from the normalization condition

$$\begin{aligned}
1 &= \int_0^\infty \mathrm{d}r r^2 \int_0^\pi \mathrm{d}\theta \sin\theta \int \mathrm{d}\phi\; \varphi_{nlm}^*(r, \theta, \phi)\varphi_{nlm}(r, \theta, \phi) \\
&= N_{nl}^2 \int_0^\infty \mathrm{d}r r^2 R_{nl}^2(r) \\
&= N_{nl}^2 \int_0^\infty \mathrm{d}r\; u_{nl}^2(2r/na_0)
\end{aligned} \tag{12.37}$$

and corresponding to an energy eigenvalue

$$E_n = -\frac{me^4}{2n^2(4\pi)^2\hbar^2} \tag{12.38}$$

The choice $c_0 = 1$ is arbitrary. A different choice would simply lead to a different normalization constant, and the wave function would end up exactly the same.

It is clear that the energy eigenvalues are degenerate. To specify a unique eigenstate $\varphi_{nlm}(r, \theta, \phi)$, it is necessary to specify three integers: n, l, m. On the other hand, the energy eigenvalue E_n depends only on one of those integers. To determine the degree of degeneracy, i.e. the number of linearly independent eigenstates with the same energy E_n, we observe that the maximum index j_s of the (non-zero) c_j coefficients satisfies

$$j_s = n - l - 1 \geqslant 0 \tag{12.39}$$

and therefore

$$l = 0, 1, 2, \ldots, n - 1 \tag{12.40}$$

Also, there are $2l + 1$ values of m,

$$m = -l, -l + 1, \ldots, l - 1, l \tag{12.41}$$

corresponding to a given value of l. Therefore, the total number of l, m combinations that can exist for a given integer n is

$$\begin{aligned} \text{degeneracy of } E_n &= \sum_{l=0}^{n-1}(2l + 1) \\ &= \left(2\sum_{l=0}^{n-1} l\right) + n \\ &= n^2 \end{aligned} \tag{12.42}$$

In atomic physics, the integer values of the l quantum number are assigned letters, namely:

$$\begin{aligned} l &= 0 \quad \text{S} \\ l &= 1 \quad \text{P} \\ l &= 2 \quad \text{D} \\ l &= 3 \quad \text{F} \end{aligned} \tag{12.43}$$

Beyond $l = 3$, the assignment is alphabetical, i.e. $l = 3, 4, 5, 6\ldots$ correspond to 'F, G, H, I, …'. respectively. (The reasons for these letter assignments go back to the history of spectroscopy, and need not concern us here.)[2] Thus the $\{nlm\} = \{100\}$ state is referred to as the '1S' state; the $\{nlm\} = \{21m\}$ states are referred to as the '2P' states, and so on, as indicated in table 12.1.

[2] 'S' stands for the 'sharp' series of spectral lines, 'P' stands for the 'principal' series, 'D' for 'diffuse', and 'F' for 'fundamental'.

Table 12.1. Hydrogen atom quantum numbers and spectroscopic notation.

n	1	2		3		
l	0	0	1	0	1	2
notation	1S	2S	2P	3S	3P	3D
m	0	0	−1, 0, 1	0	−1, 0, 1	−2, −1, 0, 1, 2
degeneracy	1	4		9		

We now use the formulas in equation (12.36) to calculate the 1S and 2S wave functions explicitly.

- **The 1S ground state**. Since $j_s = n - l - 1 \geqslant 0$, the smallest possible value for n is $n = 1$, which implies that $l = m = 0$. This is the lowest energy state, or 'ground state' of the hydrogen atom; it is the state where the binding energy is greatest. According to (12.36) we have

$$\begin{aligned}\varphi_{100} &= N\frac{u_{10}(2r/a_0)}{r}Y_{00} \\ &= \frac{N}{\sqrt{4\pi}}\frac{u_{10}(2r/a_0)}{r}\end{aligned} \tag{12.44}$$

where

$$\begin{aligned}u_{10}(\rho) &= \mathrm{e}^{-\rho/2}\rho F_{10}(\rho) \\ &= \mathrm{e}^{-\rho/2}\rho\sum_{j=0}^{1-0-1}c_j\rho^j \\ &= \mathrm{e}^{-\rho/2}\rho\end{aligned} \tag{12.45}$$

Therefore

$$\varphi_{100} = \frac{N}{\sqrt{4\pi}}\frac{2}{a_0}\mathrm{e}^{-r/a_0} \tag{12.46}$$

We must normalize the wave function to determine N:

$$\begin{aligned}1 &= \int r^2\mathrm{d}r\mathrm{d}\Omega\,|\varphi_{100}|^2 \\ &= N^2\int \mathrm{d}r r^2 u_{10}(2r/a_0)^2 \\ &= N^2\left(\frac{2}{a_0}\right)^2\int_0^\infty \mathrm{d}r\ r^2\mathrm{e}^{-2r/a_0} \\ &= N^2 a_0\end{aligned} \tag{12.47}$$

So, finally, the ground-state wave function is

$$\varphi_{100}(r, \theta, \phi) = \frac{1}{\sqrt{\pi a_0^3}} e^{-r/a_0} \tag{12.48}$$

- **The 2S excited state**. The $n = 2$ states are known as the 'first excited' states, since they correspond to the lowest energy excitation above the ground state. The $l = 0$ state is always referred to as the S-state; so the $\{nlm\} = \{200\}$ is known as the 2S state, in spectroscopic notation. Again, using equation (12.36)

$$\begin{aligned}
\phi_{200}(r, \theta, \phi) &= N \frac{u_{20}(r/a_0)}{r} Y_{00} \\
u_{200}(\rho) &= e^{-\rho/2} \rho F_{20}(\rho) \\
&= e^{-\rho/2} \rho \sum_{j=0}^{2-0-1} c_j \rho^j \\
&= e^{-\rho/2} \rho (c_0 + c_1 \rho) \\
c_1 &= \frac{0 + l + 1 - n}{(0+1)(0+2l+2)} c_0 \\
&= -\frac{1}{2}
\end{aligned} \tag{12.49}$$

so that

$$\begin{aligned}
\phi_{200} &= \frac{N}{\sqrt{4\pi}} \left(1 - \frac{1}{2}\rho\right) \frac{\rho}{r} e^{-\rho/2} \\
&= \frac{N}{a_0 \sqrt{4\pi}} \left(1 - \frac{r}{2a_0}\right) e^{-r/2a_0}
\end{aligned} \tag{12.50}$$

normalization gives $N = (2a_0)^{-1/2}$, and therefore

$$\varphi_{200} = \frac{1}{\sqrt{8\pi a_0^3}} \left(1 - \frac{r}{2a_0}\right) e^{-r/2a_0} \tag{12.51}$$

In general, the wave function for quantum numbers n, l, m has the form

$$\varphi_{nlm} = (\text{polynomial in } r \text{ of order } \leqslant n-1) \times e^{-r/na_0} Y_{lm}(\theta, \phi) \tag{12.52}$$

A polynomial can go through zero, so in general the probability distribution in the radial direction has 'bumps', whose positions depend on both n and l. In particular, if we are only interested in the probability of finding an electron between radii r_1 and r_2, then

$$\begin{aligned} \text{prob}(r_1 < r < r_2) &= \int_{r_1}^{r_2} \mathrm{d}r r^2 \int \mathrm{d}\Omega\, \varphi^*_{nlm}\varphi_{nlm} \\ &= \int_{r_1}^{r_2} \mathrm{d}r\, r^2 R^2_{nl}(r) \\ &= \int_{r_1}^{r_2} \mathrm{d}r\, P(r) \end{aligned} \tag{12.53}$$

where we define the radial probability density

$$\begin{aligned} P(r) &= r^2 R^2_{nl}(r) \\ &= (2n\text{th}-\text{order polynomial in } r) \times \mathrm{e}^{-2r/na_0} \end{aligned} \tag{12.54}$$

12.1 The scale of the world

We know that the volume of a gram of water is one cubic centimeter. Why is the volume not instead one cubic kilometer, or one cubic parsec?

If you burn a log in a fireplace, enough energy is released to warm the room. If you burn instead a stick of dynamite, the energy released will destroy the room, and probably the rest of the house. But neither the log nor the dynamite will release enough energy to flatten a city, or vaporize the planet. Why not?

Quantum mechanics answers these questions, given as input the masses of the electron and proton, and the electron charge. In fact, the solution of the hydrogen atom provides us with rough, order-of-magnitude estimates for the volume of a mole of solid anything, and the energy released by burning a mole of anything. This is because the volume of a solid depends mainly on the number of atoms, and the size of each atom. The energy released in a chemical reaction also depends on the number of molecules participating in the reaction, and the energy released by each molecule. The number of molecules in a mole is known as Avogadro's number, which is

$$\begin{aligned} N_A &= 6.02 \times 10^{23} \\ &\approx \frac{1 \text{ gram}}{m_{\text{proton}} \text{ in grams}} \end{aligned} \tag{12.55}$$

The proton mass gives slightly the wrong value, due to the proton–neutron mass difference. Anyway, N_A tells us approximately how many protons or neutrons there are in a gram, or how many molecules in a mole (= (no. of protons + neutrons per molecule) × 1 gram). Quantum mechanics then predicts, in principle, the *volume* of each molecule, and the energy released in any given chemical reaction.

Of course, using quantum mechanics to find the precise volume of any given molecule is a formidable task. But we can use the solution of the hydrogen atom to make order-of-magnitude estimates. The volume of any given molecule is certainly larger than that of the hydrogen atom, but usually not by much more than an order of magnitude. Likewise, the energy released in a chemical reaction is typically less than the hydrogen atom binding energy, but this is still a rough order-of-magnitude estimate. If we are satisfied with accuracy within a couple of orders of magnitude, then we can answer the two questions posed above:

1. What is the volume of a mole of (solid or liquid) anything?
2. How much energy is released by burning a mole of anything?

Start with the first question. We will imagine that all molecules fit inside a cube of length equal to the diameter of the hydrogen atom in its ground state. This gives us a volume

$$v \sim (2a_0)^3 \tag{12.56}$$

where a_0 is the Bohr radius

$$a_0 = \frac{4\pi\hbar^2}{me^2} = 5 \times 10^{-9}\ \text{cm} \tag{12.57}$$

so that

$$v \sim 10^{-24}\ \text{cm}^3 \tag{12.58}$$

Then the volume of a mole is

$$V_{\text{mole}} = N_A v \sim 0.6\ \text{cm}^3 \tag{12.59}$$

which is at least the right order of magnitude. The volume of a mole of anything is a few cubic centimeters, not cubic kilometers or cubic parsecs.

To estimate the energy released by burning a mole of anything, assume that the energy released per molecule is of the order of magnitude of the binding energy of an electron in the hydrogen atom ground state. This is surely not very accurate, but again, we are not going to worry about a few powers of ten. The binding energy in SI units is

$$|E_1| = \frac{me^4}{2(4\pi)^2\hbar^2} \approx 2 \times 10^{-18}\ \text{J} \tag{12.60}$$

Then burning a mole of anything would release an energy E on the order

$$E \sim N_A|E_1| \sim 1200\ \text{kJ} \tag{12.61}$$

which is sufficient to turn half a kilogram of water, at 100 °C and 1 atm pressure, completely into steam. Released in a short time this would make a good bang, but it certainly is not enough energy to flatten a city.

In this way we can use quantum mechanics to estimate the 'scale of the world'; i.e. volume per gram, energy released in chemical reactions, and so on. As explained in chapter 6, the size and binding energy of the hydrogen atom is ultimately due to an interplay between the attractive Coulomb potential and a kinetic energy coming from the uncertainty principle. It is this interplay that sets the scale of the world. Of course, the argument is a little incomplete: it required as input the masses of the electron and proton, as well as the electron charge.[3] Where do *these* numbers come from? A once-popular idea is that all of these numbers would one day be derived from an ultimate 'Theory of Everything'. Nowadays this notion is regarded with some skepticism. For one thing, it seems that if some of these numbers (quark masses, say) were only slightly different, life would be impossible. It is a lot to ask of a theory that things come out just right so that life can exist! One idea, motivated by string theory, is that our observable Universe is part of an enormous 'multiverse'. In almost all regions of this multiverse, the physical constants are such that life is impossible. Life is only possible in very rare and special regions, such as the region we live in. For a non-technical exposition of this idea, see [1].

Problems

1. For the ground state of the hydrogen atom, calculate
   a) the mean radius $\langle r \rangle$ of the ground state of the hydrogen atom (give the answer in centimeters). Compare this value to the Bohr radius;
   b) the expectation values of kinetic energy $p^2/2m$ and potential energy $-e^2/4\pi r$.
2. Find the energy levels of a particle of mass m, moving in the spherically symmetric potential

$$V(r) = \frac{A}{r^2} - \frac{B}{r} \tag{12.62}$$

3. Angular momentum is conserved in any central potential, but in the special case of a Coulomb potential, the vector quantity

$$\mathbf{A} = \frac{1}{2m}(\mathbf{p} \times \mathbf{L} - \mathbf{L} \times \mathbf{p}) - \frac{e^2}{r}\mathbf{r} \tag{12.63}$$

known as the ***Laplace–Runge–Lenz vector*** is also conserved. Show that the operator $\tilde{\mathbf{A}}$ is Hermitian, and commutes with the Hamiltonian. As a general rule, the more linearly independent operators there are which commute with the Hamiltonian but not with each other, the greater the energy degeneracy. The fact that the Hamiltonian for a Coulomb potential commutes not only with **L** but also with **A** is the underlying reason why the energy eigenvalues are more degenerate in the Coulomb case (where the energy depends only on the principal quantum number, and not on l) than in other central potentials.

[3] Nowadays it is possible to calculate the mass of the proton from the theory of the strong interactions (QCD), but this still requires input of a dimensionful constant known as 'QCD Λ'.

4. For some purposes, the electric charge distribution due to an electron in an n, l, m state of the hydrogen atom can be regarded as proportional to the electron positional probability density, i.e.

$$\rho(r, \theta, \phi) = -e\left|\phi_{nml}(r, \theta, \phi\right|^2 \qquad (12.64)$$

Treating the nucleus as a point source of charge $+e$, and using classical electrostatics, make a plot of the electric potential of the hydrogen atom in its ground state as a function of r.

5. Consider an electron in a potential

$$V(x, y, z) = \begin{cases} -\dfrac{e^2}{4\pi r} & z > 0 \\ \infty & z \leqslant 0 \end{cases} \qquad (12.65)$$

Find the energy eigenstates and eigenvalues.

Reference

[1] Susskind L 2005 *The Cosmic Landscape: String Theory and the Illusion of Intelligent Design* (Boston, MA: Little, Brown)

Part III

Aspects of spin

IOP Publishing

An Introduction to Quantum Theory

Jeff Greensite

Chapter 13

Electron spin

Let us go back, for a moment, to the classical picture of the hydrogen atom, with an electron moving in a circular orbit around the proton. A charge moving in a circle constitutes a current loop, and, according to the usual laws of electromagnetism a current loop is associated with a magnetic moment according to the rule

$$\mu = \frac{1}{c}IA \tag{13.1}$$

where μ is the magnetic moment, I is the current, and A is the area enclosed by the circular orbit. The current is the charge per unit length of loop, times the velocity, so in this case, for a circular orbit of radius r

$$\begin{aligned} I &= -\frac{e}{2\pi r}v \\ v &= \frac{p}{M} = \frac{L}{Mr} \\ A &= \pi r^2 \end{aligned} \tag{13.2}$$

where M is the electron mass. Putting it all together, and noting that as vector quantities both $\boldsymbol{\mu}$ and $\mathbf{L}$ are normal to the loop, we obtain

$$\boldsymbol{\mu} = -\frac{e}{2Mc}\mathbf{L} \tag{13.3}$$

(note that the radius r drops out of the expression). Finally, in an external magnetic field $\mathbf{B}$, the interaction energy between the magnetic dipole and the field is given by

$$\begin{aligned} E_{\text{mag}} &= -\mathbf{B}\cdot\boldsymbol{\mu} \\ &= \frac{e}{2Mc}\mathbf{B}\cdot\mathbf{L} \end{aligned} \tag{13.4}$$

doi:10.1088/978-0-7503-1167-0ch13

Suppose then that a hydrogen atom is sitting in an external constant magnetic field, directed along the z-axis. In that case, the Hamiltonian relevant for the motion of the electron is

$$H = H_0 + \frac{e}{2Mc} B_z L_z \tag{13.5}$$

where H_0 is the hydrogen atom Hamiltonian in the absence of an external field

$$H_0 = \frac{p^2}{2M} - \frac{e^2}{4\pi r} \tag{13.6}$$

Going over to quantum theory, H_0 and L_z become operators which, as it happens, commute with each other: an eigenstate φ_{nlm} of $\tilde{H}_0$ is also an eigenstate of $\tilde{L}_z$

$$\tilde{L}_z \varphi_{nlm} = m\hbar\varphi_{nlm} \tag{13.7}$$

Therefore, an eigenstate of $\tilde{H}_0$ is an eigenstate of the total Hamiltonian

$$\tilde{H}\varphi_{nlm} = \left(E_n^0 + mB_z \frac{e\hbar}{2Mc}\right)\varphi_{nlm} \tag{13.8}$$

with energy eigenvalues

$$E_{nlm} = E_n^0 + mB_z \frac{e\hbar}{2Mc} \tag{13.9}$$

where E_n^0 are the energy eigenvalues of the electron when there is no external field.

This means that if we insert a collection of hydrogen atoms between the poles of a magnet, then spectral lines corresponding to transitions between principal quantum numbers n_1 and n_2 should split into many spectral lines, corresponding to transitions between states with different values of the L_z quantum numbers m_1 and m_2. There are certain ***selection rules*** governing which transitions are possible; these rules will be derived in a later chapter. For now it will be sufficient to state the result relevant to the Paschen–Back effect discussed below: an electron can only make transitions between energy levels for which $|l_2 - l_1| = 1$, and $m_2 - m_1 = 1, 0, -1$. So let us consider transitions between the 2P and 1S levels of the hydrogen atom. Then according to (13.9) there should be three spectral lines resulting from this transition, whereas in the absence of a magnetic field we would expect only one line. This splitting of spectral lines in the presence of a magnetic field is known as the ***Zeeman effect***.

13.1 The Paschen–Back effect and the Stern–Gerlach experiment

It turns out, however, that there are actually more spectral lines in the 2P → 1S transition than would be expected from (13.9). This is known as the ***anomalous Zeeman effect***, and it led Samuel Goudsmidt and George Uhlenbeck to propose the existence of electron spin in 1925. The analysis is easiest when the magnetic field is very strong; this limit is known as the ***Paschen–Back effect***. For very strong magnetic fields there are three pairs of closely spaced spectral lines in the 2P → 1S transition, rather than just the expected three lines. This extra multiplicity is not unique to the 2P → 1S transition, but occurs in all atomic transitions in a strong magnetic field. It is found that the extra spectral lines would be accounted for if there are twice as many energy levels as one would expect, for each pair of quantum numbers n and l. If, instead of the $2l + 1$ levels one would expect (because there are $2l + 1$ values of m for each l), there would be *two* sets of $2l + 1$ levels, with energies

$$\begin{aligned} E^{+}_{nlm} &\approx E^{0}_{n} + (m+1)B_z\frac{e\hbar}{2Mc} \\ E^{-}_{nlm} &\approx E^{0}_{n} + (m-1)B_z\frac{e\hbar}{2Mc} \end{aligned} \tag{13.10}$$

then, together with the selection rules, this would explain the splitting of spectral lines. Note the approximation (≈) sign. If (13.10) were an equality, then we would have $E^{+}_{nl,\,m-1} = E^{-}_{nl,\,m+1}$, which is not quite right. In fact there are other effects which split this degeneracy slightly.[1] But what can explain the doubling of energy levels?

The most natural explanation is that, in addition to the magnetic moment due to its orbital angular momentum, an electron also has an intrinsic magnetic moment, associated with an intrinsic spin angular momentum. Classically, after all, any spinning charge acquires a magnetic moment in the direction of the spin angular momentum. So let us suppose that the intrinsic magnetic moment of the electron is

$$\boldsymbol{\mu} = -\frac{eg}{2Mc}\mathbf{S} \tag{13.11}$$

where g is some constant, known as the ***gyromagnetic ratio***. In quantum theory S must be a Hermitian operator, and we assume that it satisfies the same commutation relations as orbital angular momentum

$$\begin{aligned} \left[\tilde{S}_x, \tilde{S}_y\right] &= i\hbar\tilde{S}_z \\ \left[\tilde{S}_y, \tilde{S}_z\right] &= i\hbar\tilde{S}_x \\ [\tilde{S}_z, \tilde{S}_x] &= i\hbar\tilde{S}_y \end{aligned} \tag{13.12}$$

From these commutation relations alone, we know from the discussion in chapter 11 that the possible eigenvalues of $\tilde{S}^2$ and $\tilde{S}_z$ are

$$\begin{aligned} \tilde{S}^2 &= s(s+1)\hbar^2 \qquad & s &= 0, \frac{1}{2}, 1, \frac{3}{2}, \ldots \\ \tilde{S}_z &= s_z\hbar & -s &\leqslant s_z \leqslant s \end{aligned} \tag{13.13}$$

Taking the electron magnetic moment into account, the total Hamiltonian is then

$$\begin{aligned} \tilde{H} &= \tilde{H}_0 + \frac{e}{2Mc}B_z\tilde{L}_z - (\mu_e)_z B_z \\ \tilde{H} &= \tilde{H}_0 + \frac{e}{2Mc}B_z\left(\tilde{L}_z + g\tilde{S}_z\right) \end{aligned} \tag{13.14}$$

[1] One effect is known as ***spin–orbit coupling***, and we will come to that in due course.

The electron spin is independent of the electron position and momentum, therefore we may assume that

$$[\tilde{H}_0, \tilde{S}_z] = [L_z, \tilde{S}_z] = 0 \tag{13.15}$$

and this means that operators $\tilde{H}_0$, $\tilde{L}_z$, $\tilde{S}_z$ have a common set of eigenstates, which we denote $|nlms_z\rangle$. Then

$$\begin{aligned} \tilde{H}|nlms_z\rangle &= E_{nms_z}|nlms_z\rangle \\ E_{nms_z} &= E_n^0 + \frac{e\hbar}{2Mc}B_z(m + gs_z) \end{aligned} \tag{13.16}$$

Comparison to (13.10) shows that we get agreement if the electron has an intrinsic spin

$$s = \frac{1}{2} \quad \Rightarrow \quad s_z = \pm\frac{1}{2} \tag{13.17}$$

and gyromagnetic ratio

$$g \approx 2 \tag{13.18}$$

so that E^+ corresponds to $s_z = \frac{1}{2}$, and E^- to $s_z = -\frac{1}{2}$, and in general

$$E_{nms_z} \approx E_n^0 + \frac{e\hbar}{2Mc}B_z(m + 2s_z) \tag{13.19}$$

where we have inserted an approximation sign to allow for a small fine-structure effect, alluded to earlier, which splits levels that would otherwise be degenerate. From (13.19) the origin of the six spectral lines in the 2P → 1S transition is clear, as illustrated in figure 13.1.

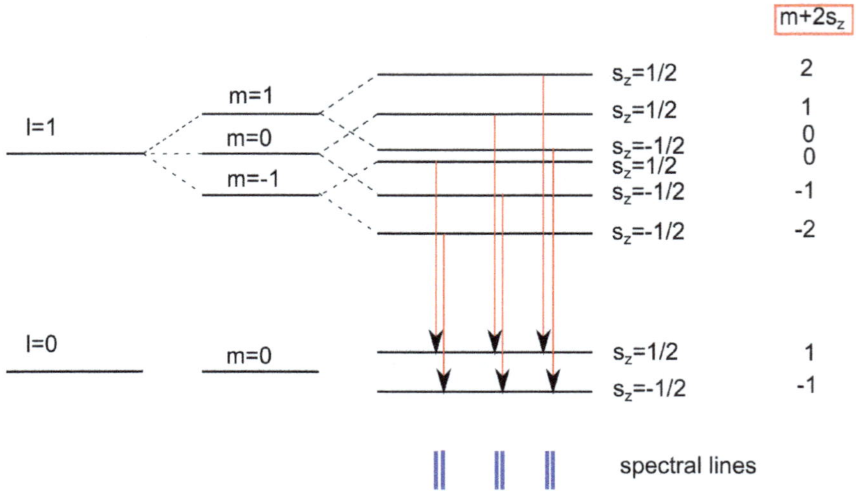

Figure 13.1. Atomic transitions in the Paschen–Back effect.

An independent check of the double-valued character of the electron magnetic moment is provided by the ***Stern–Gerlach experiment***, in which a beam of electrons is sent through a (non-uniform) magnetic field, oriented (mainly) in the z-direction, as shown in figure 13.2. Classically, the force exerted on a dipole μ in a non-uniform magnetic field is

$$\mathbf{F} = \nabla(\boldsymbol{\mu} \cdot \mathbf{B}) \tag{13.20}$$

so the z-component of the force is

$$\begin{aligned} F_z &= \frac{\partial B_z}{\partial z}\mu_z \\ &= -\frac{eg}{2Mc}\frac{\partial B_z}{\partial z}S_z \end{aligned} \tag{13.21}$$

A measurement of the deflection of the electron beam is therefore going to be a measurement of the z-component of electron spin. If $s = \frac{1}{2}$, then there are only two possibilities: $s_z = \pm\frac{1}{2}$, and the force is oriented either up or down along the z-axis, with equal magnitude. Each electron should be deflected along one of only two trajectories, as shown in figure 13.2. This is, in fact what is found experimentally. The Stern–Gerlach apparatus is therefore a way of measuring the component of electron spin along a given axis, for any given electron. If deflected upwards, the

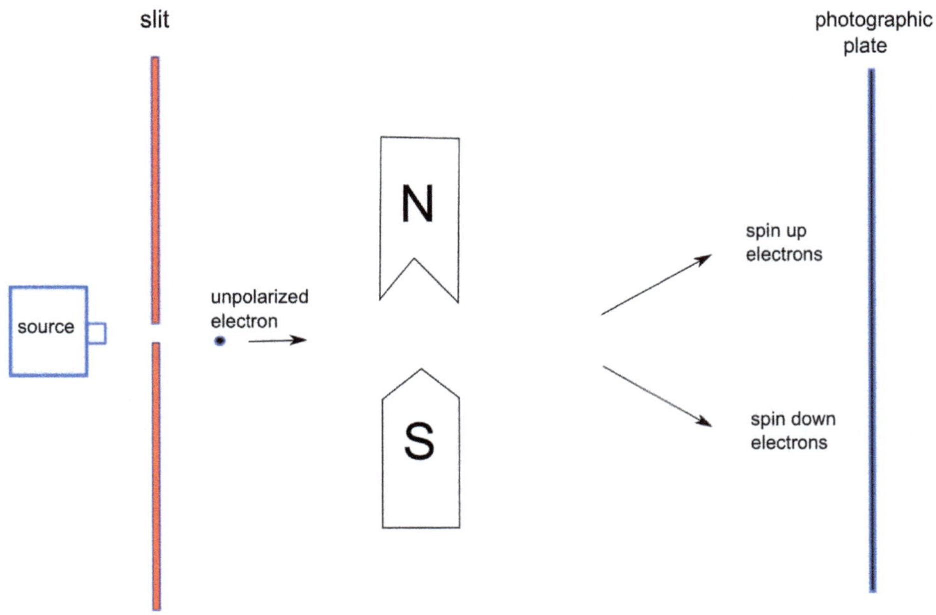

Figure 13.2. The Stern–Gerlach apparatus for measuring electron spin.

electron is in an $S_z = +\frac{1}{2}\hbar$ (or 'spin up') eigenstate. If the electron is deflected downwards, it is in an $S_z = -\frac{1}{2}\hbar$ (or 'spin down') eigenstate. A very careful analysis of the results will also show that it is necessary to set the gyromagnetic ratio to $g = 2$, although this is not quite as obvious as in the Paschen–Back effect.

As already mentioned, any spinning charged object has a magnetic moment, according to the laws of classical electromagnetism. Should we then think of the electron as being *literally* a spinning ball of charge? There are two main problems with such a picture. First, if we imagine that the electron is composed of some substance with a constant charge-to-mass ratio of e/M, then, generalizing only slightly the reasoning that applies to a current loop, the electron should have a magnetic moment

$$\boldsymbol{\mu} = -\frac{e}{2Mc}\mathbf{S} \tag{13.22}$$

instead of what is actually found, which is about twice that value

$$\boldsymbol{\mu} = -\frac{e}{2Mc}g\mathbf{S} \tag{13.23}$$

with gyromagnetic ratio $g \approx 2$. The second reason is simply the fact that, in this picture, the electron spin angular momentum is just the orbital angular momentum of electron 'substance'. But, as we have already seen, orbital angular momentum can only have values $l = 0, 1, 2, \ldots$; in particular, $l = \frac{1}{2}$ is ruled out. So the picture of an electron as a spinning volume of charged 'electron stuff' cannot easily account for the origin of the electron magnetic moment.

13.2 Spin wave functions

The next question is how to write the wave function of an eigenstate of spin angular momentum. If the electron were a spinning ball of charge, then it would be natural to represent the spin wave function by spherical harmonics $Y_{ss_z}(\theta', \phi')$, where $\partial_t\theta'$ is the classical angular velocity of the spinning ball. As we have seen, the 'ball of charge' picture is inadequate, and in any case there are no spherical harmonics with $s = \frac{1}{2}$. So, what to do?

For the moment let's forget all about the x, y, z-degrees of freedom of the particle, and concentrate just on the spin. Now, although we do not have spherical harmonics for $s = 1/2$, we can still represent the eigenstates of $\tilde{S}^2$ and $\tilde{S}_z$ by two orthonormal ket vectors

$$\left|s = \frac{1}{2}, s_z = \frac{1}{2}\right\rangle \quad \text{and} \quad \left|s = \frac{1}{2}, s_z = -\frac{1}{2}\right\rangle \tag{13.24}$$

where

$$\tilde{S}^2\left|\frac{1}{2}, \frac{1}{2}\right\rangle = \frac{3}{4}\hbar^2\left|\frac{1}{2}, \frac{1}{2}\right\rangle$$
$$\tilde{S}_z\left|\frac{1}{2}, \frac{1}{2}\right\rangle = \frac{1}{2}\hbar\left|\frac{1}{2}, \frac{1}{2}\right\rangle$$
$$\tilde{S}^2\left|\frac{1}{2}, -\frac{1}{2}\right\rangle = \frac{3}{4}\hbar^2\left|\frac{1}{2}, -\frac{1}{2}\right\rangle \tag{13.25}$$
$$\tilde{S}_z\left|\frac{1}{2}, -\frac{1}{2}\right\rangle = -\frac{1}{2}\hbar\left|\frac{1}{2}, -\frac{1}{2}\right\rangle$$

Then (if we disregard position dependence) *any* $s = \frac{1}{2}$ state can be represented as a superposition

$$|\psi\rangle = a\left|\frac{1}{2}, \frac{1}{2}\right\rangle + b\left|\frac{1}{2}, -\frac{1}{2}\right\rangle \tag{13.26}$$

Now this looks exactly like the way we would represent a vector in a two-dimensional space. So before going on, it is worth recalling a few facts of vector algebra.

Let $\mathbf{e}_1$ and $\mathbf{e}_2$ be two orthonormal vectors (i.e. orthogonal unit vectors) in a two-dimensional space. For example, $\mathbf{e}_1$ and $\mathbf{e}_2$ could be unit vectors along the x- and y-axes, respectively. Orthonormality means that

$$\mathbf{e}_1 \cdot \mathbf{e}_1 = 1$$
$$\mathbf{e}_2 \cdot \mathbf{e}_2 = 1 \tag{13.27}$$
$$\mathbf{e}_1 \cdot \mathbf{e}_2 = 0$$

If we have a set of D orthonormal vectors in a D-dimensional space, those vectors are known as a *basis* for the vector space, and any vector can be expressed as a linear combination of those basis vectors. In the case we are considering, $D = 2$, any vector can be written in the form

$$\mathbf{v} = a\mathbf{e}_1 + b\mathbf{e}_2 \tag{13.28}$$

It is useful and traditional to represent a (ket) vector in a finite dimensional space as a column of numbers, i.e.

$$\mathbf{v} \leftrightarrow \begin{bmatrix} a \\ b \end{bmatrix} \tag{13.29}$$

This is the column vector representation of the vector $\mathbf{v}$, in the *basis* $\{\mathbf{e}_1, \ \mathbf{e}_2\}$. You will notice that the components of the column vector are inner products of the ket vector $\mathbf{v}$ with the bra vectors $\{\mathbf{e}_1, \ \mathbf{e}_2\}$, i.e.

$$a = \mathbf{e}_1 \cdot \mathbf{v} = \langle e_1|v\rangle$$
$$b = \mathbf{e}_2 \cdot \mathbf{v} = \langle e_2|v\rangle \tag{13.30}$$

We can choose $\mathbf{v} = \mathbf{e}_1$ or $\mathbf{e}_2$, and find the values of a and b for the basis vectors. It is easy to see that the corresponding column vectors are

$$\mathbf{e}_1 \leftrightarrow \begin{bmatrix} 1 \\ 0 \end{bmatrix} \qquad \mathbf{e}_2 \leftrightarrow \begin{bmatrix} 0 \\ 1 \end{bmatrix} \tag{13.31}$$

In chapter 3 we discussed the notion of a linear operator: it is simply a rule for changing any vector $\mathbf{v}$ into another vector $\mathbf{v}'$

$$\mathbf{v}' = M\mathbf{v} \tag{13.32}$$

such that, for any two vectors,

$$M(a\mathbf{u} + b\mathbf{v}) = aM\mathbf{u} + bM\mathbf{v} \tag{13.33}$$

The ***matrix element*** M_{ij} of a linear operator M in the basis $\{\mathbf{e}_k\}$ is given by the inner product

$$M_{ij} = \mathbf{e}_i \cdot M\mathbf{e}_j \tag{13.34}$$

or, in bra-ket notation

$$M_{ij} = \langle e_i|M|e_j\rangle \tag{13.35}$$

Suppose we are given the matrix elements of some linear operator M in a certain basis. Then it is easy to see that the components

$$v'_i = \langle e_i|v'\rangle \tag{13.36}$$

of the transformed vector $|v'\rangle = M|v\rangle$, in terms of the components

$$v_i = \langle e_i|v\rangle \tag{13.37}$$

of the original vector, are given by the usual rule of matrix multiplication

$$\begin{aligned} v'_i &= \langle e_i|v'\rangle \\ &= \langle e_i|M|v\rangle \\ &= \Big\langle e_i\Big|M\sum_j v_j\Big|e_j\Big\rangle \\ &= \sum_j \langle e_i|M|e_j\rangle v_j \\ &= \sum_j M_{ij}v_j \end{aligned} \tag{13.38}$$

In column vector notation, this is written

$$\begin{bmatrix} v'_1 \\ v'_2 \end{bmatrix} = \begin{bmatrix} M_{11} & M_{12} \\ M_{21} & M_{22} \end{bmatrix} \begin{bmatrix} v_1 \\ v_2 \end{bmatrix} \tag{13.39}$$

After this brief excursion in vector algebra, let us return to the problem of electron spin. The idea is that we can express the spin wave function as a column vector of two components, known as a ***spinor***, and express the spin operators as 2×2 matrices. To begin with, the basis vectors are chosen to be the two eigenstates of $\tilde{S}^2$, $\tilde{S}_z$ for $s = \frac{1}{2}$,

$$|e_1\rangle = \left|\frac{1}{2}, \frac{1}{2}\right\rangle \qquad |e_2\rangle = \left|\frac{1}{2}, -\frac{1}{2}\right\rangle \tag{13.40}$$

In this basis, the eigenstates themselves can be written as the column vectors (spinors)

$$\left|\frac{1}{2}, \frac{1}{2}\right\rangle \leftrightarrow \chi_+ \equiv \begin{bmatrix} 1 \\ 0 \end{bmatrix} \qquad \left|\frac{1}{2}, -\frac{1}{2}\right\rangle \leftrightarrow \chi_- \equiv \begin{bmatrix} 0 \\ 1 \end{bmatrix} \tag{13.41}$$

and *any* spin-$\frac{1}{2}$ state can be written as a superposition, in ket notation

$$|\psi\rangle = \psi_+|e_1\rangle + \psi_-|e_2\rangle \tag{13.42}$$

or, in column vector notation,

$$|\psi\rangle \leftrightarrow \begin{bmatrix} \psi_+ \\ \psi_- \end{bmatrix} \tag{13.43}$$

Now we can figure out what the spin operators $\tilde{S}^2$, $\tilde{S}_x$, $\tilde{S}_y$, $\tilde{S}_z$ look like as matrices, using the relationships that were proved, in chapter 11, for *any* operators satisfying the angular momentum commutation relations (13.12). Writing equations (11.51) and (11.96) from chapter 11 in ket-notation, we have

$$\begin{aligned}
\tilde{S}^2|ss_z\rangle &= s(s+1)\hbar^2|ss_z\rangle \\
\tilde{S}_z|ss_z\rangle &= s_z\hbar|ss_z\rangle \\
\tilde{S}_-|ss_z\rangle &= \sqrt{(s+s_z)(s-s_z+1)}\,\hbar|s, s_z-1\rangle \\
\tilde{S}_+|ss_z\rangle &= \sqrt{(s-s_z)(s+s_z+1)}\,\hbar|s, s_z+1\rangle \\
\tilde{S}_x &= \frac{1}{2}\left(\tilde{S}_+ + \tilde{S}_-\right) \\
\tilde{S}_y &= \frac{1}{2i}\left(\tilde{S}_+ - \tilde{S}_-\right)
\end{aligned} \tag{13.44}$$

where we have just used the symbol $\tilde{S}$ instead of $\tilde{L}$, s instead of l, and s_z instead of m. For the electron, $s = \frac{1}{2}$ and $s_z = \frac{1}{2}, -\frac{1}{2}$. From these relations, we easily obtain all the matrix elements, e.g.

$$
\begin{aligned}
(S_x)_{12} &= \langle e_1|S_x|e_2\rangle \\
&= \left\langle \frac{1}{2}, \frac{1}{2}\middle| S_x \middle| \frac{1}{2}, -\frac{1}{2}\right\rangle \\
&= \frac{1}{2}\left[\left\langle \frac{1}{2}, \frac{1}{2}\middle| S_+ \middle| \frac{1}{2}, -\frac{1}{2}\right\rangle + \left\langle \frac{1}{2}, \frac{1}{2}\middle| S_- \middle| \frac{1}{2}, -\frac{1}{2}\right\rangle\right] \\
&= \frac{1}{2}\left[\left\langle \frac{1}{2}, \frac{1}{2}\middle| \hbar \middle| \frac{1}{2}, \frac{1}{2}\right\rangle + 0\right] \\
&= \frac{1}{2}\hbar
\end{aligned}
\tag{13.45}
$$

After computing all the needed components, the matrix form of the spin operators, for spin $s = \frac{1}{2}$, are

$$
\begin{aligned}
S^2 &= \frac{3}{4}\hbar^2\begin{bmatrix} 1 & 0 \\ 0 & 1 \end{bmatrix} \qquad S_z = \frac{1}{2}\hbar\begin{bmatrix} 1 & 0 \\ 0 & -1 \end{bmatrix} \\
S_x &= \frac{1}{2}\hbar\begin{bmatrix} 0 & 1 \\ 1 & 0 \end{bmatrix} \\
S_y &= \frac{1}{2}\hbar\begin{bmatrix} 0 & -i \\ i & 0 \end{bmatrix}
\end{aligned}
\tag{13.46}
$$

It is useful to extract a common factor of $\frac{\hbar}{2}$, and to write

$$
S_x = \frac{\hbar}{2}\sigma_x \qquad S_y = \frac{\hbar}{2}\sigma_y \qquad S_z = \frac{\hbar}{2}\sigma_z \tag{13.47}
$$

where we have defined

The Pauli spin matrices

$$
\sigma_x = \begin{bmatrix} 0 & 1 \\ 1 & 0 \end{bmatrix} \quad \sigma_y = \begin{bmatrix} 0 & -\mathrm{i} \\ \mathrm{i} & 0 \end{bmatrix} \quad \sigma_z = \begin{bmatrix} 1 & 0 \\ 0 & -1 \end{bmatrix} \tag{13.48}
$$

Next we must take account of the spatial degrees of freedom. Suppose the wave function ψ is an eigenstate of momenta *and* an eigenstate of S_z, with eigenvalue $s_z = \frac{1}{2}\hbar$ ('spin up'). Then we require, as usual,

$$
\begin{aligned}
-\mathrm{i}\hbar\partial_x\psi &= p_x\psi \\
-\mathrm{i}\hbar\partial_y\psi &= p_y\psi \\
-\mathrm{i}\hbar\partial_z\psi &= p_z\psi
\end{aligned}
\tag{13.49}
$$

and, in addition,

$$S_z\psi = \frac{1}{2}\hbar\psi \tag{13.50}$$

where S_z is the matrix shown in (13.46). It is easy to see that the solution is

$$\psi_{p+} = e^{i\mathbf{p}\cdot\mathbf{x}/\hbar}\begin{bmatrix}1\\0\end{bmatrix} \tag{13.51}$$

Similarly, an eigenstate of momentum and S_z with $s_z = -\frac{1}{2}$ ('spin down') will be

$$\psi_{p-} = e^{i\mathbf{p}\cdot\mathbf{x}/\hbar}\begin{bmatrix}0\\1\end{bmatrix} \tag{13.52}$$

According to our general rules, the (normalized) superposition of two physical states is also a physical state, so in general an electron wave function must have the form

$$\begin{aligned}\psi &= \int d^3p\, \left[f_+(p)\psi_{p+} + f_-(p)\psi_{p-}\right] \\ &= \int d^3p \begin{bmatrix} f_+(p)e^{i\mathbf{p}\cdot\mathbf{x}} \\ f_-(p)e^{i\mathbf{p}\cdot\mathbf{x}} \end{bmatrix} \\ &= \begin{bmatrix}\psi_+(x)\\ \psi_-(x)\end{bmatrix}\end{aligned} \tag{13.53}$$

where $\psi_+(x)$ and $\psi_-(x)$ are any two square-integrable functions satisfying a normalization condition

$$\begin{aligned}\langle\psi|\psi\rangle &= \sum_{i=+,-}\int d^3x\; \psi_i^*(\mathbf{x})\psi_i(\mathbf{x}) \\ &= \int d^3x\, \left[|\psi_+(\mathbf{x})|^2 + |\psi_-(\mathbf{x})|^2\right] \\ &= 1\end{aligned} \tag{13.54}$$

The interpretation of the two terms in the spinor wave function, $\psi_{+/-}(x)$, is easy to see from a computation of the S_z expectation value

$$\begin{aligned}\langle S_z\rangle &= \langle\psi|S_z|\psi\rangle \\ &= \int dx\, dy\, dz\, \left[\psi_+^*, \psi_-^*\right]\frac{\hbar}{2}\begin{bmatrix}1 & 0\\ 0 & -1\end{bmatrix}\begin{bmatrix}\psi_+\\ \psi_-\end{bmatrix} \\ &= \int dx\, dy\, dz\left(+\frac{1}{2}\hbar\, |\psi_+|^2 + \left(-\frac{1}{2}\hbar\right)|\psi_-|^2\right) \\ &= \left(\frac{1}{2}\hbar\right)\text{Prob(spin up)} + \left(-\frac{1}{2}\hbar\right)\text{Prob(spin down)}\end{aligned} \tag{13.55}$$

where 'spin up' and 'spin down' refer to the two possibilities of finding $s_z = \frac{1}{2}$ and $s_z = -\frac{1}{2}$. Comparing the last two lines

$$\begin{aligned} \text{Prob(spin up)} &= \int \mathrm{d}x\, \mathrm{d}y\, \mathrm{d}z \;\; |\psi_+|^2 \\ \text{Prob(spin down)} &= \int \mathrm{d}x\, \mathrm{d}y\, \mathrm{d}z \;\; |\psi_-|^2 \end{aligned} \tag{13.56}$$

Although ψ_+ and ψ_- are, in general, independent functions, there are many important cases where the spatial and spin parts of the wave function factorize, i.e.

$$\psi = \psi(x)\begin{bmatrix} a \\ b \end{bmatrix} \tag{13.57}$$

An example we have just seen are the eigenstates of momenta and S_z shown in equations (13.51) and (13.52). Another example is the set of eigenstates of $\tilde{H}, \tilde{L}^2, \tilde{L}_z, \tilde{S}^2, \tilde{S}_z$, where $\tilde{H}$ is the hydrogen atom Hamiltonian. The 'spin-up' eigenstates are

$$\varphi(r, \theta, \phi)\chi_+ = R_{nl}(r)\, Y_{lm}(\theta, \phi)\begin{bmatrix} 1 \\ 0 \end{bmatrix} \tag{13.58}$$

while the 'spin-down' eigenstates are

$$\varphi(r, \theta, \phi)\chi_- = R_{nl}(r)\, Y_{lm}(\theta, \phi)\begin{bmatrix} 0 \\ 1 \end{bmatrix} \tag{13.59}$$

In ket notation, the states are labeled by their quantum numbers:

$$\{|nlmss_z\rangle\} \qquad \text{where} \qquad s = \frac{1}{2}, \quad s_z = \pm\frac{1}{2} \tag{13.60}$$

We can get some practice in the use of spinor notation and spin matrices by studying the precession of electron spin in an external magnetic field.

13.2.1 Electron precession

Suppose an electron is in a state where $p \approx 0$, so the electron spinor, while it can depend on time, does not depend on space, i.e.

$$\psi(t) = \begin{bmatrix} \psi_+(t) \\ \psi_-(t) \end{bmatrix} \tag{13.61}$$

In addition, suppose that there is an external, constant magnetic field, of magnitude B, in the z-direction. The Hamiltonian is then

$$\begin{aligned}\tilde{H} &= \frac{p^2}{2M} - \vec{\mu}\cdot\vec{B}\\ &\approx \frac{e\hbar Bg}{4Mc}\sigma_z\\ &\approx \frac{1}{2}\hbar\Omega\sigma_z\end{aligned} \tag{13.62}$$

where

$$\Omega = \frac{eB}{Mc} \tag{13.63}$$

Ω is known as the 'cyclotron frequency'. The eigenstates of this Hamiltonian are then simply $\chi_\pm$, because $\tilde{H}$ is proportional to σ_z, and $\chi_\pm$ are the eigenstates of σ_z:

$$\begin{aligned}\tilde{H}\chi_+ &= E_+\chi_+ \quad \text{where} \quad E_+ = \frac{1}{2}\hbar\Omega\\ \tilde{H}\chi_- &= E_-\chi_- \quad \text{where} \quad E_- = -\frac{1}{2}\hbar\Omega\end{aligned} \tag{13.64}$$

Eigenstates of the Hamiltonian are stationary states; an electron in such a state will remain in that eigenstate indefinitely. But suppose instead that the electron spin is initially in an eigenstate of S_x or S_y. What we now show is that the electron spin will then tend to precess around the z-axis.

First, some notation. Define 'spin-up' states and 'spin-down' states in the x-, y-, and z-directions to be the $\pm\frac{1}{2}\hbar$ eigenstates of matrices S_x, S_y, S_z, respectively:

$$\begin{aligned}S_x\alpha_x &= \frac{1}{2}\hbar\alpha_x \qquad & S_x\beta_x &= -\frac{1}{2}\hbar\beta_x\\ S_y\alpha_y &= \frac{1}{2}\hbar\alpha_y \qquad & S_y\beta_y &= -\frac{1}{2}\hbar\beta_y\\ S_z\alpha_z &= \frac{1}{2}\hbar\alpha_z \qquad & S_z\beta_z &= -\frac{1}{2}\hbar\beta_z\end{aligned} \tag{13.65}$$

We know already that

$$\alpha_z = \chi_+ = \begin{bmatrix}1\\0\end{bmatrix} \quad \text{and} \quad \beta_z = \chi_- = \begin{bmatrix}0\\1\end{bmatrix} \tag{13.66}$$

Finding the eigenstates α_y, β_y requires finding the eigenstates of the Pauli matrix σ_y. It is instructive do this in detail.

The first thing to do is to find the eigenvalues of σ_y. The eigenvalue equation is

$$\sigma_y\mathbf{u} = \lambda\mathbf{u} \tag{13.67}$$

and the eigenvalues are determined by solving the ***secular equation***

$$\det\left[\sigma_y - \lambda\mathbb{1}\right] = 0 \tag{13.68}$$

where $\mathbb{1}$ is the unit matrix. For the Pauli matrix σ_y, this is

$$\det\begin{bmatrix} -\lambda & -\mathrm{i} \\ \mathrm{i} & -\lambda \end{bmatrix} = \lambda^2 - 1 = 0 \tag{13.69}$$

so the eigenvalues are

$$\lambda_1 = +1 \qquad \lambda_2 = -1 \tag{13.70}$$

Note that (i) σ_y is a Hermitian matrix; and (ii) the eigenvalues are both real. As shown in chapter 7, the eigenvalues of any Hermitian operator are real, and eigenstates corresponding to different eigenvalues are orthogonal. So far, we have verified the reality of the eigenvalues. Now we solve for the two corresponding eigenstates, denoted

$$\alpha_y = \begin{bmatrix} a_1 \\ a_2 \end{bmatrix} \qquad \text{for eigenvalue} \qquad \lambda = 1 \tag{13.71}$$

and

$$\beta_y = \begin{bmatrix} b_1 \\ b_2 \end{bmatrix} \qquad \text{for eigenvalue} \qquad \lambda = -1 \tag{13.72}$$

Start by solving for

$$\begin{gathered} \begin{bmatrix} 0 & -\mathrm{i} \\ \mathrm{i} & 0 \end{bmatrix}\begin{bmatrix} a_1 \\ a_2 \end{bmatrix} = \begin{bmatrix} a_1 \\ a_2 \end{bmatrix} \\ \begin{bmatrix} -\mathrm{i}a_2 \\ \mathrm{i}a_1 \end{bmatrix} = \begin{bmatrix} a_1 \\ a_2 \end{bmatrix} \end{gathered} \tag{13.73}$$

which has the solution

$$a_2 = \mathrm{i}a_1 \qquad \text{or} \qquad \alpha_y = \begin{bmatrix} a_1 \\ \mathrm{i}a_1 \end{bmatrix} \tag{13.74}$$

Finally, we determine a_1 by normalization

$$\begin{aligned} 1 &= \alpha_y \cdot \alpha_y = |a_1|^2 + |\mathrm{i}a_1|^2 = 2\,|a_1|^2 \\ &\Longrightarrow a_1 = \frac{1}{\sqrt{2}} \end{aligned} \tag{13.75}$$

so

$$\lambda_1 = +1, \qquad \alpha_y = \frac{1}{\sqrt{2}}\begin{bmatrix} 1 \\ \mathrm{i} \end{bmatrix} \tag{13.76}$$

The procedure for $\lambda_2 = -1$ is identical:

$$\begin{bmatrix} 0 & -\mathrm{i} \\ \mathrm{i} & 0 \end{bmatrix}\begin{bmatrix} b_1 \\ b_2 \end{bmatrix} = -\begin{bmatrix} b_1 \\ b_2 \end{bmatrix}$$
$$\begin{bmatrix} -\mathrm{i}b_2 \\ \mathrm{i}b_1 \end{bmatrix} = \begin{bmatrix} -b_1 \\ -b_2 \end{bmatrix} \tag{13.77}$$

And this time $b_2 = -\mathrm{i}b_1$. Normalizing to determine b_1, we find

$$\lambda_2 = -1, \qquad \beta_y = \frac{1}{\sqrt{2}}\begin{bmatrix} 1 \\ -\mathrm{i} \end{bmatrix} \tag{13.78}$$

Note that the inner product

$$\alpha_y \cdot \beta_y = \frac{1}{2}[1, -\mathrm{i}]\begin{bmatrix} 1 \\ -\mathrm{i} \end{bmatrix} = 1 + (-\mathrm{i})^2 = 0 \tag{13.79}$$

vanishes, so α_y and β_y are orthogonal, as expected.

By similar means we can find the eigenstates of σ_x, and in summary:

$$\alpha_x = \frac{1}{\sqrt{2}}\begin{bmatrix} 1 \\ 1 \end{bmatrix}, \qquad \beta_x = \frac{1}{\sqrt{2}}\begin{bmatrix} 1 \\ -1 \end{bmatrix}$$
$$\alpha_y = \frac{1}{\sqrt{2}}\begin{bmatrix} 1 \\ \mathrm{i} \end{bmatrix}, \qquad \beta_y = \frac{1}{\sqrt{2}}\begin{bmatrix} 1 \\ -\mathrm{i} \end{bmatrix} \tag{13.80}$$
$$\chi_+ = \alpha_z = \begin{bmatrix} 1 \\ 0 \end{bmatrix}, \qquad \chi_- = \beta_z = \begin{bmatrix} 0 \\ 1 \end{bmatrix}$$

To analyze the variation with time of the electron magnetic moment in an external magnetic field, we use the time-dependent Schrödinger equation with Hamiltonian (13.62):

$$\mathrm{i}\hbar\partial_t\psi = \frac{1}{2}\hbar\Omega\sigma_z\psi \tag{13.81}$$

or, in matrix form,

$$\begin{bmatrix} \partial_t\psi_+ \\ \partial_t\psi_- \end{bmatrix} = -\mathrm{i}\frac{1}{2}\Omega\begin{bmatrix} 1 & 0 \\ 0 & -1 \end{bmatrix}\begin{bmatrix} \psi_+ \\ \psi_- \end{bmatrix} = \begin{bmatrix} -\frac{1}{2}\mathrm{i}\Omega\psi_+ \\ \frac{1}{2}\mathrm{i}\Omega\psi_- \end{bmatrix} \tag{13.82}$$

These are two independent first-order differential equations, one for ψ_+ and one for ψ_-, and the general solution is easily seen to be

$$\psi(t) = \begin{bmatrix} a\mathrm{e}^{-\mathrm{i}\Omega t/2} \\ b\mathrm{e}^{\mathrm{i}\Omega t/2} \end{bmatrix} \tag{13.83}$$

where a and b are constants.

Suppose that at time $t = 0$ the electron is in a 'spin-up' state in the x-direction. This means that the constants a and b are determined to be

$$a = b = \frac{1}{\sqrt{2}} \tag{13.84}$$

Let $T = 2\pi/\Omega$ be the period corresponding to an angular frequency Ω. Then, from the solution above, we see that the wave function for the electron spin 'precesses' around the z-axis:

$$\begin{aligned}
\psi(0) &= \alpha_x \qquad \text{(assumption)} \\
\psi\left(\frac{T}{4}\right) &= \frac{1}{\sqrt{2}}\begin{bmatrix} e^{-i\pi/4} \\ e^{i\pi/4} \end{bmatrix} = e^{-i\pi/4}\alpha_y \\
\psi\left(\frac{T}{2}\right) &= \frac{1}{\sqrt{2}}\begin{bmatrix} e^{-i\pi/2} \\ e^{i\pi/2} \end{bmatrix} = e^{-i\pi/2}\beta_x \\
\psi\left(\frac{3T}{4}\right) &= \frac{1}{\sqrt{2}}\begin{bmatrix} e^{-3i\pi/4} \\ e^{3i\pi/4} \end{bmatrix} = e^{-3i\pi/4}\beta_y \\
\psi(T) &= \frac{1}{\sqrt{2}}\begin{bmatrix} e^{-i\pi} \\ e^{i\pi} \end{bmatrix} = e^{-i\pi}\alpha_x
\end{aligned} \tag{13.85}$$

Thus the spin direction of the electron precesses around the z-axis with a precession frequency $f = \Omega/2\pi$. The fact that a spin 1/2 particle precesses in the presence of an external magnetic field is of considerable practical importance for the nuclear magnetic resonance imaging technique in medicine, which takes advantage of the precession of atomic nuclei of spin 1/2.

13.2.2 Spin–orbit coupling

Even when there is no external magnetic field, the frequencies of the spectral lines of the hydrogen atom are not *exactly* as predicted by Bohr. Certain lines of the spectrum, when observed using a high-resolution spectrometer, are found to actually be two or more closely spaced spectral lines; this phenomenon is known as atomic fine structure. Atomic fine structure is due to several relativistic effects, and the effect most closely related to electron spin is known as 'spin-orbit coupling'.

To understand the origin of spin-orbit coupling, consider an electron moving with velocity $\mathbf{v}$ across a static electric field, such as the Coulomb field due to an atomic nucleus. According to the theory of special relativity, the electromagnetic field as it appears in the rest frame of the electron is no longer entirely electric, but also contains a magnetic component

$$\begin{aligned}
\mathbf{B} &= -\frac{1}{\sqrt{1 - \dfrac{v^2}{c^2}}}\frac{\mathbf{v}}{c} \times \mathbf{E} \\
&\approx -\frac{1}{Mc}\mathbf{p} \times \mathbf{E}
\end{aligned} \tag{13.86}$$

Given that the electron has a magnetic moment $\boldsymbol{\mu}$, there should be an interaction energy due to the electron magnetic moment, in the rest frame of the electron

$$H' = -\boldsymbol{\mu} \cdot \mathbf{B} \tag{13.87}$$

However, we really want the interaction energy in the rest frame of the laboratory, which is also (approximately) the rest frame of the proton. In transforming this energy back to the proton rest frame there is a slight subtlety: the electron is not moving with a uniform velocity $\mathbf{v}$, but is rather in *accelerated* motion as it circles around the proton. To take this acceleration carefully into account would take us too far afield, but suffice it to say that the interaction energy H' above is modified by a factor of $\frac{1}{2}$, called the Thomas precession factor[2]

$$\begin{aligned} H' &= -\frac{1}{2}\boldsymbol{\mu} \cdot \mathbf{B} \\ &= -\frac{\boldsymbol{\mu}}{2Mc} \cdot \left(\mathbf{E} \times \mathbf{p}\right) \\ &= -\frac{e}{2Mc}\boldsymbol{\mu} \cdot \left(\frac{\mathbf{r}}{4\pi r^3} \times \mathbf{p}\right) \\ &= -\frac{e}{2Mc}\frac{1}{4\pi r^3}\mathbf{L} \cdot \boldsymbol{\mu} \end{aligned} \tag{13.88}$$

where we have inserted, in the third line, the electric field due to the static proton. Using

$$\begin{aligned} \boldsymbol{\mu} &= -\frac{eg}{2Mc}\mathbf{S} \\ &\approx -\frac{e}{Mc}\mathbf{S} \end{aligned} \tag{13.89}$$

we obtain

$$H' = \frac{e^2}{2M^2 4\pi c^2}\frac{1}{r^3}\mathbf{L} \cdot \mathbf{S} \tag{13.90}$$

This expression is known as the 'spin–orbit' coupling, because it involves a coupling of the electron spin angular momentum with the electron orbital angular momentum. The full hydrogen atom Hamiltonian should contain this spin-dependent term.

Now the spin–orbit term involves all the x-, y-, z-components of angular momentum, and we know that there is no physical state which is an eigenstate of all of these terms simultaneously. However, let us define the *total* electron angular momentum

$$\mathbf{J} = \mathbf{L} + \mathbf{S} \tag{13.91}$$

[2] A derivation can be found in, e.g. Jackson's *Classical Electrodynamics* [1].

Then

$$J^2 = L^2 + 2\mathbf{L} \cdot \mathbf{S} + S^2 \tag{13.92}$$

or

$$\mathbf{L} \cdot \mathbf{S} = \frac{1}{2}(J^2 - L^2 - S^2) \tag{13.93}$$

The total Hamiltonian is then

$$H = H_0 + \frac{e^2}{4\pi} \frac{1}{4M^2c^2} \frac{1}{r^3}(J^2 - L^2 - S^2) \tag{13.94}$$

Now, in the quantum theory

$$[\tilde{J}^2, \tilde{L}^2] = [\tilde{J}^2, \tilde{S}^2] = 0 \tag{13.95}$$

since $\tilde{L}^2$ and $\tilde{S}^2$ commute with all the components of $\vec{J} = \vec{L} + \mathbf{S}$, and therefore

$$[H, J^2] = [H, \tilde{L}^2] = [H, \tilde{S}^2] = 0 \tag{13.96}$$

Recall that in solving for the eigenstates of H_0, we had to first find the eigenstates of $\tilde{L}^2$. For similar reasons, to find the eigenstates of H, we need first to find the eigenstates of J^2, $\tilde{L}^2$, and $\tilde{S}^2$. Finding the eigenstates of *total* angular momentum J^2, given the eigenstates

$$Y_{lm}\chi_{\pm} \tag{13.97}$$

of orbital ($\tilde{L}^2$, $\tilde{L}_z$) and spin ($\tilde{S}^2$, $\tilde{S}_z$) angular momentum is carried out by a systematic procedure known as the ***addition of angular momentum***, which is our next order of business.

Gyromagnetic ratio and g − 2
We pause, however, to mention that the gyromagnetic ratio of the electron, also called the ***g-factor***, is very close but not exactly equal to 2. For the electron, the g-factor has been measured to astonishing accuracy:

$$g = 2.00231930436182(52) \tag{13.98}$$

where the numbers in parentheses give the error estimate of the last two digits. This makes the g-factor of the electron one of the most precisely measured constants in Nature. One of the recent experiments which has achieved this amazing precision is reported in [2]. For the muon, the result is

$$g = 2.0023318414(12) \tag{13.99}$$

There is no obvious explanation for why $g \approx 2$ in non-relativistic quantum mechanics; for this, relativity is required. In 1928 Dirac proposed a simple relativistic wave equation for the electron, and derived from this equation that $g = 2$ exactly. The deviation of the g-factor from 2, known as ***g-2*** was explained in 1949 from the relativistic quantum field theory known as quantum electrodynamics. It is one of the triumphs of theoretical physics that the calculated g-factor for the electron agrees with the experimental number to within the extraordinary 12-digit precision of experiment. The theoretical calculation is beyond the scope of this book. At the present time, however, there is still a very tiny discrepancy between the experimental value and the theoretical prediction for the muon g-2, a difference residing in the last three digits. Whether this is an indication of physics beyond the Standard Model, or simply an error in the theoretical treatment of contributions associated with the strong nuclear force, is not yet known.

It should also be noted that the theoretical prediction for the g-factor holds for point-like particles, with no internal structure. Particles such as the proton and neutron, which are bound states of quarks, have quite different g-factors, e.g. the g-factor of the proton is $g \approx 5.6$.

Problems

1. Consider a particle with spin $s = 1$. Find the matrix representation of $\tilde{S}^2$, $\tilde{S}_x$, $\tilde{S}_y$, $\tilde{S}_z$ in the basis of eigenstates of $\tilde{S}^2$, $\tilde{S}_z$, denoted $|ss_z\rangle = |11\rangle$, $|10\rangle$, $|1\ -1\rangle$.
2. Obtain the eigenstates α_x and β_x of the matrix S_x, for $s = 1/2$.
3. Find the 3×3 matrix form of the operator $\tilde{L}_x$ for orbital angular momentum $l = 1$, in the $\tilde{L}^2$, $\tilde{L}_z$ representation. Obtain the eigenvalues and eigenvectors of this matrix. Let $\tilde{L}^2$, $\tilde{L}_x$ be the differential operators corresponding to observables L^2, L_x. Using only your solutions for the eigenvalues and eigenvectors of the 3×3 matrix (i.e. do not try to solve any differential equations!), write down the function $\Phi(\theta, \phi)$ which satisfies

$$\tilde{L}^2\Phi(\theta, \phi) = 2\hbar^2\Phi(\theta, \phi) \qquad \tilde{L}_x\Phi(\theta, \phi) = -\hbar\Phi(\theta, \phi)$$

expressed as a linear combination of spherical harmonics.
4. We have given a matrix representation to the spin angular momentum operators in the $|ss_z\rangle$ basis, but there is nothing sacred about the z-axis. If $\tilde{S}_{x,y,z} = \frac{\hbar}{2}\sigma_{x,y,z}$ for spin 1/2, and the $\tilde{S}_{x,y,z}$ are operators, then we should also regard the $\sigma_{x,y,z}$ as Pauli *operators* $\tilde{\sigma}_{x,y,z}$, rather than Pauli matrices. The matrix representation of these Pauli operators must depend on the basis, i.e. in a basis $\mathbf{e}_1$, $\mathbf{e}_2$, the (i, j) matrix components are $\langle e_i|\sigma_{x,y,z}|e_j\rangle$. From this fact, compute the matrix representation of the Pauli operators in the basis of eigenstates of $\tilde{S}_x$. (Hint: while the components of vectors and matrices depend on the basis, inner products such as $\langle e_i|\sigma_{x,y,z}|e_j\rangle$ do not, and can be evaluated in any convenient basis.)
5. Each time a button is pushed, a spin-1/2 particle emerges in an unknown (but always the same) spin state

$$\psi = a\alpha_z + b\beta_z \tag{13.100}$$

Bob can measure the spin component along the x-, y-, and z-axes. Explain how Bob can determine this state, i.e. the constants a and b, from spin up/down probabilities along the x-, y-, z-axes. Since ψ and $e^{i\delta}\psi$ are the same physical state, you may choose constant a to be real and positive. But in general b is a complex number.

6. Let A and B denote two Hermitian operators acting on a two-dimensional Hilbert space. The operators satisfy the eigenvalue equations

$$\begin{aligned} A|\phi_1\rangle &= 3|\phi_1\rangle \\ A|\phi_2\rangle &= 7|\phi_2\rangle \\ B|\omega_1\rangle &= 2|\omega_1\rangle \\ B|\omega_2\rangle &= 3|\omega_2\rangle \end{aligned} \tag{13.101}$$

where the two sets of eigenstates are related by

$$\begin{aligned} |\omega_1\rangle &= \frac{1}{\sqrt{2}}\left[|\phi_1\rangle + |\phi_2\rangle\right] \\ |\omega_2\rangle &= \frac{1}{\sqrt{2}}\left[|\phi_1\rangle - |\phi_2\rangle\right] \end{aligned} \tag{13.102}$$

Write A and B as 2×2 matrices in the basis spanned by $\{|\phi_1\rangle, |\phi_2\rangle\}$, and also in the basis spanned by $\{|\omega_1\rangle, |\omega_2\rangle\}$.

References

[1] Jackson J D 1991 *Classical Electrodynamics* 3rd edn (Hoboken, NJ: Wiley)

[2] Odom B C, Hanneke D, D'Urso B and Gabrielse G 2006 New measurement of the electron magnetic moment using a one-electron quantum cyclotron *Phys. Rev. Lett.* **97** 030801
Odom B C, Hanneke D, D'Urso B and Gabrielse G 2007 *Phys. Rev. Lett.* **99** 039902

IOP Publishing

An Introduction to Quantum Theory

Jeff Greensite

Chapter 14

The addition of angular momentum

Consider the lilies of the field, how they grow. They toil not, neither do they spin...

The Book of Matthew

In the previous chapter we found that there was a contribution to the hydrogen atom Hamiltonian proportional to the operator

$$\tilde{\mathbf{L}} \cdot \tilde{\mathbf{S}} = \frac{1}{2}\left(\tilde{J}^2 - \tilde{L}^2 - \tilde{S}^2\right) \tag{14.1}$$

In order to find eigenstates of $\tilde{\mathbf{L}} \cdot \tilde{\mathbf{S}}$, we need to find eigenstates of $\tilde{L}^2$, $\tilde{S}^2$, and the total angular momentum $\tilde{J}^2$. Now the usual eigenstates of $\tilde{L}^2$, $\tilde{L}_z$, $\tilde{S}^2$, $\tilde{S}_z$ that we wrote down in the last chapter, namely

$$\begin{gathered} \varphi(r, \theta, \phi)\chi_+ = R_{nl}(r)Y_{lm}(\theta, \phi)\begin{bmatrix}1\\0\end{bmatrix} \\ \text{and} \\ \varphi(r, \theta, \phi)\chi_- = R_{nl}(r)Y_{lm}(\theta, \phi)\begin{bmatrix}0\\1\end{bmatrix} \end{gathered} \tag{14.2}$$

are (mostly) *not* eigenstates of $\tilde{\mathbf{L}} \cdot \tilde{\mathbf{S}}$. This is because $\tilde{\mathbf{L}} \cdot \tilde{\mathbf{S}} = \tilde{L}_x\tilde{S}_x + \tilde{L}_y\tilde{S}_y + \tilde{L}_z\tilde{S}_z$, and while the wave functions above are eigenstates of $\tilde{L}_z$ and $\tilde{S}_z$, they are (mostly) not eigenstates of $\tilde{L}_x$, $\tilde{L}_y$ and $\tilde{S}_x$, $\tilde{S}_y$. Because these wave functions are not eigenstates of $\tilde{\mathbf{L}} \cdot \tilde{\mathbf{S}}$, they are not eigenstates of the total angular momentum $\tilde{J}^2$, either.

Defining $\tilde{J}_x = \tilde{L}_x + \tilde{S}_x$ etc, and using the fact that the orbital angular momentum operators commute with the spin angular momentum operators (they act on different degrees of freedom), it is easy to see that the components of total angular momentum satisfy the now-familiar commutation relations

doi:10.1088/978-0-7503-1167-0ch14

$$[\tilde{J}_x, \tilde{J}_y] = i\hbar\tilde{J}_z$$
$$[\tilde{J}_y, \tilde{J}_z] = i\hbar\tilde{J}_x \qquad (14.3)$$
$$[\tilde{J}_z, \tilde{J}_x] = i\hbar\tilde{J}_y$$

So this means that $\tilde{J}_z = \tilde{L}_z + \tilde{S}_z$ must commute with $\tilde{J}^2$. What we want to do is to construct eigenstates of $\tilde{J}^2$, $\tilde{J}_z$, $\tilde{L}^2$, $\tilde{S}^2$, denoted $|jj_z\,ls\rangle$, as linear combinations of the eigenstates of $\tilde{L}^2$, $\tilde{L}_z$, $\tilde{S}^2$, $\tilde{S}_z$ denoted by $|lmss_z\rangle$, whose corresponding eigenfunctions are shown in (14.2). In other words, we want to find the set of ***Clebsch–Gordon coefficients*** $C^{lmss_z}_{jj_z\,ls}$ such that

$$|jj_z\,ls\rangle = \sum_{s_z, m=j_z-s_z} C^{lmss_z}_{jj_z\,ls}|lmss_z\rangle \qquad (14.4)$$

or, in wave function notation

$$\Phi_{jj_z} = \sum_{s_z, m=j_z-s_z} C^{lmss_z}_{jj_z\,ls} Y_{lm}\chi_{s_z} \qquad (14.5)$$

(In the case of the hydrogen atom, $s = \frac{1}{2}$. But the technique we are going to use will work for any value of s.)

The trick is to notice that one of the $|lmss_z\rangle$ states, known as the ***highest weight state***, is also an eigenstate of $\tilde{J}^2$. Suppose we ask which state is the eigenstate with the highest eigenvalue of $\tilde{J}_z$. Now $\tilde{J}_z = \tilde{L}_z + \tilde{S}_z$, and any eigenstate of $\tilde{L}_z$, $\tilde{S}_z$ is also an eigenstate of $\tilde{J}_z$, although it is usually not an eigenstate of $\tilde{J}^2$. The eigenstate with the highest eigenvalue of $\tilde{J}_z$ has to be the state with the highest eigenvalue of $\tilde{L}_z$ and $\tilde{S}_z$, and there is only one such state

$$\text{'highest weight state'} = Y_{ll}\chi_+ \qquad (14.6)$$

which has $j_z = l + \frac{1}{2}$. Now if the highest j_z is $l + \frac{1}{2}$, then this must also be the highest possible value of j that can be constructed from the $Y_{lm}\chi_{s_z}$. But there is only one state with $j_z = l + s$, so this must also be the eigenstate of the operator $\tilde{J}^2$ with $j = l + s$. So we conclude that

$$\Phi_{jj} = Y_{ll}\chi_+ \qquad (j = l + s) \qquad (14.7)$$

By exactly the same reasoning, there is also a lowest weight state

$$\Phi_{j,-j} = Y_{l,-l}\chi_- \qquad (j = l + s) \qquad (14.8)$$

with $j_z = -(l + s)$. For a direct proof that the highest/lowest weight states are eigenstates of $\tilde{J}^2$, see problem 2 at the end of the chapter.

From one eigenstate of $\tilde{J}^2$, i.e. the highest weight state, it is simple to construct all the other eigenstates of $\tilde{J}^2$ and $\tilde{J}_z$, by successively applying the ladder operator

$$\tilde{J}_- = \tilde{L}_- + \tilde{S}_- \tag{14.9}$$

and the relations

$$\begin{aligned}
\tilde{J}_- |jj_z\rangle &= \sqrt{j(j+1) - j_z(j_z - 1)}\,\hbar |j, j_z - 1\rangle \\
\tilde{L}_- |lm\rangle &= \sqrt{l(l+1) - m(m-1))}\,\hbar |l, m - 1\rangle \\
\tilde{S}_- |ss_z\rangle &= \sqrt{s(s+1) - s_z(s_z - 1)}\,\hbar |s, s_z - 1\rangle
\end{aligned} \tag{14.10}$$

To see how it works, let us pick a particular example to work out, with $l = 1$. In this case, the highest weight state is

$$\Phi_{\frac{3}{2}\frac{3}{2}} = Y_{11}\chi_+ \tag{14.11}$$

Now apply the $\tilde{J}_-$ operator on both sides of this equation, and use equation (14.10)

$$\begin{aligned}
\tilde{J}_-\Phi_{\frac{3}{2}\frac{3}{2}} &= (\tilde{L}_- + \tilde{S}_-)\,Y_{11}\chi_+ \\
\hbar\sqrt{\frac{3}{2}\left(\frac{3}{2}+1\right) - \frac{3}{2}\left(\frac{3}{2}-1\right)}\,\Phi_{\frac{3}{2}\frac{1}{2}} &= (\tilde{L}_-Y_{11})\chi_+ + Y_{11}(\tilde{S}_-\chi_+) \\
\hbar\sqrt{3}\,\Phi_{\frac{3}{2}\frac{1}{2}} &= \hbar\Big(\sqrt{1(1+1) - 0}\;Y_{10}\chi_+ \\
&\quad + Y_{11}\cdot\sqrt{\frac{1}{2}\left(\frac{1}{2}+1\right) - \frac{1}{2}\left(\frac{1}{2}-1\right)}\chi_-\Big) \\
&= \hbar\left(\sqrt{2}\,Y_{10}\chi_+ + Y_{11}\chi_-\right)
\end{aligned} \tag{14.12}$$

Dividing both sides by $\sqrt{3}\,\hbar$, we have found the eigenstate of $\tilde{J}^2$, $\tilde{J}_z$ with $j = \frac{3}{2}$ and $j_z = \frac{1}{2}$

$$\Phi_{\frac{3}{2}\frac{1}{2}} = \sqrt{\frac{2}{3}}\,Y_{10}\chi_+ + \sqrt{\frac{1}{3}}\,Y_{11}\chi_- \tag{14.13}$$

To get the state with $j_z = -\frac{1}{2}$, we just apply $\tilde{J}_-$ again, to both sides of (14.13)

$$\begin{aligned}
\tilde{J}_- \Phi_{\frac{3}{2}\frac{1}{2}} &= \left(\tilde{L}_- + \tilde{S}_-\right)\left(\sqrt{\frac{2}{3}}\, Y_{10}\chi_+ + \sqrt{\frac{1}{3}}\, Y_{11}\chi_-\right) \\
\hbar\sqrt{\frac{3}{2}\left(\frac{3}{2}+1\right) - \frac{1}{2}\left(\frac{1}{2}-1\right)}\,\Phi_{\frac{3}{2}-\frac{1}{2}} &= \hbar\sqrt{\frac{1}{3}}\Big(\sqrt{2}\left(\tilde{L}_- Y_{10}\right)\chi_+ + \left(\tilde{L}_- Y_{11}\right)\chi_- \\
&\quad + \sqrt{2}\, Y_{10}\left(\tilde{S}_-\chi_+\right) + Y_{11}\left(\tilde{S}_-\chi_-\right)\Big) \\
2\hbar\Phi_{\frac{3}{2}-\frac{1}{2}} &= \hbar\sqrt{\frac{1}{3}}\Big(2\, Y_{1-1}\chi_+ + \sqrt{2}\, Y_{10}\,\chi_- \\
&\quad + \sqrt{2}\, Y_{10}\chi_- + 0\Big)
\end{aligned} \tag{14.14}$$

and we obtain

$$\Phi_{\frac{3}{2}-\frac{1}{2}} = \sqrt{\frac{2}{3}}\, Y_{10}\chi_- + \sqrt{\frac{1}{3}}\, Y_{1-1}\chi_+ \tag{14.15}$$

The remaining state with $j = \frac{3}{2}$ has to be the lowest weight state (14.8). But just as a check of the algebra, we obtain it by applying the ladder operator one more time

$$\begin{aligned}
\tilde{J}_-\Phi_{\frac{3}{2}-\frac{1}{2}} &= \left(\tilde{L}_- + \tilde{S}_-\right)\left(\sqrt{\frac{2}{3}}\, Y_{10}\chi_- + \sqrt{\frac{1}{3}}\, Y_{1-1}\chi_+\right) \\
\hbar\sqrt{3}\,\Phi_{\frac{3}{2}-\frac{3}{2}} &= \hbar\sqrt{\frac{1}{3}}\Big(\sqrt{2}\left(\tilde{L}_- Y_{10}\right)\chi_- + \left(\tilde{L}_- Y_{1-1}\right)\chi_+ + \sqrt{2}\, Y_{10}\left(\tilde{S}_-\chi_-\right) \\
&\quad + Y_{1-1}\left(\tilde{S}_-\chi +\right)\Big) \\
&= \hbar\sqrt{\frac{1}{3}}\left(2\, Y_{1-1}\chi_- + 0 + 0 + Y_{1-1}\chi_-\right) \\
&= \hbar\sqrt{3}\, Y_{1-1}\chi_-
\end{aligned} \tag{14.16}$$

and therefore

$$\Phi_{\frac{3}{2}-\frac{3}{2}} = Y_{1-1}\chi_- \tag{14.17}$$

as it should.

But are these all the eigenstates of $\tilde{J}^2$, $\tilde{J}_z$ that we can form from the $l = 1$, $s = \frac{1}{2}$ states? So far we have constructed *four* orthonormal states

$$\left\{\Phi_{\frac{3}{2}j_z},\ j_z = \frac{3}{2}, \frac{1}{2}, -\frac{1}{2}, -\frac{3}{2}\right\} \tag{14.18}$$

which are eigenstates of $\tilde{J}^2$, $\tilde{J}_z$, out of *six* orthonormal states

$$\begin{matrix} Y_{11}\chi_+ & Y_{10}\chi_+ & Y_{1-1}\chi_+ \\ Y_{11}\chi_- & Y_{10}\chi_- & Y_{1-1}\chi_- \end{matrix} \tag{14.19}$$

So out of these six states, we ought to be able to build two more states which are orthogonal to the four $\{\Phi_{\frac{3}{2}j_z}\}$. Now these extra two states cannot have $j = 3/2$, because we have already constructed all of those states, and they cannot be $j > 3/2$, basically because a state with $j_z > \frac{3}{2}$ is not available. So let us try to find states with $j = 1/2$. There are exactly two such states

$$\Phi_{\frac{1}{2}\frac{1}{2}} \quad \text{and} \quad \Phi_{\frac{1}{2},-\frac{1}{2}} \tag{14.20}$$

so this would bring the total to six. But how do we obtain them? The trick is to first find $\Phi_{\frac{1}{2}\frac{1}{2}}$ from the fact that it has to be orthogonal to $\Phi_{\frac{3}{2}\frac{1}{2}}$, and also normalized. Then we can use the ladder operator to obtain the other state.

We begin from the fact that $\Phi_{\frac{1}{2}\frac{1}{2}}$ is an eigenstate of $\tilde{J}_z$, with $j_z = \frac{1}{2}$. There are two states in the set of six (14.19) which have this value of j_z, namely

$$Y_{11}\chi_- \quad \text{and} \quad Y_{10}\chi_+ \tag{14.21}$$

but these are not eigenstates of $\tilde{J}^2$. So we look for a linear combination

$$\Phi_{\frac{1}{2}\frac{1}{2}} = aY_{11}\chi_- + bY_{10}\chi_+ \tag{14.22}$$

which has the orthonormality properties

$$\begin{aligned} \left\langle \Phi_{\frac{3}{2}\frac{1}{2}} \middle| \Phi_{\frac{1}{2}\frac{1}{2}} \right\rangle &= 0 \\ \left\langle \Phi_{\frac{1}{2}\frac{1}{2}} \middle| \Phi_{\frac{1}{2}\frac{1}{2}} \right\rangle &= 1 \end{aligned} \tag{14.23}$$

Substituting (14.13) and (14.22) into the first equation or (14.23)

$$\begin{aligned} 0 &= \left[\sqrt{\frac{1}{3}}\langle Y_{11}\chi_-| + \sqrt{\frac{2}{3}}\left\langle Y_{10}\chi_+\right|\right]\left[a\left|Y_{11}\chi_-\right\rangle + b\left|Y_{10}\chi_+\right\rangle\right] \\ &= \sqrt{\frac{1}{3}}\left[a + \sqrt{2}b\right] \end{aligned} \tag{14.24}$$

Therefore

$$\Phi_{\frac{1}{2}\frac{1}{2}} = a\left[Y_{11}\chi_- - \sqrt{\frac{1}{2}}\,Y_{10}\chi_+\right] \tag{14.25}$$

Impose the normalization condition on this state

$$
\begin{aligned}
1 &= a^2\left[\langle Y_{11}\chi_-| - \sqrt{\frac{1}{2}}\langle Y_{10}\chi_+|\right]\left[|Y_{11}\chi_-\rangle - \sqrt{\frac{1}{2}}|Y_{10}\chi_+\rangle\right] \\
&= a^2\left(1 + \frac{1}{2}\right) \qquad \Rightarrow \qquad a = \sqrt{\frac{2}{3}}
\end{aligned}
\tag{14.26}
$$

So

$$
\Phi_{\frac{1}{2}\frac{1}{2}} = \sqrt{\frac{2}{3}}\, Y_{11}\chi_- - \sqrt{\frac{1}{3}}\, Y_{10}\chi_+ \tag{14.27}
$$

The last thing to do is to apply the ladder operator to both sides of this equation to obtain the last of the six states

$$
\begin{aligned}
\tilde{J}_-\Phi_{\frac{1}{2}\frac{1}{2}} &= \left(\tilde{L}_- + \tilde{S}_-\right)\left(\sqrt{\frac{2}{3}}\, Y_{11}\chi_- - \sqrt{\frac{1}{3}}\, Y_{10}\chi_+\right) \\
\hbar\Phi_{\frac{1}{2},-\frac{1}{2}} &= \sqrt{\frac{2}{3}}\left(\tilde{L}_- Y_{11}\chi_- - \sqrt{\frac{1}{3}}\left(\tilde{L}_- Y_{10}\right)\chi_+ + \sqrt{\frac{2}{3}}\, Y_{11}\left(\tilde{S}_-\chi_-\right) - \sqrt{\frac{1}{3}}\, Y_{10}\left(\tilde{S}_-\chi_+\right)\right. \\
&= \frac{\hbar}{\sqrt{3}}\left[2\, Y_{10}\chi_- \sqrt{2}\, Y_{1-1}\chi_+ + 0 - Y_{10}\chi_-\right]
\end{aligned}
\tag{14.28}
$$

and finally

$$
\Phi_{\frac{1}{2},-\frac{1}{2}} = \sqrt{\frac{1}{3}}\, Y_{10}\chi_- - \sqrt{\frac{2}{3}}\, Y_{1-1}\chi_+ \tag{14.29}
$$

The set of all six eigenstates of $\tilde{J}^2$, $\tilde{J}_z$ that can be formed from the six states $Y_{1m}\chi_\pm$ is displayed below (the arrow indicates that the next state is obtained from the state above by applying the lowering operator $\tilde{J}_-$ to the left-hand side, and $\tilde{L}_- + \tilde{S}_-$ to the right-hand side):

$$
\underline{\mathrm{j} = \frac{3}{2}} \qquad\qquad\qquad\qquad \underline{\mathrm{j} = \frac{1}{2}}
$$

$$
\Phi_{\frac{3}{2}\frac{3}{2}} = Y_{11}\chi_+
$$

$$
\Downarrow
$$

$$
\Phi_{\frac{3}{2}\frac{1}{2}} = \sqrt{\frac{2}{3}}\, Y_{10}\chi_+ + \sqrt{\frac{1}{3}}\, Y_{11}\chi_- \quad \text{orthonormality}\rightarrow \quad \Phi_{\frac{3}{2}\frac{1}{2}} = \sqrt{\frac{1}{3}}\, Y_{10}\chi_+ - \sqrt{\frac{2}{3}}\, Y_{11}\chi_-
$$

$$
\Downarrow \qquad\qquad\qquad\qquad\qquad\qquad \Downarrow
$$

$$
\Phi_{\frac{3}{2}-\frac{1}{2}} = \sqrt{\frac{2}{3}}\, Y_{10}\chi_- + \sqrt{\frac{1}{3}}\, Y_{1-1}\chi_+ \qquad\qquad \Phi_{\frac{3}{2}-\frac{1}{2}} = \sqrt{\frac{1}{3}}\, Y_{10}\chi_+ - \sqrt{\frac{2}{3}}\, Y_{1-1}\chi_+
$$

$$
\Downarrow
$$

$$
\Phi_{\frac{3}{2}-\frac{3}{2}} = Y_{1-1}\chi_-
$$

But there is nothing special about $l = 1$, of course, and the procedure shown above works for arbitrary l. Note that, since $-l \leqslant m \leqslant l$, there are $(2l + 1)$ values for m, and two possible values of $s_z = \pm\frac{1}{2}$. This means there are $2(2l + 1)$ orthogonal $Y_{lm}\chi_\pm$ states for each fixed l, $s = \frac{1}{2}$. From these, one constructs a set of eigenstates of $\tilde{J}^2$ with $j_{\max} = l + \frac{1}{2}$, and $j_{\min} = l - \frac{1}{2}$. The total number of $\tilde{J}^2$ eigenstates is the same as the number of $Y_{lm}\chi_\pm$ states, i.e.

$$
\begin{aligned}
N &= \left(2j_{\max} + 1\right) + \left(2j_{\min} + 1\right) \\
&= \left(2\left(l + \frac{1}{2}\right) + 1\right) + \left(2\left(l - \frac{1}{2}\right) + 1\right) \\
&= 2\,(2l + 1)
\end{aligned}
\tag{14.30}
$$

Now let us go back to the hydrogen atom. Let $\tilde{H}_0$ be the hydrogen atom Hamiltonian without the spin–orbit contribution, i.e.

$$
\tilde{H}_0 = -\frac{\hbar^2}{2M}\nabla^2 - \frac{e^2}{4\pi r}
\tag{14.31}
$$

Including electron spin, the eigenstates of $\tilde{H}_0$, $\tilde{L}^2$, $\tilde{L}_z$, $\tilde{S}^2$, $\tilde{S}_z$ are the set of states

$$
\begin{aligned}
\{|nlmss_z\rangle\} \Leftrightarrow \Big\{ R_{nl}(r)\,Y_{lm}(\theta, \phi)\chi_{s_z}, \quad & n = 1, 2, \ldots; \quad l = 0, 1 .., n - 1; \\
m = -l, \ldots, l, \quad & s_z = \frac{1}{2}, -\frac{1}{2} \Big\}
\end{aligned}
\tag{14.32}
$$

What has been done in this chapter is to show how to construct eigenstates of $\tilde{H}_0$, $\tilde{J}^2$, $\tilde{J}_z$, $\tilde{L}^2$, $\tilde{S}^2$

$$
\begin{aligned}
\{|njj_zls\rangle\} \Leftrightarrow \Big\{ R_{nl}(r)\Phi_{jj_z}, \quad & n = 1, 2, \ldots; \quad l = 0, 1 ,.., n - 1, \\
j = l + \frac{1}{2}, l - \frac{1}{2}; \quad & j_z = -j, \ldots, j \Big\}
\end{aligned}
\tag{14.33}
$$

Now, taking into account the spin–orbit term, the Hamiltonian of the hydrogen atom is really

$$
\tilde{H} = \tilde{H}_0 + \tilde{H}'
\tag{14.34}
$$

where

$$
\tilde{H}' = \frac{e^2}{4\pi}\frac{1}{4M^2c^2}\frac{1}{r^3}\left(\tilde{J}^2 - \tilde{L}^2 - \tilde{S}^2\right)
\tag{14.35}
$$

Since $\tilde{H}$ still commutes with $\tilde{L}^2$, $\tilde{S}^2$, $\tilde{J}^2$, $\tilde{J}_z$, the eigenfunctions of $\tilde{H}$ will still have the form

$$
R'_{nlj}(r)\Phi_{jj_z}
\tag{14.36}
$$

but now $R'(r)$ will be a little different from $R_{nl}(r)$, and also the new energy eigenvalues E'_{njl} will be a little different from the Bohr energies E_n. Since spin–orbit coupling is a small effect, there is a powerful method known as time-independent perturbation theory which allows one to 'sneak up' on the eigenstates and eigenvalues of $\tilde{H}$, given the eigenstates and eigenvalues of $\tilde{H}_0$. This method will be presented in all its glory in chapter 17, but let us borrow, ahead of time, one of the results, which is that, if $|njj_z ls\rangle$ is an eigenstate of $\tilde{H}_0$, the energy eigenvalue of $\tilde{H}$ is approximately

$$E'_{njj_z ls} \approx E_n + \langle njj_z ls|H'|njj_z ls\rangle \tag{14.37}$$

Then the spin–orbit coupling introduces a correction to the atomic energies which can be calculated:

$$\begin{aligned}
\Delta E &= \langle njj_z ls|H'|njjzls\rangle \\
&= \frac{e^2}{4\pi}\frac{1}{4M^2c^2}\int \mathrm{d}r\; r^2 R_{nl}(r)\frac{1}{r^3}R_{nl}(r)\left\langle \Phi_{jj_z}\middle|\left(\tilde{J}^2 - \tilde{L}^2 - \tilde{S}^2\right)\middle|\Phi_{jj_z}\right\rangle \\
&= \frac{e^2}{4\pi}\frac{1}{4M^2c^2}\int \mathrm{d}r\; r^2 R_{nl}(r)\frac{1}{r^3}R_{nl}(r) \\
&\quad \times \left[j(j+1) - l(l+1) - \frac{1}{2}\left(\frac{1}{2}+1\right)\right]\hbar^2 \\
&= \left[j(j+1) - l(l+1) - \frac{1}{2}\left(\frac{1}{2}+1\right)\right] \\
&\quad \times \hbar^2 \frac{e^2}{4\pi}\frac{1}{M^2c^2a_0^3n^3l(l+1)(2l+1)}
\end{aligned} \tag{14.38}$$

where a_0 is the Bohr radius and we have used without proof the fact that

$$\int_0^\infty \mathrm{d}r\frac{R_{nl}^2}{r} = \frac{1}{a_0^3n^3l(l+1/2)(l+1)}$$

For a given l, we have seen that the j quantum number can have one of two values, $j = l + \frac{1}{2}$ or $j = l - \frac{1}{2}$, in which case

$$j(j+1) - l(l+1) - \frac{3}{4} = \begin{cases} l & \text{for} \quad j = l + \dfrac{1}{2} \\ -(l+1) & \text{for} \quad j = l - \dfrac{1}{2} \end{cases} \tag{14.39}$$

Also using the expression for the Bohr energy

$$E_n = -\frac{Me^4}{2\hbar^2(4\pi)^2n^2} \tag{14.40}$$

and defining the ***fine structure constant***

$$\alpha \equiv \frac{e^2}{4\pi\hbar c} \approx \frac{1}{137} \tag{14.41}$$

the final result for the energies E' of atomic orbits is

$$\begin{aligned} E'_{n,j=l+\frac{1}{2}} &= E_n + \frac{1}{(2l+1)(l+1)}|E_n|\frac{\alpha^2}{n} \\ E'_{n,j=l-\frac{1}{2}} &= E_n - \frac{1}{(2l+1)l}|E_n|\frac{\alpha^2}{n} \end{aligned} \tag{14.42}$$

The fact that the electron energy levels now depend (slightly) on l and $j = l \pm \frac{1}{2}$, in addition to the principal quantum number n, causes a small splitting of the atomic spectral lines associated with transitions between the various electron states. The splitting is very small; we can see that the correction to atomic energies is on the order of the square of the fine structure constant, α^2, which is less than one part in 10 000.

At this point you may feel a little let down. Did we go to all the trouble of learning how to add angular momenta, just to compute a tiny, arcane splitting of hydrogen atom spectral lines?[1] The answer, of course, is no. The technique for adding angular momentum is one of the most important tools there is in atomic, molecular, nuclear, particle, and condensed matter physics, and the reason is simple: things are made of other things, and those other things spin. Atoms, for example, consist of electrons and a nucleus. Like the electron, atomic nuclei also spin. So the possible spin angular momenta of a given atom is determined by adding the total angular momentum of the electron ($J = L + S$) to the spin angular momentum of the nucleus. Nuclei are made of protons and neutrons, each of which has spin $\frac{1}{2}$. The energy levels of nuclei depend on the total spin and orbital angular momentum of the protons and neutrons, and this requires adding the spins of the protons and neutrons, together with their orbital angular momentum in the nucleus. Likewise, protons and neutrons (and many other short-lived particles) are composed of three quarks. Each quark has spin $\frac{1}{2}$. To determine which combination of three quarks would give spin $\frac{1}{2}$ protons and neutrons, and which would lead to higher spin objects like hyperons, it is necessary to be able to add the spin angular momenta of the quarks. In short, it is important to learn how to add angular momentum, because all physical objects are composed of spinning particles. Viewed at the atomic level *everything* spins, including the lilies of the field.

14.1 The general method

Consider a composite system consisting of two subsystems, and suppose each subsystem has a certain angular momentum. It could be that the system is an atom, one subsystem is all the electrons, and the other subsystem is the nucleus. Or perhaps the system is a deuterium nucleus, consisting of one proton, or one neutron.

[1] In fact, we haven't even finished computing the splitting. There are other relativistic corrections (e.g. to the kinetic energy) of the same order of magnitude that have been ignored here.

Or maybe the system is a π-meson, consisting of one quark, and one antiquark. Depending on the particular case, the angular momentum of each subsystem might be the orbital angular momentum of the subsystem, the spin of the subsystem, or maybe the total (spin plus orbital) angular momentum of the subsystem. The point is, adding angular momentum is an algebraic technique, and it really does not matter whether we are adding spin to spin, orbital to orbital, orbital to spin, etc. We will just denote the angular momentum of one system by J_1, and the angular momentum of the other system by J_2, with the understanding that these can be any type of angular momentum.

It is assumed that we know the eigenstates of $\tilde{J}_1^2$, $\tilde{J}_{1z}$, $\tilde{J}_2^2$, $\tilde{J}_{2z}$, which we denote

$$\left\{\phi^1_{j_1m_1}\phi^2_{j_2m_2}\right\} \tag{14.43}$$

or, in ket notation

$$\left\{\left|j_1, m_1, j_2, m_2\right\rangle\right\} \tag{14.44}$$

with eigenvalues

$$\begin{aligned}
\tilde{J}_1^2\phi^1_{j_1m_1}\phi^2_{j_2m_2} &= j_1(j_1+1)\hbar^2\phi^1_{j_1m_1}\phi^2_{j_2m_2} \\
\tilde{J}_{1z}\phi^1_{j_1m_1}\phi^2_{j_2m_2} &= m_1\hbar\phi^1_{j_1m_1}\phi^2_{j_2m_2} \\
\tilde{J}_2^2\phi^1_{j_1m_1}\phi^2_{j_2m_2} &= j_2(j_2+1)\hbar^2\phi^1_{j_1m_1}\phi^2_{j_2m_2} \\
\tilde{J}_{2z}\phi^1_{j_1m_1}\phi^2_{j_2m_2} &= m_2\hbar\phi^1_{j_1m_1}\phi^2_{j_2m_2}
\end{aligned} \tag{14.45}$$

The eigenstates of $\tilde{J}_1^2$, $\tilde{J}_{1z}$, $\tilde{J}_2^2$, $\tilde{J}_{2z}$ are known as the ***uncoupled representation***. If, for example, $\tilde{J}_1$ and $\tilde{J}_2$ were both orbital angular momenta, then

$$\phi^1_{j_1m_1}\phi^2_{j_2m_2} = Y_{j_1m_1}(\theta_1, \phi_1)Y_{j_2m_2}(\theta_2, \phi_2) \tag{14.46}$$

Denote the sum of the two angular momenta by

$$\mathbf{J} = \mathbf{J}_1 + \mathbf{J}_2 \tag{14.47}$$

Given the eigenstates of $\tilde{J}_1^2$, $\tilde{J}_{1z}$, $\tilde{J}_2^2$, $\tilde{J}_{2z}$, we want to find the eigenstates of Φ_{jm} of the total angular momentum $\tilde{J}^2$, $\tilde{J}_z$, $\tilde{J}_1^2$, $\tilde{J}_2^2$. The eigenstates of $\tilde{J}^2$, $\tilde{J}_z$, $\tilde{J}_1^2$, $\tilde{J}_2^2$ are known as the ***coupled representation***.

Once again, the idea is to start from the highest-weight state, namely

$$\Phi_{j_mj_m} = \phi^1_{j_1j_1}\phi^2_{j_2j_2} \qquad (j_m = j_1 + j_2) \tag{14.48}$$

and get all the other states with $j = j_m = j_1 + j_2$ by applying the ladder operators. Then we construct states with the next lowest value of angular momentum, $j = j_1 + j_2 - 1$, and then states with $j = j_1 + j_2 - 2$, and so on. The first question is: where does this procedure end? If $j = j_1 + j_2$ is the highest possible value of total angular momentum, what is the lowest possible value?

This question can be answered by counting states. For a given j_1 and j_2 there are a total of $2j_1 + 1$ possible values of m_1, and $2j_2 + 1$ possible values of m_2, so the total number of orthonormal states $\psi^1_{j_1 m_1}\psi^2_{j_2 m_2}$ is

$$N = (2j_1 + 1)(2j_2 + 1) \tag{14.49}$$

Denote by $j_{\min}$ the lowest possible value of total angular momentum that can be constructed from states of given j_1 and j_2. Then the total number of eigenstates of total angular momentum will be

$$N' = \sum_{j=j_{\min}}^{j_1+j_2} (2j + 1) \tag{14.50}$$

and $j_{\min}$ is determined from the requiring $N = N'$. The solution to this condition is that

$$j_{\min} = |j_1 - j_2| \tag{14.51}$$

which we now quickly verify. Suppose, e.g. that $j_1 > j_2$, so that $j_{\min} = j_1 - j_2$. Then

$$\begin{aligned} N' &= \sum_{j=j_1-j_2}^{j_1+j_2} (2j + 1) \\ &= \sum_{j=1}^{j_1+j_2} (2j + 1) - \sum_{j=1}^{j_1-j_2-1} (2j + 1) \\ &= 2\sum_{j=1}^{j_1+j_2} j - 2\sum_{j=1}^{j_1-j_2-1} j + (2j_2 + 1) \end{aligned} \tag{14.52}$$

and using the fact that

$$\sum_{l=1}^{L} l = \frac{L(L + 1)}{2} \tag{14.53}$$

we find

$$\begin{aligned} N' &= (j_1 + j_2)(j_1 + j_2 + 1) - (j_1 - j_2)(j_1 - j_2 - 1) + (2j_2 + 1) \\ &= (2j_1 + 1)(2j_2 + 1) \\ &= N \end{aligned} \tag{14.54}$$

Starting from the highest weight state, equation (14.48), we apply $\tilde{J}_-$ to the left-hand side and $\tilde{J}_{1-} + \tilde{J}_{2-}$ to the right-hand side to get $\Phi_{j,j-1}$. Proceeding in this way, one finds all the states with $j_{\max} = j_1 + j_2$. Next, the state with $j = m = j_{\max} - 1$ must have the form

$$\Phi_{j_{\max}-1,\,j_{\max}-1} = a\phi^1_{j_1 j_1}\phi^2_{j_2,\,j_2-1} + b\phi^1_{j_1,\,j_1-1}\phi^2_{j_2,\,j_2} \tag{14.55}$$

because these are the only states with $j_z = j_{\max} - 1$. The two constants a and b are determined by orthogonality

$$\left\langle \Phi_{j_{\max}, j_{\max}-1} | \Phi_{j_{\max}-1, j_{\max}-1} \right\rangle = 0 \tag{14.56}$$

and normalization

$$\left\langle \Phi_{j_{\max}-1, j_{\max}-1} | \Phi_{j_{\max}-1, j_{\max}-1} \right\rangle = 1 \tag{14.57}$$

Having determined a and b, apply $\tilde{J}_-$ to the left-hand side, and $\tilde{J}_{1-} + \tilde{J}_{2-}$ to the right-hand side, to find $\Phi_{j_{\max}-1, j_{\max}-2}$, and continue applying the ladder operator successively to find all states with $j_{\max} - 1$. Then determine the three constants a, b, c in the state

$$\Phi_{j_{\max}-2, j_{\max}-2} = a\phi^1_{j_1 j_1}\phi^2_{j_2, j_2-2} + b\phi^1_{j_1, j_1-1}\phi^2_{j_2, j_2-1} + c\phi^1_{j_1 j_1-2}\phi^2_{j_2, j_2} \tag{14.58}$$

from the two orthogonality conditions

$$\begin{aligned} \left\langle \Phi_{j_{\max}, j_{\max}-2} | \Phi_{j_{\max}-2, j_{\max}-2} \right\rangle &= 0 \\ \left\langle \Phi_{j_{\max}-1, j_{\max}-2} | \Phi_{j_{\max}-2, j_{\max}-2} \right\rangle &= 0 \end{aligned} \tag{14.59}$$

and the normalization condition

$$\left\langle \Phi_{j_{\max}-2, j_{\max}-2} | \Phi_{j_{\max}-2, j_{\max}-2} \right\rangle = 1 \tag{14.60}$$

Ladder operators are applied again, and the whole procedure is continued until all of the states with total angular momentum $j = |j_1 - j_2|$ have been found. At that point, stop. A summary of the procedure is here:

$$\begin{array}{ccccc}
\underline{j_m = j_1 + j_2} & \underline{j_m - 1} & \underline{j_m - 2} & \cdots\ \cdots & \underline{|j_1 - j_2|} \\
\Phi_{j_m j_m} = \phi^1_{j_1 j_1}\phi^2_{j_2 j_2} & & & & \\
\Downarrow & & & & \\
\Phi_{j_m j_m - 1} & \Phi_{j_m - 1, j_m - 1} & & & \\
\Downarrow & \Downarrow & & & \\
\Phi_{j_m j_m - 2} & \Phi_{j_m - 1, j_m - 2} & \Phi_{j_m - 2, j_m - 2} & & \\
\Downarrow & \Downarrow & \Downarrow & & \\
\vdots & \vdots & \vdots & & \Phi_{|j_1 - j_2|, |j_1 - j_2|} \\
\Downarrow & \Downarrow & \Downarrow & & \Downarrow \\
\vdots & \vdots & \vdots & \vdots & \vdots \\
\vdots & \vdots & \vdots & & \Phi_{|j_1 - j_2|, -|j_1 - j_2|} \\
\vdots & \vdots & \vdots & & \\
\Phi_{j_m, -(j_m - 2)} & \Phi_{j_m - 1, -(j_m - 2)} & \Phi_{j_m - 2, -(j_m - 2)} & & \\
\Downarrow & \Downarrow & & & \\
\Phi_{j_m, -(j_m - 1)} & \Phi_{j_m - 1, -(j_m - 1)} & & & \\
\Downarrow & & & & \\
\Phi_{j_m, -j_m} & & & &
\end{array}$$

An important application of this procedure is the addition of the spin angular momentum of two spin $\frac{1}{2}$ particles; a case that turns up again and again in atomic, nuclear, and particle physics. In this case there are four eigenstates of $\tilde{S}_1^2$, $\tilde{S}_{1z}$, $\tilde{S}_2^2$, $\tilde{S}_{2z}$ in the uncoupled representation, denoted

$$\chi_\pm^1 \chi_\pm^2 \tag{14.61}$$

We want to find all possible eigenstates of $\tilde{S}^2$, $\tilde{S}_z$, $\tilde{S}_1^2$, $\tilde{S}_2^2$ in the coupled representation. In this case

$$s_{\max} = \frac{1}{2} + \frac{1}{2} = 1 \qquad s_{\min} = \left|\frac{1}{2} - \frac{1}{2}\right| = 0 \tag{14.62}$$

The highest weight state is

$$\Phi_{11} = \chi_+^1 \chi_+^2 \tag{14.63}$$

Applying the ladder operators

$$\begin{aligned} \tilde{S}_-\Phi_{11} &= \left(\tilde{S}_{1-} + \tilde{S}_{2-}\right)\chi_+^1\chi_+^2 \\ \sqrt{2}\,\hbar\Phi_{10} &= \left(\tilde{S}_{1-}\chi_+^1\right)\chi_+^2 + \chi_+^1\left(\tilde{S}_{2-}\chi_+^2\right) \\ &= \hbar\left[\chi_-^1\chi_+^2 + \chi_+^1\chi_-^2\right] \end{aligned} \tag{14.64}$$

to find

$$\Phi_{10} = \frac{1}{\sqrt{2}}\left[\chi_+^1\chi_-^2 + \chi_-^1\chi_+^2\right] \tag{14.65}$$

Again apply the ladder operators

$$\begin{aligned} \tilde{S}_-\Phi_{10} &= \frac{1}{\sqrt{2}}\left(\tilde{S}_{1-} + \tilde{S}_{2-}\right)\left[\chi_+^1\chi_-^2 + \chi_-^1\chi_+^2\right] \\ \hbar\sqrt{2}\,\Phi_{1-1} &= \frac{1}{\sqrt{2}}\left[\left(\tilde{S}_{1-}\chi_+^1\right)\chi_-^2 + \left(\tilde{S}_{1-}\chi_-^1\right)\chi_+^2 + \chi_+^1\left(\tilde{S}_{2-}\chi_-^2\right) + \chi_-^1\left(\tilde{S}_{2-}\chi_+^2\right)\right] \\ &= \frac{\hbar}{\sqrt{2}}\left[\chi_-^1\chi_-^2 + 0 + 0 + \chi_-^1\chi_-^2\right] \end{aligned} \tag{14.66}$$

which gives, as it should, the lowest weight state

$$\Phi_{1-1} = \chi_-^1\chi_-^2 \tag{14.67}$$

There are three states with $j = 1$, which are known as the ***triplet spin states***. The one remaining state at $j = 0$ is known, for obvious reasons, as the ***singlet spin state***. The singlet state

$$\Phi_{00} = a\chi_+^1\chi_-^2 + b\chi_-^1\chi_+^2 \tag{14.68}$$

is determined by the orthogonality and normalization conditions

$$\begin{aligned} \langle\Phi_{10}|\Phi_{00}\rangle &= 0 \\ \langle\Phi_{00}|\Phi_{00}\rangle &= 1 \end{aligned} \tag{14.69}$$

From the orthogonality condition

$$\begin{aligned} 0 &= \frac{1}{\sqrt{2}}\left[\langle\chi_+^1\chi_-^2| + \langle\chi_-^1\chi_+^2|\right]\left[a|\chi_+^1\chi_-^2\rangle + b|\chi_-^1\chi_+^2\rangle\right] \\ &= \frac{1}{\sqrt{2}}(a+b) \end{aligned} \tag{14.70}$$

which tells us that $b = -a$. Substituting into the normalization condition

$$\begin{aligned} 1 &= a^2\left[\langle\chi_+^1\chi_-^2| - \langle\chi_-^1\chi_+^2|\right]\left[|\chi_+^1\chi_-^2\rangle - |\chi_-^1\chi_+^2\rangle\right] \\ &= 2a^2 \qquad \Leftarrow \qquad a = \frac{1}{\sqrt{2}} \end{aligned} \tag{14.71}$$

we find that the spin singlet state is

$$\Phi_{00} = \frac{1}{\sqrt{2}}\left[\chi_+^1\chi_-^2 - \chi_-^1\chi_+^2\right] \tag{14.72}$$

One of the many applications of addition of spin $\frac{1}{2}$ angular momentum is to the so-called 'hyperfine splitting' of the hydrogen atom ground state; this has important consequences for radio astronomy. Now the ground state of the hydrogen atom has zero orbital angular momentum, so there is no $L \cdot S$ splitting of the energy level. However, the proton, like the electron, is a spin $\frac{1}{2}$ particle, and, like the electron, it has a certain magnetic moment

$$\boldsymbol{\mu}_p = \frac{g_p e}{2m_p c}\mathbf{S}_p \tag{14.73}$$

where, experimentally, $g_p = 5.59$. For the electron, the corresponding 'g-factor' is very close to $g_e = 2$, and the electron magnetic moment is

$$\boldsymbol{\mu}_e = -\frac{e}{2m_e c}\mathbf{S}_e \tag{14.74}$$

Now, according to classical electrodynamics, the magnetic field at point **r** due to a magnetic dipole at the origin is given by

$$\mathbf{B} = \frac{1}{4\pi r^3}[3(\boldsymbol{\mu} \cdot \mathbf{e}_r)\mathbf{e}_r - \boldsymbol{\mu}] + \frac{2}{3}\boldsymbol{\mu}\delta^3(r) \tag{14.75}$$

So the contribution to the Hamiltonian for the electron, due to the interaction of the electron magnetic dipole moment with the magnetic field due to the proton dipole moment is

$$\tilde{H}' = e^2\frac{g_p}{8\pi c^2 m_p m_e}\frac{3\left(\tilde{\mathbf{S}}_p \cdot \mathbf{e}_r\right)(\tilde{\mathbf{S}}_e \cdot \mathbf{e}_r) - \tilde{\mathbf{S}}_p \cdot \tilde{\mathbf{S}}_e}{r^3} + e^2\frac{g_p}{3c^2 m_p m_e}\tilde{\mathbf{S}}_p \cdot \tilde{\mathbf{S}}_e\delta^3(r) \tag{14.76}$$

The $\tilde{\mathbf{S}}_p$ operators in $\tilde{H}'$ act on the spin degrees of freedom of the proton. Therefore we have to enlarge, a little bit, our expression for the hydrogen wave function, to include these extra quantized degrees of freedom. The ground state wave function(s), which are eigenstates of H_0, $\tilde{S}_e^2$, $\tilde{S}_{ez}$, $\tilde{S}_p^2$, $\tilde{S}_{pz}$, are the four states

$$R_{n0}(r)Y_{00}(\theta, \phi)\chi^e_\pm\chi^p_\pm \tag{14.77}$$

where $\chi^e_\pm$ refers to the spin state of the electron, and $\chi^p_\pm$ refers to the spin state of the proton. We can also, following the procedure above, reorganize these four states into four other states, which are eigenstates of H_0, $\tilde{S}^2$, $\tilde{S}_z$, $\tilde{S}_e^2$, $\tilde{S}_p^2$, where $\mathbf{S} = \mathbf{S}_e + \mathbf{S}_p$, i.e.

$$\psi^{\text{triplet}}_{s=1,\, s_z} = R_{n0}(r)Y_{00}(\theta, \phi)\begin{cases} \chi^e_+\chi^p_+ & (s_z = 1) \\ \frac{1}{\sqrt{2}}\left(\chi^e_+\chi^p_- + \chi^e_-\chi^p_+\right) & (s_z = 0) \\ \chi^e_-\chi^p_- & (s_z = -1) \end{cases}$$

$$\psi^{\text{singlet}}_{s=0,\, s_z=0} = R_{n0}(r)Y_{00}(\theta, \phi)\frac{1}{\sqrt{2}}\left(\chi^e_+\chi^p_- - \chi^e_-\chi^p_+\right) \tag{14.78}$$

Then, once again making use of a result (to be shown) from first order perturbation theory

$$\begin{aligned} \Delta E^{\text{triplet}}_{1s_z} &= \left\langle\psi^{\text{triplet}}_{ss_z}|H'|\psi^{\text{triplet}}_{ss_z}\right\rangle \\ \Delta E^{\text{singlet}} &= \langle\psi^{\text{singlet}}|H'|\psi^{\text{singlet}}\rangle \end{aligned} \tag{14.79}$$

The expectation values above involve an integration over angles θ, ϕ. The Y_{00} spherical harmonic has no angular dependence, but the first term in $\tilde{H}'$ does have such dependence. When the integral over solid angle is carried out, the first term in $\tilde{H}'$ averages to zero. The integration over r, θ, ϕ for the second term is very easily carried out, since it involves a delta function, and we find

$$\begin{aligned}
\Delta E &= e^2 \frac{g_p}{3c^2 m_e m_p} \left\langle \psi \middle| \delta^3(r) \tilde{S}_p \cdot \tilde{S}_e \middle| \psi \right\rangle \\
&= e^2 \frac{g_p}{3c^2 m_e m_p} \left\langle \psi \middle| \delta^3(r) \frac{1}{2}\left(\tilde{S}^2 - \tilde{S}_p^2 - \tilde{S}_e^2\right) \middle| \psi \right\rangle \\
&= e^2 \frac{g_p}{3c^2 m_e m_p} \frac{1}{2} \hbar^2 \left[s(s+1) - \frac{3}{4} - \frac{3}{4} \right) \langle \psi | \delta^3(r) | \psi \rangle \\
&= e^2 \frac{g_p \hbar^2}{3c^2 m_e m_p} |\psi(r=0)|^2 \frac{1}{2} \left[s(s+1) - \frac{3}{2} \right]
\end{aligned} \tag{14.80}$$

Then, using the fact that for the hydrogen atom ground state

$$|\psi(r=0)|^2 = \frac{1}{\pi a_0^3} \tag{14.81}$$

with Bohr radius $a_0 = 4\pi\hbar^2/m_e e^2$, we obtain

$$\Delta E = \frac{4 g_p \hbar^4}{3 m_p m_e^2 c^2 a_0^4} \begin{cases} \frac{1}{4} & \text{(triplet)} \\ -\frac{3}{4} & \text{(singlet)} \end{cases} \tag{14.82}$$

It is then possible for the electron–proton system to make a transition from the higher energy triplet state, to the lower energy singlet state, emitting a photon, whose energy is the difference in energies between the triplet/singlet states. According to the de Broglie relation, the frequency of this emitted photon will be

$$\begin{aligned}
f &= \frac{\Delta E_{\text{triplet}} - \Delta E_{\text{singlet}}}{h} \\
&= \frac{4 g_p \hbar^4}{3 m_p m_e^2 c^2 a_0^4 h} \\
&= 1420 \text{ MHz}
\end{aligned} \tag{14.83}$$

with a corresponding wavelength of 21 cm.

Atomic hydrogen is the most common element found in interstellar space, and transitions back and forth between the triplet and singlet states (separated by only about 6 millionths of an electron volt), occur very easily. For that reason, the 21 cm spectral line due to transitions between the hydrogen triplet and singlet states is one of the most dominant features of the interstellar radio spectrum.

Problems

1. a. Suppose ψ is an eigenstate of each of two non-commuting operators $\tilde{A}$ and $\tilde{B}$. Show that this requires

$$[\tilde{A}, \tilde{B}]\psi = 0 \tag{14.84}$$

b. Let $\mathbf{J} = \mathbf{L} + \mathbf{S}$. Compute $[\tilde{J}^2, \tilde{L}_z]$ and $[\tilde{J}^2, \tilde{S}_z]$ in terms of raising and lowering operators $\tilde{L}_\pm$ and $\tilde{S}_\pm$. Show that

$$\begin{aligned} \left[\tilde{J}^2, \tilde{L}_z\right] Y_{ll}(\theta, \phi)\chi_+ &= 0 \\ \left[\tilde{J}^2, \tilde{S}_z\right] Y_{ll}(\theta, \phi)\chi_+ &= 0 \end{aligned} \tag{14.85}$$

so that $Y_{ll}\chi_+$ can be an eigenstate of $\tilde{J}^2, \tilde{J}_z, \tilde{L}_z, \tilde{S}_z$, even though $\tilde{J}^2, \tilde{L}_z, \tilde{S}_z$ do not commute.
Is it possible to find a wave function that is an eigenstate of both the position ($\tilde{x}$) and momentum ($\tilde{p}$) operators?

2. Using

$$\begin{aligned} \tilde{J}^2 &= \tilde{L}^2 + \tilde{S}^2 + 2\tilde{L} \cdot \tilde{S} \\ \tilde{J}_z &= \tilde{L}_z + \tilde{S}_z \end{aligned} \tag{14.86}$$

and expressing $\tilde{L}_x, \tilde{L}_y, \tilde{S}_x, \tilde{S}_y$ in terms of ladder operators, show explicitly that the highest weight state is an eigenstate of $\tilde{J}^2$, i.e.

$$\begin{aligned} &\tilde{J}^2 Y_{ll}\chi_+ = j_{\max}(j_{\max} + 1)\hbar^2 Y_{ll}\chi_+ \quad \text{where} \quad j_{\max} = l + \frac{1}{2} \\ &\tilde{J}_z Y_{ll}\chi_+ = j_{\max}\, \hbar\, Y_{ll}\chi_+ \end{aligned} \tag{14.87}$$

3. Let $\hat{n}$ be a unit vector pointing in the (θ, ϕ) direction in spherical coordinates. Find the eigenstates and eigenvalues of the operator

$$\sigma_n = \hat{n} \cdot \vec{\sigma} = n_x\sigma_x + n_y\sigma_y + n_z\sigma_z \tag{14.88}$$

where the σ are the Pauli matrices. For an electron in each of these spin eigenstates, find the expectation value of the z-component of spin angular momentum.

4. Find all the eigenstates of the total angular momentum operators $\tilde{J}^2, \tilde{J}_z$ for an electron with orbital angular momentum $l = 2$ and spin $s = 1/2$. Express your answers both in wave function notation (i.e. ϕ_{jj_z}= linear combination of $Y_{lm}\chi_\pm$) and bra-ket notation ($|jj_z ls\rangle$= linear combination of $|lmss_z\rangle$). Also write a table of the corresponding Clebsch–Gordon coefficients

$$C_j^{ms_z} = \langle lmss_z | jj_z ls \rangle \tag{14.89}$$

for each j.

5. The total angular momentum of a system of two particles is $\mathbf{J} = \mathbf{J}_1 + \mathbf{J}_2$. Suppose $j_1 = j_2 = 1$. Find the eigenstates of $\tilde{J}^2$, $\tilde{J}_z$, denoted $|jmj_1j_2\rangle$ of the coupled representation, as linear combinations of the $\tilde{J}_1^2$, $\tilde{J}_{1z}$, $\tilde{J}_2^2$, $\tilde{J}_{2z}$ eigenstates, denoted $|j_1m_1j_2m_2\rangle$, of the uncoupled representation.
6. Imagine an 'atom' consisting of a single electron and, in place of the proton, a massive, positively charged, spin-2 particle.
   a) What 'spin' (i.e. total angular momentum) can the atom have, if the electron is in an orbital S-wave state? Or in a P-wave state?
   b) In the S-wave state, express the $J = 5/2$, $J_z = 3/2$ state in the coupled representation as a linear combination of states in the uncoupled representation.
7. Find all the eigenstates and eigenvalues of the operator

$$Q = a\tilde{\mathbf{S}}_3 \cdot \left(\tilde{\mathbf{S}}_1 + \tilde{\mathbf{S}}_2\right) + b\left(\tilde{\mathbf{S}}_1 + \tilde{\mathbf{S}}_2\right)^2 + c\left(\tilde{\mathbf{S}}_{1z} + \tilde{\mathbf{S}}_{2z} + \tilde{\mathbf{S}}_{3z}\right) \tag{14.90}$$

where $\tilde{\mathbf{S}}_{1,2,3}$ are the spin angular momentum operators of three *non*-identical spin $\frac{1}{2}$ particles. Denote the spin up states of particles 1, 2, and 3 by α_1, α_2, α_3, respectively, and the spin down states by β_1, β_2, β_3. (Hint: begin with the eigenstates of $\tilde{S}'^2$, $\tilde{S}'_z$, where $\mathbf{S}' \equiv \mathbf{S}_1 + \mathbf{S}_2$. You then need to find the eigenstates of $\tilde{S}^2_{\text{tot}}$, $\tilde{S}_{\text{tot},z}$ where $\mathbf{S}_{\text{tot}} = \mathbf{S}' + \mathbf{S}_3$.)

IOP Publishing

An Introduction to Quantum Theory

Jeff Greensite

Chapter 15

Identical particles

I wouldn't belong to any club that would have <u>*me*</u> *as a member!*

Groucho Marx

In ordinary experience, no two objects are *exactly* identical. 'Identical' twins are similar, rather than identical; 'identical' ping-pong balls can always be distinguished (at least under a microscope) by some small difference in size, shape, or surface markings.

It is different with elementary particles. As far as we can tell there are no scratches, marks, or personality traits that would distinguish one electron from another electron, or one proton from another. When we speak of identical particles in quantum physics, we mean the particles are *identical*, not just very very similar. The fact that there are objects in the world which seem to be exactly identical leads to a very interesting (almost metaphysical) question of principle regarding physical states.

To start out, imagine that we have two identical spin-0 particles. The position of one particle is denoted by x_1, and that of the other particle by x_2. Consider the state where particle 1 is very close to the point a, and particle 2 is very close to the point b, where 'very close' to point a means that the particle wave function is strongly peaked near point a. For example, the particle could be in a state described by a Gaussian wave function

$$\psi_a(x) = \frac{1}{(\pi\epsilon^2)^{1/4}} \exp\left[-\frac{(x-a)^2}{2\epsilon^2}\right] \tag{15.1}$$

where ϵ is very small. So the two-particle state would be

$$\psi(1, 2) = \psi_a(x_1)\psi_b(x_2) \tag{15.2}$$

doi:10.1088/978-0-7503-1167-0ch15

Now suppose someone comes along and interchanges the two particles. The particles would then be in a different physical state

$$\begin{aligned}\psi'(1, 2) &= \psi_a(x_2)\psi_b(x_1) \\ &= \psi(2, 1)\end{aligned} \tag{15.3}$$

However, if the particles are *exactly* identical, then there is no way of ever distinguishing between the states $\psi(1, 2)$, and $\psi'(1, 2)$. A measurement could only determine that one particle was found near point a, and one particle was found near point b. The particles themselves do not carry labels telling us which is particle 1, and which is particle 2. But if there are different physical states which are experimentally indistinguishable, then the difference between them amounts to a kind of 'hidden' information, which is not accessible to human knowledge.

Now there is really no obvious reason, *a priori*, that wave functions *cannot* carry hidden information. Perhaps this information is really there, but only accessible to some supernatural being. In a similar vein, the example of the Heisenberg microscope in chapter 2 suggested that there is no experimental way to measure position and momentum simultaneously to arbitrary precision. That does not logically imply that a particle *cannot* have a definite position and momentum, at any given time, but we found in fact that there are no physical states with this property. Therefore, let us tentatively take the point of view that physical states do not contain any experimentally inaccessible information, and see where it leads.

In order to even write down a two-particle wave function, we have to label the particles in some way. We could call them particle 1 and particle 2. We could just as well call them 'Mary' and 'Billy'. It does not matter; the point is that there has to be a place in the wave function where the labels go. For instance, in the previous example, we could write, instead of labels 1 and 2,

$$\psi(\text{Mary, Billy}) = \psi_a(x_{\text{Mary}})\psi_b(x_{\text{Billy}}) \tag{15.4}$$

We can be fairly confident, however, that particles do not know their names. Since there is no way of distinguishing between truly identical particles, we should have the same physical state, however we choose to label the particles. In other words, the state ψ(Mary, Billy), and the state we obtain by changing the labels of the particles, i.e. ψ(Billy, Mary), should be the same physical state. Let us elevate this idea to a fundamental principle:

The interchange condition

The only physical states allowed in Nature are those for which interchanging the labels of identical particles, in the wave function which represents the state, does not result in a different physical state.

Clearly a state of the form (15.4) does not fulfill this requirement, because

$$\psi_a(x_{\text{Mary}})\psi_b(x_{\text{Billy}}) \neq \psi_a(x_{\text{Billy}})\psi_b(x_{\text{Mary}}) \tag{15.5}$$

so according to the interchange condition this is not a physical state. On the other hand, in quantum mechanics we are allowed to build up states by superposition, so let us consider instead the state

$$\psi_S(\text{Mary, Billy}) = \frac{1}{\sqrt{2}}\big(\psi_a(x_{\text{Mary}})\psi_b(x_{\text{Billy}}) + \psi_a(x_{\text{Billy}})\psi_b(x_{\text{Mary}})\big) \tag{15.6}$$

where the subscript 'S' stands for 'symmetric'. In this state it clearly does not matter if we interchange the names of the two particles, because

$$\psi_S(\text{Mary, Billy}) = \psi_S(\text{Billy, Mary}) \tag{15.7}$$

Therefore ψ_S does qualify as a physical state.

However, there is another state that can be constructed from states $\psi_{a,\,b}(x)$ which *also* satisfies the interchange condition, namely

$$\psi_A(\text{Mary, Billy}) = \frac{1}{\sqrt{2}}\big(\psi_a(x_{\text{Mary}})\psi_b(x_{\text{Billy}}) - \psi_a(x_{\text{Billy}})\psi_b(x_{\text{Mary}})\big) \tag{15.8}$$

(the 'A' stands for 'antisymmetric'). In this case, the change in labels gives a change in sign

$$\psi_A(\text{Mary, Billy}) = -\psi_A(\text{Billy, Mary}) \tag{15.9}$$

This change in sign is actually not a violation of the interchange condition, since two wave functions which agree on every possible expectation value of every possible observable actually represent the same physical state. This means that two functions which differ only by an overall phase, i.e.

$$\psi \qquad \text{and} \qquad e^{i\delta}\psi \tag{15.10}$$

correspond to the same physical state, because the constant $e^{i\delta}$ factor drops out of all expectation values and probabilities. Such quantities are always bilinear in ψ, ψ^*, so an extra phase $e^{i\delta}$ always cancels against its complex conjugate $e^{-i\delta}$. In particular, an extra minus sign, corresponding to $\delta = \pi$, always cancels out. This means that Ψ_A and $-\Psi_A$ correspond to the same physical state. So in general, we are looking for two-particle wave functions which satisfy (switching back to gender and culture neutral labels 1 and 2 instead of Mary and Billy)

$$\psi(1, 2) = e^{i\delta}\psi(2, 1) \tag{15.11}$$

What possible values can there be, in equation (15.11), for the phase $e^{i\delta}$?

Let us define an ***exchange operator*** that operates on wave functions by interchanging labels

$$P_E\psi(1, 2) = \psi(2, 1) \tag{15.12}$$

Then states satisfying the interchange condition must satisfy

$$P_E\psi(1, 2) = e^{i\delta}\psi(1, 2) \tag{15.13}$$

It is easy to see that P_E is a linear operator, and equation (15.13) is an eigenvalue equation. Then, by the same logic we used to deduce the eigenstates of parity, we apply equation (15.12) twice to obtain

$$\begin{aligned} P_E P_E \psi(1, 2) &= P_E \psi(2, 1) \\ &= \psi(1, 2) \end{aligned} \tag{15.14}$$

Next apply equation (15.13) twice, to find

$$\begin{aligned} P_E P_E \psi(1, 2) &= \epsilon^{i\delta} P_E \psi(1, 2) \\ &= (\epsilon^{i\delta})^2 \psi(1, 2) \end{aligned} \tag{15.15}$$

Comparing (15.14) and (15.15), we deduce that the only possible eigenvalues are

$$e^{i\delta} = \pm 1 \tag{15.16}$$

Therefore, the only two-particle physical states which satisfy the interchange condition are the ***symmetric states*** $\psi_S(1, 2)$, with the property

$$\psi_S(2, 1) = +\psi_S(1, 2) \tag{15.17}$$

and the ***antisymmetric states*** $\psi_A(1, 2)$ with

$$\psi_A(2, 1) = -\psi_A(1, 2) \tag{15.18}$$

All of this easily generalizes to systems containing any number of identical particles. Suppose there are N such particles, and the corresponding wave function is denoted $\psi(1, 2, \ldots, N)$. Denote by P_E^{ij} the operator which exchanges the particle labels i and j, i.e.

$$P_E^{ij}\psi(1, 2, \ldots, i, \ldots, j, \ldots N) = \psi(1, 2, \ldots j, \ldots, i, \ldots N) \tag{15.19}$$

Then the states which are allowed by the interchange hypothesis are the symmetric states, satisfying

$$\psi_S(1, 2, \ldots, i, \ldots, j, \ldots N) = +\psi_S(1, 2, \ldots j, \ldots, i, \ldots N) \tag{15.20}$$

and the antisymmetric states, satisfying

$$\psi_A(1, 2, \ldots, i, \ldots, j, \ldots N) = -\psi_A(1, 2, \ldots j, \ldots, i, \ldots N) \tag{15.21}$$

for any choice of i, j with $i \neq j$.

So far, all this reasoning has been very hypothetical. *If* quantum states carry no experimentally inaccessible 'hidden information', *then* the wave functions must be either symmetric or antisymmetric with respect to an interchange of labels. Now comes an absolutely astounding fact, known as

The spin statistics theorem

Systems of identical particles with integer spin $s = 0, 1, 2, \ldots$ are described by wave functions that are *symmetric* under the interchange of particle label. Systems of identical particles with half-integer spin $s = \frac{1}{2}, \frac{3}{2}, \frac{5}{2}, \ldots$ are described by wave functions that are *antisymmetric* under the interchange of particle label.

Particles with integer spin are known as ***bosons***. Examples are π-mesons (spin zero), photons (spin 1), and gravitons (spin 2). Particles with half-integer spin are known as ***fermions***. Examples include electrons, protons, and neutrinos (spin 1/2), the Omega hyperon (spin 3/2), and the (hypothetical) gravitino (spin 3/2) predicted by a theory known as supergravity.

The spin statistics theorem is called a theorem because it can be proven from a handful of axioms (causality, locality, Lorentz invariance,...) in the context of a relativistic extension of quantum mechanics known as quantum field theory. We will touch briefly on quantum field theory at the very end of the book but the proof of the spin statistics theorem is not easy, and is simply beyond the scope of this book. Let us just note here that its implications have been tested experimentally in many ways. The most important consequence for atomic physics is known as

The Pauli exclusion principle

No two electrons can be in exactly the same quantum state.

This is the fact which makes chemistry possible, and leads to the remarkable repetition of chemical properties among the elements which is summarized in the periodic table. But in order to understand the exclusion principle, and how it follows from the spin statistics theorem, it is useful to first look at some simple examples of two-electron wave functions.

15.1 Two-electron states

Suppose we have two spin-zero identical particles (such as pions) which are prepared in such a way that one particle has momentum p_a, and the other particle has momentum p_b. What is the wave function of the system?

Were it not for the spin statistics theorem, the wave function could be (up to a normalization constant)

$$\psi(x_1, x_2) = e^{ip_a \cdot x_1/\hbar} e^{ip_b \cdot x_2/\hbar} \quad \text{or} \quad e^{ip_a \cdot x_2/\hbar} e^{ip_b \cdot x_1/\hbar} \tag{15.22}$$

or any linear combination

$$\psi(x_1, x_2) = A e^{ip_a \cdot x_1/\hbar} e^{ip_b \cdot x_2/\hbar} + B e^{ip_a \cdot x_2/\hbar} e^{ip_b \cdot x_1/\hbar} \tag{15.23}$$

However, the spin statistics theorem tells us that the only possible combination is the one which is symmetric with respect to the interchange of coordinate labels $1 \leftrightarrow 2$. Thus there is only one possible state in this case

$$\psi_S(x_1, x_2) = e^{ip_a \cdot x_1/\hbar} e^{ip_b \cdot x_2/\hbar} + e^{ip_a \cdot x_2/\hbar} e^{ip_b \cdot x_1/\hbar} \tag{15.24}$$

Next suppose there are two electrons, one with momentum p_a, the other with momentum p_b. This time, we also have to say something about the spin states. First, suppose the particle with momentum p_a has spin up, and the particle with momentum p_b has spin down. Again, without knowing the spin statistics theorem, the states could be

$$\psi = e^{ip_a \cdot x_1/\hbar} \chi_+^1 e^{ip_b \cdot x_2/\hbar} \chi_-^2 \quad \text{or} \quad e^{ip_a \cdot x_2/\hbar} \chi_+^2 e^{ip_b \cdot x_1/\hbar} \chi_-^1 \tag{15.25}$$

or any linear combination (χ_+^1 means particle 1 has spin up, etc). But the spin statistics theorem tells us that, up to a normalization constant, only the antisymmetric combination

$$\psi_A = e^{ip_a \cdot x_1/\hbar} \chi_+^1 e^{ip_b \cdot x_2/\hbar} \chi_-^2 - e^{ip_a \cdot x_2/\hbar} \chi_+^2 e^{ip_b \cdot x_1/\hbar} \chi_-^1 \tag{15.26}$$

will occur in Nature.

The two-electron state (15.26) is not an eigenstate of the total electron spin

$$S^2 = (S_1 + S_2)^2 = S_1^2 + S_2^2 + 2S_1 \cdot S_2 \tag{15.27}$$

because $\chi_+^1 \chi_-^2$ is not an eigenstate of $S_1 \cdot S_2$. But we saw, in the last chapter, that it is possible to combine the spin angular momentum of two spin $\frac{1}{2}$ particles into eigenstates of total spin $s = 0$ and $s = 1$. What is then the wave function if one particle has momentum p_a, the other p_b, and the total spin angular momentum is $s = 1$?

Let us begin with the spin part of the wave function. In the last chapter, it was found that spin $\frac{1}{2}$ combinations that give $s = 1$ are

$$\begin{aligned} \Phi_{11}^S &= \chi_+^1 \chi_+^2 \\ \Phi_{10}^S &= \frac{1}{\sqrt{2}} \left[\chi_+^1 \chi_-^2 + \chi_-^1 \chi_+^2 \right] \\ \Phi_{1-1}^S &= \chi_-^1 \chi_-^2 \end{aligned} \tag{15.28}$$

The superscript 'S' has been added to draw attention to the fact that these states are symmetric with respect to the interchange $1 \leftrightarrow 2$. But the spin statistics theorem only says that the *total* wave function, including the spin and the spatial parts, must change sign under the $1 \leftrightarrow 2$ interchange. So if the spin part is symmetric, the spatial part must be antisymmetric, i.e.

$$e^{ip_a \cdot x_1/\hbar} e^{ip_b \cdot x_2/\hbar} - e^{ip_a \cdot x_2/\hbar} e^{ip_b \cdot x_1/\hbar} \tag{15.29}$$

and the possible wave functions for $s = 1$ are

$$\psi^{\text{triplet}} = \left(e^{ip_a\cdot x_1/\hbar}e^{ip_b\cdot x_2/\hbar} - e^{ip_a\cdot x_2/\hbar}e^{ip_b\cdot x_1/\hbar}\right)\begin{cases} \Phi^S_{11} = \chi^1_+\chi^2_+ \\ \Phi^S_{10} = \frac{1}{\sqrt{2}}\left[\chi^1_+\chi^2_- + \chi^1_-\chi^2_+\right] \\ \Phi^S_{1-1} = \chi^1_-\chi^2_- \end{cases} \tag{15.30}$$

The other possibility is $s = 0$. In this case, the spin wave function is

$$\Phi^A_{00} = \frac{1}{\sqrt{2}}\left[\chi^1_+\chi^2_- - \chi^1_-\chi^2_+\right] \tag{15.31}$$

Since this part of the wave function is antisymmetric under the $1 \leftrightarrow 2$ interchange, the spatial part must be symmetric, and therefore

$$\begin{aligned} \psi^{\text{singlet}} &= (e^{ip_a\cdot x_1/\hbar}e^{ip_b\cdot x_2/\hbar} + e^{ip_a\cdot x_2/\hbar}e^{ip_b\cdot x_1/\hbar})\Phi^A_{00} \\ &= (e^{ip_a\cdot x_1/\hbar}e^{ip_b\cdot x_2/\hbar} + e^{ip_a\cdot x_2/\hbar}e^{ip_b\cdot x_1/\hbar}) \times \frac{1}{\sqrt{2}}\left[\chi^1_+\chi^2_- - \chi^1_-\chi^2_+\right] \end{aligned} \tag{15.32}$$

Finally, let us ask if it is possible to prepare two electrons in a state where both electrons have the same momentum $p_a = p_b = p$, and both particles have the same spin, e.g. both are spin up. It is not hard to see that the only wave function which would describe such a system is

$$\psi = e^{ip\cdot x_1/\hbar}e^{ip\cdot x_2/\hbar}\chi^1_+\chi^2_+ \tag{15.33}$$

But this is a symmetric state, therefore, it cannot exist in Nature! Two electrons can never be prepared in a state, or found to be in a state, with the same momenta and the same spin. Such a state, incidentally, is an $s = 1$ state, and one can see from equation (15.30) that the $s = 1$ states vanish for $p_a = p_b$. This is one example of the Pauli exclusion principle stated above. And of course the argument holds in general. Suppose two electrons were each in the same quantum state denoted ψ_a. Then the wave function represented such a two-particle state could only be

$$\psi(1, 2) = \psi_a(1)\psi_a(2) \tag{15.34}$$

But this is a symmetric state $\psi(1, 2) = +\psi(2, 1)$, and is therefore not allowed by the spin statistics theorem. This is how Pauli exclusion follows from spin statistics.

What about more complicated states, involving N electrons? Begin from the fact that any state of N electrons is a superposition of products of N one-particle states, with each one-particle state of the form

$$\psi_a(i) = \begin{bmatrix} \psi_{a+}(x_i) \\ \psi_{a-}(x_i) \end{bmatrix} \tag{15.35}$$

For example, for the hydrogen atom the index a could be the set of quantum numbers $a = (nlmss_z)$, so that, e.g.

$$\psi_{nlmss_z}(i) = R_{nl}(r_i) Y_{lm}(\theta_i, \phi_i) \chi^i_{s_z} \tag{15.36}$$

which is an eigenstate of H, L^2, L_z, S^2, S_z. A one-particle state can, in general, be expressed as a superposition of some set of eigenstates

$$\psi(1) = \sum_a c_a \psi_a(1) \tag{15.37}$$

In the same way, a state consisting of two identical particles can be expressed as

$$\psi(1, 2) = \sum_a \sum_b c_{ab} \psi_a(1) \psi_b(2) \tag{15.38}$$

where the indices a, b run over the same set of quantum numbers. But according to the spin statistics theorem $\psi(1, 2) = -\psi(2, 1)$, which means that

$$\sum_a \sum_b c_{ab} \psi_a(1) \psi_b(2) = -\sum_a \sum_b c_{ab} \psi_a(2) \psi_b(1) \tag{15.39}$$

Now, on the right-hand side of this equation, switch the labels on the indices $a \to b$ and $b \to a$. We can always do that, since it does not matter what letter we use to denote the index of a sum. So then we have

$$\begin{aligned} \sum_a \sum_b c_{ab} \psi_a(1) \psi_b(2) &= -\sum_b \sum_a c_{ba} \psi_b(2) \psi_a(1) \\ &= -\sum_a \sum_b c_{ba} \psi_a(1) \psi_b(2) \end{aligned} \tag{15.40}$$

This means that

$$c_{ab} = -c_{ba} \tag{15.41}$$

and in particular $c_{aa} = 0$. In other words, the superposition can never include in the sum any states in which both electrons are in the same state. So let us go on and consider a state of N electrons

$$\begin{aligned} \psi(1, 2, \ldots, i, \ldots, j, \ldots, N) = &\sum_a \sum_b \cdots \sum_s \cdots \sum_w \cdots \sum_z c_{ab\ldots s\ldots w\ldots z} \\ &\times \psi_a(1) \psi_b(2) \ldots \psi_s(i) \ldots \psi_w(j) \ldots \psi_z(N) \end{aligned} \tag{15.42}$$

Now interchange labels $i \leftrightarrow j$

$$\begin{aligned} \psi(1, 2, \ldots, j, \ldots, i, \ldots, N) = &\sum_a \sum_b \cdots \sum_s \cdots \sum_w \cdots \sum_z c_{ab\ldots s\ldots w\ldots z} \\ &\times \psi_a(1) \psi_b(2) \ldots \psi_s(j) \ldots \psi_w(i) \ldots \psi_z(N) \end{aligned} \tag{15.43}$$

and relabel the indices $s \leftrightarrow w$ in the sums. This brings us to

$$\begin{aligned}\psi(1, 2, \ldots, j, \ldots, i, \ldots, N) &= \sum_a \sum_b \cdots \sum_w \cdots \sum_s \cdots \sum_z c_{ab\ldots w\ldots s\ldots z} \\ &\quad \times \psi_a(1)\psi_b(2)\ldots\psi_s(i)\ldots\psi_w(j)\ldots\psi_z(N) \\ &= -\psi(1, 2, \ldots, i, \ldots, j, \ldots, N)\end{aligned} \tag{15.44}$$

where the last line follows from the spin statistics theorem. Comparing to (15.42), we see that the previous equation can only be true if

$$c_{ab\ldots w\ldots s\ldots z} = -c_{ab\ldots s\ldots w\ldots z} \tag{15.45}$$

which means in particular that $c_{ab\ldots s\ldots s\ldots z} = 0$. What this means is that inside the sum over states in an N-particle state, no two electrons i, j can be in the same state s. There is zero probability, in an N-particle state, to find two electrons in the same quantum state. This is the essence of the Pauli exclusion principle, whose consequences for atomic physics are profound.

15.2 The helium atom

Before talking about atoms in general, let us discuss one in particular: helium. This is obviously the next step up in complexity from hydrogen and, since it has two electrons rather than one, is a good place to apply our knowledge of two-electron wave functions. The Hamiltonian for the system is

$$\begin{aligned}H &= \left\{\frac{p_1^2}{2m} - \frac{2e^2}{4\pi r_1}\right\} + \left\{\frac{p_2^2}{2m} - \frac{2e^2}{4\pi r_2}\right\} + \frac{e^2}{4\pi r_{12}} \\ &= H_0(1) + H_0(2) + V'\end{aligned} \tag{15.46}$$

where r_1 (r_2) is the distance of electron 1 (electron 2) to the nucleus, and r_{12} is the distance between the two electrons. We have neglected all other terms in the Hamiltonian that involve spin; nevertheless, due to the spin statistics theorem, electron spin turns out to be very important in determining the possible electron states.

The first thing to do is neglect V' also, and construct eigenstates of

$$H_0(12) = H_0(1) + H_0(2) \tag{15.47}$$

This is easy because the variables are separable, and the solutions of

$$H_0\Psi(x_1, x_2) = E\Psi(x_1, x_2) \tag{15.48}$$

are just

$$\begin{aligned}\Psi_{nlm,\,n'l'm'} &= \phi_{nlm}(r_1, \theta_1, \phi_1)\phi_{n'l'm'}(r_2, \theta_2, \phi_2) \\ E_{nlm,\,n'l'm'} &= E_n + E_{n'}\end{aligned} \tag{15.49}$$

where the $\phi_{nlm}(r, \theta, \phi)$ and E_n are just the hydrogen atom wave functions, with the one modification that the charge e^2 in those expressions is replaced by $2e^2$, since the helium nucleus has two protons. Of course, if the quantum numbers (nlm) and $(n'l'm')$ are different, then any linear combination

$$\Psi = A\phi_{nlm}(r_1, \theta_1, \phi_1)\phi_{n'l'm'}(r_2, \theta_2, \phi_2) + B\phi_{nlm}(r_2, \theta_2, \phi_2)\phi_{n'l'm'}(r_1, \theta_1, \phi_1) \quad (15.50)$$

is also an eigenstate of $H_0(12)$, with energy $E_n + E_{n'}$. In particular, we can form normalized, symmetric, and antisymmetric combinations

$$\begin{aligned}\Psi^S_{nlm,\,n'm'l'} &= \frac{1}{\sqrt{2}}\Big[\phi_{nlm}(r_1, \theta_1, \phi_1)\phi_{n'l'm'}(r_2, \theta_2, \phi_2) \\ &\quad + \phi_{nlm}(r_2, \theta_2, \phi_2)\phi_{n'l'm'}(r_1, \theta_1, \phi_1)\Big] \\ \Psi^A_{nlm,\,n;l'm'} &= \frac{1}{\sqrt{2}}\Big[\phi_{nlm}(r_1, \theta_1, \phi_1)\phi_{n'l'm'}(r_2, \theta_2, \phi_2) \\ &\quad - \phi_{nlm}(r_2, \theta_2, \phi_2)\phi_{n'l'm'}(r_1, \theta_1, \phi_1)\Big]\end{aligned} \quad (15.51)$$

So far we have only written down the spatial part of the energy eigenstates, and ignored the spin part. But H commutes with the total spin operators S^2, S_z, so we should be able to construct energy eigenstates which are also total spin eigenstates. The only possibilities for total spin is $s = 1$, which has a triplet of symmetric spin wave functions, and $s = 0$, which has a single antisymmetric spin wave function. To satisfy spin statistics, $s = 1$ must go with an antisymmetric spatial wave function, and $s = 0$ with a symmetric spatial wave function, so all in all

$$\begin{aligned}\Psi^{\text{triplet}}_{nlm,\,n'l'm'} &= \frac{1}{\sqrt{2}}\Big[\phi_{nlm}(r_1, \theta_1, \phi_1)\phi_{n'l'm'}(r_2, \theta_2, \phi_2) \\ &\quad - \phi_{nlm}(r_2, \theta_2, \phi_2)\phi_{n'l'm'}(r_1, \theta_1, \phi_1)\Big]\begin{cases}\Phi_{11} \\ \Phi_{10} \\ \Phi_{1-1}\end{cases} \\ \Psi^{\text{singlet}}_{nlm,\,n'l'm'} &= \frac{1}{\sqrt{2}}\Big[\phi_{nlm}(r_1, \theta_1, \phi_1)\phi_{n'l'm'}(r_2, \theta_2, \phi_2) \\ &\quad + \phi_{nlm}(r_2, \theta_2, \phi_2)\phi_{n'l'm'}(r_1, \theta_1, \phi_1)\Big]\Phi_{00}\end{aligned} \quad (15.52)$$

if (nlm) and $(n'm'l')$ are different, and

$$\Psi^{\text{singlet}}_{nlm,\,nlm} = \phi_{nlm}(r_1, \theta_1, \phi_1)\phi_{nlm}(r_2, \theta_2, \phi_2)\Phi_{00} \quad (15.53)$$

if (nlm) and $(n'm'l')$ are the same. Note that in the ground state

$$\Psi^{\text{singlet}}_{100,\,100} = \phi_{100}(1)\phi_{100}(2) \times \frac{1}{\sqrt{2}}\Big[\chi^1_+\chi^2_- - \chi^1_-\chi^2_+\Big] \quad (15.54)$$

the electrons are in spin up/spin down combinations. It is impossible to put two spin up electrons (or two spin down electrons) into the helium ground state; this is another illustration of the exclusion principle.

Now let us consider the case where one electron is in the (100) state, and the other electron is in the (nlm) state. The $s = 0$ singlet combination is known as ***parahelium***, and the $s = 1$ triplet combination is known as ***orthohelium***. Neglecting the electrostatic repulsive potential V' between the two electrons, orthohelium and parahelium are degenerate in energy.

In order to see how the electron repulsion removes the degeneracy between ortho- and parahelium, we will treat V' as though it were a small perturbing potential, and use the formula (to be derived in chapter 17)

$$\Delta E = \langle \Psi | V' | \Psi \rangle \tag{15.55}$$

which is valid if V' is a small correction to the rest of the Hamiltonian. Actually, in this case V' is not so small; it is the same order of magnitude in strength as the attractive potential of the electrons to the nucleus. But we will go ahead and use equation (15.55) anyway, trusting that it will produce results that are at least qualitatively, if not quantitatively, correct. So the correction to the zeroth-order energy

$$E^{O,P} = E_0 + \Delta E \tag{15.56}$$

for ortho- and parahelium will be

$$\begin{aligned} \Delta E^{O,P} = \frac{1}{2} \int & d^3x_1 d^3x_2 \{\phi^*_{100}(x_1)\phi^*_{nlm}(x_2) \pm \phi^*_{100}(x_2)\phi^*_{nlm}(x_1)\} \\ & \times \frac{e^2}{4\pi|\mathbf{x}_1 - \mathbf{x}_2|}\{\phi_{100}(x_1)\phi_{nlm}(x_2) \pm \phi_{100}(x_2)\phi_{nlm}(x_1)\} \\ & \times \langle \Phi_{s=0,1} | \Phi_{s=0,1} \rangle \end{aligned} \tag{15.57}$$

where the plus sign is for parahelium, and the minus sign for orthohelium. The spin states are normalized, so $\langle \Phi | \Phi \rangle = 1$, and therefore, collecting like terms,

$$\begin{aligned} \Delta E^P &= A + B \\ \Delta E^O &= A - B \end{aligned} \tag{15.58}$$

where

$$A = \int d^3x_1 d^3x_2 |\phi_{100}(x_1)|^2 \frac{e^2}{4\pi r_{12}} |\phi_{nlm}(x_2)|^2 \tag{15.59}$$

and

$$B = \int d^3x_1 d^3x_2 [\phi^*_{nlm}(x_1)\phi_{100}(x_1)] \frac{e^2}{4\pi r_{12}} [\phi_{nlm}(x_2)\phi^*_{100}(x_2)] \tag{15.60}$$

Term A is associated with a simple, intuitive picture. If we think of the electron in state (100) as having a charge density proportional to its probability density

$$\rho_{100}(x) = -e|\phi_{100}(x)|^2 \tag{15.61}$$

and likewise for the electron in state (nlm)

$$\rho_{nlm}(x) = -e|\phi_{nlm}(x)|^2 \tag{15.62}$$

then A is simply the Coulomb interaction energy between those two charge distributions, i.e.

$$A = \int d^3x_1 d^3x_2 \frac{\rho_{100}(x_1)\rho_{nlm}(x_2)}{4\pi|\mathbf{x}_1 - \mathbf{x}_2|} \tag{15.63}$$

On the other hand, there is really no classical picture for the origin of term B.

Both the quantities A and B are positive numbers, so we can conclude that

$$E \text{ of parahelium } (s = 0) > E \text{ of orthohelium } (s = 1) \tag{15.64}$$

The intuitive reason for this fact is something like the following: The $s = 1$ state is symmetric in spins, it is therefore antisymmetric in space. In spatially antisymmetric states, the electrons tend to be further apart, on average, than for the corresponding spatially symmetric states. For example, in an antisymmetric state the wave function vanishes at $x_1 = x_2$. Being further apart means that the repulsive potential is somewhat smaller, and hence the energy of the state is smaller. This is an example of a rule in atomic physics known as Hund's first rule (see next section), which says that, other things being equal, the state of highest spin is the state of lowest energy.

15.3 The periodic table

One learns in a beginning chemistry course that chemical reactions among the elements involve electrons hopping from one atom to another, either in the form of an ionic bond (one atom captures the electron of another, as in NaCl), or a covalent bond (one or more electrons are shared between atoms, as in H_2). But, why should electrons hop back and forth between atoms? Why do they not simply sit in the lowest possible energy state of each atom, and refuse to budge unless seriously molested? The key to that question, and to understanding the curious periodicity of chemical properties among the elements, is the Pauli exclusion principle.

In principle, everything that can be known about an atom is contained in the wave functions of its energy eigenstates. For purposes of chemistry it is the atomic ground state, and particularly the ionization energy, which are important (the ionization energy is the minimum energy required to remove a single electron from the atom). In principle the procedure for finding the ground state of an atom of atomic number Z is very straightforward: First, write down Schrödinger's equation for the Z-electron state

$$\left[\sum_{n=1}^{Z}\left\{-\frac{\hbar^2}{2m}\nabla_n^2 - \frac{Ze^2}{4\pi r_n}\right\} + \sum_{n>m}\frac{e^2}{4\pi|\mathbf{x}_n - \mathbf{x}_m|}\right]\Psi = E\Psi \tag{15.65}$$

Second, find the solution, antisymmetric with respect to particle interchange, which has the lowest energy. The antisymmetry restriction is very important; the lowest energy eigenstates are not antisymmetric in general.

Unfortunately, the second part of this suggested procedure is very hard to carry out in practice. The problem is that, due to the interactions among electrons, there is no way of separating variables in the many-electron Schrödinger equation. What is more, the electron interaction term is too large to be treated as a small perturbation, for which there exist systematic approximation methods.

But in the case that the number of electrons is fairly large, the theoretical physicist has another trick up his or her sleeve. In condensed matter, nuclear, and particle physics this trick is known variously as the 'mean field', or 'self-consistent field', or 'independent particle', or 'random phase' approximation. In atomic physics the method is known as the 'Hartree approximation'. Roughly the idea is this: when each particle interacts with many other particles, the potential due to those many other particles can be treated, on average, as though it were a classical potential, in which each particle moves independently. If we knew this average potential, we could calculate the wave function of each particle by solving a one-particle Schrödinger equation; and, knowing the wave function of each particle, we could calculate the average potential. Therefore one calculates the wave functions 'self-consistently', i.e. finding solutions which lead to a potential which gives those solutions. That is the general idea. The weakness of the approximation is that it ignores all correlations in the positions of the various particles. Still, when each particle interacts with many other particles, ignoring these subtle correlations is often justified. We will not actually do any calculations of multi-electron atomic wave functions using this method (the computations are quite involved), but it is still interesting and useful to see how one gets started.

The first part of the Hartree approximation is to imagine that each electron, numbered 1, 2, ..., Z is in an individual state, denoted $\phi_1, \phi_2, \ldots, \phi_Z$, so the total wave function would be

$$\Phi(1, 2, \ldots, Z) = \phi_1(1)\phi_2(2)\ldots\phi_Z(Z) \tag{15.66}$$

In other words, the electron labeled 1 is in state ϕ_1, and so on. This is of course already in conflict with the spin statistics theorem because Φ should be antisymmetric with respect to interchange of particle labels. In fact, the approximation can be improved by antisymmetrizing Φ (it is called the 'Hartree–Fock' approximation), but the improvement is on the order of 10%–20%, so we will ignore this additional complication. The Pauli exclusion principle, however, must be respected! This is imposed by requiring that the ϕ_n are all orthogonal, or nearly so, so that no two electrons are in the same quantum state. We will also suppose, for now, that the ϕ_n are of the form

$$\phi_n(n) = \phi_n(x_n)\chi_\pm^n \tag{15.67}$$

The essence of the Hartree approximation is that the electron in the kth state 'sees' the electron in the jth state as being a cloud of charge, with charge density given by

$$\rho(x) = -e\left|\phi_j(x)\right|^2 \tag{15.68}$$

In that case, we can write down a Schrödinger equation for the wave function of the k-th electron, treating the other electrons as though they were simply classical charge distributions of the form (15.68), i.e.

$$\left\{\frac{-\hbar^2}{2m}\nabla^2 - \frac{Ze^2}{4\pi r} + \sum_{n\neq k}\int \mathrm{d}^3y \frac{e^2\left|\phi_n(y)\right|^2}{4\pi|\mathbf{x}-\mathbf{y}|}\right\}\phi_k(x) = \mathcal{E}\phi_k(x) \tag{15.69}$$

To make things even simpler, the last term is approximated by its angular average, i.e.

$$\left\{\frac{-\hbar^2}{2m}\nabla^2 - \frac{Ze^2}{4\pi r} + V_k(r)\right\}\phi_k(x) = \mathcal{E}\phi_k(x) \tag{15.70}$$

where[1]

$$V_k(r) = \frac{1}{4\pi}\int \mathrm{d}\Omega \sum_{n\neq k}\int \mathrm{d}^3y \frac{e^2\left|\phi_n(y)\right|^2}{4\pi|\mathbf{x}-\mathbf{y}|} \tag{15.71}$$

There are two important observations to make about equation (15.70). First, it is an equation with a spherically symmetric potential. This means that the Hamiltonian commutes with L^2, L_z, and the solutions can be labeled by the same quantum numbers as the hydrogen atom wave functions, i.e. (nlm). Second, the potential is no longer exactly $1/r$. When a given electron is close to the nucleus, it sees a potential $-Ze^2/4\pi r$, due to the full charge of the nucleus. However, when the electron is near the edge of the atom, it will only see a potential of strength $-e^2/4\pi r$, because the other $Z-1$ electrons are interposed, and screen the nuclear charge. As a result, energies are no longer independent of the l quantum number. In fact, for a given n, the larger values of l have a higher energy (i.e. are more weakly bound). The reason is that the higher angular momentum states are further, on average, from the nucleus. Therefore an electron in such states sees a smaller attractive potential, and therefore has a smaller binding energy.[2] Equation (15.70) is actually a set of Z coupled partial differential equations, since the potential $V_k(r)$ in the equation for electron k depends on the solution of all the other $j \neq k$ equations. Solutions are obtained by an iterative, numerical approach.

[1] This is the Coulomb term. The Hartree approximation, because it has not been properly antisymmetrized, misses the exchange term.

[2] You might wonder why a spherically symmetric Hamiltonian with a $1/r$ potential has a greater degeneracy of eigenstates than Hamiltonians with other central potentials $V(r)$. The answer is that Hamiltonians with $1/r$ potentials actually have a symmetry which includes, but is greater than, spherical symmetry. A full discussion of this point, however, is beyond the scope of this text.

Without solving anything, a ground-state wave function in the Hartree approximation is denoted by listing the energy levels that the Z electrons are in. The notation for each level is

$$n\ \mathbf{l}^N \quad \begin{cases} n = 1, 2, 3, \ldots & \text{principal quantum number} \\ \mathbf{l} = s, p, d, \ldots & \text{spectroscopic symbol for } l = 0, 1, 2, \ldots, n-1 \\ N = 1, 2, 3, \ldots & \text{number of electrons in this level} \leqslant 2(2l+1) \end{cases} \tag{15.72}$$

The Pauli principle tells us that the maximum value for N, at any given l, is

$$N_{\max}(l) = 2(2l + 1) \tag{15.73}$$

Each electron has to be in a different state, and there are $2l + 1$ possible values of m for each l. Therefore there can be $2l + 1$ electrons with spin up, and $2l + 1$ electrons with spin down; a total of $2(2l + 1)$ electrons in all. When there are exactly this many electrons in a given level, the level is said to be 'filled'.

The last thing needed, in order to list the electron configuration of each atomic ground state, is the actual order of the energies. In the hydrogen atom, the energies depend only on the n quantum number, but in multi-electron atoms, as just mentioned, the binding energy decreases as l increases. This listing can only be obtained by actually carrying out the Hartree approximation, and is as follows:

$$1s,\ 2s,\ 2p,\ 3s,\ 3p,\ [4s,\ 3d],\ 4p,\ [5s,\ 4d],\ 5p,\ [6s,\ 4f,\ 5d],\ 6p,\ \ldots \tag{15.74}$$

where the levels in brackets are very close in energy. Note that there are cases, such as $4s$ and $3d$, where a level with smaller n and larger l has a higher energy than the level with higher n and smaller l. Again, this is due to the l dependence of the energy: the $4s$-orbital tends to be closer to the nucleus, and hence sees a larger effective positive charge, than the $3d$-orbital, which tends to be further from the nucleus, whose charge is therefore screened to a greater extent by intervening electrons.

Knowledge of the atomic number Z of an atom, and the order of energies (15.74), allows us to immediately write down the configuration of electrons in the ground state. As an example, consider sodium, with an atomic number of $Z = 11$. Of the 11 electrons, the first two go into the $1s$ state, which is filled. Two more electrons go into the $2s$ state, which is also filled. Of the remaining seven electrons, six fill the $2p$ level, and the remaining electron occupies the $3s$ level. So the electron configuration is written[3]

$$\text{Na:}\ \ 1s^2\ 2s^2\ 2p^6\ 3s \tag{15.75}$$

Electrons in the highest energy level are known as ***valence electrons***. It is these electrons that determine the chemical properties of the atom. Sodium has one valence electron, in the $3s$ state.

[3] When there is only one electron in a given level, one omits the superscript, i.e. $3s \equiv 3s^1$.

A quick tour of the periodic table

The periodic table begins with hydrogen, with a $1s$ electron configuration in the ground state, and an ionization energy of 13.6 eV or 1312 kJ/mol. The next element, helium, has two electrons in a $1s^2$ configuration, but (as we see in figure 15.1(a)), the ionization energy has almost doubled, to 24.6 eV (2372 kJ/mol). The reason for this increase in ionization energy is quite simple: the charge of the nucleus is twice as big as for hydrogen, so each of the $1s$ levels is more tightly bound to the nucleus. Helium will not easily give up an electron to form a molecule; moreover, since the

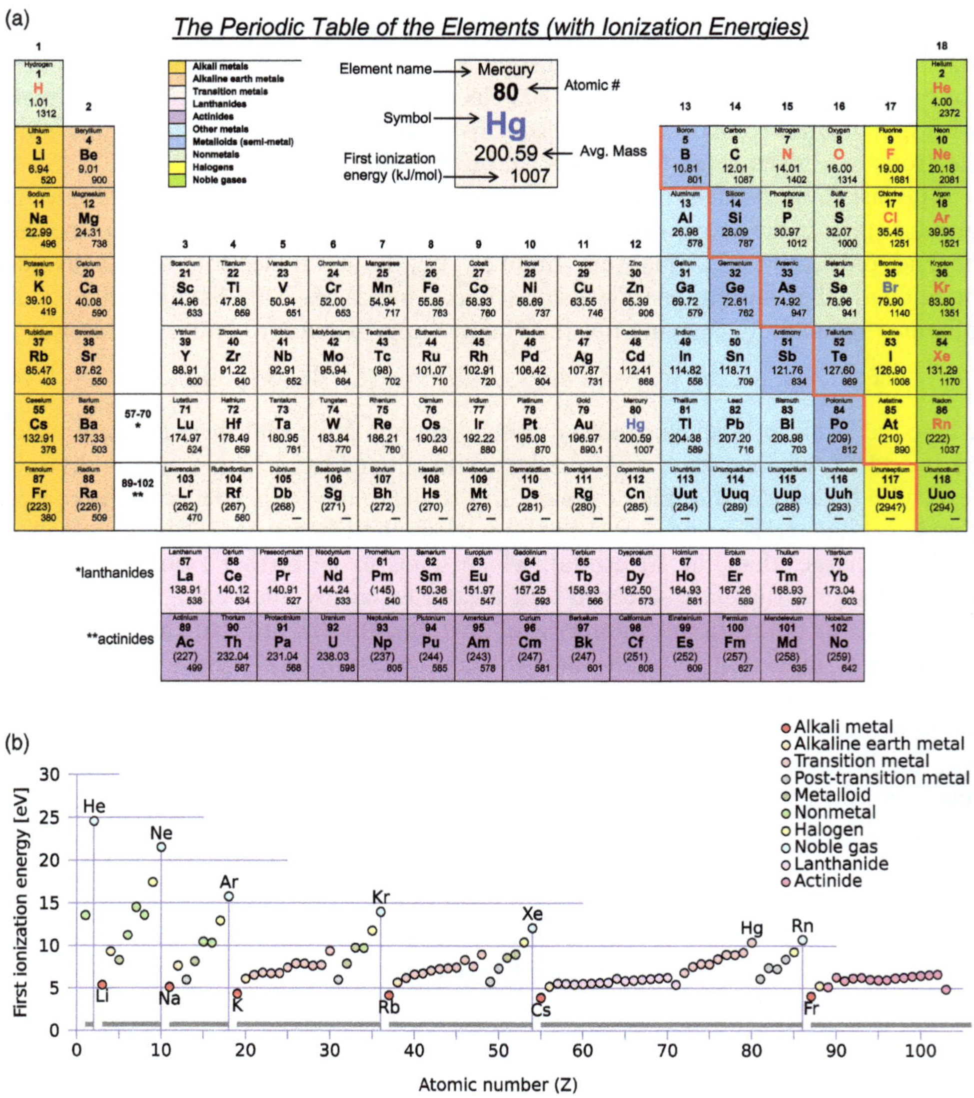

Figure 15.1. (a) Periodic Table with ionization energies in kJ mole^{-1}. Figure courtesy of Andy Allan, www.sciencegeek.net. (b) Ionization energies in eV versus atomic number.

$1s$ level is filled, it will not capture an electron either. This element is therefore chemically inert.

The electron configuration of the third element, lithium, is $1s^2\ 2s$, and the ionization energy is much less than helium, being only 5.4 eV (520.2 kJ/mol). Naively, since the charge of the nucleus is now $Z = 3$, one might expect the ionization energy to increase, as it did with helium, so that lithium would be even more inactive. It is at this point that the exclusion principle comes into play, with a constraint that allows chemistry to be interesting, and life to exist. Since the $1s$ level is filled, the third electron must occupy the $2s$ level. But an electron in the $2s$ level is further away from the nucleus, on average, than the two electrons in the $1s$ level, so the valence electron sees a shielded charge, which is only about $Z = 1$. Having seen this, we can estimate that the binding energy of the $2s$ electron in lithium should be about the same as that of a $2s$ electron in hydrogen, namely 3.4 eV. The actual binding energy (which is also the ionization energy) is in fact slightly higher than that (5.4 eV, as mentioned above), and this is because the electron in the $2s$ state has some amplitude for being close to the nucleus, where it would see a larger effective nuclear charge. Moving to the next element in line, beryllium, the configuration is $1s^2\ 2s^2$, and again the ionization energy increases because $Z = 4$, and the binding energy of each of the $2s$ states is greater. But although the $2s$ shell is filled, the $2p$ shell is quite close in energy, and the presence of a nearby atom may be enough to cause a rearrangement in which one of the $2s$ electrons goes into the $2p$ state. So beryllium is not as chemically inert as helium. In general, the most stable electron configurations are those for which: (i) the valence level is filled; and (ii) there is a large jump in energy to the next level; e.g. when all l levels for a given n are filled.

Roughly then, this is the picture. The various electron levels can be lumped together into groups which have about the same energy. These groups, and the maximum number of electrons allowed in each group, are

$$\begin{array}{ll} 1s & 2 \text{ electrons} \\ 2s,\ 2p & 8 \quad '' \\ 3s,\ 3p & 8 \quad '' \\ 4s,\ 3d,\ 4p & 18 \quad '' \\ 5s,\ 4d,\ 5p & 18 \quad '' \\ 6s,\ 4f,\ 5d,\ 6p & 32 \quad '' \\ 7s,\ 6d,\ 5f, \ldots & \end{array} \tag{15.76}$$

As one goes up in Z in each group, the binding energy tends to increase (figure 15.1(b)), due to the fact that charge of the nucleus increases. Thus the ionization energy builds to a maximum, which is the case when all of the energy levels in a group are filled, as seen in figure 15.1. These maxima occur for the ***inert gases***: helium, neon, argon, krypton. However, as Z increases further and the levels of the next higher group begin to be filled, the ionization energy drops sharply from its value at the inert gas. The reason is that the sudden decrease in binding energy of the next higher shell predominates over the increase in Z. It should be noted that unless a shell is completely filled, a high ionization energy does not mean that an element is necessarily very inert. Chlorine, for example, has an ionization energy 13 eV (1251.2 kJ/mol) which is nearly as high as the inert gas argon. But the valence level of chlorine is not filled; there is room for one more electron. And because the binding energy of the valence level is so high, it is possible for chlorine to capture an electron, thereby filling its valence level. This is why, when sodium and chlorine are brought into proximity, it is energetically favorable for sodium to lose its valence electron (this only costs 5.1 eV), which is more than compensated by a lowering of the chlorine energy by the valence electron bound state energy (−13 eV). The positively charged sodium ion is then bound to the negatively charged chlorine ion by ordinary electrostatic attraction. The resulting compound is ordinary table salt: NaCl.

The groups shown above in (15.76) are associated with rows in the periodic table. The first group, corresponding to the 1s level, is occupied only by hydrogen and helium. The next two groups are the 'short' rows of the periodic table, and constitute the eight elements of lithium to neon in the upper row, and sodium to argon in the lower row. Next come two long rows of the periodic table, with 18 elements each: potassium to krypton in the upper row, and rubidium to xenon in the lower row. Then comes a long group of 32 elements, containing the rare earths, whose names I can never remember. The last group of the heaviest elements, containing thorium, uranium, and some other elements that have been created in the laboratory but which do not appear in Nature, is not completely filled. The limitation on the size of the periodic table comes not from atomic physics, but from nuclear physics, since nuclei beyond some limiting Z tend to be highly radioactive, and decay an instant after they are produced.

Term notation

The Hartree approximation ignores all correlations between electrons, and in particular it ignores the correlations that lead to the atomic ground state being an eigenstate of *total* spin, orbital, and spin+orbital angular momenta

$$\begin{aligned} \mathbf{S} &= \mathbf{S}_1 + \mathbf{S}_2 + \cdots + \mathbf{S}_Z \\ \vec{L} &= \vec{L}_1 + \vec{L}_2 + \cdots + \vec{L}_Z \\ \mathbf{S} &= \vec{J}_1 + \vec{J}_2 + \cdots + \vec{J}_Z \end{aligned} \tag{15.77}$$

In fact, the Hamiltonian commutes with $\vec{J}^2$, L^2, S^2, and the ground state in particular has definite values of the quantum numbers J, L, S, where we use capitals letters to emphasize that these are quantum numbers for the ensemble of electrons. The ***term notation*** or ***spectroscopic description*** of a many-electron state with definite quantum numbers J, L, S is given by the expression

$${}^{2S+1}\mathcal{L}_J \tag{15.78}$$

where

$$\begin{aligned} L &= 0\ 1\ 2\ 3\ 4\ 5\ \ldots \\ \mathcal{L} &= S\ P\ D\ F\ G\ H\ \ldots \end{aligned} \tag{15.79}$$

There exists some folklore known as ***Hund's rules*** which allow one to guess, sometimes, the term description for the ground state of a given atom. In these rules, the phrase 'highest (lowest) possible' means: highest (lowest) possible consistent with the exclusion principle. The rules are as follows:

Hund's rules

1. Other things being equal, the lowest energy state of atomic electrons is the state with the highest S.
2. For a given total spin S, the lowest energy state is the state with the largest L.
3. For a given total S and L, if the valence level is not more than half-filled, the lowest energy state has the minimum $J = |L - S|$; if the shell is more than half-filled, the lowest energy state has $J = L + S$.

As an example of these rules we consider carbon, which has an atomic configuration $1s^2\ 2s^2\ 2p^2$. The filled shells have zero spin and zero orbital angular momentum (because there is no state one can make with $m^{\text{total}} \neq 0,\ \ S_z^{\text{total}} \neq 0$), so only the two valence electrons are relevant for determining the overall spin, orbital, and total angular momentum. The two spin $\frac{1}{2}$ electrons in the $2p$ states can combine into $S = 0$ or $S = 1$. By Hund's first rule, the $S = 1$ value is chosen. This means (for $S_z = +1$) both electrons are spin up. They are also both in an $l = 1$ orbital, so addition of angular momentum could lead to any of the values $L = 2, 1, 0$. But because of the exclusion principle, the two spin-up electrons cannot possibly have the same m values. This means that the highest and lowest weight states for adding $L_1 = 1$ and $L_2 = 1$ are excluded, since we cannot have $m_1 = m_2 = \pm 1$. So this rules out $L = 2$. The highest L left, by Hund's second rule, is $L = 1$. By Hund's third rule, since the valence shell is less than half filled, $J = |L - S| = 0$. So the term notation (or 'spectroscopic description') for the ground state of carbon is ${}^3P_0$.

There is much more that can be deduced about the periodic table, atomic physics, and chemical reactions from quantum-mechanical principles. This is the subject matter of theoretical chemistry. But there are two points that, once again, deserve special emphasis. First, the existence of chemical reactions is a consequence of the exclusion principle. Were it not for the exclusion principle, all atomic electrons would sit in the $1s$ state, becoming more and more tightly bound as Z increases. The second point is that the observed repetition of chemical properties is connected to the $2(2l + 1)$ occupancy of the s, p, d, f orbitals, and this in turn is connected to the peculiar algebra of the angular momentum operators. In short, the chemical properties of the elements are due essentially to the exclusion principle, and to the fact that two rotations, done in different orders, end up in different places. Life is a byproduct of chemistry; it is a curious thought that we owe our existence to the absolute identity of electrons, and to the non-commutative property of rotations.

Problems

1. Two equal mass particles are in the first excited state of a one-dimensional harmonic oscillator: that means that one particle is in the ground state ϕ_0, and the other is in the first excited state ϕ_1. Find the average square distance

$$D = \langle (x_1 - x_2)^2 \rangle \tag{15.80}$$

between the particles, in the case that:
   a. The particles are non-identical;
   b. The particles are identical and spin 0;
   c. The particles are identical and spin 1/2, in a total spin S = 1 state.

   In which case are the particles closest, and in which case furthest apart, on average?
2. Two particles are moving with equal and opposite momentum $\pm p$ along the x-axis. The Hamiltonian of the two-particle system is given by

$$H = \frac{p_1^2}{2m} + \frac{p_2^2}{2m} + a\mathbf{S}_1 \cdot \mathbf{S}_2 + b(S_{1z} + S_{2z})$$

where $\mathbf{S}_1$ and $\mathbf{S}_2$ are the spin operators of the two particles. Find the eigenvalues and eigenstates of this Hamiltonian in the case that:
   a. The particles are identical and spinless;
   b. The particles are identical and each of spin 1/2;
   c. The particle moving to the right has spin 1/2, and the particle moving to the left has spin 3/2.
3. Using the ordering of atomic energy levels given in (15.76), apply Hund's rules to determine the term notation for the ground states of the elements with atomic number $Z = 14, 15, 24, 28$.

IOP Publishing

An Introduction to Quantum Theory

Jeff Greensite

Chapter 16

Live wires and dead stars

Stars die. Their long lives are interesting and complex, occasionally culminating in fantastic explosions (supernovas) that can briefly outshine entire galaxies. The details are the subject matter of some other book. For us it is sufficient to know that every 'living' star balances its immense gravitational force, which tends to crush all the atoms of the star inward to a single point, with an equally enormous outward pressure, due to the heat produced by nuclear fusion. Eventually any star will exhaust its nuclear fuel, and then the gravitational force, unopposed, crushes the atoms of the star to a fantastic density. This is especially true for the remnants of stars following a supernova explosions. For remnants greater than about one solar mass, the atomic electrons are absorbed by protons in the nuclei, via the process

$$e^- + p \quad \rightarrow \quad n + \nu \tag{16.1}$$

and the massless neutrinos ν, due to their weak interaction with all other matter, escape into space. At this stage, all of the particles composing the star are neutrons, the density of the star approximates that of atomic nuclei, and the dead star is known as a 'neutron star'. For stars with masses less than about four solar masses, that is the end of the story: the cold dead star remains as a neutron star until the end of time. But this brings up the question: what force can there possibly be, within the cold neutron star, that is capable of opposing the mighty gravitational force, which would otherwise crush all the matter of the star to a single point?

It seems incredible that all this astrophysical drama should have anything at all to do with the apparently more mundane question of why some materials conduct electricity, and some do not. Nevertheless, the physics of dead stars and that of quite ordinary solids are related in certain unexpected ways. Both neutron stars and the electrons in cold metals are examples of what are known as ***degenerate Fermi gases***. We begin by taking up the question of how is it possible that certain solids conduct electricity, while others are insulators.

16.1 The Kronig–Penny model

A crystalline solid is a regular array of atoms, and, at first sight, conduction of electricity is a mystery: if electrons are bound to atoms, how is it possible for them to move through the solid under the influence of a small electric field? The answer is that in a crystal, not all of the electrons are actually bound to the atoms; in fact, some of the electrons in the metal behave more like a gas of free particles, albeit with some peculiar characteristics which are due to the exclusion principle.

To understand how electrons in a crystal can act as a gas, it is useful to solve for the electron energy eigenstates in a highly idealized model of a solid, known as the Kronig–Penny model, which makes the following simplifications:

S1 The solid is one-dimensional, rather than three-dimensional. The N atoms are spaced a distance a from one another. In order that there are no special effects at the boundaries, we consider a solid that has no boundary at all, by arranging the atoms in a circle.

S2 Instead of a Coulomb potential, the potential of the nth atom is represented by a delta-function attractive potential well

$$V_n(x) = -g\delta(x - x_n) \tag{16.2}$$

where x_n is the position of the nth atom.

S3 Interactions between electrons in the one-dimensional solid are ignored.

Obviously, these are pretty drastic simplifications. An important feature of this model, which it shares with realistic solids, is that the electrons are moving in a periodic potential. Arranging the atoms in a circle means that the position variable is periodic, like an angle. Just as $\theta + 2\pi$ is the same angle as θ, so the position $x + L$ is the same position as x, where

$$L = Na \tag{16.3}$$

is the length of the solid. Let the position of the nth particle be $x_n = na$, $n = 0, 1, \ldots, N - 1$, the potential then has the form

$$V(x) = -g \sum_{n=0}^{N-1} \delta(x - na) \tag{16.4}$$

It is clear that the potential satisfies

$$V(x + a) = V(x) \tag{16.5}$$

providing that we impose the periodicity requirement that $x + L$ denotes the same point as x. The 'periodic delta-function' which incorporates this requirement is given by

$$\delta(x) = \frac{1}{L} \sum_{m=-\infty}^{\infty} \exp[2\pi i m x/L] \tag{16.6}$$

which can be verified by expanding the periodic delta function in a Fourier series. The time-independent Schrödinger equation for the motion of any one electron in this potential has the usual form

$$H\psi_k(x) = \left[-\frac{\hbar^2}{2m}\frac{\partial^2}{\partial x^2} - g\sum_{n=0}^{N-1}\delta(x-na)\right]\psi_k(x) = E_k\psi_k(x) \tag{16.7}$$

Because $V(x)$ has the symmetry (16.5), it is useful to consider the translation operator first introduced in chapter 10,

$$\begin{aligned} T_a f(x) &= \exp[\mathrm{i}a\tilde{p}/\hbar]f(x) \\ &= f(x+a) \end{aligned} \tag{16.8}$$

Likewise,

$$\begin{aligned} T_{-a} f(x) &= \exp[-\mathrm{i}a\tilde{p}/\hbar]f(x) \\ &= f(x-a) \end{aligned} \tag{16.9}$$

Because $\tilde{p}$ is a Hermitian operator, it is easy to see (just expand the exponentials in a power series) that

$$\begin{aligned} T_a^\dagger &= T_{-a} \\ [T_a, T_{-a}] &= 0 \\ T_{-a}T_a &= 1 \end{aligned} \tag{16.10}$$

Due to the periodicity of the potential $V(x)$, the Hamiltonian commutes with the translation operators, which, as we have seen, also commute with each other, i.e.

$$[T_a, H] = [T_{-a}, H] = [T_a, T_{-a}] = 0 \tag{16.11}$$

This means (see chapter 10) that we can choose energy eigenstates to also be eigenstates of $T_{\pm a}$, i.e.

$$\begin{aligned} T_a\psi_E(x) &= \lambda_E\psi_E(x) \\ T_{-a}\psi_E(x) &= \lambda'_E\psi_E(x) \end{aligned} \tag{16.12}$$

Therefore,

$$\begin{aligned} \lambda_E &= \langle\psi_E|T_a|\psi_E\rangle \\ &= \langle T_a^\dagger\psi_E|\psi_E\rangle \\ &= \langle T_{-a}\psi_E|\psi_E\rangle \\ &= (\lambda'_E)^* \end{aligned} \tag{16.13}$$

But also

$$\begin{aligned} \psi_E(x) &= T_a T_{-a} \psi_E \\ &= \lambda_E \lambda'_E \psi_E \\ &= \psi_E(x) \end{aligned} \tag{16.14}$$

This means that $\lambda'_E = (\lambda_E)^{-1}$. Insert that fact into (16.13) and we conclude that $\lambda^*_E = (\lambda_E)^{-1}$, i.e.

$$\lambda_E = \exp(\mathrm{i}Ka) \tag{16.15}$$

for some K. In this way we arrive at:

Bloch's theorem

For potentials with the periodicity property $V(x + a) = V(x)$, each energy eigenstate of the Schrödinger equation satisfies

$$\psi(x + a) = \mathrm{e}^{\mathrm{i}Ka}\psi(x) \tag{16.16}$$

for some value of K.

It is also easy to work out the possible values of K, from the fact that

$$\begin{aligned} \psi(x) &= \psi(x + L) \\ &= (T_a)^N \psi(x) \\ &= \exp[\mathrm{i}NKa]\psi(x) \end{aligned} \tag{16.17}$$

which implies

$$K = \frac{2\pi}{Na} j \qquad j = 0, 1, 2, \ldots, N - 1 \tag{16.18}$$

The limiting value of j is $N - 1$, simply because

$$\exp\left[\mathrm{i}2\pi \frac{j + N}{N}\right] = \exp\left[\mathrm{i}2\pi \frac{j}{N}\right] \tag{16.19}$$

so $j \geqslant N$ does not lead to any further eigenvalues (j and $N - j$ are equivalent).

According to Bloch's theorem, if we can solve for the energy eigenstates in the region $0 \leqslant x \leqslant a$, then we have also solved for the wave function at all other values of x. Now the periodic delta function potential $V(x)$ vanishes in the region $0 < x < a$, so in this region (call it region I) the solution must have the free particle form

$$\psi_I(x) = A \sin(kx) + B \cos(kx) \qquad E = \frac{\hbar^2 k^2}{2m} \tag{16.20}$$

But according to Bloch's theorem, in the region $-a < x < 0$ (region II),

$$\begin{aligned} \psi_{\mathrm{II}}(x) &= \mathrm{e}^{-\mathrm{i}Ka} \psi_I(x + a) \\ &= \mathrm{e}^{-\mathrm{i}Ka}[A \sin k(x + a) + B \cos k(x + a)] \end{aligned} \tag{16.21}$$

Now we apply continuity of the wave function at $x = 0$ to obtain

$$B = e^{-iKa}[A\sin(ka) + B\cos(ka)] \tag{16.22}$$

For a delta function potential, we found in section 8.6 that there is a discontinuity in the slope of the wave function at the location ($x = 0$) of the delta function spike, namely

$$\left(\frac{d\psi}{dx}\right)_{|\epsilon} - \left(\frac{d\psi}{dx}\right)_{|-\epsilon} = -\frac{2mg}{\hbar^2}\psi(0) \tag{16.23}$$

and this condition gives us

$$kA - e^{-iKa}k[A\cos(ka) - B\sin(ka)] = -\frac{2mg}{\hbar^2}B \tag{16.24}$$

Solving equation (16.22) for A,

$$A = \frac{e^{iKa} - \cos(ka)}{\sin(ka)}B \tag{16.25}$$

inserting into equation (16.24) and cancelling B on both sides of the equation leads finally, after a few manipulations, to

$$\cos(Ka) = \cos(ka) - \frac{mg}{\hbar^2 k}\sin(ka) \tag{16.26}$$

This equation determines the possible values of k, and thereby, via $E = \hbar^2k^2/2m$, the possible energy eigenvalues of an electron in a periodic potential.

Now comes the interesting point. The parameter K can take on the values $2\pi n/Na$, and $\cos(Ka)$ varies from $\cos(Ka) = +1$ ($n = 0$) down to $\cos(Ka) = -1$ ($n = N/2$), and back up to $\cos(Ka) \approx +1$ ($n = N - 1$). So the left-hand side is always in the range $[-1, 1]$. On the other hand, the right-hand side is *not* always in this range, and that means *there are gaps in the allowed energies of an electron in a periodic potential.* This is shown in figure 16.1, where the right-hand side of (16.26) is plotted for a certain choice of parameters. Values of k for which the curve is outside the range $[-1, 1]$ correspond to regions of forbidden energies, known as ***energy gaps***, while the values where the curve is inside the $[-1, 1]$ range correspond to allowed energies, known as ***energy bands***. The structure of bands and gaps is indicated in figure 16.2; each of the closely spaced horizontal lines is associated with a definite value of K.

In the case that we have $M > N$ non-interacting electrons, each of the electrons must be in an energy state corresponding to a line in one of the allowed energy bands. The lowest energy state would naively be that of all electrons in the lowest energy level, but at this point we must invoke the exclusion principle: there can be no more than one electron in any given quantum state. Thus there can a maximum of two electrons (spin up and spin down) at any allowed energy in an energy band.

At the lowest possible temperature ($T = 0$ K), the electrons' configuration is the lowest possible energy consistent with the exclusion principle. A perfect ***insulator*** is a

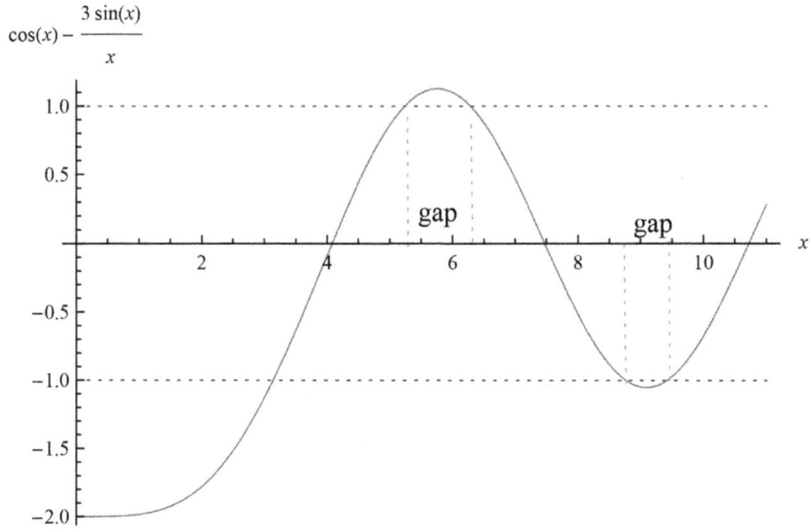

Figure 16.1. A plot of the right-hand side of equation (16.26), with $x = ka$, and $mg/\hbar^2 = 3a$. Intervals where the curve is greater than 1 or less than −1 are intervals where equation (16.26) cannot be satisfied. These regions correspond to 'gaps' in the electron energy levels.

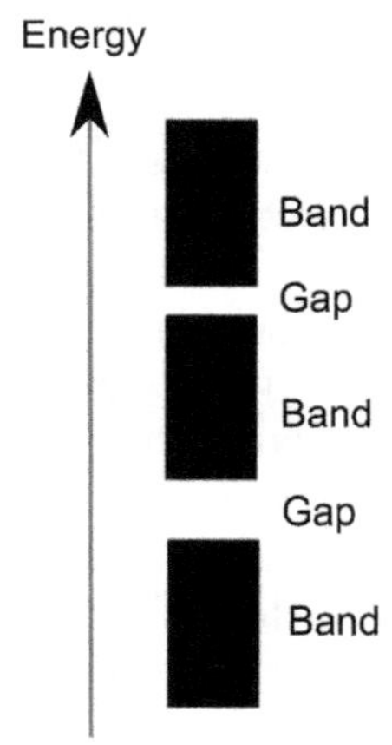

Figure 16.2. Energy bands and gaps. The black regions actually correspond to discrete but very closely spaced energy levels.

crystal in which the electrons completely fill one or more energy bands, and there is a gap in energy from the most energetic electron to the next unoccupied energy level. In a ***conductor***, the highest energy band containing electrons is only partially filled.

In an applied electric field the electrons in a crystal will tend to accelerate, and increase their energy. But they can only increase their energy if there are (nearby) higher energy states available, for electrons to occupy. If there are no nearby higher energy states, as in an insulator, no current will flow (unless the applied field is so enormous that electrons can 'jump' across the energy gap). In a conductor, there is an enormous number of nearby energy states for electrons to move into. Electrons are therefore free to accelerate, and a current flows through the material.

The actual physics of conduction, in a real solid, is of course far more complex than this little calculation would indicate. Still, the Kronig–Penny model does a remarkable job of isolating the essential effect, namely, the formation of energy bands, separated by gaps in which no electron states exist.

16.2 The free electron gas

In the Kronig–Penney model, the electron wave functions have a free particle form in the interval between the atoms; there is just a discontinuity in slope at precisely the position of the atoms. In passing to the three-dimensional case, we will simplify the situation just a bit more, by ignoring even the discontinuity in slope. The electron wave functions are then entirely of the free particle form, with only some boundary conditions that need to be imposed at the surface of the solid. Tossing away the atomic potential means losing the energy gaps; there is only one 'band', whose energies are determined entirely by the boundary conditions. For some purposes (such as thermodynamics of solids, or computing the bulk modulus), this is not such a terrible approximation.

We consider the case of N electrons in a cubical solid of length L on a side. Since the electrons are constrained to stay within the solid, but we are otherwise ignoring atomic potentials and inter-electron forces, the problem maps directly into a gas of non-interacting electrons in a cubical box. Inside the box, the Schrödinger equation for each electron has the free particle form

$$-\frac{\hbar^2}{2m}\nabla^2\psi(x, y, z) = E\psi(x, y, z) \tag{16.27}$$

While, at the boundaries, we impose the 'box' conditions

$$\begin{aligned} \psi(0, y, z) &= \psi(L, y, z) = 0 \\ \psi(x, 0, z) &= \psi(x, L, z) = 0 \\ \psi(x, y, 0) &= \psi(x, y, L) = 0 \end{aligned} \tag{16.28}$$

The solution of the particle in a box is a simple generalization of the particle in a tube (chapter 5), and the particle in a square (chapter 10). The eigenstates are

$$\psi_{n_1n_2n_3}(x, y, z) = \left(\frac{2}{L}\right)^{3/2} \sin\left(\frac{\pi n_1}{L}x\right)\sin\left(\frac{\pi n_2}{L}y\right)\sin\left(\frac{\pi n_3}{L}z\right) \tag{16.29}$$

with energy eigenvalues

$$E_{n_1n_2n_3} = \frac{\pi^2\hbar^2}{2mL^2}(n_1^2 + n_2^2 + n_3^2) \tag{16.30}$$

Suppose the solid is at temperature $T = 0$. We ask the questions: (i) what is the total energy E_T of the electrons in the solid; and (ii) what is the energy E_F of the most energetic electron in the solid? The energy of the most energetic electron in a cold solid, E_F, is known as the ***Fermi energy***.

Start with E_F. The method for calculating the Fermi energy is to first suppose that we know it, and, assuming that every energy $E_{n_1n_2n_3} < E_F$ is filled with two electrons (spin up/down), figure out the total number of electrons in the solid. By setting this number to N, we can determine E_F.

To each eigenstate there corresponds a set of integers n_1, n_2, n_3, so each possible energy state can be represented by a point in a three-dimensional space, with positive integer coordinates. Denote the maximum value of $n_1^2 + n_2^2 + n_3^2$, for the most energetic state, by R^2. The Fermi energy is therefore

$$E_F = \frac{\pi^2\hbar^2}{2mL^2}R^2 \tag{16.31}$$

Now all of the states with $E \leqslant E_F$ are occupied. So we need to count the number of points with integer coordinates $(n_1n_2n_3)$ such that

$$n_1^2 + n_2^2 + n_3^2 \leqslant R^2 \tag{16.32}$$

But since there is one site with integer coordinates per unit volume, the number of sites satisfying (16.32) is simply the volume of a one octant of a sphere of radius R (see figure 16.3). Since there can be no more than two electrons per site $(n_1n_2n_3)$, the total number of electrons with energies less than E_F, with all levels filled, is

$$N = 2 \times \frac{1}{8} \times \frac{4}{3}\pi R^3 = \frac{1}{3}\pi R^3 \tag{16.33}$$

According to (16.31),

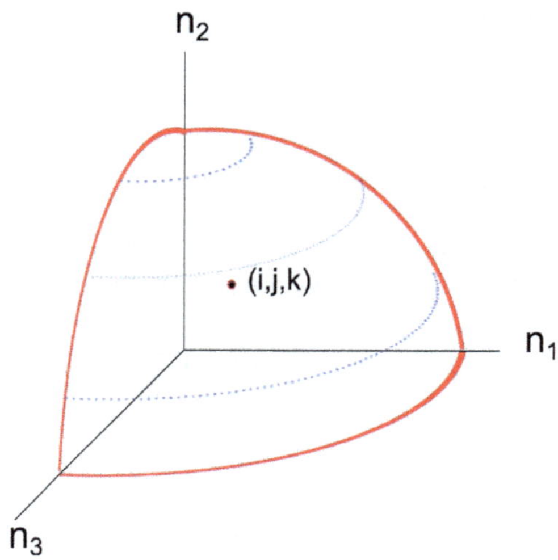

Figure 16.3. Energy eigenstates in the free electron gas are specified by three integers. Counting the number of states up to a maximum energy amounts to counting the number of points, with integer coordinates, in an octant of a sphere.

$$R = \left(\frac{2mL^2}{\hbar^2\pi^2}E_F\right)^{1/2} \tag{16.34}$$

so the number of electrons, in terms of E_F, becomes

$$N = \frac{\pi}{3}L^3\left(\frac{2\,mE_F}{\hbar^2\pi^2}\right)^{3/2} \tag{16.35}$$

Solving for the Fermi energy, we obtain

$$E_F = \frac{\hbar^2\pi^2}{2m}\left(\frac{3N}{\pi L^3}\right)^{2/3} \tag{16.36}$$

But the number density of electrons, n_e, in the solid is just

$$n_e = \frac{N}{L^3} \tag{16.37}$$

Therefore, we find that the Fermi energy just depends on the electron mass and density (and not on the size of the box)

$$E_F = \frac{\hbar^2\pi^2}{2m}\left(\frac{3}{\pi}n_e\right)^{2/3} \tag{16.38}$$

Next, approximate the sum over all states with energies less than E_F by an integral

$$\begin{aligned} E_T &= \sum_{n_1}\sum_{n_2}\sum_{n_3} 2E_{n_1n_2n_3} \\ &\approx \int_{\vec{n}\cdot\vec{n}\leqslant R^2} \mathrm{d}^3n\; 2E_{n_1n_2n_3} \\ &= \frac{\hbar^2\pi^2}{mL^2}\frac{1}{8}\int \mathrm{d}^3n\;\vec{n}\cdot\vec{n} \\ &= \frac{\hbar^2\pi^2}{8mL^2}4\pi\int_0^R \mathrm{d}n\; n^4 \\ &= \frac{\hbar^2\pi^3}{10mL^2}R^5 \end{aligned} \tag{16.39}$$

where the factor of 1/8 in front of the integral comes, again, because the sites lie in the octant of the sphere with *positive* integer coordinates. Next from (16.33)

$$R = \left(\frac{3N}{\pi}\right)^{1/3} \tag{16.40}$$

Using $L = V^{1/3}$, where V is the volume of the solid, we obtain finally

$$E_T = \frac{\hbar^2\pi^3}{10m}\left(\frac{3N}{\pi}\right)^{5/3} V^{-2/3} \tag{16.41}$$

The pressure with which any object resists compression is given by

$$p = -\frac{\mathrm{d}E_T}{\mathrm{d}V} \tag{16.42}$$

where E_T is the energy of the object. For a cold electron gas, this is known as the ***degeneracy pressure***. The bulk modulus B describes the change in pressure per fractional decrease in volume, i.e.

$$\begin{aligned} B &= -V\frac{\mathrm{d}p}{\mathrm{d}V} \\ &= V\frac{\mathrm{d}^2E_T}{\mathrm{d}V^2} \\ &= \frac{\pi^3\hbar^2}{9m}\left(\frac{3N}{\pi}\right)^{5/3} V^{-5/3} \\ &= \frac{\pi^3\hbar^2}{9m}\left(\frac{3n_e}{\pi}\right)^{5/3} \end{aligned} \tag{16.43}$$

Rather surprisingly, given all the unrealistic assumptions that have gone into this expression, it gives answers that compare reasonably well (within 30% in some cases) to the experimental results.

16.2.1 Neutron stars

Finally, we turn to the neutron star. The radius of the dead star is determined by a balance between the inward gravitational pressure and the outward degeneracy pressure, which is due to the fact that neutrons, like electrons, are spin $\frac{1}{2}$ objects. The degeneracy pressure is the same expression for neutron stars as it is for ordinary solids, the only difference being that the electron mass m should be replaced by the neutron mass m_n, the electron number density n_e by the neutron number density n_n, and N is the number of neutrons. The degeneracy pressure is

$$\begin{aligned} p_D &= -\frac{\mathrm{d}E_T}{\mathrm{d}V} \\ &= \frac{\hbar^2\pi^3}{15m_n}\left(\frac{3N}{\pi}\right)^{5/3} V^{-5/3} \\ &= \frac{\hbar^2\pi^3}{15m_n}\left(\frac{3n_n}{\pi}\right)^{5/3} \end{aligned} \tag{16.44}$$

where E_T is the total kinetic energy of the neutrons up to the Fermi level. On the other hand, the gravitational potential energy of a sphere of constant density ρ and radius $\mathcal{R}$ is given by

$$E_G = -\frac{(4\pi)^2}{15} G\rho^2 \mathcal{R}^5 \tag{16.45}$$

Using $\rho = M/V$, where M is the mass of the star, and $V = \frac{4}{3}\pi\mathcal{R}^3$, we obtain an expression for E_G in terms of V:

$$E_G = -\frac{3}{5}\left(\frac{4\pi}{3}\right)^{1/3} GM^2 V^{-1/3} \tag{16.46}$$

Then the inward gravitational pressure is

$$\begin{aligned} p_G &= -\frac{\mathrm{d}E_G}{\mathrm{d}V} \\ &= -\frac{1}{5}\left(\frac{4\pi}{3}\right)^{1/3} GM^2 V^{-4/3} \end{aligned} \tag{16.47}$$

Finally, requiring that the inward (p_G) and outward (p_D) pressures balance,

$$\frac{1}{5}\left(\frac{4\pi}{3}\right)^{1/3} GM^2 V^{-4/3} = \frac{\hbar^2\pi^3}{15m_n}\left(\frac{3N}{\pi}\right)^{5/3} V^{-5/3} \tag{16.48}$$

solving for V and then for $\mathcal{R} = (3V/4\pi)^{1/3}$, we obtain

$$\mathcal{R} = \left(\frac{81\pi^2}{16}\right)^{1/3} \frac{\hbar^2}{Gm_n^3} N^{-1/3} \tag{16.49}$$

For a two solar mass star, $N = M/m_n$, R turns out to be about 10 km.

It seems from equation (16.49) that no matter how massive the star (i.e. no matter how large N may be) the degeneracy pressure will always end the collapse at a finite value of R. Actually this is not quite right, because we have ignored relativistic effects. The free Schrödinger equation is based on $E = p^2/2m$, i.e. E grows quadratically with p. In fact, the correct expression in relativity is $E = \sqrt{p^2c^2 + m^2c^4}$, and at velocities very close to the speed of light, the energy is approximately $E \approx pc$; i.e. the energy grows only linearly with p. This means that the degeneracy pressure for relativistic neutrons is less than one would expect from the non-relativistic calculation. In fact, for stars at about four solar masses, the neutrons in the star *do* become relativistic, and the degeneracy pressure is insufficient to counter the gravitational pressure. For stars of that mass and greater, no known physical process can stop the continuing collapse, and the star becomes a black hole.

For a star with a mass somewhat less than that of the Sun, the cold star never reaches the neutron star stage, since the electron degeneracy pressure is sufficient to stabilize the star at radii comparable to planetary radii. For masses somewhat greater than that of the Sun, the electrons becomes relativistic, and the collapse continues until, for masses less than four solar masses, the star becomes a neutron star.

It is clear that electron spin (and the spin statistics theorem) explains a lot. It enables us to understand the qualitative features of the periodic table and the existence of conduction bands in crystals. It allows us to estimate the bulk modulus of solids, and the radii of dead stars. And this is not even half the story! In addition to degenerate fermion gases, there are also degenerate boson gases, in which a large fraction of the total number of particles occupies the lowest energy level. Aggregates of that kind are of crucial importance in understanding such startling phenomena as superconductivity and superfluidity, which are discussed in most textbooks on condensed matter physics.

Problems

1. Suppose that in copper metal there is one electron, from each copper atom, which can be viewed as belonging to a free electron gas. Under this assumption, calculate the Fermi energy for copper, and the classical velocity of an electron with kinetic energy equal to the Fermi energy (compare to the speed of light). (The density of copper is 8.96 $\mathrm{g/cm^3}$ and its atomic weight is 63.5 amu.)
2. Verify that the expression for a periodic delta function (16.6) is both periodic, and satisfies the defining property of a delta function (4.38), when the integration range runs from $x = 0$ to $x = L$.
3. Carry out the algebra to verify equation (16.26).
4. A white dwarf star is a dead star stabilized by electron degeneracy pressure. Imagine a star completely composed of hydrogen (i.e. equal numbers of electrons and protons), with a mass equal to 1/10 that of the Sun, stabilized by electron degeneracy pressure. What would be the radius of such a star? At about what mass would electrons at the Fermi surface be moving at half the speed of light?

Part IV

Approximation methods

IOP Publishing

An Introduction to Quantum Theory

Jeff Greensite

Chapter 17

Time-independent perturbation theory

There are only a handful of potentials for which the time-independent Schrödinger equation can be solved exactly. In one dimension there are rectangular potentials, the harmonic oscillator, linear potentials, and a few others. In three dimensions the time-independent Schrödinger equation is a partial differential equation, and these are almost impossible to solve exactly unless the separation of variables technique can be employed. Separation of variables usually requires a symmetry, typically spherical symmetry, as in a central potential. Even then there are only a few potentials for which the radial equation of section 11.5 can be solved exactly. Fortunately these include cases of great importance to physics and chemistry, such as the Coulomb potential. But apart from those special cases, where eigenstates of the Hamiltonian turn out to be Bessel functions, Hermite polynomials, Airy functions, hypergeometric functions, etc, most Hamiltonians encountered in real situations (as opposed to idealized examples in textbooks), do not have exact solutions in terms of special functions.

We have already seen that even the hydrogen atom, meaning the *real* hydrogen atom with spin effects included, is not exactly soluble. There is an $L \cdot S$ coupling, there is a coupling between the electron and proton spins, there are other relativistic effects, and these introduce additional potentials that defy an exact treatment. But we have also seen that these contributions to the potential are extremely small compared to the Coulomb potential. When this is the case, i.e. when there is a very small part of the potential, a ***perturbation***, which spoils the exact solubility, then there is a very powerful technique which allows us to 'sneak up' on the solution, starting from the solution of the problem which ignores that troublesome piece of the potential. The technique is known as ***perturbation theory***, and in this chapter we will develop the theory for perturbations which do not depend on time. Perturbations which do depend on time, such as the potential energy of electrons interacting with light, will be dealt with in the next two chapters.

doi:10.1088/978-0-7503-1167-0ch17

17.1 Solution by power series

Consider a Hamiltonian which has this form:

$$H = H_0 + \text{a little bit extra potential} \tag{17.1}$$

where H_0, with potential energy V_0, is a Hamiltonian whose eigenvalue equation is already solved:

$$H_0\phi_n^0 = E_n^0\phi_n^0 \tag{17.2}$$

It then makes sense that the solutions to the eigenvalue equation

$$H\phi_n = E_n\phi_n \tag{17.3}$$

can be written, for each n, as

$$\begin{aligned} \phi_n &= \phi_n^0 + \text{a little bit extra function} \\ E_n &= E_n^0 + \text{a little bit extra constant} \end{aligned} \tag{17.4}$$

An example,

$$H = -\frac{\hbar^2}{2m}\frac{d}{\mathrm{d}x^2} + \frac{1}{2}kx^2 + \lambda x^4 \tag{17.5}$$

In this case

$$H_0 = -\frac{\hbar^2}{2m}\frac{d}{\mathrm{d}x^2} + \frac{1}{2}kx^2 \tag{17.6}$$

is the Hamiltonian of a harmonic oscillator, whose energy eigenvalues and eigenfunctions are well known, and

$$\text{a little bit extra potential} = \lambda x^4 \tag{17.7}$$

Another example is the case of a hydrogen atom in an external electric field, directed (say) along the z-axis

$$H = -\frac{\hbar^2}{2m}\nabla^2 - \frac{e^2}{4\pi r} + eEz \tag{17.8}$$

This time

$$H_0 = -\frac{\hbar^2}{2m}\nabla^2 - \frac{e^2}{4\pi r} \tag{17.9}$$

is the usual hydrogen atom Hamiltonian, and

$$\text{a little bit extra potential} = eEz \tag{17.10}$$

Of course, in each case these extra potentials can only be characterized as 'a little bit extra' if the constants λ and eE are small, in some appropriate units (to be discussed further below). Let us write in general that

$$\text{a little bit extra potential} = V'(x) = \lambda V(x) \tag{17.11}$$

where λ is some constant which controls whether $V'(x)$ is small or not. The total potential is $V_{\text{tot}} = V_0 + V' = V_0 + \lambda V$. Then the solutions of the eigenvalue equation (17.3) are, obviously, functions of λ, which can be expanded in a Taylor series around $\lambda = 0$, i.e.

$$\begin{aligned} \phi_n &= \phi_n(x, \lambda) \\ &= \phi_n^{(0)}(x) + \lambda\phi_n^{(1)}(x) + \lambda^2\phi_n^{(2)}(x) + \cdots \\ E_n &= E_n(\lambda) \\ &= E_n^{(0)} + \lambda E_n^{(1)} + \lambda^2 E_n^{(2)} + \cdots \end{aligned} \tag{17.12}$$

where

$$\begin{aligned} E_n^{(k)} &\equiv \frac{1}{k!}\left(\frac{d^k}{d\lambda^k}E_n(\lambda)\right)_{\lambda=0} \\ \phi_n^k(x) &\equiv \frac{1}{k!}\left(\frac{d^k}{d\lambda^k}\phi(x, \lambda)\right)_{\lambda=0} \end{aligned} \tag{17.13}$$

and of course

$$H = H_0 + \lambda V \tag{17.14}$$

Then the eigenvalue equation (17.3) becomes

$$\begin{aligned} &(H_0 + \lambda V)\left(\phi_n^{(0)} + \lambda\phi_n^{(1)} + \lambda^2\phi_n^{(2)} + \cdots\right) \\ &\quad = \left(E_n^{(0)} + \lambda E_n^{(1)} + \lambda^2 E_n^{(2)} + \cdots\right)\left(\phi_n^{(0)}(x) + \lambda\phi_n^{(1)}(x) + \lambda^2\phi_n^{(2)}(x) + \cdots\right) \end{aligned} \tag{17.15}$$

Collecting like powers of λ on each side we have

$$\begin{aligned} &H_0\phi_n^{(0)} + \lambda\left(H_0\phi_n^{(1)} + V\phi_n^{(0)}\right) + \cdots + \lambda^N\left(H_0\phi_n^{(N)} + V\phi_n^{(N-1)}\right) + \cdots \\ &\quad = E_n^{(0)}\phi_n^{(0)} + \lambda\left(E_n^{(0)}\phi_n^{(1)} + E_n^{(1)}\phi_n^{0}\right) + \lambda^2\left(E_n^{(0)}\phi_n^{(2)} + E_n^{(1)}\phi_n^{(1)} + E_n^{(2)}\phi_n^{(0)}\right) + \cdots \\ &\qquad + \lambda^N \sum_{j=0}^{N} E_n^{(j)}\phi_n^{(N-j)} + \cdots \end{aligned} \tag{17.16}$$

or

$$
\begin{aligned}
0 &= \left(H_0\phi_n^{(0)} - E_n^{(0)}\phi_n^{(0)}\right) + \lambda\left(H_0\phi_n^{(1)} + V\phi_n^{(0)} - E_n^{(0)}\phi_n^{(1)} - E_n^{(1)}\phi_n^{0}\right) + \cdots \\
&\quad + \lambda^N\left(H_0\phi_n^{(N)} + V\phi_n^{(N-1)} - \sum_{j=0}^{N} E_n^{(j)}\phi_n^{(N-j)}\right) + \cdots \\
&= F^{(0)} + \lambda F^{(1)} + \lambda^2 F^2 + \cdots
\end{aligned}
\tag{17.17}
$$

Now this equation has to be true for *every* choice of λ. But since, by definition, the $E_n^{(k)}$ and $\phi_n^{(k)}(x)$ are independent of λ, we must have

$$
F^{(0)} = F^{(1)} = F^{(2)} = \cdots = 0 \tag{17.18}
$$

which gives us an infinite set of coupled equations:

$$
\begin{aligned}
&\left(H_0 - E_n^{(0)}\right)\phi_n^{(0)} = 0 \\
&\left(H_0 - E_n^{(0)}\right)\phi_n^{(1)} = \left(E_n^{(1)} - V\right)\phi_n^{(0)} \\
&\left(H_0 - E_n^{(0)}\right)\phi_n^{(2)} = -\,V\phi_n^{(1)} + E_n^{(1)}\phi_n^{(1)} + E_n^{(2)}\phi_n^{(0)} \\
&\qquad\vdots \\
&\left(H_0 - E_n^{(0)}\right)\phi_n^{(N)} = -\,V\phi^{(N-1)} + \sum_{j=1}^{N-1} E_n^{(j)}\phi_n^{(N-j)} + E_n^{N}\phi_n^{(0)} \\
&\qquad\vdots
\end{aligned}
\tag{17.19}
$$

We already know the solution of $F^{(0)} = 0$, which is simply the set of zeroth-order (in λ) eigenstates and eigenfunctions $\{\phi_n^{(0)}, E_n^{(0)}\}$. Then the idea is input these solutions to solve the equation $F^{(1)} = 0$ for the set $\{\phi_n^{(1)}, E_n^{(1)}\}$. This provides the solution to ϕ_n, E_n up to first order in λ. We can then continue to solve these equations iteratively, and obtain the eigenstates and eigenvalues to any order in λ desired.

But we first have to deal with a slight ambiguity. Notice that if $\phi_n^{(1)}$ is a solution of

$$
\left(H_0 - E_n^{(0)}\right)\phi_n^{(1)} = \left(E_n^{(1)} - V\right)\phi_n^{(0)} \tag{17.20}
$$

then so is

$$
\phi'^{(1)}_n = \phi_n^{(1)} + a\phi_n^{(0)} \tag{17.21}
$$

where a is any constant. There is a similar ambiguity for any $\phi_n^{(N)}$. We can get rid of this ambiguity in a simple way. We first note that the eigenvalue equation $H\phi_n = E_n\phi_n$ is a linear equation, which does not set the overall normalization $\langle\phi_n|\phi_n\rangle$ of the eigenstates. So let us temporarily choose the normalization such that the overlap

$$
\left\langle\phi_n^{0}|\phi_n\right\rangle = 1 \tag{17.22}
$$

is unity, for any λ. It's easy to see, from the expansion (17.12), that this implies for each $N > 0$,

$$\left\langle \phi_n^0 | \phi_n^{(N)} \right\rangle = 0 \tag{17.23}$$

and this condition eliminates the ambiguity mentioned above. Of course, after having computed $\phi_n(x)$ by the process of iteratively solving the set of $F^{(N)} = 0$ equations, we will finally have to normalize $\phi_n(x)$ in the usual way to obtain a physical state.

Let us see how this iterative process works. Suppose we have solved all the $F^{(k)} = 0$ equations up to $k = N - 1$. The $k = N$ equation, in bra-ket notation, is

$$\left(H_0 - E_n^{(0)}\right)\left|\phi_n^{(N)}\right\rangle = -V\left|\phi_n^{(N-1)}\right\rangle + \sum_{j=1}^{N-1} E_n^{(j)}\left|\phi_n^{(N-j)}\right\rangle + E_n^N\left|\phi_n^{(0)}\right\rangle \tag{17.24}$$

To solve for $\phi_n^{(N)}$, first express it in terms of a complete set of states spanning the Hilbert space, namely, the zeroth-order eigenstates of H_0

$$\phi_n^{(N)} = \sum_{i \neq n} c_{ni}^N \phi_i^{(0)} \tag{17.25}$$

where we can neglect $i = n$ because of (17.23). Now multiply equation (17.24) on both sides by the bra vector $\langle \phi_i^{(0)}|$, with $i \neq n$

$$\begin{aligned}
\left\langle \phi_i^{(0)}\right|\left(H_0 - E_n^{(0)}\right)\left|\phi_n^{(N)}\right\rangle &= -\left\langle \phi_i^{(0)} | V | \phi_n^{(N-1)}\right\rangle + \sum_{j=1}^{N-1} E_n^{(j)}\left\langle \phi_i^{(0)} | \phi_n^{(N-j)}\right\rangle \\
&\quad + E_n^N\left\langle \phi_i^{(0)} | \phi_n^{(0)}\right\rangle \\
\left(E_i^{(0)} - E_n^{(0)}\right)\left\langle \phi_i^{(0)} | \phi_n^{(N)}\right\rangle &= -\left\langle \phi_i^{(0)} | V | \phi_n^{(N-1)}\right\rangle + \sum_{j=1}^{N-1} E_n^{(j)}\left\langle \phi_i^{(0)} | \phi_n^{(N-j)}\right\rangle \\
\left(E_i^{(0)} - E_n^{(0)}\right)c_{ni}^N &= -\left\langle \phi_i^{(0)} | V | \phi_n^{(N-j)}\right\rangle + \sum_{j=1}^{N-1} E_n^{(j)} c_{ni}^{N-j}
\end{aligned} \tag{17.26}$$

so that

$$c_{ni}^N = \frac{1}{E_n^{(0)} - E_i^{(0)}}\left[\left\langle \phi_i^{(0)} | V | \phi_n^{(N-1)}\right\rangle - \sum_{j=1}^{N-1} E_n^{(j)} c_{ni}^{N-j}\right] \tag{17.27}$$

Therefore

$$\phi_n^{(N)} = \sum_{i \neq n} \frac{1}{E_n^{(0)} - E_i^{(0)}}\left[\left\langle \phi_i^{(0)} | V | \phi_n^{(N-1)}\right\rangle - \sum_{j=1}^{N-1} E_n^{(j)} c_{ni}^{N-j}\right]\phi_n^{(0)} \tag{17.28}$$

is the Nth-order correction to the zeroth-order wave function, expressed in terms of the lower order $E_n^{(j)}$ and c_{ni}^{N-1}.

Next we have to get the Nth-order correction to the energy, and for this we multiply equation (17.24) on both sides by the bra vector $\langle\phi_n^{(0)}|$

$$\begin{aligned}\left\langle\phi_n^{(0)}\middle|\left(H_0 - E_n^{(0)}\right)\middle|\phi_n^{(N)}\right\rangle &= -\left\langle\phi_n^{(0)}|V|\phi_n^{(N-1)}\right\rangle + \sum_{j=1}^{N-1} E_n^{(j)}\left\langle\phi_n^{(0)}|\phi_n^{(N-j)}\right\rangle \\ &\quad + E_n^N\left\langle\phi_n^{(0)}|\phi_n^{(0)}\right\rangle \\ 0 &= -\left\langle\phi_n^{(0)}|V|\phi_n^{(N-1)}\right\rangle + \sum_{j=1}^{N-1} E_n^{(j)}\left\langle\phi_n^{(0)}|\phi_n^{(N-j)}\right\rangle + E_n^N\end{aligned} \tag{17.29}$$

and making use of the constraint (17.23), we have

$$0 = -\left\langle\phi_n^{(0)}|V|\phi_n^{(N-1)}\right\rangle + E_n^{(N)} \tag{17.30}$$

Putting everything together, we have the iterative equations

$$\begin{aligned}\phi_n^{(N)} &= \sum_{i\neq n} c_{ni}^N\phi_i^{(0)} \\ c_{ni}^N &= \frac{1}{E_n^{(0)} - E_i^{(0)}}\left[\left\langle\phi_i^{(0)}|V|\phi_n^{(N-1)}\right\rangle - \sum_{j=1}^{N-1} E_n^{(j)}c_{ni}^{N-1}\right] \\ E_n^{(N)} &= \left\langle\phi_n^{(0)}|V|\phi_n^{(N-1)}\right\rangle\end{aligned} \tag{17.31}$$

which allow us to compute the solutions

$$\begin{aligned}\phi_n &= \phi_n^{(0)} + \lambda\phi_n^{(1)} + \lambda^2\phi_n^{(2)} + \cdots \\ E_n &= E_n^{(0)} + \lambda E_n^{(1)} + \lambda^2 E_n^{(2)} + \cdots\end{aligned} \tag{17.32}$$

to any desired accuracy. Incidentally, since every power of λ in these expressions comes with one factor of $V(x)$, the expressions can just as well be regarded as power series in $V'(x)$. In fact, if the expansion constant λ is not obvious but $V'(x) \ll V_0(x)$ according to a criterion explained below, then one could just as well set $\lambda = 1$ in these series expansions, and replace $V(x)$ by $V'(x)$.

Let us now work out the corrections to the energies and wave functions to second order in λ (it is usually not necessary to go further than this). First of all, setting $N = 1$ in (17.31)

$$c_{ni}^1 = \frac{\left\langle\phi_i^{(0)}|V|\phi_n^{(0)}\right\rangle}{E_n^{(0)} - E_i^{(0)}} \tag{17.33}$$

so that

$$\phi_n^{(1)} = \sum_{i \neq n} \frac{\left\langle \phi_i^{(0)} | V | \phi_n^{(0)} \right\rangle}{E_n^{(0)} - E_i^{(0)}} \phi_i^0 \tag{17.34}$$

and

$$E_n^{(1)} = \left\langle \phi_n^{(0)} | V | \phi_n^{(0)} \right\rangle \tag{17.35}$$

We can substitute these formulas into the $N = 2$ equations to obtain the second-order correction to the wave function,

$$\begin{aligned} \phi_n^{(2)} &= \sum_{i \neq n} \frac{1}{E_n^{(0)} - E_i^{(0)}} \left[\left\langle \phi_i^{(0)} | V | \phi_n^{(1)} \right\rangle - E_n^{(1)} c_{ni}^1 \right] \phi_n^{(0)} \\ &= \sum_{i \neq n} \frac{1}{E_n^{(0)} - E_i^{(0)}} \left[\langle \phi_i^{(0)} | V | \left\{ \sum_{k \neq n} c_{nk}^1 | \phi_k^{(0)} \right\rangle \right\} - E_n^{(1)} c_{ni}^1 \right] \phi_n^{(0)} \\ &= \sum_{i \neq n} \sum_{k \neq n} \frac{1}{E_n^{(0)} - E_i^{(0)}} \frac{1}{E_n^{(0)} - E_k^{(0)}} \left\langle \phi_i^{(0)} | V | \phi_k^{(0)} \right\rangle \left\langle \phi_k^{(0)} | V | \phi_n^{(0)} \right\rangle \phi_i^{(0)} \\ &\quad - \sum_{i \neq n} \frac{\left\langle \phi_i^{(0)} | V | \phi_n^{(0)} \right\rangle \left\langle \phi_i^{(0)} | V | \phi_n^{(0)} \right\rangle}{\left(E_n^{(0)} - E_i^{(0)} \right)^2} \phi_i^{(0)} \end{aligned} \tag{17.36}$$

and the second-order correction to the energy,

$$\begin{aligned} E_n^{(2)} &= \left\langle \phi_n^{(0)} | V | \phi_n^1 \right\rangle \\ &= \left\langle \phi_n^{(0)} \middle| V \sum_{i \neq n} \frac{\left\langle \phi_i^{(0)} | V | \phi_n^{(0)} \right\rangle}{E_n^{(0)} - E_i^{(0)}} \middle| \phi_i^{(0)} \right\rangle \\ &= \sum_{i \neq n} \frac{\left| \left\langle \phi_n^{(0)} | V | \phi_i^{(0)} \right\rangle \right|^2}{E_n^{(0)} - E_i^{(0)}} \end{aligned} \tag{17.37}$$

You may wonder how much of this you really have to remember. The average quantum physicist can recite from memory the result for the wave function to first order, and the energy to second order in λ. So these are the results to remember:

$$\phi_n = \phi_n^{(0)} + \lambda \sum_{i \neq n} \frac{\langle \phi_i^{(0)} | V | \phi_n^{(0)} \rangle}{E_n^{(0)} - E_i^{(0)}} \phi_i^{(0)}$$
$$E_n = E_n^{(0)} + \lambda \langle \phi_n^{(0)} | V | \phi_n^{(0)} \rangle + \lambda^2 \sum_{i \neq n} \frac{|\langle \phi_n^{(0)} | V | \phi_i^{(0)} \rangle|^2}{E_n^{(0)} - E_i^{(0)}} \tag{17.38}$$

17.2 Validity of perturbation theory

Perturbation theory works when the perturbing potential V' is small compared to H_0. But what do we mean by 'small'? A good rule of thumb is that V' is small if the first-order correction $\lambda\phi_n^{(1)}$ is much less than the zeroth-order wave function, which requires (at least) that

$$\lambda |c_{in}^1| \ll 1 \quad \Longrightarrow \quad \lambda \left| \frac{\langle \phi_i^{(0)} | V | \phi_n^{(0)} \rangle}{E_n^{(0)} - E_i^{(0)}} \right| \ll 1 \tag{17.39}$$

In other words, the 'matrix element' V'_{in} of the perturbing potential is much smaller than the corresponding energy difference

$$|V'_{in}| \equiv \lambda \left| \langle \phi_i^{(0)} | V | \phi_n^{(0)} \rangle \right| \ll |E_n^{(0)} - E_i^0| \tag{17.40}$$

This is, of course, for $i \neq n$ (otherwise the energy difference is trivially zero). For $i = n$, we require that the first-order correction to the energy is small compared to the zeroth-order energy, i.e.

$$E_n^{(1)} \ll E_n^{(0)} \quad \Longrightarrow \quad \lambda \left| \langle \phi_n^{(0)} | V | \phi_n^{(0)} \rangle \right| \ll E_n^{(0)} \tag{17.41}$$

These conditions should be regarded as *only* a rule of thumb. Even if they are satisfied, there is no guarantee that the perturbation series actually converges. What sometimes happens is that the number of terms at nth order grows like $n!$, eventually overwhelming the suppression by λ^n, no matter how small λ may be. That does not mean, however, that such a series is necessarily useless. What often happens is that the perturbation series gets closer and closer to the right answer as n increases up to some order n_{best}, after which the series diverges away from the correct result. A series of this kind is known as an ***asymptotic expansion***. As an example, consider evaluating the integral

$$I = \int_{-\infty}^{\infty} \mathrm{d}x \mathrm{e}^{-ax^2 - \lambda x^4} \tag{17.42}$$

for very small λ by expanding the exponential factor in a Taylor series,

$$I = \int_{-\infty}^{\infty} \mathrm{d}x \sum_{n=0}^{\infty} \lambda^n x^{4n} \mathrm{e}^{-ax^2} \tag{17.43}$$

and then using the known results for Gaussian integrals

$$\begin{aligned} I_{2n} &= \int_{-\infty}^{\infty} \mathrm{d}x \; x^{2n} \mathrm{e}^{-ax^2} \\ &= (-1)^n \frac{\mathrm{d}^n}{\mathrm{d}a^n} \sqrt{\frac{\pi}{a}} \\ &= \sqrt{\pi} \left(\frac{a^{-(n+1/2)}}{2^n} \right) 1 \cdot 3 \cdot 5 \cdot 7 \ldots \cdot (2n-1) \end{aligned} \tag{17.44}$$

This gives the series

$$I = \sum_{n=0}^{\infty} \lambda^n I_{4n} \tag{17.45}$$

which diverges, no matter how small λ may be. Nevertheless, for small λ the series initially converges towards the correct answer, before diverging at large n. Many perturbative solutions have this feature. An example is the Feynman diagram expansion in quantum electrodynamics. Feynman diagrams are a graphical representation of terms in a perturbation expansion, and they are used to describe the scattering of charged particles by a power series in the fine structure constant $e^2/4\pi\hbar c \approx 1/137$. The theory at zeroth-order is the non-interacting theory, which is exactly soluble. The Feynman diagram expansion eventually diverges. Nevertheless, if we keep only the lower order terms in the expansion, this expansion delivers results which are in exceptionally close agreement with experiment.

A general discussion of asymptotic expansions can be found in Bender and Orszag [1].

17.3 Example: the anharmonic oscillator

Consider the Hamiltonian

$$\begin{aligned} H &= -\frac{\hbar^2}{2m} \frac{d^2}{\mathrm{d}x^2} + \frac{1}{2} kx^2 + \lambda x^4 \\ &= H_0 + \lambda x^4 \end{aligned} \tag{17.46}$$

The shift in the energy, to first order in λ, is

$$\begin{aligned} \Delta E_n &= \lambda E_n^{(1)} \\ &= \lambda \langle n | x^4 | n \rangle \end{aligned} \tag{17.47}$$

where we introduce the notation, for the harmonic oscillator, that

$$|n\rangle \equiv \left|\phi_n^{(0)}\right\rangle \tag{17.48}$$

As usual in harmonic oscillator problems, it helps to express the position operator x in terms of raising and lowering operators, keeping only the terms in x^4 which will contribute to $\langle n|x^4|n\rangle$:

$$\begin{aligned}
\tilde{x} &= \sqrt{\frac{\hbar}{2m\omega}}(\tilde{a} + \tilde{a}^\dagger) \\
\tilde{x}^4 &= \left(\frac{\hbar}{2m\omega}\right)^2(\tilde{a} + \tilde{a}^\dagger)^4 \\
&= \left(\frac{\hbar}{2m\omega}\right)^2[\tilde{a}^2(\tilde{a}^\dagger)^2 + \tilde{a}\tilde{a}^\dagger\tilde{a}\tilde{a}^\dagger + \tilde{a}(\tilde{a}^\dagger)^2\tilde{a} \\
&\quad + \tilde{a}^\dagger\tilde{a}^2\tilde{a}^\dagger + \tilde{a}^\dagger\tilde{a}\tilde{a}^\dagger\tilde{a} + (\tilde{a}^\dagger)^2\tilde{a}^2] \\
&\quad + \text{non-contributing terms}
\end{aligned} \tag{17.49}$$

Using the raising/lowering operator properties

$$\begin{aligned}
\tilde{a}^\dagger|n\rangle &= \sqrt{n+1}\,|n+1\rangle \\
\tilde{a}|n\rangle &= \sqrt{n}\,|n-1\rangle
\end{aligned} \tag{17.50}$$

we find, e.g., that

$$\begin{aligned}
\langle n|\tilde{a}^2(\tilde{a}^\dagger)^2|n\rangle &= \sqrt{n+1}\,\langle n|\tilde{a}^2\tilde{a}^\dagger|n+1\rangle \\
&= \sqrt{n+1}\,\sqrt{n+2}\,\langle n|\tilde{a}^2|n+2\rangle \\
&= \sqrt{n+1}\,(n+2)\langle n|\tilde{a}|n+1\rangle \\
&= (n+1)(n+2)
\end{aligned} \tag{17.51}$$

Evaluating all of the relevant terms in this way, we find

$$\begin{aligned}
\langle n|x^4|n\rangle &= \left(\frac{\hbar}{2m\omega}\right)^2[(n+2)(n+1) + (n+1)^2 + n(n+1) + n(n+1) \\
&\quad + n^2 + n(n-1)] \\
&= 3\left(\frac{\hbar}{2m\omega}\right)^2[1 + 2n(n+1)]
\end{aligned} \tag{17.52}$$

Then, to first order in λ

$$E_n = \hbar\omega\left(n + \frac{1}{2}\right) + 3\lambda\left(\frac{\hbar}{2m\omega}\right)^2[1 + 2n(n+1)] \tag{17.53}$$

which we can also express as

$$E_n = \hbar(\omega + \delta\omega)\left(n + \frac{1}{2}\right) + n^2\hbar\delta\omega \tag{17.54}$$

where

$$\delta\omega = 6\lambda\frac{\hbar}{4m^2\omega^3} \tag{17.55}$$

So the effect of the λx^4 perturbation is essentially a shift $\delta\omega$ in the angular frequency of the oscillator, together with an increase in the spacing of energy levels. It is interesting to note that no matter how small λ may be, the perturbative expansion for the energy must break down completely when n is large enough. This is because the zeroth-order energy increases only linearly with n, whereas the first-order correction increases quadratically. This breakdown of perturbation theory has a simple physical explanation. Whatever the value of λ, it is always true that $\lambda x^4 > \frac{1}{2}kx^2$ when x is large enough. But highly excited harmonic oscillator states, which have large n, spread out far from the origin $x = 0$, and therefore probe the region where the perturbing potential is larger than the zeroth-order potential. In such regions one cannot expect perturbation theory to work, and in fact it does not.

17.4 Perturbation theory in matrix notation

Let us define the matrix elements of an operator, in the basis of Hilbert space spanned by $\{\phi_n^{(0)}\}$, as

$$O_{ij} = \left\langle\phi_i^{(0)}|O|\phi_j^{(0)}\right\rangle \tag{17.56}$$

In this basis, H_0 is a diagonal matrix:

$$\begin{aligned}[H_0]_{ij} &= \left\langle\phi_i^{(0)}\middle|H_0\middle|\phi_j^{(0)}\right\rangle \\ &= \delta_{ij}E_j^{(0)} \\ H_0 &= \begin{bmatrix} E_1^0 & 0 & 0 & \cdots \\ 0 & E_2^0 & 0 & \cdots \\ 0 & 0 & E_3^0 & \cdots \\ \vdots & \vdots & \vdots & \cdots \\ \vdots & \vdots & \vdots & \cdots \end{bmatrix}\end{aligned} \tag{17.57}$$

For diagonal matrices, solving the eigenvalue problem

$$H_0\phi_n^{(0)} = E_n^{(0)}\phi_n^{(0)} \tag{17.58}$$

is very easy. The eigenvectors are

$$\vec{\phi}_1^{(0)} = \begin{bmatrix} 1 \\ 0 \\ 0 \\ \vdots \end{bmatrix}, \quad \vec{\phi}_2^{(0)} = \begin{bmatrix} 0 \\ 1 \\ 0 \\ \vdots \end{bmatrix}, \quad \vec{\phi}_3^{(0)} = \begin{bmatrix} 0 \\ 0 \\ 1 \\ \vdots \end{bmatrix} \cdots \tag{17.59}$$

and the eigenvalues are the diagonal elements of the matrix.

So now the problem is to find the eigenvectors of the (non-diagonal) matrix

$$H = H_0 + V' = \begin{bmatrix} E_1^{(0)} + V'_{11} & V'_{12} & V'_{13} & \cdots \\ V'_{21} & E_2^{(0)} + V'_{22} & V'_{23} & \cdots \\ V'_{31} & V'_{32} & E_3^{(0)} + V'_{33} & \cdots \\ \vdots & \vdots & \vdots & \cdots \end{bmatrix} \tag{17.60}$$

where

$$H_{ij} = \left\langle \phi_i^{(0)} | H_0 + V' | \phi_j^{(0)} \right\rangle = E_j \delta_{ij} + V'_{ij} \tag{17.61}$$

But this problem is already solved! To first order (in V') for the eigenvectors, and second order in the eigenvalues, the solution is

$$\begin{aligned} \vec{\phi}_n &= \vec{\phi}_n^{(0)} + \sum_{i \neq n} \frac{V'_{in}}{E_n^{(0)} - E_i^{(0)}} \vec{\phi}_i^{(0)} + O(V'^2) \\ E_n &= E_n^0 + V'_{nn} + \sum_{i \neq n} \frac{|V'_{in}|^2}{E_n^{(0)} - E_i^{(0)}} + O(V'^3) \end{aligned} \tag{17.62}$$

You can see that we can apply these equations to finding the eigenstates and eigenvalues of any matrix of the form

$$M = M_0 + Q \tag{17.63}$$

where M_0 is a diagonal matrix ($M_{ij} = m_i \delta_{ij}$), and Q is a 'perturbation' matrix such that

$$|Q_{ij}| \ll |m_i - m_j| \quad , \quad |Q_{ii}| \ll |m_i| \tag{17.64}$$

It is just a matter of changing notation (replace H_0 by M_0, $E_n^{(0)}$ by m_n, etc).

Getting back to the original problem, suppose we have found the eigenstates $\{\vec{\phi}_n\}$ by the perturbative method (equation (17.62)), or by some other method, and then normalized the states so that

$$\vec{\phi}_n \cdot \vec{\phi}_n = 1 \tag{17.65}$$

Now suppose we use the states $\{\phi_n\}$ to span the Hilbert space, instead of the set $\{\phi_n^{(0)}\}$. Then we can easily see that in this new basis, it is the matrix H, rather than H_0, which is diagonal

$$H = \begin{bmatrix} E_1 & 0 & 0 & \cdots \\ 0 & E_2 & 0 & \cdots \\ 0 & 0 & E_3 & \cdots \\ \cdot & \cdot & \cdot & \cdots \\ \cdot & \cdot & \cdot & \cdots \end{bmatrix} \tag{17.66}$$

because

$$H_{ij} = \langle \phi_i | H | \phi_j \rangle = \delta_{ij} E_j \tag{17.67}$$

It is for this reason that the process of solving the Hamiltonian eigenvalue equation $H\phi_n = E_n\phi_n$ is often referred to as *'diagonalizing the Hamiltonian'*.

17.5 Degenerate perturbation theory

The energy eigenvalues of the hydrogen atom are degenerate. The energy eigenstates ϕ_{nlm} depend on quantum numbers n, l, m, but the energy eigenvalues depend only on n. As we found back in chapter 10, this degeneracy is typical when the Hamiltonian is invariant with respect to some symmetry operations (e.g. rotation around the x-, y-, or z-axes) which do not commute with each other.

But now if we try to apply perturbation theory to a Hamiltonian with degenerate energy eigenvalues, we can immediately see that there is a problem. Consider, e.g. the first-order correction to the wave function

$$\phi_n^{(1)} = \sum_{i \neq n} \frac{V_{in}}{E_n^{(0)} - E_i^{(0)}} \phi_i^{(0)} \tag{17.68}$$

Obviously, if there is some energy $E_k^{(0)}$ such that $E_n^{(0)} = E_k^{(0)}$, then the right-hand side of this equation is not small, it is infinite. That is not exactly a small perturbation! What can be done in this case, which is quite typical in atomic physics?

First of all, to simplify things a little bit, let us suppose that there is just one set of states with degenerate energy eigenvalues, and that these are the states $\phi_n^{(0)}$ labeled by $n = 1, 2, \ldots, q$, i.e.

$$E_1^{(0)} = E_2^{(0)} = \cdots = E_q^{(0)} \tag{17.69}$$

The first-order correction to the wave function is derived from equation (17.26) with $N = 1$ and $i \neq n$, and requires

$$\left(E_i^{(0)} - E_n^{(0)}\right) c_{ni}^1 = -\langle \phi_i^{(0)} | V | \phi_n^{(0)} \rangle \tag{17.70}$$

Now if $i, n \leqslant q$, then $E_i^{(0)} = E_n^{(0)}$ and the left-hand side of the equation is zero. But this is impossible unless

$$\langle \phi_i^{(0)} | V | \phi_n^{(0)} \rangle = 0 \tag{17.71}$$

for all $i, n < q$ with $i \neq n$, which is generally not the case.

The way out of this dilemma is to first notice that the set of $\{\phi_n^{(0)}, n = 1, 2, \ldots, q\}$ spans a finite, q-dimensional subspace of the full Hilbert space, and *any* state in this subspace is an eigenstate of H_0, with the same degenerate energy $E_1^{(0)} = E_2^{(0)} = \cdots$. Then let $\bar{V}$ be the Hermitian, $q \times q$ matrix, whose matrix elements are

$$\bar{V}_{ij} = \left\langle \phi_i^{(0)} | V | \phi_j^{(0)} \right\rangle \qquad i, j \leqslant q \tag{17.72}$$

By the usual theorems about Hermitian operators (which also apply to Hermitian matrices), the eigenstates $\{\varphi_n, n = 1, 2, \ldots, q\}$ of $\bar{V}$

$$\bar{V}\varphi_n = \mathcal{E}_n \varphi_n \tag{17.73}$$

span the *same* q-dimensional Hilbert space as the $\{\phi_n^{(0)}, n = 1, 2, \ldots, q\}$, and moreover, for $i \neq n$

$$\left\langle \varphi_i^{(0)} | V | \varphi_n^{(0)} \right\rangle = 0 \tag{17.74}$$

providing all the $\mathcal{E}_n$ are different (which we will assume).

This means that if we replace the initial set of eigenstates $\{\phi_n^{(0)}\}$ of H_0 with a new set of zeroth-order eigenstates

$$\phi'^{(0)}_n = \begin{cases} \varphi_n & n \leqslant q \\ \phi_n^{(0)} & n > q \end{cases} \tag{17.75}$$

then equation (17.26) is automatically satisfied for $i, n \leqslant q$

$$0 = \left(E_i^{(0)} - E_n^{(0)}\right) c_{ni}^1 = -\left\langle \phi'^{(0)}_i | V | \phi'^{(0)}_n \right\rangle = 0 \tag{17.76}$$

We can then consistently take

$$c_{ni}^1 = 0 \qquad \text{for} \quad i, n \leqslant q \tag{17.77}$$

Given the new set of zeroth-order wave functions $\{\phi'^{(0)}_n\}$, perturbation theory can be applied exactly as in the non-degenerate case, and we find for the first-order corrections that

$$\begin{aligned} \phi_n^{(1)} &= \sum_{i>q} \frac{\left\langle \phi_i^{(0)} | V | \varphi_n \right\rangle}{E_n^{(0)} - E_i^{(0)}} \phi_i^{(0)} \quad \text{for} \quad n \leqslant q \\ \phi_n^{(1)} &= \sum_{i \neq n} \frac{\left\langle \phi_i'^{(0)} | V | \varphi_n \right\rangle}{E_n^{(0)} - E_i^{(0)}} \phi'^{(0)}_i \quad \text{for} \quad n > q \\ E_n^{(1)} &= \left\langle \phi'^{(0)}_n | V | \phi'^{(0)}_n \right\rangle \\ &= \begin{cases} \mathcal{E}_n & n \leqslant q \\ \left\langle \phi_n^{(0)} | V | \phi_n^{(0)} \right\rangle & n > q \end{cases} \end{aligned} \tag{17.78}$$

17.5.1 Diagonalizing the submatrix

Equations (17.78) give the perturbative solutions to $H\phi_n = E_n\phi_n$ to first order in λ, when the first q energy eigenvalues are degenerate, providing one can solve the eigenvalue equation

$$\bar{V}\ \varphi_n = \mathcal{E}_n\varphi_n \tag{17.79}$$

Since the first q eigenstates $\{\phi_n^{(0)},\ n = 1, 2, \ldots, q\}$ are a basis of the q-dimensional subspace, we can write

$$\varphi_n = \sum_{i=0}^{q} a_{ni}\phi_i^{(0)} \tag{17.80}$$

Writing the $\{\phi_n^{(0)}, n = 1, 2, \ldots, q\}$ as basis vectors, as in equation (17.59), we can write the eigenvectors φ_n in the q-dimensional subspace in the form

$$\vec{\varphi}_n = \begin{bmatrix} a_{n1} \\ a_{n2} \\ a_{n3} \\ \cdot \\ \cdot \\ \cdot \\ a_{nq} \end{bmatrix} \tag{17.81}$$

(In the full Hilbert space φ_n is an infinite-component vector, with zeros following a_{nq}.) Then, in matrix notation, the eigenvalue problem (17.79) is simply

$$\begin{bmatrix} V_{11} & V_{12} & V_{13} & \cdot & \cdot & V_{1q} \\ V_{21} & V_{22} & V_{23} & \cdot & \cdot & V_{2q} \\ V_{31} & V_{32} & V_{33} & \cdot & \cdot & V_{3q} \\ \cdot & \cdot & \cdot & \cdot & \cdot & \cdot \\ \cdot & \cdot & \cdot & \cdot & \cdot & \cdot \\ V_{q1} & V_{q2} & V_{q3} & \cdot & \cdot & V_{qq} \end{bmatrix} \vec{\varphi}_n = \mathcal{E}_n\vec{\varphi}_n \tag{17.82}$$

We have already discussed how to solve the eigenvalue problem for a 2×2 matrix; for example, in solving for the eigenstates of the Pauli spin matrix σ_y in chapter 13. In general, one first solves for the eigenvalues $\mathcal{E}_n$ by solving the ***secular equation***

$$\det[\bar{V} - \mathcal{E}\mathbb{1}] = 0 \tag{17.83}$$

where $\mathbb{1}$ is the $q \times q$ unit matrix. The secular equation is a qth-order polynomial which (in general) has q roots. These roots are the set of eigenvalues of the matrix $\bar{V}$, which we have denoted $\{\mathcal{E}_n\}$. Then, for each eigenvalue, we solve for the vector $\vec{\varphi}_n$ in equation (17.82) algebraically (as illustrated for the eigenstates of σ_y in chapter 13).

17.5.2 Example: the two-dimensional harmonic oscillator

The Hamiltonian for a harmonic oscillator in two dimensions is

$$H_0 = -\frac{\hbar^2}{2M}\left[\frac{\partial^2}{\partial x^2} + \frac{\partial^2}{\partial y^2}\right] + \frac{1}{2}k(x^2 + y^2) \tag{17.84}$$

This system is symmetric with respect to rotations (around the z-axis), reflections ($x \to -x$ and $y \to -y$), and interchange $x \to y,\ \ y \to x$. It is not hard to verify that these symmetries do not all commute with each other, and therefore at least some of the energy eigenvalues must be degenerate. In fact, it is easy to solve for the eigenvalues and eigenstates of H_0 by the method of separation of variables. Writing

$$H_0 = h[x] + h[y] \tag{17.85}$$

where $h[x]$ and $h[y]$ are one-dimensional harmonic oscillator Hamiltonians in the x- and y-coordinates, respectively, and

$$\phi_{mn}^{(0)}(x, y) = \phi_m(x)\phi_n(y) \tag{17.86}$$

we end up with the one-dimensional harmonic oscillator eigenvalue equations

$$\begin{aligned} h[x]\phi_m(x) &= E_m\phi_m(x) \\ h[y]\phi_n(y) &= E_n\phi_n(y) \end{aligned} \tag{17.87}$$

with the total energy eigenvalue of $\phi_{mn}^{(0)}(x, y)$ being

$$E_{nm}^{(0)} = E_m + E_n \tag{17.88}$$

The eigenvalue equations (17.87) were solved back in chapter 9 using raising/lowering operators. In this case, we should introduce separate raising/lowering operators in the x- and y-coordinates, i.e.

$$\begin{aligned} \tilde{a} &= \frac{1}{\sqrt{2\hbar}}\left(\sqrt{M\omega}\,x + \mathrm{i}\frac{\tilde{p}_x}{\sqrt{M\omega}}\right) \\ \tilde{b} &= \frac{1}{\sqrt{2\hbar}}\left(\sqrt{M\omega}\,y + \mathrm{i}\frac{\tilde{p}_y}{\sqrt{M\omega}}\right) \end{aligned} \tag{17.89}$$

with commutator relations

$$[\tilde{a}, \tilde{a}^\dagger] = [\tilde{b}, \tilde{b}^\dagger] = 1 \tag{17.90}$$

and

$$[\tilde{a}, \tilde{b}] = [\tilde{a}, \tilde{b}^\dagger] = [\tilde{a}^\dagger, \tilde{b}] = [\tilde{a}^\dagger, b^\dagger] = 0 \tag{17.91}$$

In terms of these operators, the two-dimensional harmonic oscillator Hamiltonian is simply

$$H_0 = \hbar\omega(\tilde{a}^\dagger\tilde{a} + \tilde{b}^\dagger\tilde{b} + 1) \tag{17.92}$$

and the $\phi_m(x)$ and $\phi_n(y)$ are one-dimensional harmonic oscillator eigenstates. The total energy eigenvalue is the sum of $E_m + E_m$

$$\begin{aligned} \phi^{(0)}_{mn}(x, y) &= \phi_m(x)\phi_n(y) \\ E^{(0)}_{mn} &= \hbar\omega(m + n + 1) \end{aligned} \tag{17.93}$$

It is easy to see that the energy $E^{(0)}_{00}$ is unique, that $E^{(0)}_{01} = E^{(0)}_{10}$ is two-fold degenerate, $E^{(0)}_{20} = E^{(0)}_{11} = E^{(0)}_{02}$ is three-fold degenerate, and so on (in general, the energy E^{0}_{mn} is $(m + n + 1)$-fold degenerate).

Now introduce a perturbing potential $V' = \lambda xy$,

$$H = H_0 + \lambda xy \tag{17.94}$$

As an exercise in degenerate perturbation theory, we will now compute the correction to the energies $E^{(0)}_{10} = E^{(0)}_{01}$ of the first excited states.

The first two excited states with degenerate energy eigenvalues, ϕ_{10} and ϕ_{01}, span a two-dimensional subspace of the Hilbert space. Our first task is to find the eigenstates of the 2×2 matrix

$$\overline{V} = \begin{bmatrix} \langle\phi_{10}|xy|\phi_{10}\rangle & \langle\phi_{10}|xy|\phi_{01}\rangle \\ \langle\phi_{01}|xy|\phi_{10}\rangle & \langle\phi_{01}|xy|\phi_{01}\rangle \end{bmatrix} \tag{17.95}$$

Using

$$\tilde{x}\tilde{y} = \frac{1}{2\beta^2}(\tilde{a} + \tilde{a}^\dagger)(\tilde{b} + \tilde{b}^\dagger) \qquad \beta \equiv \sqrt{\frac{M\omega}{\hbar}} \tag{17.96}$$

we obtain

$$\begin{aligned} \langle\phi_{10}|xy|\phi_{10}\rangle &= \frac{1}{2\beta^2}\langle\phi_1(x)|(\tilde{a} + \tilde{a}^\dagger)|\phi_1(x)\rangle\langle\phi_0(y)|(\tilde{b} + \tilde{b}^\dagger)|\phi_0(y)\rangle = 0 \\ \langle\phi_{01}|xy|\phi_{01}\rangle &= \frac{1}{2\beta^2}\langle\phi_0(x)|(\tilde{a} + \tilde{a}^\dagger)|\phi_0(x)\rangle\langle\phi_1(y)|(\tilde{b} + \tilde{b}^\dagger)|\phi_1(y)\rangle = 0 \\ \langle\phi_{10}|xy|\phi_{01}\rangle &= \frac{1}{2\beta^2}\langle\phi_1(x)|(\tilde{a} + \tilde{a}^\dagger)|\phi_0(x)\rangle\langle\phi_0(y)|(\tilde{b} + \tilde{b}^\dagger)|\phi_1(y)\rangle = \frac{1}{2\beta^2} \\ \langle\phi_{01}|xy|\phi_{10}\rangle &= \frac{1}{2\beta^2}\langle\phi_0(x)|(\tilde{a} + \tilde{a}^\dagger)|\phi_1(x)\rangle\langle\phi_1(y)|(\tilde{b} + \tilde{b}^\dagger)|\phi_0(y)\rangle = \frac{1}{2\beta^2} \end{aligned} \tag{17.97}$$

and therefore

$$\bar{V} = \frac{1}{2\beta^2}\begin{bmatrix} 0 & 1 \\ 1 & 0 \end{bmatrix} \tag{17.98}$$

The problem of finding the eigenvectors and eigenvalues ('diagonalizing') the matrix $\bar{V}$ is pretty much the same as solving the eigenvalue problem for the Pauli matrix σ_x. First we solve the secular equation

$$\det\left[\bar{V} - \mathcal{E}I\right] = \mathcal{E}^2 - \left(\frac{1}{2\beta^2}\right)^2 = 0 \tag{17.99}$$

to find two roots

$$\mathcal{E}_{\pm} = \pm\frac{1}{2\beta^2} \tag{17.100}$$

corresponding to eigenstates

$$\begin{aligned} \phi_+ &= a\phi_{10} + \beta\phi_{01} \\ \phi_- &= c\phi_{10} + \mathrm{d}\phi_{01} \end{aligned} \tag{17.101}$$

We find these eigenstates by solving

$$\begin{aligned} \frac{1}{2\beta^2}\begin{bmatrix} 0 & 1 \\ 1 & 0 \end{bmatrix}\begin{bmatrix} a \\ b \end{bmatrix} &= \frac{1}{2\beta^2}\begin{bmatrix} a \\ b \end{bmatrix} \\ \frac{1}{2\beta^2}\begin{bmatrix} 0 & 1 \\ 1 & 0 \end{bmatrix}\begin{bmatrix} c \\ d \end{bmatrix} &= -\frac{1}{2\beta^2}\begin{bmatrix} c \\ d \end{bmatrix} \end{aligned} \tag{17.102}$$

subject to the condition that

$$\begin{aligned} \langle\phi_+|\phi_+\rangle &= a^2 + b^2 = 1 \\ \langle\phi_-|\phi_-\rangle &= c^2 + \mathrm{d}^2 = 1 \end{aligned} \tag{17.103}$$

The solutions are

$$\begin{aligned} \phi_+ &= \frac{1}{\sqrt{2}}\begin{bmatrix} 1 \\ 1 \end{bmatrix} \\ &= \frac{1}{\sqrt{2}}\left[\phi_{10} + \phi_{01}\right] \\ \phi_- &= \frac{1}{\sqrt{2}}\begin{bmatrix} 1 \\ -1 \end{bmatrix} \\ &= \frac{1}{\sqrt{2}}\left[\phi_{10} - \phi_{01}\right] \end{aligned} \tag{17.104}$$

Instead of two degenerate energies, $E_{10} = E_{01}$, we now have energy eigenvalues to first order in λ

$$
\begin{aligned}
H\phi_+ &= E_+\phi_+ \\
H\phi_- &= E_-\phi_-
\end{aligned}
\tag{17.105}
$$

where $E = E^{(0)} + \lambda\mathcal{E}$, i.e.

$$
\begin{aligned}
E_+ &= E_{10} + \frac{\lambda}{2\beta^2} \\
E_- &= E_{10} - \frac{\lambda}{2\beta^2}
\end{aligned}
\tag{17.106}
$$

We say that the degeneracy is 'lifted at first order' by the perturbation.

17.5.3 Example: the Stark effect

The Stark effect is a splitting of spectral lines due to the (partial) lifting of atomic energy level degeneracy by an external electric field.

Let us suppose that the electric field is directed along the z-axis. The Hamiltonian of the hydrogen atom is then

$$
\begin{aligned}
H_0 &= \left(-\frac{\hbar^2}{2m}\nabla^2 - \frac{e^2}{4\pi r}\right) - eE_z z \\
&= H_0 + \lambda V
\end{aligned}
\tag{17.107}
$$

where this time

$$\lambda = eE_z \qquad \text{and} \qquad V = z \tag{17.108}$$

Since the ground state energy is non-degenerate (only the ϕ_{100} state has this energy at zeroth order), the lifting of degeneracy first occurs at $n = 2$. There are four states at $n = 2$ with the same energy $E_2^{(0)}$

$$|\phi_{nlm}\rangle = |nlm\rangle = |200\rangle, \quad |211\rangle, \quad |210\rangle, \quad |21-1\rangle \tag{17.109}$$

which span a 4×4 subspace of Hilbert space. We first have to compute the 4×4 $\bar{V}$ matrix with matrix elements $\langle 2l_1m_1|z|2l_2m_2\rangle$.

Consider the case $m_1 \neq m_2$. Then

$$\langle 2l_1m_1|z|2l_2m_2\rangle \sim \int_0^{2\pi} \mathrm{d}\phi \mathrm{e}^{\mathrm{i}(m_2-m_1)} = 0 \tag{17.110}$$

since $z = r\cos(\theta)$ does not depend on ϕ. Therefore only terms with $m_1 = m_2$ are non-zero. Next, consider $l_1 = l_2$, where we find

$$\langle 2lm|z|2lm\rangle \sim \int \mathrm{d}\Omega |Y_{lm}|^2 \cos(\theta) = 0 \tag{17.111}$$

The integral is zero because $|Y_{lm}|^2 > 0$ is an even function around $\theta = \pi/2$, while $\cos(\theta)$ is an odd function for reflections around $\pi/2$.

Thus, the only non-zero matrix elements in this subspace are

$$\begin{aligned}\langle 210|z|200\rangle &= \langle 200|z|210\rangle \\ &= \int \mathrm{d}r r^2 \int \mathrm{d}\Omega \phi_{200} z \phi_{210} \\ &= \int \mathrm{d}r r^2 \int \mathrm{d}\Omega \frac{2}{(2a_0)^{3/2}}\left(1 - \frac{r}{2a_0}\right)\mathrm{e}^{-r/2a_0} Y_{00} \\ &\quad \times r\cos(\theta)\frac{1}{\sqrt{3}(2a_0)^{3/2}}\frac{r}{a_0}\mathrm{e}^{-r/2a_0} Y_{10}(\theta,\phi) \\ &= 3a_0\end{aligned} \tag{17.112}$$

where $a_0 = 4\pi\hbar^2/me^2$ is the Bohr radius. The secular equation is then

$$\begin{aligned}&\det[\overline{V} - \mathcal{E}I] \\ &= \det\begin{bmatrix} \langle 200|z|200\rangle - \mathcal{E} & \langle 200|z|211\rangle & \langle 200|z|210\rangle & \langle 200|z|21-1\rangle \\ \langle 211|z|200\rangle & \langle 211|z|211\rangle - \mathcal{E} & \langle 211|z|210\rangle & \langle 211|z|21-1\rangle \\ \langle 210|z|200\rangle & \langle 210|z|211\rangle & \langle 210|z|210\rangle - \mathcal{E} & \langle 210|z|21-1\rangle \\ \langle 21-1|z|200\rangle & \langle 21-1|z|211\rangle & \langle 21-1|z|210\rangle & \langle 21-1|z|21-1\rangle - \mathcal{E} \end{bmatrix} \\ &= \det\begin{bmatrix} -\mathcal{E} & 0 & 3a_0 & 0 \\ 0 & -\mathcal{E} & 0 & 0 \\ 3a_0 & 0 & -\mathcal{E} & 0 \\ 0 & 0 & 0 & -\mathcal{E} \end{bmatrix} \\ &= \mathcal{E}^4 - (3a_0)^2\mathcal{E}^2 \\ &= 0\end{aligned} \tag{17.113}$$

The roots of the secular equation are

$$\mathcal{E} = 0, \quad 3a_0, -3a_0 \tag{17.114}$$

Therefore the four-fold degenerate E_2 energy level splits into three (not four) levels

$$E_2 \Longrightarrow \begin{cases} E_2 + 3a_0 eE_z \\ E_2 \\ E_2 - 3a_0 eE_z \end{cases} \tag{17.115}$$

Because the subspace is four-dimensional, but the perturbation results only in three distinct energies, the degeneracy is not entirely lifted; a subset of states in this four-dimensional subspace still has degenerate energies.

Next we figure out the eigenstates corresponding to the $n = 2$ energy eigenvalues at first order. We can always express a general state in the subspace spanned by the zeroth-order $n = 2$ state as a superposition

$$|\psi\rangle = a|\phi_{200}\rangle + b|\phi_{211}\rangle + c|\phi_{210}\rangle + d|\phi_{21-1}\rangle \tag{17.116}$$

In vector notation, we can write

$$\psi = \begin{bmatrix} a \\ b \\ c \\ d \end{bmatrix} \tag{17.117}$$

and the eigenvalue equation

$$\bar{V}\psi = \mathcal{E}\psi \tag{17.118}$$

becomes

$$\begin{bmatrix} 0 & 0 & 3a_0 & 0 \\ 0 & 0 & 0 & 0 \\ 3a_0 & 0 & 0 & 0 \\ 0 & 0 & 0 & 0 \end{bmatrix}\begin{bmatrix} a \\ b \\ c \\ d \end{bmatrix} = \mathcal{E}\begin{bmatrix} a \\ b \\ c \\ d \end{bmatrix} \tag{17.119}$$

There are three values of $\mathcal{E}$ to consider:

Case I. $\mathcal{E} = 3a_0$
Matrix multiplication on the left-hand side of (17.119) gives us

$$3a_0\begin{bmatrix} c \\ 0 \\ a \\ 0 \end{bmatrix} = 3a_0\begin{bmatrix} a \\ b \\ c \\ d \end{bmatrix} \tag{17.120}$$

from which we conclude that

$$c = a \ , \qquad b = d = 0 \tag{17.121}$$

Normalization of

$$\vec{\phi} = \begin{bmatrix} a \\ 0 \\ a \\ 0 \end{bmatrix} \tag{17.122}$$

gives us $a = 1/\sqrt{2}$. So to first order in $\lambda = eE_z$ we have an eigenstate

$$|\phi\rangle_{E_{2+}} = \frac{1}{\sqrt{2}}\left[|\phi_{200}\rangle + |\phi_{210}\rangle\right] \tag{17.123}$$

for energy eigenvalue

$$E_{2+} = E_2^{(0)} + 3ea_0E_z \tag{17.124}$$

Case II. $\mathcal{E} = -3a_0$

$$3a_0\begin{bmatrix} c \\ 0 \\ a \\ 0 \end{bmatrix} = -3a_0\begin{bmatrix} a \\ b \\ c \\ d \end{bmatrix} \tag{17.125}$$

and therefore

$$c = -a \ , \qquad b = d = 0 \tag{17.126}$$

Normalization of

$$\vec{\phi} = \begin{bmatrix} a \\ 0 \\ -a \\ 0 \end{bmatrix} \tag{17.127}$$

still gives $a = 1/\sqrt{2}$, leading to the eigenstate

$$|\phi\rangle_{E_{2-}} = \frac{1}{\sqrt{2}}\left[|\phi_{200}\rangle - |\phi_{210}\rangle\right] \tag{17.128}$$

for energy eigenvalue

$$E_{2-} = E_2^{(0)} - 3ea_0E_z \tag{17.129}$$

Case III. $\mathcal{E} = 0$

Finally, we consider $\mathcal{E} = 0$, and equation (17.119) becomes

$$3a_0\begin{bmatrix} c \\ 0 \\ a \\ 0 \end{bmatrix} = 0 \tag{17.130}$$

This time, we can only conclude that $a = c = 0$, and that any normalized eigenstate of the form

$$|\phi\rangle_{E_2} = b|\phi_{211}\rangle + d|\phi_{21-1}\rangle \tag{17.131}$$

has an energy eigenvalue

$$E_2 = E_2^{(0)} \tag{17.132}$$

The normalization condition in this case only tells us that

$$b^2 + d^2 = 1 \tag{17.133}$$

and therefore the linearly independent eigenstates with energy E_2 span a two-dimensional subspace of the full Hilbert space. Thus the perturbation does not remove *all* of the degeneracy in energy eigenvalues; there is still some left.

The reason that there is still some degeneracy left, as seen in the $\mathcal{E} = 0$ case, is that there is still some symmetry left after the external electric field is applied. For an electric field directed along the z-axis, the Hamiltonian is still invariant with respect to rotations around the z-axis, as well as reflections along the x- and y-axes; the result is some remnant degeneracy in the energy eigenvalues.

Problems

1. A one-dimensional harmonic oscillator is perturbed by an extra potential term λx^3.
   a) Calculate the change in each energy level to second order in λ.
   b) Perturbation theory can sometimes be misleading. Consider the full potential $\frac{1}{2}kx^2 + \lambda x^3$ as $x \to -\infty$, and make a rough estimate of the energy expectation value of a narrow wavepacket of width Δx centered at large negative x. Based on your finding, does this system really have a ground state?
2. A particle moves in an infinite square well potential of the form

$$V(x) = \begin{cases} \lambda \sin\left(\dfrac{\pi x}{L}\right) & 0 < x < L \\ \infty & \text{otherwise} \end{cases} \tag{17.134}$$

   Calculate all the energy eigenvalues, and the ground state wave function, to first order in λ.
3. Consider the nucleus of a hydrogen atom as a charge of $+e$ uniformly distributed in a sphere of radius $R = 1$ fm $=10^{-15}$ m. The difference between the potential of the uniformly charged sphere and the Coulomb potential can be considered a perturbation of the Coulomb potential (it is only non-zero for radii $r < R$, of course). Calculate the change in the energy of the hydrogen atom ground state due to this perturbation.
4. Consider a harmonic oscillator perturbed by a potential V', having matrix elements $V_{ij} = \langle \phi_i | V' | \phi_j \rangle$ where the ϕ_n are zeroth-order energy eigenstates, and indices i, j run from zero to infinity. Suppose that for $i, j = 0, 1, 2, 3, 4$ the V' matrix elements are given by

$$V' = \lambda \begin{pmatrix} 1 & 0 & -\sqrt{1/2} & 0 & \sqrt{3/8} \\ 0 & 0 & 0 & 0 & 0 \\ -\sqrt{1/2} & 0 & 1/2 & 0 & -\sqrt{3/16} \\ 0 & 0 & 0 & 0 & 0 \\ \sqrt{3/8} & 0 & -\sqrt{3/16} & 0 & 3/8 \end{pmatrix} \tag{17.135}$$

   All other matrix elements are negligible. Give the energy eigenvalues of all eigenstates to first order in λ. In addition, compute the energy eigenvalue of the ground state to second order in λ, and the ground state wave function (in terms of the ϕ_n) to first order.

5. A rigid rotor (moment of inertia I) rotates in the x–y plane about the z-axis. The Hamiltonian is $H_0 = L_z^2/(2I)$. Write down the (zeroth-order) eigenstates and energy eigenvalues. The rotor has an electric dipole moment $\vec{\mu}$ oriented along the line of the rotor, and is placed in a uniform electric field $\vec{\epsilon} = \epsilon\hat{\mathbf{i}}$ directed along the x-axis. The resulting additional potential is

$$V'(\theta) = -\vec{\epsilon} \cdot \vec{\mu}$$

Compute the energy eigenstates of the rotor to first order, and the energy eigenvalues to second order, in the additional potential V'.

6. Consider the Hamiltonian

$$H = AS_z^2 + B\left(S_x^2 - S_y^2\right) \tag{17.136}$$

for a system of spin 1, and $|B \gg |A|$. (Such a Hamiltonian is actually relevant to a spin-1 ion located in a crystal with rhombic symmetry.)
   a) In the $|s = 1, s_z\rangle$ basis, write this Hamiltonian as a matrix.
   b) Find the eigenvalues of the Hamiltonian, using degenerate perturbation theory.

7. Two identical spin $\frac{1}{2}$ particles move in a two-dimensional rectangular region of sides L and $\frac{1}{2}L$; the potential outside this region is infinite. Inside the rectangular region, the potential is just $V = \alpha\mathbf{S}_1 \cdot \mathbf{S}_2$, where $\mathbf{S}_1$, $\mathbf{S}_2$ are the spin angular momentum operators for each particle.
   a) Find the lowest energy, and lowest energy eigenstates, for each of the possible values of total spin angular momentum S^2.
   b) Suppose there is also a repulsive potential between the particles, given by the potential $\beta\delta(x_1 - x_2)\delta(y_1 - y_2)$. Treating this as a perturbation, how do the energy eigenvalues in part (a) change, to first order in β?

8. A certain system is characterized by a Hamiltonian H_0, having only three orthogonal eigenstates ϕ_1, ϕ_2, ϕ_3 whose energies are 3, 7, 7 eV, respectively. The system is perturbed by a potential V with matrix elements

$$\langle\phi_1|V|\phi_1\rangle = 0.1 \text{ eV}$$

$$\langle\phi_2|V|\phi_3\rangle = (1 + \mathrm{i})/10 \text{ eV}$$

$$\langle\phi_3|V|\phi_2\rangle = (1 - \mathrm{i})/10 \text{ eV}$$

with all other $\langle\phi_i|V|\phi_j\rangle = 0$. Find the eigenstates and eigenvalues of the perturbed system via degenerate perturbation theory.

9. Consider a particle of mass m confined to a two-dimensional infinite square well in the x–y plane, of length L on each side. The particle is subject to a small perturbing potential

$$V(x, y) = g \sin\left(\frac{2\pi x}{L}\right) \sin\left(\frac{2\pi y}{L}\right) \tag{17.137}$$

Calculate, to first order in g, the energy eigenvalues and eigenstates of the first two excited states above the ground state.

10. Consider a particle of mass m and charge q confined in a box with sides of length a, a, b in the x-, y-, z-directions, respectively, with $b \ll a$. Inside the box there is also a weak electric field, with components

$$E_x = \mathcal{E}\frac{y}{a}, \quad E_y = \mathcal{E}\frac{x}{a}, \quad E_z = 0 \tag{17.138}$$

Find the corresponding potential energy and then calculate, to first order in $\mathcal{E}$, the energies of the ground state and first excited states.

11. Two identical spin 1/2 particles are contained in a box with sides of length $L_x > L_y > L_z$. The particles interact via the potential

$$V = V_0\delta(x_1 - x_2)\delta(y_1 - y_2)\delta(z_1 - z_2) \tag{17.139}$$

Calculate, in first-order perturbation theory, the lowest energy possible for the cases $s = 0$ and $s = 1$, where s is the total spin angular momentum quantum number.

12. A certain system is characterized by a Hamiltonian H_0, having three orthogonal eigenstates ϕ_1, ϕ_2, ϕ_3 whose energies are 3, 3, 4, respectively (in units of eV). The system is perturbed by a potential V' with matrix elements

$$\begin{aligned} \langle\phi_1|V'|\phi_2\rangle = \langle\phi_2|V'|\phi_1\rangle = 0.1 \\ \langle\phi_1|V'|\phi_3\rangle = \langle\phi_3|V'|\phi_1\rangle = 0.2 \end{aligned} \tag{17.140}$$

with all other matrix elements equal to zero.

a) Write H_0 and V' as 3×3 matrices in the basis ϕ_1, ϕ_2, ϕ_3.

b) Use degenerate perturbation theory to find the energy eigenvalues to first order. Denote the eigenstates that you obtain from this procedure as ϕ'_1, ϕ'_2, ϕ'_3.

c) Write H_0 and V' as 3×3 matrices in the new basis ϕ'_1, ϕ'_2, ϕ'_3.

d) Using the new basis as the zeroth-order eigenstates, calculate the energy eigenvalues to second order in the perturbing potential. (Note that in the sum over states that arises at second order, any terms in the sum with zero in the numerator should be dropped.)

Reference

[1] Bender C M and Orszag S A 1978 *Advanced Mathematical Methods for Scientists and Engineers* (New York: McGraw-Hill)

IOP Publishing

An Introduction to Quantum Theory

Jeff Greensite

Chapter 18

Time-dependent potentials

The time dependence of any observable is associated with an uncertainty in the energy of the system, and this association is quantified in the time–energy uncertainty relation. So suppose an object has settled into its ground state. Then nothing about the system can change in time, neither its energy nor the expectation value of any other observable, until someone or something comes along and gives the object a kick. Then the energy and other observables *can* change. But a 'kick' that occurs at a certain time or period of time means, from the quantum-mechanical point of view, that the object has been exposed to a time-dependent potential. We have not yet discussed how to deal with this situation.[1]

There are a number of special cases to be considered. First, assume that the time-dependent potential is very weak, compared to time-independent part of the Hamiltonian. This calls for a time-dependent version of perturbation theory, which will be developed in the next section. A second case is that the time-dependent potential is not necessarily weak, but varies very slowly in time. Finally there is the opposite situation, where the potential changes very suddenly, e.g. by someone flipping a switch. Each case requires a different approach.

Some of the most interesting and technologically important aspects of atomic physics concern the interaction of atoms with electromagnetic radiation, and electromagnetic waves consist of orthogonal electric and magnetic fields which oscillate, at any point, harmonically in time. If the field strength is small enough, then the associated electrostatic potential can be viewed as a small, harmonic perturbation to the usual Hamiltonian of the atom. We will derive the effect of

[1] When the potential is time-dependent, the energy of an object is not conserved. This does not violate any sacred principle, but only reflects the fact that we are neglecting the back-reaction of the object on the source of the potential. In a closed system, where there are only interactions among objects within the system and no potentials external to the system, then of course the total energy must be conserved. Then there can be no time-dependent potentials per se, although there can be a time dependence in the wave functions of the sources which produce those potentials.

doi:10.1088/978-0-7503-1167-0ch18

harmonic perturbations in this chapter, and then apply the results, in the next chapter, to the interaction of atoms and light.

18.1 Time-dependent perturbation theory

Let us consider time-dependent Hamiltonians of the form

$$H = H_0 + \lambda V(x, t) \tag{18.1}$$

where we assume that the eigenstates of H_0, now denoted φ_n,

$$H_0\varphi_n(x) = E_n\varphi_n(x) \tag{18.2}$$

are known, and λV is 'weak' compared to H_0 in the sense of the last chapter. Denote the corresponding energy-eigenstate solutions of the time-dependent Schrödinger equation by

$$\psi_n(x, t) = \varphi_n(x)e^{-i\omega_n t} \qquad \omega_n = \frac{E_n}{\hbar} \tag{18.3}$$

Just to have a definite picture in mind, we might imagine that the φ_n are the energy eigenstates of the hydrogen atom, and the time-varying potential $\lambda V(x, t)$ is the electrostatic potential, at point x, of an incident electromagnetic wave. The question we are interested in answering is this: if the electron starts out at time t_0 in a given orbital φ_n, what is the probability that the incident radiation will result in the electron being found in the orbital φ_m at some later time t?

To begin with, since the states $\{\psi_n(x, t)\}$ span the Hilbert space at any time t, we can always expand the solution of the time-dependent Schrödinger equation $\psi(x, t)$, at any time t, in that basis:

$$\psi(x, t) = \sum_n c_n(t)\psi_n(x, t) \tag{18.4}$$

The initial value problem, i.e. 'given $\psi(x, 0)$, find $\psi(x, t)$', can be phrased as: given $\{c_n(0)\}$, find $\{c_n(t)\}$. Substituting (18.4) into the time-dependent Schrödinger equation

$$i\hbar\frac{d}{dt}\psi = (H_0 + \lambda V)\psi \tag{18.5}$$

to obtain

$$i\hbar\sum_n\left(\frac{dc_n}{dt} - i\omega_n c_n\right)\psi_n = \sum_n(\hbar\omega_n + \lambda V)c_n\psi_n \tag{18.6}$$

Cancelling terms on each side proportional to $\hbar\omega_n$, and going to ket notation

$$i\hbar\sum_n\frac{dc_n}{dt}|\psi_n\rangle = \sum_n\lambda V c_n|\psi_n\rangle \tag{18.7}$$

Multiply both sides by $\langle\psi_k|$

$$i\hbar\frac{\mathrm{d}c_k}{\mathrm{d}t} = \lambda\sum_n\langle\psi_k|V|\psi_n\rangle c_n \tag{18.8}$$

Obviously, the solutions $c_n = c_n(t)$ are actually functions of both time *and* λ. We can therefore expand the c_n in a Taylor series around $\lambda = 0$

$$\begin{aligned} c_k(t) &= c_k^{(0)} + \lambda c_k^{(1)}(t) + \lambda^2 c_k^{(2)}(t) + \cdots \\ c_k^{(n)}(t) &= \frac{1}{n!}\left(\frac{\mathrm{d}^n c_k}{\mathrm{d}\lambda^n}\right)_{\lambda=0} \end{aligned} \tag{18.9}$$

Substituting this expansion into (18.8)

$$\begin{aligned} &i\hbar\left[\frac{\mathrm{d}c_k^{(0)}}{\mathrm{d}t} + \lambda\frac{\mathrm{d}c_k^{(1)}}{\mathrm{d}t} + \lambda^2\frac{\mathrm{d}c_k^{(2)}}{\mathrm{d}t} + \cdots\right] \\ &\quad = \sum_n\langle\psi_k|V|\psi_n\rangle\left[\lambda c_n^{(0)} + \lambda^2 c_n^1 + \lambda^3 c_n^{(2)} + \cdots\right] \end{aligned} \tag{18.10}$$

and equating equal powers of λ on the right- and left-hand sides gives an infinite set of equations

$$\begin{aligned} i\hbar\frac{\mathrm{d}c_k^{(0)}}{\mathrm{d}t} &= 0 \\ i\hbar\frac{\mathrm{d}c_k^{(1)}}{\mathrm{d}t} &= \sum_n\langle\psi_k|V|\psi_n\rangle c_n^{(0)} \\ i\hbar\frac{\mathrm{d}c_k^{(2)}}{\mathrm{d}t} &= \sum_n\langle\psi_k|V|\psi_n\rangle c_n^{(1)} \\ \ldots &= \ldots \\ i\hbar\frac{\mathrm{d}c_k^{(m+1)}}{\mathrm{d}t} &= \sum_n\langle\psi_k|V|\psi_n\rangle c_n^{(m)} \\ \ldots &= \ldots \end{aligned} \tag{18.11}$$

This set can be solved iteratively, starting from knowledge of the set $\{c_n^{(0)}\}$.

Suppose that initially (say at $t \to -\infty$), before the perturbation is turned on, the system was in an energy eigenstate of the unperturbed Hamiltonian φ_l. For example, in the case we have in mind, perhaps the electron of a hydrogen atom is in some definite orbital before a laser, aimed at the atom, is turned on. Then we have

$$c_n(t \to -\infty) = c_n^{(0)} = \delta_{nl} \tag{18.12}$$

Substituting into the equation for $\mathrm{d}c_k^{(1)}/\mathrm{d}t$,

$$i\hbar\frac{\mathrm{d}c_k^{(1)}}{\mathrm{d}t} = \langle\psi_k|V|\psi_l\rangle \tag{18.13}$$

which is integrated to yield

$$c_k^{(1)}(t) = \frac{1}{\mathrm{i}\hbar}\int_{-\infty}^{t} \mathrm{d}t_1 \langle \psi_k(t_1)|V(t_1)|\psi_l(t_1)\rangle \tag{18.14}$$

Likewise,

$$\mathrm{i}\hbar\frac{\mathrm{d}c_k^{(2)}}{\mathrm{d}t} = \sum_n \langle \psi_k|V|\psi_n\rangle c_n^{(1)} \tag{18.15}$$

is integrated to

$$c_k^{(2)}(t) = \frac{1}{\mathrm{i}\hbar}\sum_{n_1}\int_{-\infty}^{t} \mathrm{d}t_1 \left\langle \psi_k(t_1)|V(t_1)|\psi_{n_1}(t_1)\right\rangle c_{n_1}^{(1)}(t_1) \tag{18.16}$$

inserting (18.14)

$$c_k^{(2)}(t) = \left(\frac{1}{\mathrm{i}\hbar}\right)^2 \sum_{n_1}\int_{-\infty}^{t} \mathrm{d}t_1 \left\langle \psi_k(t_1)|V(t_1)|\psi_{n_1}(t_1)\right\rangle \int_{-\infty}^{t_1} \mathrm{d}t_2 \left\langle \psi_{n_1}(t_2)|V(t_2)|\psi_l(t_2)\right\rangle \tag{18.17}$$

By induction, it is not hard to see that

$$\begin{aligned} c_k^{(m)}(t) = &\left(\frac{1}{\mathrm{i}\hbar}\right)^m \sum_{n_1,n_2,\ldots,n_{m-1}} \int_{-\infty}^{t}\mathrm{d}t_1 \int_{-\infty}^{t_1}\mathrm{d}t_2 \int_{-\infty}^{t_2}\mathrm{d}t_3 \\ &\ldots \int_{-\infty}^{t_{m-1}} \mathrm{d}t_m \left\langle \psi_k(t_1)|V(t_1)|\psi_{n_1}(t_1)\right\rangle \\ &\times \left\langle \psi_{n_1}(t_2)|V(t_2)|\psi_{n_2}(t_2)\right\rangle \left\langle \psi_{n_2}(t_3)|V(t_3)|\psi_{n_3}(t_3)\right\rangle \\ &\ldots \left\langle \psi_{n_{m-1}}(t_m)|V(t_m)|\psi_l(t_m)\right\rangle \end{aligned} \tag{18.18}$$

or, using

$$\begin{aligned} |\psi_n(t)\rangle &= \mathrm{e}^{-\mathrm{i}\omega_n t}|\varphi_n\rangle \\ \langle \psi_n(t)| &= \mathrm{e}^{\mathrm{i}\omega_n t}\langle \varphi_n| \end{aligned} \tag{18.19}$$

we have in general

$$\begin{aligned} c_k^{(m)}(t) = &\left(\frac{1}{\mathrm{i}\hbar}\right)^m \sum_{n_1,n_2,\ldots,n_{m-1}} \int_{-\infty}^{t}\mathrm{d}t_1 \int_{-\infty}^{t_1}\mathrm{d}t_2 \int_{-\infty}^{t_2}\mathrm{d}t_3 \ldots \int_{-\infty}^{t_{m-1}}\mathrm{d}t_m \\ &\mathrm{e}^{\mathrm{i}(\omega_k-\omega_{n_1})t_1}\mathrm{e}^{\mathrm{i}(\omega_{n_1}-\omega_{n_2})t_2}\ldots \mathrm{e}^{\mathrm{i}(\omega_{n_{m-1}}-\omega_l)t_m} \\ &\times \left\langle \varphi_k|V(t_1)|\varphi_{n_1}\right\rangle \left\langle \varphi_{n_1}|V(t_2)|\varphi_{n_2}\right\rangle \left\langle \varphi_{n_2}|V(t_3)|\varphi_{n_3}\right\rangle \\ &\ldots \left\langle \varphi_{n_{m-1}}|V(t_m)|\varphi_l\right\rangle \end{aligned} \tag{18.20}$$

In what follows, we really only need the coefficients $c_k(t)$ to first order in λ:

$$c_k^{(1)}(t) = \frac{1}{\mathrm{i}\hbar}\int_{-\infty}^{t} \mathrm{d}t' \mathrm{e}^{\mathrm{i}(\omega_k-\omega_l)t'}\langle \varphi_k|V(t')|\varphi_l\rangle \tag{18.21}$$

It is also often the case that the perturbing potential factorizes into time and space-dependent pieces,

$$\lambda V(x, t) = \lambda v(x) f(t) \tag{18.22}$$

so that to first order

$$c_k(t) = \delta_{kl} + \lambda c_k^{(1)}(t) = \delta_{kl} + \lambda\langle\varphi_k|v|\varphi_l\rangle\frac{1}{\mathrm{i}\hbar}\int_{-\infty}^{t} \mathrm{d}t' \mathrm{e}^{\mathrm{i}(\omega_k-\omega_l)t'} f(t') \tag{18.23}$$

Then the transition probability $P_{l\to k}$ for the system, initially in state ψ_l at time $t \to -\infty$, to be found in a different state ψ_k, $k \neq l$ at time t, is just

$$\begin{aligned} P_{l\to k} &= |\langle\psi_k(t)|\psi(t)\rangle|^2 \\ &= |c_k(t)|^2 \\ &= \lambda^2\frac{1}{\hbar^2}|\langle\varphi_k|v|\varphi_l\rangle|^2\left|\int_{-\infty}^{t} \mathrm{d}t' \mathrm{e}^{\mathrm{i}(\omega_k-\omega_l)t'} f(t')\right|^2 \end{aligned} \tag{18.24}$$

18.2 Harmonic perturbations

As already mentioned, the electric potential associated with electromagnetic radiation varies harmonically in time. Let us then consider a time-dependent potential of the form

$$\lambda V(x, t) = \begin{cases} 0 & t \leqslant 0 \\ \lambda v(x) 2\cos(\omega t) & t > 0 \end{cases} \tag{18.25}$$

Again we imagine that the system is in the state ψ_l at $t \leqslant 0$, so that $c_k(0) = c_k^0 = \delta_{kl}$. Substituting (18.25) into (18.23) gives us, for $k \neq l$,

$$\begin{aligned} c_k(t) &= \lambda c_k^{(1)}(t) \\ &= \lambda\langle\varphi_k|v|\varphi_l\rangle\frac{1}{\mathrm{i}\hbar}\int_0^t \mathrm{d}t' \mathrm{e}^{\mathrm{i}(\omega_k-\omega_l)t'}\left[\mathrm{e}^{\mathrm{i}\omega t'} + \mathrm{e}^{-\mathrm{i}\omega t'}\right] \\ &= -\lambda\langle\varphi_k|v|\varphi_l\rangle\frac{1}{\mathrm{i}\hbar}\left[\frac{\mathrm{e}^{\mathrm{i}(\omega_{kl}-\omega)t} - 1}{\omega_{kl} - \omega} + \frac{\mathrm{e}^{\mathrm{i}(\omega_{kl}+\omega)t} - 1}{\omega_{kl} + \omega}\right] \\ &= -2\mathrm{i}\lambda\langle\varphi_k|v|\varphi_l\rangle\frac{1}{\hbar}\left[\frac{\mathrm{e}^{\mathrm{i}(\omega_{kl}-\omega)t/2}}{\omega_{kl} - \omega}\sin\left[\frac{1}{2}(\omega_{kl} - \omega)t\right]\right. \\ &\quad \left. + \frac{\mathrm{e}^{\mathrm{i}(\omega_{kl}+\omega)t/2}}{\omega_{kl} + \omega}\sin\left[\frac{1}{2}(\omega_{kl} + \omega)t\right]\right] \end{aligned} \tag{18.26}$$

where we have defined

$$\omega_{kl} = \omega_k - \omega_l = \frac{E_k - E_l}{\hbar} \tag{18.27}$$

Clearly, the transition probability $P_{lk} = |c_k|^2$ is going to be largest when the angular frequency ω of the perturbing potential is equal to either ω_{kl} or $-\omega_{kl}$, i.e.

$$\begin{aligned} \hbar\omega &= E_k - E_l \quad \text{'absorption'} \quad (E_k > E_l) \\ \hbar\omega &= E_l - E_k \text{ 'stimulated emission' } (E_l > E_k) \end{aligned} \tag{18.28}$$

The terms 'absorption' and 'stimulated emission' refer to atomic electrons in the field of an electromagnetic wave. According to the Einstein–Bohr picture, the incoming radiation consists of photons, each of energy $E = hf = \hbar\omega$. An electron, initially in an orbital with energy E_l, absorbs a photon of energy $\hbar\omega$ from the incoming electromagnetic radiation, and jumps into a higher energy state $E_k = E_l + \hbar\omega$. This is the absorption case. The result (18.26) has its most common application in atomic physics, and this application will be discussed in some detail in the next chapter. But equation (18.26) of course applies in general to any perturbation of any quantum-mechanical system, when the perturbation varies harmonically in time.

A plot of P_{lk} versus $\omega - \omega_{kl}$ is shown in figure 18.1. Of course the height depends on the strength of the perturbing potential, but the main feature to notice is that the width $4\pi/t$ of the central peak becomes narrower and the height becomes higher as t increases. This means that after a long time there is only a transition if $\hbar\omega$ is very close to $E_k - E_l$. The implications of this fact will be discussed in the next chapter.

18.3 Potentials that vary slowly in time

Let H be a Hamiltonian with a time-dependent potential, and denote the Hamiltonian operator at a given time by $\tilde{H}_t = \tilde{p}^2/2m + V(x, t)$. This operator is still a Hermitian operator at any time t, with an eigenvalue equation

$$\tilde{H}_t\phi_n(x, t) = E_n(t)\phi_n(x, t) \tag{18.29}$$

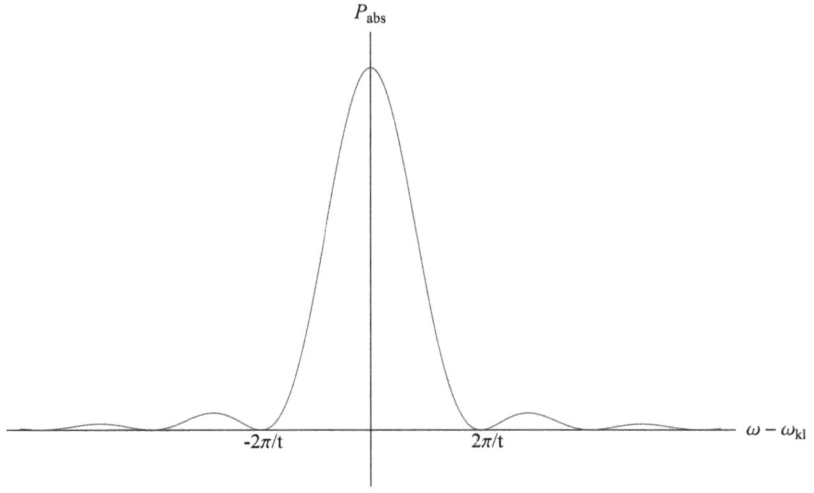

Figure 18.1. Transition (absorption) probability P_{abs} versus angular frequency $\omega - \omega_{kl}$ at a fixed time t. The central peak becomes narrower (width $= 4\pi/t$) and higher as $t \to \infty$.

whose solutions depend on the time t. We will assume for simplicity that the eigenvalues are discrete and non-degenerate. If the potential varies very slowly, according to a criterion that will be described, then the time-evolution of the system is given by

The adiabatic theorem

If a system is in the nth eigenstate of $\tilde{H}_{t_0}$ at time $t = t_0$, and if the potential varies sufficiently slowly in time, then at any later time t the wave function will equal the nth eigenstate of operator $\tilde{H}_t$, multiplied by a certain phase:

$$\psi(x, t) = \phi_n(x, t)e^{i\theta(t)}$$

We will now prove this theorem and compute the phase $\theta(t)$.

At any time t the eigenstates of H_t form a basis in Hilbert space, so the solution to the time-dependent Schrödinger equation can always be expanded as

$$\psi(x, t) = \sum_n c_n(t)\phi_n(x, t) \tag{18.30}$$

Plugging this expression into the time-dependent Schrödinger equation

$$i\hbar\frac{\partial\psi}{\partial t} = \tilde{H}_t\psi \tag{18.31}$$

gives

$$i\hbar\sum_m\left(\frac{\partial c_m}{\partial t}\phi_m(x, t) + c_m\frac{\partial\phi_m}{\partial t}\right) = \sum_m c_m(t)E_m(t)\phi_m(x, t) \tag{18.32}$$

Take the inner product on both sides with $\phi_n(x, t)$:

$$i\hbar\frac{\partial c_n}{\partial t} + i\hbar\sum_m c_m\langle\phi_n|\partial_t|\phi_m\rangle = c_nE_n(t) \tag{18.33}$$

For ***adiabatic*** (= slowly varying) potentials, all terms in the sum are negligible except for $m = n$, and then the first-order differential equation can be solved for c_n. To show that those terms are negligible, we take the time derivative of both sides of (18.29)

$$(\partial_t\tilde{H}_t)\phi_m + \tilde{H}_t(\partial_t\phi_m) = (\partial_tE_m)\phi_m + E_m\partial_t\phi_m \tag{18.34}$$

and then take the inner product of both sides with ϕ_n, with $n \neq m$. This eliminates the term proportional to ∂_tE_m, and using the Hermiticity of $\tilde{H}_t$ we have

$$\langle\phi_n|\partial_t\tilde{H}_t|\phi_m\rangle + E_n\langle\phi_n|\partial_t|\phi_m\rangle = E_m\langle\phi_n|\partial_t|\phi_m\rangle \tag{18.35}$$

or

$$\langle\phi_n|\partial_t|\phi_m\rangle = \frac{\langle\phi_n|(\partial_t H_t)|\phi_m\rangle}{E_n - E_m} \tag{18.36}$$

The adiabatic theorem holds when the time-dependent potential varies so slowly that the right-hand side of equation (18.36) is negligible. When that is the case, only the $m = n$ contribution to the sum in (18.33) remains, and (18.33) can be expressed as

$$\begin{aligned} \mathrm{i}\hbar\frac{\partial c_n}{\partial t} + \mathrm{i}\hbar c_n\langle\phi_n|\partial_t|\phi_n\rangle = c_n E_n(t) \\ \frac{1}{c_n}\partial_t c_n = -\frac{\mathrm{i}}{\hbar}E_n - \langle\phi_n|\partial_t|\phi_n\rangle \\ \frac{\partial}{\partial t}\log(c_n) = -\frac{\mathrm{i}}{\hbar}E_n - \langle\phi_n|\partial_t|\phi_n\rangle \\ \log\left(\frac{c_n(t)}{c_n(t_0)}\right) = \int_{t_0}^{t}\mathrm{d}t'\left(\frac{-\mathrm{i}}{\hbar}E_n(t') + \mathrm{i}\langle\phi_n|\mathrm{i}\partial_t|\phi_n\rangle\right) \end{aligned} \tag{18.37}$$

so that finally

$$c_n(t) = c_n(t_0)\mathrm{e}^{-\mathrm{i}\mathcal{E}_n(t)/\hbar}\mathrm{e}^{\mathrm{i}\gamma(t)} \tag{18.38}$$

where

$$\begin{aligned} \mathcal{E}_n(t) = \int_{t_0}^{t}\mathrm{d}t'\ E_n(t') \\ \gamma_n(t) = \langle\phi_n|\mathrm{i}\partial_t|\phi_n\rangle \end{aligned} \tag{18.39}$$

The fact that $\gamma_n(t)$ is real is seen from the fact that $\partial_t\langle\phi_n|\phi_n\rangle = 0$, which implies that $\langle\phi_n|\partial_t|\phi_n\rangle$ is imaginary. The adiabatic theorem follows for $c_n(t_0) = 1$, $c_{m\neq n}(t_0) = 0$, and the overall phase is $\theta_n(t) = \mathcal{E}_n(t) + \gamma_n(t)$. In more generality, the solution of the time-dependent Schrödinger equation for an adiabatic potential is given by

$$\psi(x, t) = \sum_n c_n(t_0)\phi_n(x, t)\mathrm{e}^{-\mathrm{i}\mathcal{E}_n(t)/\hbar}\mathrm{e}^{\mathrm{i}\gamma_n(t)} \tag{18.40}$$

The factor $\gamma_n(t)$ is known as the ***Berry phase***. Its existence has some deep consequences, which will not be pursued here. For an introduction, see [1].

18.3.1 Example

Consider a particle of mass m, initially (time $t = 0$) in its ground state in a one-dimensional box of length L; i.e.

$$\begin{aligned} \varphi_1(x) = \sqrt{\frac{2}{L}}\sin\left(\frac{\pi x}{L}\right) \\ E_1 = \frac{\pi^2\hbar^2}{2mL^2} \end{aligned} \tag{18.41}$$

Suppose that the walls of the box move apart from each other very slowly, so that the adiabatic assumption is justified, and that after some long time t the walls are a distance αL apart.

We do not need perturbation theory to solve this problem. All that is required is to solve the time-independent Schrödinger equation (exactly) at time t. The result is

$$\begin{aligned} \varphi_1(x, t) &= \sqrt{\frac{2}{\alpha L}} \sin\left(\frac{\pi x}{\alpha L}\right) \\ E_1(t) &= \frac{\pi^2 \hbar^2}{2m\alpha^2 L^2} \end{aligned} \tag{18.42}$$

You can see that the particle has an energy loss

$$\Delta E = \frac{\pi^2 \hbar^2}{2\, mL^2}\left(1 - \frac{1}{\alpha^2}\right) \tag{18.43}$$

which has a simple classical interpretation. The particle in the box exerts a pressure on the walls. As the walls move, the particle does positive work, and loses energy. This is a good thing; otherwise steam and automobile engines would not work! Gas molecules ultimately obey the laws of quantum mechanics. In pushing a piston, they had better give up some of their energy to the piston, otherwise it would be hard to understand how quantum mechanics would be consistent with the principles of heat engine design that were worked out in the 19th century.

18.4 Potentials that change suddenly in time

Finally, let us consider the other extreme of time-dependent perturbation, namely, the case where the potential changes instantly (or nearly so, e.g. by someone flipping a switch):

$$V'(x, t) = \begin{cases} 0 & t < 0 \\ V'(x) & t \geqslant 0 \end{cases} \tag{18.44}$$

We will suppose that the possible energy eigenstates and eigenvalues both before and after $t = 0$ are known:

$$\begin{aligned} H_0 \phi_n &= E_n \phi_n \\ H \phi'_n &= E'_n \phi'_n \\ H &= H_0 + V'(x, t) \end{aligned} \tag{18.45}$$

Possibly the set $\{\phi'_n, E'_n\}$ can be determined by time-independent perturbation theory, or is even known exactly (for some simple forms of $V'(x)$).

Consider a system which is in an initial state

$$\psi_{in}(x, t = 0^-) \tag{18.46}$$

at a moment ($t = 0^-$) just before the perturbation is switched on. With this initial condition, we want to know what the wave function will be at any time $t > 0$, and in particular the probability of finding any particle energy eigenvalue E'_k.

This is actually quite simple. First of all, at any time $t > 0$, the solution to the time-dependent Schrödinger equation has the general form

$$\psi(x, t > 0) = \sum_n c_n \phi'_n(x) e^{-iE_n t/\hbar} \tag{18.47}$$

Then continuity of the wave function in time, and particularly continuity at time $t = 0$, requires that

$$\psi_{in}(x, t = 0^-) = \sum_n c_n \phi'_n(x) \tag{18.48}$$

This equation determines all the c_n in the usual way (go to bra-ket notation, and multiply on the left by $\langle \phi'_k |$)

$$c_k = \langle \phi'_k | \psi_{in}(t = 0^-) \rangle \tag{18.49}$$

The probability of measuring energy E'_k at any time $t > 0$ is then

$$\begin{aligned} P(E_k) &= |\langle \phi'_k | \psi(t) \rangle|^2 \\ &= |c_k|^2 \\ &= |\langle \phi'_k | \psi_{in}(t = 0^-) \rangle|^2 \end{aligned} \tag{18.50}$$

Example

We consider the potential

$$V(x, t) = \begin{cases} \frac{1}{2} k x^2 & t < 0 \\ \frac{1}{2} k' x^2 & t \geqslant 0 \end{cases} \tag{18.51}$$

So for any time t the Hamiltonian is that of a harmonic oscillator (which is exactly soluble) but with spring constant k at $t < 0$, and k' at $t > 0$.

Suppose the particle is in its ground state at times $t < 0$. What is the probability that the particle is in an excited state at $t > 0$?

Clearly the probability of being in an excited state is related to the probability $P(E'_0)$ of remaining in the ground state, i.e.

$$P_{ex} = 1 - P(E'_0) = 1 - |c_0|^2 \tag{18.52}$$

The ground states before and after $t = 0$ are

$$\begin{aligned} \phi_0(x) &= \left(\frac{mk}{\pi^2 \hbar^2} \right)^{1/8} \exp\left[-\frac{1}{2} \sqrt{mk}\, x^2 / \hbar \right] \\ \phi'_0(x) &= \left(\frac{mk'}{\pi^2 \hbar^2} \right)^{1/8} \exp\left[-\frac{1}{2} \sqrt{mk'}\, x^2 / \hbar \right] \end{aligned} \tag{18.53}$$

Then

$$\begin{aligned} c_0 &= \langle \phi'_0 | \phi_0 \rangle \\ &= \left(\frac{m}{\pi^2\hbar^2}\right)^{1/4} (kk')^{1/8} \int \mathrm{d}x \exp\left[-\frac{1}{2}(\sqrt{mk} + \sqrt{mk'})x^2/\hbar\right] \\ &= \sqrt{2}\frac{(kk')^{1/8}}{(\sqrt{k} + \sqrt{k'})^{1/2}} \end{aligned} \tag{18.54}$$

so that

$$P(E'_0) = 2\frac{(kk')^{1/4}}{\sqrt{k} + \sqrt{k'}} \tag{18.55}$$

It is interested to plug in some numbers. Suppose that the spring constant changes drastically at $t = 0$, by a factor of 16, i.e.

$$k' = \frac{k}{16} \tag{18.56}$$

We might expect that such a large change would be very likely to kick the particle out of the ground state. But in fact, one finds in this case that $P(E'_0) = \frac{4}{5}$, so that there is actually only a 20% chance that the particle is found in an excited state after time $t = 0$.

Problems

1. A harmonic oscillator of mass m, charge e, and angular frequency ω is in its ground state.
   a) A uniform electric field $\mathcal{E}$ is turned on at $t = 0$, and turned off at $t = \tau$. Estimate the probability that the system is excited to its nth state, for times $t > \tau$.
   b) Same as (a), only for a sinusoidal field $\mathcal{E} = \mathcal{E}_0 \sin(\omega_0 t)$, turned on at $t =$ 0 and off at $t = \tau$.

   In this one-dimensional problem, the direction of the electric field is aligned with the direction of vibration of the oscillator.

2. Consider a one-dimensional harmonic oscillator of mass m and spring constant k, perturbed by a time-dependent potential

$$V(x, t) = \frac{\lambda}{\sqrt{\pi}\tau} x e^{-t^2/\tau^2} \tag{18.57}$$

   If the oscillator was in the ground state at time $t = -\infty$, find (to lowest order in λ) the probability that it will be in an excited state at time $t = +\infty$.

3. The electron of a hydrogen atom is in its ground state $\phi_{100}(r)$. At time $t = 0$ a uniform electric field in the z-direction is turned on (so the potential is linear in z), and then decays to zero according to

$$E_z(t) = E_0 e^{-\Gamma t} \tag{18.58}$$

Find the probability, to first order in E_0, that the electron is found to be in any one of the excited $n = 2$ states of the hydrogen atom, after a very long time.

The wave functions for the ϕ_{100} and ϕ_{200} states are found in chapter 12. You will also need

$$\phi_{210}(r, \theta, \phi) = \frac{1}{\sqrt{32\pi a_0^3}} \frac{r}{a_0} e^{-r/2a_0} \cos(\theta) \tag{18.59}$$

4. There is an old saying that a watched pot never boils. To investigate this concept, let us consider a quantum-mechanical pot.
   a) The pot is assumed to have only two eigenstates, φ_a ('not boiling') and φ_b ('boiling'), of energies E_a and E_b, and it is perturbed by a harmonic potential
   $$V(x, t) = F(x)\cos(\omega t) \tag{18.60}$$
   where $\omega \approx (E_b - E_a)/\hbar$. Suppose the pot is initially in state φ_a at time $t = 0$, and it is observed again at time $t = \tau$. Assuming τ is very small, i.e. $(\omega_b - \omega_a - \omega)\tau \ll 1$, find the probability that the pot has made a transition from state φ_a to φ_b. Assume that you know the matrix element $F_{ab} = \langle\varphi_a|F|\varphi_b\rangle$.
   b) Now assume the pot is observed every τ seconds, beginning at time $t = 0$, when it is found to be in the 'not boiling' state. Find the probability that the pot does not boil (i.e. remains in state φ_a) after t seconds, in the limit that the pot is 'watched' continuously, i.e. $\tau \to 0$.

Reference

[1] Berry M V 1988 The geometric phase *Sci. Am.* **259** 26–36

IOP Publishing

An Introduction to Quantum Theory

Jeff Greensite

Chapter 19

Atoms and light

In the early 20th century, Planck, Einstein, and Bohr were concerned with the interaction of atoms with light.[1] Their great insights, namely the Planck black-body spectrum, the photon composition of light, and the Bohr atom, preceded the discovery, in 1924–25, of the laws of quantum mechanics. But from the perspective of quantum mechanics, some aspects of those early ideas seem hard to understand, however successful they might be. Take, for example, Bohr's idea that an electron in an excited state of an atom will eventually, all by itself and without any particular inducement, jump spontaneously to a lower energy state, emitting a photon in the process. This is called ***spontaneous emission***. Classically a charged particle in orbital motion will lose energy by radiation; that makes perfect sense. But in quantum mechanics this process should be forbidden, from everything learned so far! Excited atomic states of definite energy are eigenstates of the Hamiltonian and therefore stationary, meaning nothing ever changes. How, then, can the transitions envisaged by Bohr ever take place? Also, Einstein's theory of the photoelectric effect is based on the absorption of photons by electrons. Somehow it is not hard to imagine an electron gobbling up a photon, and gaining energy as a result. But a laser works on the principle that electromagnetic radiation, of the appropriate frequency, can also induce an electron to *lose* energy by emission of a photon. This ***stimulated emission*** is a little harder to visualize, and it is not obvious why such a thing should occur.

In 1917 Albert Einstein showed that absorption of light by atoms implies not only stimulated emission, which is the basis of the laser, but also spontaneous emission. Furthermore, given the absorption rate, the rates of stimulated and spontaneous emission were also determined. It was an amazing feat of deduction, given that a full understanding of the interaction between atoms and light requires relativistic

[1] In this chapter, the word 'light' is shorthand for 'electromagnetic radiation', not necessarily in the visible range.

quantum field theory, and at the time even non-relativistic quantum mechanics was still seven years in the future.

This chapter will begin with Einstein's argument, and the relation he deduced between the rates of absorption, stimulated emission, and spontaneous emission of light by atoms. The rest of the chapter will be devoted to actually computing the rates of absorption and stimulated emission from the principles of quantum mechanics, as an application of time-dependent perturbation theory. Spontaneous emission is a little more mysterious, and the mechanism cannot really be understood without quantizing the electromagnetic field (a topic to be introduced in chapter 27).

19.1 The Einstein *A* and *B* coefficients

Imagine that we have a very large number of atoms in a cavity, whose walls are held at temperature T. For simplicity we will imagine that these are hydrogen atoms, and the electrons of these atoms interact with the thermal black-body radiation which fills the cavity. Let us focus on electrons which are in the ground state with energy E_1, and the first excited state, with energy E_2. It is assumed that

1. An electron in ground state E_1 can absorb, from the thermal radiation, a photon of energy $\hbar\omega = E_2 - E_1$, and jump to the higher energy state E_2. The rate at which this occurs is proportional to the energy density per unit frequency $\mathcal{E}(\omega)$ of the black-body radiation, i.e.

$$\text{Absorption rate} = B_{1\to 2}\mathcal{E}(\omega) \tag{19.1}$$

2. The thermal radiation can stimulate an electron in an exited state E_2 to emit a photon of energy $\hbar\omega = E_2 - E_1$, and fall into the lower energy state E_1, with a rate proportional to the energy density per unit frequency $\mathcal{E}(\omega)$ of the black-body radiation,

$$\text{Stimulated emission rate} = B_{2\to 1}\mathcal{E}(\omega) \tag{19.2}$$

3. Even in the absence of thermal radiation, an electron in excited state E_2 may spontaneously emit a photon of energy $\hbar\omega = E_2 - E_1$, and fall into the lower energy state E_1, with a rate independent of the black-body energy density

$$\text{Spontaneous emission rate} = A \tag{19.3}$$

The atoms are supposed to be in thermal equilibrium with the radiation, and this means that the number of electrons absorbing a photon per second, and making the transition from E_1 to E_2, must be the same as the number of electrons emitting a photon per second, and going from energy E_2 to E_1. If this were not true, then the radiation field would be either losing or gaining energy with time, which means there is no equilibrium. So let N_1, N_2 be the number of electrons at energy levels E_1, E_2, respectively. Equating the overall $1 \to 2$ and $2 \to 1$ transition rates means that

$$N_1 B_{1\to 2}\mathcal{E}(\omega) = N_2(B_{2\to 1}\mathcal{E}(\omega) + A) \tag{19.4}$$

On the other hand, according to the standard principles of statistical mechanics (namely, the Boltzmann probability distribution)

$$\frac{N_2}{N_1} = e^{-(E_2-E_1)/kT} = e^{-\hbar\omega/kT} \tag{19.5}$$

Combining (19.4) and (19.5) and solving for $\mathcal{E}(\omega)$, we find

$$\mathcal{E}(\omega) = \frac{A}{e^{\hbar\omega/kT} B_{1\to2} - B_{2\to1}} \tag{19.6}$$

But the energy density per unit frequency of black-body radiation is also given by the Planck formula of equation (2.4), and if we express this formula in terms of energy density per unit angular frequency, we have

$$\mathcal{E}(\omega) = \frac{\hbar\omega^3}{\pi^2 c^3 (e^{\hbar\omega/kT} - 1)} \tag{19.7}$$

Comparing (19.6) and (19.7), we conclude that

$$\begin{aligned} B_{2\to1} &= B_{1\to2} \\ A &= \frac{\hbar\omega^3}{\pi^2 c^3} B_{2\to1} \end{aligned} \tag{19.8}$$

If there were no emission, then after some time we would have $N_1 = 0$. This contradicts the Boltzmann distribution (19.5). But if there were only spontaneous emission, or only stimulated emission, then (19.6) would contradict the Planck formula for black-body radiation. Both spontaneous emission, independent of thermal radiation, and stimulated emission, proportional to the energy density of thermal radiation, are required, and Einstein's argument relates both quantities to the coefficient $B_{1\to2}$ for the absorption rate.

Einstein's argument is a wonderful shortcut. It shows that the laws of physics, properly applied to the interaction of atoms and light, must predict stimulated and spontaneous emission, given that they allow for absorption. But this is still a consistency argument. We would like to show that those laws really do make the required predictions, and in the process actually compute the A and B coefficients from first principles. The calculation of spontaneous emission rates requires quantization of the electromagnetic field; this calculation can be found in Weinberg [1]. But even without a full quantum treatment of radiation, we can already demonstrate the existence of absorption and stimulated emission, and work out the B coefficient, with the tools at hand.

19.2 The *B* coefficient from quantum theory

The equations of motion of a particle of mass m and charge q in an electromagnetic field is given (in the Heaviside–Lorentz units used in this book) by the Lorentz force law

$$\mathbf{F} = q\mathbf{E} + \frac{q}{c}\mathbf{v} \times \mathbf{B} \tag{19.9}$$

plus Newton's second law $\mathbf{F} = m\mathbf{a}$. It is not difficult to show that these equations of motion follow from Hamilton's equations applied to the following Hamiltonian:

$$H = \frac{1}{2m}\left(\mathbf{p} - \frac{q}{c}\mathbf{A}\right) \cdot \left(\mathbf{p} - \frac{q}{c}\mathbf{A}\right) + q\phi \tag{19.10}$$

where $\mathbf{A}$ and ϕ are the vector and electric potentials, related to $\mathbf{E}$, $\mathbf{B}$ via

$$\begin{aligned} \mathbf{E} &= -\frac{1}{c}\frac{\partial \mathbf{A}}{\partial t} - \nabla\phi \\ \mathbf{B} &= \nabla \times \mathbf{A} \end{aligned} \tag{19.11}$$

Now we suppose that the $\mathbf{A}$ field we are concerned with is a propagating electromagnetic wave, and that it moreover satisfies the Coulomb gauge condition

$$\nabla \cdot \mathbf{A} = 0 \tag{19.12}$$

Then Maxwell's equations tell us that in the absence of charges and currents, $\phi = 0$ and $\mathbf{A}$ satisfies the wave equation

$$\frac{1}{c^2}\frac{\partial^2 \mathbf{A}}{\partial t^2} - \nabla^2 \mathbf{A} = 0 \tag{19.13}$$

This equation has the general solution (allowing for the fact that $\mathbf{A}(x, t)$ is real rather than complex)

$$\begin{aligned} \mathbf{A}(x, t) = \sum_{\lambda=1,2} \int \frac{\mathrm{d}^3 k}{(2\pi)^3 (2\omega_k)} &(a(k, \lambda)\boldsymbol{\epsilon}(\mathbf{k}, \lambda)\mathrm{e}^{\mathrm{i}(\mathbf{k}\cdot\mathbf{r} - \omega_k t)} \\ &+ (a^*(k, \lambda)\boldsymbol{\epsilon}(\mathbf{k}, \lambda)\mathrm{e}^{-\mathrm{i}(\mathbf{k}\cdot\mathbf{r} - \omega_k t)}) \end{aligned} \tag{19.14}$$

where $\omega_k = c|\mathbf{k}|$. The two unit vectors $\boldsymbol{\epsilon}(\mathbf{k}, 1)$, $\boldsymbol{\epsilon}(\mathbf{k}, 2)$ are known as polarization vectors, and they point in two directions which are orthogonal to the propagation vector $\mathbf{k}$ and to each other:

$$\begin{aligned} \mathbf{k} \cdot \boldsymbol{\epsilon}(\mathbf{k}, \lambda) &= 0 \\ \boldsymbol{\epsilon}(\mathbf{k}, \lambda) \cdot \boldsymbol{\epsilon}(\mathbf{k}, \lambda') &= \delta_{\lambda\lambda'} \end{aligned} \tag{19.15}$$

For example, if $\mathbf{k}$ is in the z-direction, then $\boldsymbol{\epsilon}(\mathbf{k}, 1)$, $\boldsymbol{\epsilon}(\mathbf{k}, 2)$ could be chosen to be unit vectors in the x- and y-directions respectively. The condition that $\mathbf{k} \cdot \boldsymbol{\epsilon}(\mathbf{k}, \lambda) = 0$ is imposed by the Coulomb gauge condition $\nabla \cdot \mathbf{A} = 0$.

It is interesting to compare (19.14) with the free particle solution (5.44) of the Schrödinger equation, which of course is also a wave equation. In that expression we see that the integrand is proportional to $\exp(-\mathrm{i}E_p t/\hbar)$, with $E_p = p^2/2\,m > 0$. But the solutions of the Maxwell wave equations have factors proportional to both $\exp(-\mathrm{i}\omega t)$ and $\exp(+\mathrm{i}\omega t)$, sometimes referred to as positive and negative frequencies. The reason for having both cases is that equation (19.13) is second order in the time derivative, and the reason for *that* has to do with special relativity. We will soon see that one factor is associated with absorption, and the other with stimulated emission.

Let us begin by considering a plane electromagnetic wave propagating in the $\mathbf{k}$-direction, with amplitude A_0 and polarization λ

$$\mathbf{A}(x, t) = 2A_0\boldsymbol{\epsilon}(\mathbf{k}, \lambda)\cos(\mathbf{k}\cdot\mathbf{r} - \omega t) \tag{19.16}$$

Visible light has a wavelength on the order of 500 nm, while the diameter of the hydrogen atom is about 0.1 nm. Consequently, the spatial variation of the A-field across an atom is rather small, and to a good approximation (known as the 'dipole approximation') the $\mathbf{k}\cdot\mathbf{r}$ terms can be treated as being a constant within the volume of the atom. Taking the nucleus to be at the origin, $\mathbf{r} = 0$, we have in the neighborhood of the atom

$$\mathbf{A}(x, t) = 2A_0\boldsymbol{\epsilon}(\mathbf{k}, \lambda)\cos(\omega t) \tag{19.17}$$

From the Hamiltonian (19.10) we then have a time-dependent interaction potential, which is the operator

$$V_{\text{int}} = 2\frac{eA_0}{mc}\epsilon(\mathbf{k}, \lambda)\cdot\mathbf{p}A_0\cos(\omega t) + O(A_0^2) \tag{19.18}$$

Since we are going to be using perturbation theory, it is assumed that the term quadratic in A_0 is much smaller than the term linear in A_0, and can be neglected.[2]

Let us denote the initial state of the atom by $i \equiv (n_1l_1m_1)$ and the final state by $f \equiv (n_2l_2m_2)$. Assume that the atom is exposed to light of frequency ω and wave-vector $\mathbf{k}$ for a time t (perhaps by turning a laser on and off). Then, by the rules of time-dependent perturbation theory outlined in the previous chapter, the transition amplitude M_{if} corresponding to $c_k(t)$ of equation (18.26) is

$$\begin{aligned} M_{if}(\omega, \hat{\mathbf{k}}, \lambda; t) = A_0W_{if}(\mathbf{k}, \lambda)&\left[\frac{e^{i(\omega_{fi}-\omega)t/2}}{\omega_{fi}-\omega}\sin\left[\frac{1}{2}(\omega_{fi}-\omega)t\right]\right.\\ &\left.+\frac{e^{i(\omega_{fi}+\omega)t/2}}{\omega_{fi}+\omega}\sin\left[\frac{1}{2}(\omega_{fi}+\omega)t\right]\right] \end{aligned} \tag{19.19}$$

where

$$W_{if}\left(\hat{\mathbf{k}}, \lambda\right) = -2i\frac{e}{m\hbar c}\left\langle n_2l_2m_2|\epsilon(\mathbf{k}, \lambda)\cdot\mathbf{p}|n_1l_1m_1\right\rangle \tag{19.20}$$

and

$$\omega_{fi} = \omega_f - \omega_i = \frac{E_f - E_i}{\hbar} \tag{19.21}$$

Suppose first that $E_f > E_i$, so $\omega_{fi} > 0$. The angular frequency of visible light is on the order of 10^{15} Hz, so the term with $\omega_{fi} + \omega$ in the denominator is negligible. Then only the term coming from the positive frequency $\exp(-i\omega t)$ part of the plane wave

[2] Note that in Coulomb gauge the operator order of $\mathbf{p}$, $\mathbf{A}$ is irrelevant, since $\mathbf{A}\cdot\mathbf{p} - \mathbf{p}\cdot\mathbf{A} = i\hbar\nabla\cdot\mathbf{A} = 0$.

$$M_{if}^{\text{abs}}\left(\omega, \hat{\mathbf{k}}, \lambda; t\right) = A_0 W_{if}\left(\hat{\mathbf{k}}, \lambda\right) \frac{\mathrm{e}^{\mathrm{i}(\omega_{fi}-\omega)t/2}}{\omega_{fi} - \omega} \sin\left[\frac{1}{2}(\omega_{fi} - \omega)t\right] \tag{19.22}$$

contributes, and even that term is only significant when the angular frequency ω of the plane wave is very close to ω_{fi}. In this case the electron has absorbed some energy in making a transition to a higher energy level, which justifies the term 'absorption', for this process, and the corresponding superscript 'abs' on the transition amplitude.

Next, suppose that $E_f < E_i$, so $\omega_{fi} < 0$. Then the term in (19.19) with denominator $\omega_{fi} - \omega$ is negligible, and the other term, with denominator $\omega_{fi} + \omega$ is only significant if $\hbar\omega \approx E_i - E_f$. In this case there is no question of the electron absorbing energy; instead the radiation field has stimulated the electron to lose energy. Although we are treating the radiation field as classical, in reality it is not, and in a more accurate treatment, where the radiation field itself is quantized, the energy lost by the electron appears in the radiation field in the form of an emitted photon. The amplitude for stimulated emission is coming almost entirely from the negative frequency $\exp(\mathrm{i}\omega t)$ part of the plane wave

$$M_{if}^{\text{stim}}\left(\omega, \hat{\mathbf{k}}, \lambda; t\right) = A_0 W_{if}\left(\hat{\mathbf{k}}, \lambda\right) \frac{\mathrm{e}^{\mathrm{i}(\omega_{fi}+\omega)t/2}}{\omega_{fi} + \omega} \sin\left[\frac{1}{2}(\omega_{fi} + \omega)t\right] \tag{19.23}$$

The transition probabilities for absorption and stimulated emission are given by the squared modulus of the corresponding amplitudes, which are proportional to A_0^2. But according to standard electromagnetism, the energy density $u(\omega)$ of the plane wave (19.16) is

$$u(\omega) = \frac{2\omega^2}{c^2} A_0^2 \quad \Longrightarrow \quad A_0^2 = \frac{c^2 u(\omega)}{2\omega^2} \tag{19.24}$$

which means that the transition probabilities are proportional to energy density. In the case of absorption, the transition probability $|M_{if}|^2$ is

$$P_{if}^{\text{abs}}\left(\omega, \hat{\mathbf{k}}, \lambda; t\right) = \frac{c^2 u(\omega)}{2\omega^2} \left|W_{if}\left(\hat{\mathbf{k}}, \lambda\right)\right|^2 \frac{\sin^2\left[\frac{1}{2}(\omega_{fi} - \omega)t\right]}{(\omega_{fi} - \omega)^2} \tag{19.25}$$

Now in thermal radiation the radiant energy is distributed continuously over a range of frequencies, and is distributed evenly among all possible directions $\hat{\mathbf{k}}$ and polarizations λ. Let us then consider radiation with a distribution of energy density per unit angular frequency denoted $\mathcal{E}(\omega)$. Averaging over directions $\hat{\mathbf{k}}$ and polarizations λ, and integrating over frequencies, the absorption probability at time t is

$$P_{if}^{\text{abs}}(t) = \int \mathrm{d}\hat{\mathbf{k}}\, \frac{1}{2} \sum_{\lambda} \left|W_{if}\left(\hat{\mathbf{k}}, \lambda\right)\right|^2 \int_0^\infty \mathrm{d}\omega \frac{c^2 \mathcal{E}(\omega)}{2\omega^2} \frac{\sin^2\left[\frac{1}{2}(\omega_{fi} - \omega)t\right]}{(\omega_{fi} - \omega)^2} \tag{19.26}$$

The time-dependent expression in the integrand

$$F(t) = \frac{\sin^2\left[\frac{1}{2}(\omega_{fi} - \omega)t\right]}{(\omega_{fi} - \omega)^2} \tag{19.27}$$

is negligible for frequencies ω outside a narrow range $\omega_{fi} \pm \Delta\omega$, where $\Delta\omega = 1/t$. Clearly, for $t = 1$ s and $1/t = 1$ Hz, this is a tiny range compared to ω_{fi}, which is on the order of 10^{15} Hz. It is generally safe to treat $\mathcal{E}(\omega)/\omega^2$ as very nearly constant in that range, with the constant equal to $\mathcal{E}(\omega_{fi})/\omega_{fi}^2$. Then we have

$$P_{if}^{\text{abs}}(t) = \int \mathrm{d}\hat{\mathbf{k}} \frac{1}{2} \sum_{\lambda} \left|W_{if}(\mathbf{k}, \lambda)\right|^2 \frac{c^2\mathcal{E}(\omega_{fi})}{2\omega_{fi}^2} \int_0^\infty \mathrm{d}\omega \frac{\sin^2\left[\frac{1}{2}(\omega_{fi} - \omega)t\right]}{(\omega_{fi} - \omega)^2} \tag{19.28}$$

Now, making the change of variables $x = \frac{1}{2}(\omega_{fi} - \omega)t$ in the ω-integration

$$\begin{aligned} \int_0^\infty \mathrm{d}\omega \frac{\sin^2\left[\frac{1}{2}(\omega_{fi} - \omega)t\right]}{(\omega_{fi} - \omega)^2} &= \frac{1}{2}t \int_{-\omega_{fi}t}^\infty \mathrm{d}x \frac{\sin^2(x)}{x^2} \\ &\approx \frac{1}{2}t \int_{-\infty}^\infty \mathrm{d}x \frac{\sin^2(x)}{x^2} \\ &= \frac{1}{2}\pi t \end{aligned} \tag{19.29}$$

Dividing the probability by elapsed time t gives us a transition rate of absorption

$$\begin{aligned} R_{if}^{\text{abs}} &= \frac{\pi c^2\mathcal{E}(\omega_{fi})}{4\omega_{fi}^2} \int \mathrm{d}\hat{\mathbf{k}} \ \frac{1}{2} \sum_{\lambda} \left|W_{if}(\mathbf{k}, \lambda)\right|^2 \\ &= B_{1\to2}\mathcal{E}(\omega_{fi}) \end{aligned} \tag{19.30}$$

By the same logic, we obtain a similar expression for the rate of stimulated emission

$$\begin{aligned} R_{if}^{\text{stim}} &= \frac{\pi c^2\mathcal{E}(\omega_{if})}{4\omega_{if}^2} \int \mathrm{d}\hat{\mathbf{k}} \ \frac{1}{2} \sum_{\lambda} \left|W_{if}(\mathbf{k}, \lambda)\right|^2 \\ &= B_{2\to1}\mathcal{E}(\omega_{if}) \end{aligned} \tag{19.31}$$

and we see that the Einstein B coefficients for absorption and stimulated emission are the same, and equal to

$$B = \frac{\pi c^2}{4\omega_{fi}^2} \int \mathrm{d}\hat{\mathbf{k}} \ \frac{1}{2} \sum_{\lambda} \left|W_{if}(\mathbf{k}, \lambda)\right|^2 \tag{19.32}$$

To complete the calculation of B, we have to evaluate

$$\begin{aligned}
T_{if} &= \int \mathrm{d}\hat{\mathbf{k}}\ \frac{1}{2}\sum_{\lambda} \left|W_{if}(\mathbf{k}, \lambda)\right|^2 \\
&= \frac{4e^2}{m^2\hbar^2c^2}\int \mathrm{d}\hat{\mathbf{k}}\ \frac{1}{2}\sum_{\lambda} |\langle n_2l_2m_2|\epsilon(\mathbf{k}, \lambda)\cdot \mathbf{p}|n_1l_1m_1\rangle|^2 \\
&= \frac{4e^2}{m^2\hbar^2c^2}\int \mathrm{d}\hat{\mathbf{k}}\ \frac{1}{2}\sum_{\lambda} \epsilon_i\left(\hat{\mathbf{k}}, \lambda\right)\epsilon_j\left(\hat{\mathbf{k}}, \lambda\right) \\
&\quad \left\langle n_2l_2m_2\middle|p_i\middle|n_1l_1m_1\right\rangle\left\langle n_1l_1m_1\middle|p_j\middle|n_2l_2m_2\right\rangle
\end{aligned} \tag{19.33}$$

where subscripts i, j denote the vector components of the momentum and polarization vectors, and we use the Einstein summation convention, meaning that repeated indices are summed over. Now note from (19.15) that the sum over polarizations must have the following properties:

$$\begin{aligned}
&k_i\sum_{\lambda}\epsilon_i\left(\hat{\mathbf{k}}, \lambda\right)\epsilon_j\left(\hat{\mathbf{k}}, \lambda\right) = k_j\sum_{\lambda}\epsilon_i\left(\hat{\mathbf{k}}, \lambda\right)\epsilon_j\left(\hat{\mathbf{k}}, \lambda\right) = 0 \\
&\delta_{ij}\sum_{\lambda}\epsilon_i\left(\hat{\mathbf{k}}, \lambda\right)\epsilon_j\left(\hat{\mathbf{k}}, \lambda\right) = 2
\end{aligned} \tag{19.34}$$

From this we deduce that

$$\sum_{\lambda}\epsilon_i\left(\hat{\mathbf{k}}, \lambda\right)\epsilon_j\left(\hat{\mathbf{k}}, \lambda\right) = \delta_{ij} - \hat{k}_i\hat{k}_j \tag{19.35}$$

It is not hard to see that

$$\int \mathrm{d}\hat{\mathbf{k}}\ \hat{k}_i\hat{k}_j = \frac{1}{3}\delta_{ij} \tag{19.36}$$

The reason is that if $i \neq j$, the integral is zero because the integrand is odd under $k_i \to -k_i$. On the other hand, if $i = j$ and we sum over i, that answer must equal 1, since $\hat{\mathbf{k}}\cdot\hat{\mathbf{k}} = 1$, and the average over directions is (by definition) normalized to 1. Then if $i = j$ but we do not sum over i, we must obtain 1/3 the answer we would get if we summed over all three components, which gives us (19.36). So

$$\int \mathrm{d}\hat{\mathbf{k}}\ \sum_{\lambda}\epsilon_i\left(\hat{\mathbf{k}}, \lambda\right)\epsilon_j\left(\hat{\mathbf{k}}, \lambda\right) = \frac{2}{3}\delta_{ij} \tag{19.37}$$

and therefore, after averaging over radiation directions $\hat{\mathbf{k}}$, we have

$$T_{if} = \frac{4e^2}{3m^2\hbar^2c^2}\left\langle n_2l_2m_2|\mathbf{p}|n_1l_1m_1\right\rangle\cdot\left\langle n_1l_1m_1|\mathbf{p}|n_2l_2m_2\right\rangle \tag{19.38}$$

Selection rules

The next step is to note that, by equation (7.134),

$$\frac{\mathrm{d}}{\mathrm{d}t}\left\langle n_2 l_2 m_2 |\mathbf{r}| n_1 l_1 m_1 \right\rangle = \frac{1}{\mathrm{i}\hbar}\left\langle n_2 l_2 m_2 |[H, \mathbf{r}]| n_1 l_1 m_1 \right\rangle \tag{19.39}$$

We are calculating the transition rate to lowest order in e^2, which means that H should be replaced by H_0, which is just the Hamiltonian for the hydrogen atom in the absence of radiation. Then

$$\frac{\mathrm{d}}{\mathrm{d}t}\left\langle n_2 l_2 m_2 |\mathbf{r}| n_1 l_1 m_1 \right\rangle = \left(E_f - E_i\right)\frac{1}{\mathrm{i}\hbar}\left\langle n_2 l_2 m_2 |\mathbf{r}| n_1 l_1 m_1 \right\rangle \tag{19.40}$$

On the other hand, since $[H_0, \mathbf{r}] = -\mathrm{i}\hbar p/m$, we also have

$$\frac{\mathrm{d}}{\mathrm{d}t}\left\langle n_2 l_2 m_2 |\mathbf{r}| n_1 l_1 m_1 \right\rangle = -\frac{1}{m}\left\langle n_2 l_2 m_2 |\mathbf{p}| n_1 l_1 m_1 \right\rangle \tag{19.41}$$

Equating the right-hand sides of (19.40) and (19.41), we have

$$\left\langle n_2 l_2 m_2 |\mathbf{p}| n_1 l_1 m_1 \right\rangle = im\omega_{fi}\left\langle n_2 l_2 m_2 |\mathbf{r}| n_1 l_1 m_1 \right\rangle \tag{19.42}$$

Then

$$T_{if} = \frac{4e^2\omega_{fi}^2}{3c^2\hbar^2}\,|\langle n_2 l_2 m_2 |\mathbf{r}| n_1 l_1 m_1\rangle|^2 \tag{19.43}$$

Now in spherical coordinates

$$\mathbf{r} = r\left(\sin\theta\cos\phi\;\hat{\mathbf{e}}_x + \sin\theta\sin\phi\;\hat{\mathbf{e}}_y + \cos\theta\;\hat{\mathbf{e}}_z\right) \tag{19.44}$$

where $\hat{\mathbf{e}}_x$, $\hat{\mathbf{e}}_y$, $\hat{\mathbf{e}}_z$ are unit vectors in the x-, y-, z-directions. The angle-dependent coefficients of these unit vectors can be expressed in terms of the $l = 1$ spherical harmonics listed in (11.99), and we have

$$\begin{aligned}\mathbf{r} = r\Bigg(&-\sqrt{\frac{2\pi}{3}}\,Y_{11}(\theta, \phi)\left(\hat{\mathbf{e}}_x - i\hat{\mathbf{e}}_y\right) + \sqrt{\frac{2\pi}{3}}\,Y_{1,-1}(\theta, \phi)\left(\hat{\mathbf{e}}_x + i\hat{\mathbf{e}}_y\right)\\ &+\sqrt{\frac{4\pi}{3}}\,Y_{10}(\theta, \phi)\hat{\mathbf{e}}_z\Bigg)\end{aligned} \tag{19.45}$$

This means that the expression $\langle n_2 l_2 m_2 |\mathbf{r}| n_1 l_1 m_1\rangle$ involves angular integrations

$$\int \mathrm{d}\Omega\, Y^*_{l_2, m_2} Y_{1m} Y_{l_1 m_1} \tag{19.46}$$

where $m = -1, 0, 1$. If we look just at the integration over the spherical coordinate ϕ, this is

$$\int_0^{2\pi} \mathrm{d}\phi \mathrm{e}^{-\mathrm{i}m_2\phi}\mathrm{e}^{\mathrm{i}m\phi}\mathrm{e}^{\mathrm{i}m_1\phi} = 2\pi\delta_{m_2, m_1+m} \tag{19.47}$$

From this we deduce that m_2 can only equal $m_1 - 1$, m_1, $m_1 + 1$, i.e. the possible difference between m values in the initial and final atomic states can only be $\Delta m = m_2 - m_1 = -1, 0, 1$.

The next point is parity. The operator $\mathbf{r}$ is obviously odd parity, and the parity of a spherical harmonic Y_{lm} is $(-1)^l$. From this it follows that $l_1 + l_2$ must be an odd number, if the transition rate is to be non-zero; for otherwise the integrand has odd parity, and the integral vanishes. Finally, we need to use the addition theorem for spherical harmonics, which states that

$$Y_{l_am_a}(\theta, \phi)Y_{l_bm_b}(\theta, \phi) = \sum_{l_c=|l_a-l_b|}^{l_a+l_b} C^{l_c}_{l_am_a,\, l_bm_b} Y_{l_c,\, m_a+m_b}(\theta, \phi) \tag{19.48}$$

where $C^{l_c}_{l_am_a,\, l_bm_b}$ is the Clebsch–Gordon coefficient already discussed in chapter 14. Taking $(l_am_a) = (l_1m_1)$, $(l_bm_b) = (1m)$ together with the orthogonality of spherical harmonics, tells us that $l_2 = l_1 - 1, l_1, l_1 + 1$. But since we already know that $l_2 + l_1$ is an odd number, the only allowed possibilities are $l_2 = l_1 \pm 1$. In summary, we have:

The selection rules for radiative transitions

In the dipole approximation, transitions from a state with quantum numbers $n_1l_1m_1$ to a state with quantum numbers $n_2l_2m_2$ can only occur if the angular momentum quantum numbers satisfy

$$\Delta l = l_2 - l_1 = \pm 1 \qquad \text{and} \qquad \Delta m = m_2 - m_1 = -1, 0, 1$$

The dipole approximation ignores the variation of the radiation field within the volume of the atom, i.e. it approximates $\exp(i\mathbf{k} \cdot \mathbf{r})$ as a constant. One can go beyond the dipole approximation by expanding $\exp(i\mathbf{k} \cdot \mathbf{r})$ in a Taylor series around $\mathbf{r} = \mathbf{r}_0$, where $\mathbf{r}_0$ is the location of the atomic nucleus, and then certain transitions which violate the above selection rules can occur. However, such transitions are greatly suppressed, simply because the wavelength of light which can cause atomic transitions (frequencies in the UV range and below) is far greater than the size of an atom.

The 1S → 2P radiative transition

For the hydrogen atom (with a_0 the Bohr radius)

$$\begin{aligned} \varphi_{100} &= R_{10}(r)Y_{00}(\theta, \phi), \qquad & R_{10}(r) &= \frac{2}{a_0^{3/2}} e^{-r/a_0} \\ \varphi_{21m} &= R_{21}(r)Y_{1m}(\theta, \phi), \qquad & R_{21}(r) &= \frac{1}{2\sqrt{6}a_0^{5/2}} r e^{-r/2a_0} \end{aligned} \tag{19.49}$$

so

$$\begin{aligned}\langle 21\, m|\mathbf{r}|100\rangle &= \int_0^\infty \mathrm{d}r\ r^2 \int \mathrm{d}\Omega R_{21} Y_{1m}^* \mathbf{r} R_{10} Y_{00} \\ &= \int_0^\infty \mathrm{d}r\ r^3 R_{21}(r) R_{10}(r) \frac{1}{\sqrt{4\pi}} \int \mathrm{d}\Omega\, Y_{1m}^* \frac{\mathbf{r}}{r}\end{aligned} \tag{19.50}$$

The r-integration is straightforward,

$$\begin{aligned}\int_0^\infty \mathrm{d}r\ r^3 R_{21}(r) R_{10}(r) &= \frac{1}{a^4\sqrt{6}} \int_0^\infty \mathrm{d}r\ r^4 \mathrm{e}^{-3r/2a_0} \\ &= \frac{256 a_0}{81\sqrt{6}}\end{aligned} \tag{19.51}$$

while for the integration over solid angle we just use the orthogonality of the spherical harmonics

$$\begin{aligned}\frac{1}{\sqrt{4\pi}} \int \mathrm{d}\Omega\, Y_{1m}^* \frac{\mathbf{r}}{r} &= \frac{1}{\sqrt{4\pi}} \int \mathrm{d}\Omega\, Y_{1m}^* \Bigg(-\sqrt{\frac{2\pi}{3}}\, Y_{11}(\theta, \phi)\big(\hat{\mathbf{e}}_x - \mathrm{i}\hat{\mathbf{e}}_y\big) \\ &\quad + \sqrt{\frac{2\pi}{3}}\, Y_{1,-1}(\theta, \phi)\big(\hat{\mathbf{e}}_x + \mathrm{i}\hat{\mathbf{e}}_y\big) + \sqrt{\frac{4\pi}{3}}\, Y_{10}(\theta, \phi)\hat{\mathbf{e}}_z \Bigg) \\ &= \begin{cases} \big(-\hat{\mathbf{e}}_x + \mathrm{i}\hat{\mathbf{e}}_y\big)/\sqrt{6} & m = 1 \\ \hat{\mathbf{e}}_z/\sqrt{3} & m = 0 \\ \big(\hat{\mathbf{e}}_x + \mathrm{i}\hat{\mathbf{e}}_y\big)/\sqrt{6} & m = -1 \end{cases}\end{aligned} \tag{19.52}$$

The square modulus of the integral above is equal to 1/3, regardless of m. Together with (19.51) this gives

$$|\langle 21\, m|\mathbf{r}|100\rangle|^2 = \frac{2^{15} a_0^2}{3^{10}} \tag{19.53}$$

From (19.32), (19.33), and (19.43) we now have the B coefficient for either absorption or stimulated emission

$$\begin{aligned}B &= \frac{\pi c^2}{4\omega_{fi}^2} T_{if} = \frac{\pi c^2}{4\omega_{fi}^2} \frac{4e^2\omega_{fi}^2}{3\hbar^2 c^2} \frac{2^{15} a_0^2}{3^{10}} \\ &= \frac{e^2 \pi}{3\hbar^2} \frac{2^{15} a_0^2}{3^{10}}\end{aligned} \tag{19.54}$$

Inserting this expression into Einstein's relation for the A and B coefficient, we find for the rate of spontaneous emission

$$A = \left(\frac{e^2}{4\pi\hbar c}\right)\frac{4\omega_{fi}^3}{3c^2}\frac{2^{15}}{3^{10}}a_0^2 \tag{19.55}$$

We recognize the quantity in parentheses as the fine-structure constant $\approx 1/137$ in Heaviside–Lorentz units. The mean lifetime of the excited state, i.e. the average time it takes for an electron in the excited 2P state to fall back spontaneously to the 1S ground state is simply the inverse of the rate of emission A. Plugging in the numbers, the average lifetime is $\tau = A^{-1} = 1.6 \times 10^{-9}$ s. This implies, of course, that the 2P state is not an eigenstate of the Hamiltonian including radiation. According to the energy–time uncertainty relation, there must be an uncertainty $\Delta E \geqslant \hbar/2\tau$, which means that the frequency of emitted light lies in a frequency range of width $\Delta f \approx 1/(4\pi\tau)$. Since the light emitted in the $2P \to 1S$ transition is $f = 2.5 \times 10^{15}$ Hz, we see that the fractional spread of frequencies $\delta f/f \sim 10^{-8}$ is minute. In principle, atomic lifetimes could be measured from a spectroscopic measurement of δf, but this direct approach is foiled by the usual optical broadening of the spectral lines which swamps the broadening due to finite lifetime. In practice, the method for determining the lifetime of an excited state is to excite these states, at some position $x = 0$, in atoms moving at a velocity v in the x-direction, and then measure the intensity of emitted radiation as a function of position x. This gives a plot of intensity versus time of emission x/v, and from this plot the average lifetime can be extracted.

19.3 Subtleties

This is about as far as one can go, without actually quantizing the electromagnetic field. It is not possible to understand spontaneous emission, whose rate is equal to the A coefficient, without quantizing radiation as well as electron motion. But even for the B coefficients, a close look at the procedure we have followed raises two questions, whose answers require going beyond the framework of non-relativistic quantum mechanics:

1. Is energy exactly conserved in radiative transitions, when the interaction lasts for a fixed and finite time t?
2. In obtaining the transition probability, we have taken the modulus square of the amplitude, and then summed over frequency, direction, and polarization. If a general solution of Maxwell's equations has the form (19.14), should we not integrate the amplitude over frequency, direction, and polarization, and *then* square the modulus to obtain the probability?

Both of these questions bring up the issue of quantizing radiation.

Start with the first question, concerning conservation of energy. Obviously if the atomic electron makes a transition from one energy level to another, the energy of the atom is not conserved. But if we are treating the radiation field classically, just as a time-dependent potential, then there is no paradox. Energy is not conserved in a time-dependent potential, just as momentum is not conserved when a wavepacket reflects off a 'brick wall', represented by the infinite barrier in section 8.2. On the other hand, total momentum is conserved in Nature, and so is total energy. To see conservation of momentum in the brick-wall case, one has to consider the reaction of the brick wall to the collision with the particle. The way to do this is to denote the center of mass of the brick wall as X, the position of the particle as x, and the potential between the particle and the brick wall as $V(X - x)$. Then there is a wave function for the particle and the brick wall, $\psi(x, X)$, satisfying the Schrödinger equation

$$i\hbar\frac{\partial}{\partial t}\psi(x, X, t) = \left\{-\frac{\hbar^2}{2m}\frac{\partial^2}{\partial x^2} - \frac{\hbar^2}{2M_{\text{wall}}}\frac{\partial^2}{\partial X^2} + V(X - x)\right\}\psi(x, X, t) \qquad (19.56)$$

The Hamiltonian operator on the right-hand side of this equation is invariant under translations $x \to x + a$, $X \to X + a$, and this leads, by the arguments of chapter 10, to conservation of total momentum. The point is that to see this conservation, it is necessary to quantize the motion of the brick wall.

A similar approach is necessary to see conservation of energy in the interaction of atoms with light. We can consider the incident radiation to be in the form of a wavepacket, propagating in the z-direction, with a width Δz. Assuming an atom lies in the path of propagation, it will interact with the radiation for a time $t = \Delta z/c$, and will have some amplitude to either absorb or lose some energy. But then it must be that the radiation field has either lost or gained an equal amount of energy, and this is something which cannot be seen if we treat the radiation as though it were just a time-dependent potential, unaffected by its interaction with the electron. In fact, conservation of energy implies that when the effect of the electron on the radiation is taken into account, the state of the radiation must be correlated with the electron state. There is an amplitude, after time t, for the electron to have gained(lost) energy, in which case the radiation has lost(gained) an equivalent amount of energy. But there is also an amplitude for the electron to remain in its initial state, in which case the energy of the radiation field is unchanged. In other words, the states of the electron and the radiation have become ***entangled***, in a sense which will be made precise in chapter 24. But if the radiation can be part of a quantum superposition, then there is no alternative to introducing a wave function for the radiation as well as the electron, and treating the entire electron + radiation system quantum mechanically.

Although it is not really part of the formalism used so far, let us imagine that the incoming wavepacket is somehow made up of photons. Suppose that the time of interaction is infinite, so only a frequency $\omega = |\omega_{fi}|$ exactly will cause the atom to absorb or emit energy. Then the interpretation is obvious: the change of energy in the atom is $\Delta E = E_2 - E_1$, and this is caused by either absorption or emission of a photon of energy $\hbar|\omega_{fi}| = |E_2 - E_1|$. But what if the interaction time t is finite? In this case the incoming wavepacket is of finite length $\Delta z = ct$. Then by the uncertainty principle there is an uncertainty in the momentum of each photon of order $\Delta p_z = \hbar/\Delta z = \hbar/ct$, and therefore an uncertainty in photon energy of at least $\Delta E = c\Delta p_z = \hbar/t$. Likewise, if an atom emits a photon during a period of length t, then the photon is localized to within a distance ct, and again by the uncertainty principle the emitted photon cannot have an energy of *exactly* $E_2 - E_1$. It must have an energy uncertainty of order $\Delta E = \hbar/t$. Now if we look at the plot of absorption/emission probability versus angular frequency in figure 18.1 of the previous chapter, we see that the width of the central peak is $\Delta\omega = 4\pi/t$, and if we relate this spread in angular frequency of radiation to a spread in photon energy, it corresponds to a spread of $\delta E \approx h/t$. This is, as just seen, the order of magnitude of uncertainty in energy of the incident or emitted photons. While this is hardly a complete explanation, it does indicate that the spread in frequencies of figure 18.1 is related to the finite time of interaction, and the inevitable uncertainty in energy which follows from the time–energy uncertainty relation. The spread in frequencies is not in any way an indication of energy non-conservation, although it requires a more sophisticated treatment, in which photons emerge as quantum excitations of the electromagnetic field, to actually prove that statement.

Likewise, the Planck spectrum of black-body radiation does not represent a single electromagnetic wave solving Maxwell's equations, of the form (19.14), rather it is based on a statistical (i.e. probabilistic) distribution of the number of photons occupying each available frequency. The proper procedure is to treat the photon occupation numbers at different frequencies as statistically independent, compute the contribution to the absorption/emission probability from each possible occupation number at a given frequency, multiplied by the appropriate statistical weight, and then sum over frequencies. This is essentially equivalent to the path we have followed; i.e. compute the transition/emission probability at each frequency and then integrate over frequencies (also averaging over directions and polarizations), rather than integrate the amplitude over frequencies and directions, and then take the square modulus to obtain the probability.

As for spontaneous emission, the fact that excited atomic energy levels have a finite lifetime means that they cannot truly be eigenstates of a Hamiltonian operator which includes the electromagnetic field as quantized degrees of freedom, rather than treating it as just a fixed Coulomb potential.

Problems

1. Using classical electromagnetism and Heaviside–Lorentz units, verify the time-averaged energy density in (19.24).
2. Do the algebra to verify equation (19.45).
3. Calculate the spontaneous transition rates in the hydrogen atom for the Balmer 3P → 2S and Lyman 3P → 1S transitions which emit photons of wavelengths 656.3 nm and 102.6 nm respectively. Plug in the physical constants, and give numerical answers in inverse seconds. From these rates, compute the average lifetime of the 3P state.

   In addition to the wave functions listed in chapter 19, you will also need the radial wave functions

$$R_{20}(r) = \frac{2}{(2a_0)^{3/2}}\left(1 - \frac{r}{2a_0}\right)\mathrm{e}^{-r/2a_0}$$

$$R_{31}(r) = \frac{4\sqrt{2}}{3(3a_0)^{3/2}}\frac{r}{a_0}\left(1 - \frac{r}{6a_0}\right)\mathrm{e}^{-r/3a_0}$$

Reference

[1] Weinberg S 2013 *Lectures on Quantum Mechanics* (Cambridge: Cambridge University Press)

Chapter 20

The WKB approximation

Planck's constant is absent in classical physics, yet classical physics does a very good job of describing, e.g. the motion of the Moon around the Earth. So it must be that when quantum mechanics is applied to the motion of the Moon, the classical result is obtained in some way, but the precise value of Planck's constant is irrelevant, since it doesn't appear in the final answer. In this situation, Planck's constant is tiny compared to all relevant scales, and could just as well be zero. Two questions arise. First, how does the center-of-mass motion of the Moon, which is essentially particle motion, come out of a wave equation, even in the limit that $\hbar \to 0$? Ehrenfest's Principle is part of the answer, but not the whole answer. Secondly, is there some transitional region between classical and quantum behavior, a 'semiclassical' regime, where quantum deviations from classical laws are not large but are still non-negligible, and the finite value of $\hbar$, while small by comparison to other scales, cannot be overlooked?

With these questions in mind, let us try to systematically solve Schrödinger's equation via an expansion in powers of $\hbar$. The idea is that classical physics should derive from the leading order solution, with quantum effects appearing at higher orders, and the whole scheme should break down if, when dealing with systems very much smaller than the Moon, $\hbar$ cannot be treated as a small parameter.

Now the de Broglie wave of a macroscopic object has a fantastically small wavelength; an object with a mass of one gram, moving at one meter per second, has a de Broglie wavelength of 6.6×10^{-31} m. Rather than try a power series expansion of this very rapidly oscillating function of position, it might make more sense to make a power expansion of the phase, which may have a much smoother dependence on position. With this motivation, we make the ansatz

$$\begin{aligned} \psi(x) &= e^{iF/\hbar} = e^{-Im(F)/\hbar} e^{iRe(F)/\hbar} \\ F(x) &= F_0(x) + F_1(x)\hbar + F_2(x)\hbar^2 + \cdots \end{aligned} \tag{20.1}$$

The ***WKB approximation*** keeps only the first two terms, F_0 and F_1, and these are easy to compute.

doi:10.1088/978-0-7503-1167-0ch20

Begin by writing the Schrödinger equation in the form

$$\left(\hbar^2 \frac{\mathrm{d}^2}{\mathrm{d}x^2} + p^2(x)\right)\psi = 0, \qquad p(x) = \pm\sqrt{2m(E - V(x))} \tag{20.2}$$

Here $p(x)$ is the classical momentum at point x of a right or left-moving particle (positive or negative sign, respectively) which has a total energy E, moving in a potential $V(x)$. Substitute $\mathrm{e}^{\mathrm{i}F}$ for ψ in the Schrödinger equation, and group together terms with the same power of $\hbar$

$$\left\{-(F_0')^2 + p^2(x)\right\} + \hbar\left\{\mathrm{i}F_0'' - 2F_1'F_0'\right\} + O(\hbar^2) = 0 \tag{20.3}$$

The left-hand side of this equation must equal zero for any small value of $\hbar$, and this means that the zeroth order and first order terms must be separately equal to zero. So we have, for leading and next-to-leading terms in $\hbar$

$$\begin{aligned} -(F_0')^2 + p^2(x) &= 0 \\ \mathrm{i}F_0'' - 2F_1'F_0' &= 0 \end{aligned} \tag{20.4}$$

The solution to the first of these equations is

$$F_0(x) = \pm\int_{x_0}^{x} \mathrm{d}x' p(x') \tag{20.5}$$

where x_0 is a constant. For the second equation, we have

$$F_1' = \frac{\mathrm{i}}{2}\frac{F_0''}{F_0'} \tag{20.6}$$

or

$$\frac{\mathrm{d}}{\mathrm{d}x}F_1 = \frac{\mathrm{i}}{2}\frac{\mathrm{d}}{\mathrm{d}x}\log(F_0') \tag{20.7}$$

which implies, since $F_0' = \pm p(x)$ that

$$F_1 = \mathrm{i}\log(p^{1/2}) + c \tag{20.8}$$

where c is an irrelevant constant that can be absorbed into a normalization factor. This means that the solution to the time-independent Schrödinger equation in the WKB approximation, up to a normalization constant, is simply

$$\begin{aligned} \psi_\pm &= \frac{1}{\sqrt{p(x)}}\exp\left[\pm\frac{\mathrm{i}}{\hbar}\int_{x_0}^{x} \mathrm{d}x' p(x')\right] \\ p(x) &= \sqrt{2m(E - V(x))} \end{aligned} \tag{20.9}$$

or any superposition of ψ_+ and ψ_-.

Condition for validity

A condition for the validity of any power series expansion is that higher order terms are smaller than lower order terms, whatever the expansion parameter may be. In the WKB approximation there are only two terms, $F_0 + \hbar F_1$, so a first guess is that we should require $|F_0| \gg \hbar|F_1|$. However, the magnitude of F_0 in equation (20.5) depends on the choice of the lower limit of integration x_0, and in any case what counts is not the overall magnitude of ψ, which is adjusted by a normalization constant, but rather how rapidly $\psi(x)$ varies with x. Since the variation is controlled by $F(x)$, the inequality should really refer to the derivatives

$$|F_0'(x)| \gg \hbar|F_1'(x)| \tag{20.10}$$

Inserting the expressions (20.5) and (20.8) for F_0, F_1 on the left- and right-hand sides

$$|p(x)| \gg \hbar\left|\frac{\mathrm{d}}{\mathrm{d}x}\frac{1}{2}\log(p(x))\right| \tag{20.11}$$

or

$$\begin{aligned}\frac{|p(x)|}{\hbar} &\gg \left|\frac{1}{2}\frac{1}{p(x)}\frac{\mathrm{d}}{\mathrm{d}x}p(x)\right| \\ &\gg \frac{1}{4}\left|\frac{1}{KE}\frac{\mathrm{d}}{\mathrm{d}x}KE\right|\end{aligned} \tag{20.12}$$

where $KE = E - V(x)$ is the classical kinetic energy. Since $p(x)$ is the classical momentum, we may introduce an x-dependent 'de Broglie' wavelength

$$\lambda(x) = \frac{h}{p(x)} \tag{20.13}$$

which is a rough measure of the length over which the phase of the wave function changes by 2π. Now let ΔKE be the magnitude of the change in kinetic energy over one de Broglie wavelength. Then from equation (20.12) the fractional change $\Delta KE/KE$ satisfies

$$\frac{2\pi}{\lambda(x)} \gg \frac{1}{4}\frac{1}{\lambda(x)}\frac{\Delta KE}{KE} \tag{20.14}$$

and this can only be true if the fractional change in kinetic energy is negligible, i.e. $\Delta KE/KE \ll 1$, over one de Broglie wavelength. Since the change in the kinetic energy is minus the change in the potential energy, the conclusion is that *the WKB approximation should be valid when the wave function varies much more rapidly than the potential.*

This rapid variation of the wave function relative to the potential is already a clue about classical behavior. It is not obvious that light has wave properties, and the reason it took some time to discover those properties is that the wavelength of light is very small compared to the size of ordinary macroscopic objects. When light or any other plane wave encounters an obstacle whose size is much larger than the wavelength, it casts a shadow. But so does a stream of particles. On the other hand, when the size of an obstacle is comparable to or smaller than the wavelength, then waves 'bend around' the object, and this is something that a stream of particles cannot do. In particle motion, it is the potential which corresponds to an obstacle. So one characteristic of the classical limit is that the interference and diffraction effects associated with the de Broglie wavelength of a particle are impossible to detect, when the de Broglie wavelength is tiny compared to scale of variation of the potential.

Classical motion

For macroscopic objects, the uncertainty Δx in position is usually negligible, and remains negligible for billions of years, as we saw back in section 5.5. When this is the case then $\langle V(x)\rangle = V(\langle x\rangle)$, and also

$$\left\langle \frac{\mathrm{d}V(x)}{\mathrm{d}x} \right\rangle = \frac{\mathrm{d}V(\langle x\rangle)}{d\langle x\rangle} \tag{20.15}$$

Then Ehrenfest's principle is identical to Hamilton's equations, with $\langle x\rangle$, $\langle p\rangle$ replacing x, p, and $\langle x\rangle$ as a function of time traces out a trajectory which is a solution of the classical equations of motion. We have now found the energy eigenstates of particles moving in a potential, to the leading and next-to-leading orders in $\hbar$. How is classical motion manifest in these wave functions?

We are considering, for simplicity, motion in one dimension, and suppose that a particle with energy E_0 is moving in the direction of positive x in a potential $V(x)$, with $E_0 > V(x)$ for all x. Then the increment of distance traveled in time $\mathrm{d}t$ is $\mathrm{d}x = v\mathrm{d}t$, or

$$\frac{\mathrm{d}x}{v(x)} = \mathrm{d}t \qquad \text{with} \qquad v(x) = \sqrt{\frac{2}{m}(E_0 - V(x))} \tag{20.16}$$

and it is assumed that the particle is at position x_0 at $t = 0$. Integrating both sides of the equation we have

$$\int_{x_0}^{x_t} \frac{\mathrm{d}x}{v(x)} = t \tag{20.17}$$

which in principle can be solved, given time lapse t, for the position x_t at time t.

An energy eigenstate does not much resemble the motion of a classical particle, first because the wave function is spread out all over space, so Δx is not small at all, and secondly because an energy eigenstate is a stationary state, which means that nothing (e.g. position) can change in time. To have motion we need an uncertainty in energy; this is a consequence of the time-energy uncertainty relation. So to construct a well-localized wavepacket propagating in the positive x direction, with small uncertainty Δx and an average position $\langle x\rangle$ which can change in time, it is necessary to contruct a superposition of energy eigenstates

$$\psi(x, t) = \int \mathrm{d}E\ f(E) \frac{1}{\sqrt{p(x)}} \exp\left[\frac{\mathrm{i}}{\hbar}\left(\int_{x_0}^{x} \mathrm{d}x' \sqrt{2m(E - V(x'))} - Et\right)\right] \quad (20.18)$$

The exact form of $f(E)$ is not important; it is only assumed that it is peaked around some average value E_0. Now the motion of any oscillating wavepacket, whether it arises in quantum mechanics, electromagnetism, or fluids, can be understood in terms of constructive and destructive wave interference. The wavepacket is largest, at time t, in the neighborhood of a point x_t where the component waves of different frequencies (different E values, in the example above), interfere constructively; i.e. they have about the same phase. Elsewhere the phases of the component waves vary wildly, and the waves interfere destructively. Equation (20.18) is a sum of waves, and we ask: at what value $x = x_t$ does the phase $\phi(E)/\hbar$, with

$$\phi(E) = \int_{x_0}^{x} \mathrm{d}x' \sqrt{2m(E - V(x'))} - Et \quad (20.19)$$

vary very slowly, for $E \approx E_0$? In other words, for what value of x is $\phi(E)$ stationary at $E = E_0$? At this value of x, denoted $x = x_t$, the component waves will interfere contructively, and $\psi(x_t, t)$ will have its largest magnitude. The stationarity condition is

$$\begin{aligned} 0 &= \left(\frac{\mathrm{d}\phi(E)}{\mathrm{d}E}\right)_{E=E_0} \\ &= \int_{x_0}^{x_t} \mathrm{d}x' \frac{1}{\sqrt{2(E_0 - V(x'))/m}} - t \end{aligned} \quad (20.20)$$

Observing that $v(x) = \sqrt{2(E_0 - V(x))/m}$ is the classical velocity, the stationarity condition is exactly the classical result (20.17) above. Hence, when the WKB approximation is valid, the maximum of the wavepacket follows the same trajectory as the center-of-mass of a particle, starting at $x = x_0$ at $t = 0$ and energy E_0, in classical physics.

20.1 Bound states

So far we are considering unbound particles, where E is greater than the maximum value of the potential. But the WKB approximation can also be applied to bound states, providing that the validity condition (20.14) applies both in the classically allowed region, where $E > V(x)$, as well as the classically forbidden region, where $E < V(x)$. This condition is typically satisfied for highly excited energy eigenstates, which vary rapidly inside a potential well. In the classically forbidden regime, where $p(x)$ is imaginary and the wave function falls off exponentially, $\lambda = h/|p|$ is not really a wavelength. Nevertheless, if the fractional change in $\sqrt{V(x) - E}$ is small over a distance λ in the classically forbidden region, the WKB method still applies.

Generally we are interested in energy eigenstates when dealing with bound state problems, and those eigenstates are not well-localized wavepackets; instead the wave function extends from one classical turning point to the other, and also somewhat into the classically forbidden regime. Let's begin with an infinite potential well

$$V(x) = \begin{cases} u(x) & a < x < b \\ \infty & \text{otherwise} \end{cases} \tag{20.21}$$

Inside the well, the solution should be a superposition of $\psi_{\pm}$, with the condition that $\psi(x) = 0$ at $x = a$, b. Suppose we choose $x_0 = a$. Then the solution should have the WKB form, in the region $a < x < b$

$$\begin{aligned} \psi(x) &= \frac{A}{\sqrt{|p(x)|}} \cos\left[\frac{1}{\hbar}\int_a^x \mathrm{d}x' p(x') + \pi/2\right] \\ p(x) &= \sqrt{2m(E - u(x))} \end{aligned} \tag{20.22}$$

where the phase $\pi/2$ was chosen to guarantee $\psi(a) = 0$. In order to guarantee that also $\psi(b) = 0$, it is necessary to impose the condition

$$\int_a^b \mathrm{d}x' p(x') = n\pi\hbar \tag{20.23}$$

Since $p(x) = \sqrt{2m(E - u(x))}$, this condition implicitly determines the energy eigenvalue $E = E_n$ at each n.

Next we consider the more general case where the classically allowed region is $a < x < b$, but $V(x)$ is everywhere finite, as shown in figure 20.1. In the classically

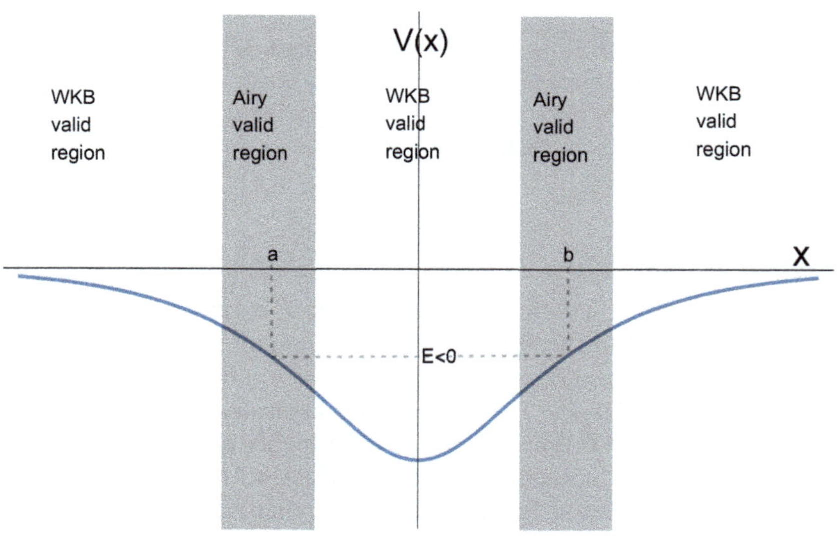

Figure 20.1. The WKB approximation can determine bound state energies in an attractive potential. The approximation works away from the classical turning points a, b. In the regions near the turning points the solution is well represented by Airy functions, which are matched to the WKB solutions at the boundaries of the shaded regions.

forbidden region, away from the turning points at $x \ll a$ and $x \gg b$, the WKB wavefunction falls off exponentially as $x \to \pm\infty$:

$$\psi(x) \propto \frac{1}{\sqrt{|p|}} \exp\left[\pm\frac{1}{\hbar}\int^x dx'|p(x')|\right] \tag{20.24}$$

where the plus sign is for $x \ll a$, and the minus sign is for $x \gg b$. However, for a given value of E, the WKB approximation breaks down at the classical turning points $x = a, b$, where $E = V(a) = V(b)$ and $p(x) = 0$. The strategy is then to solve the Schrödinger equation *exactly* in the immediate neighborhood of points a and b, and match that solution to the WKB solution at the boundaries of the shaded area shown in figure 20.1. The trick that makes an analytic solution possible near the turning points is that in the immediate region of a turning point, say the left hand turning point $x = a$, the potential can be approximated by the first two terms in its Taylor expansion:

$$V(x) = V(a) - \sigma(x - a) \qquad \text{where} \qquad \left(\frac{dV}{dx}\right)_{x=a} = -\sigma \tag{20.25}$$

Since $V(a) = E$, the Schrödinger equation in the neighborhood of $x = a$ has the form:

$$-\frac{\hbar^2}{2m}\frac{d^2\psi}{dx^2} - \sigma(x - a) = 0 \tag{20.26}$$

This equation describes a particle of zero total energy in a linear potential which vanishes at $x = a$. The region $x < a$ is classically forbidden, the region $x > a$ is classically allowed. The exact solutions are known as Airy functions, and the only important fact we'll need is that in the classically allowed region, at $x \gg a$, the Airy function solution is approximately

$$\psi(x \gg a) = \frac{A}{\sqrt{|p|}} \cos\left[\frac{1}{\hbar}\int_a^x dx'p(x') - \frac{\pi}{4}\right], \qquad p(x) = \sqrt{2m\sigma(x - a)} \tag{20.27}$$

while in the classically allowed region at $x \ll b$, the Airy function solution is

$$\psi(x \ll b) = \frac{A'}{\sqrt{|p|}} \cos\left[\frac{1}{\hbar}\int_b^x dx'p(x') + \frac{\pi}{4}\right] \tag{20.28}$$

Away from the turning points, in the classically allowed region, the WKB wavefunction must be a superposition of $\psi_\pm$ (see equation(20.9)), and because the wavefunction is real this superposition must have the form

$$\psi(x) = \frac{A}{\sqrt{p}} \cos\left[\frac{1}{\hbar}\int^x dx'p(x') + \delta\right] \tag{20.29}$$

So the Airy functions and the WKB functions have the same form near but not too close to the turning points. By matching the WKB and Airy functions in the regions where both are valid[1] we can then conclude that equations (20.28) and (20.30) hold throughout the classically allowed region (except in the immediate neighborhood of the turning points) if we take $p(x) = \sqrt{2m(E - V(x))}$. Then, since the two expressions for $\psi(x)$ must agree at $a < x < b$,

$$\frac{1}{\hbar}\int_a^x \mathrm{d}x' p(x') - \frac{\pi}{4} = \frac{1}{\hbar}\int_b^x p(x') + \frac{\pi}{4} + n\pi \tag{20.30}$$

or

$$\int_a^b \mathrm{d}x\, p(x) = \left(n + \frac{1}{2}\right)\pi\hbar \tag{20.31}$$

The integral is sometimes expressed as a 'round-trip' of a classical particle in the well, $a \to b \to a$,

$$\oint \mathrm{d}x\ p(x) = 2\pi\hbar\left(n + \frac{1}{2}\right) \tag{20.32}$$

and it determines the energy eigenvalues E_n. With $n + \frac{1}{2}$ replaced by n, this condition resembles the ***Bohr–Sommerfeld quantization rule*** of the 'old' quantum mechanics, which preceded quantum mechanics based on the Schrödinger equation. Note that unless the potential well is rectangular, the classical turning points a, b, as well as the integrand, in general depend on E.

Example: the harmonic oscillator

The harmonic oscillator is the only potential (apart from free motion) in which the classical equations of motion are linear, and it is the only potential for which the Ehrenfest equations reduce exactly to Hamilton's equations. It is also the only non-trivial example where the WKB approximation gives the exact energy levels. For spring constant k, mass m, and angular frequency $\omega = \sqrt{k/m}$ the turning points are at $x = \pm\sqrt{2E/k}$, so

$$\int_{-\sqrt{2E/k}}^{\sqrt{2E/k}} \sqrt{2m\left(E - \frac{1}{2}kx^2\right)} = \left(n + \frac{1}{2}\right)\pi\hbar \tag{20.33}$$

Change variables $x = \sqrt{2E/k}\,y$,

$$\begin{aligned}\left(n + \frac{1}{2}\right)\pi\hbar &= \sqrt{2m(2E/k)E}\int_{-1}^{1} \mathrm{d}y\sqrt{1 - y^2}\\ &= \frac{2E}{\omega}\frac{\pi}{2}\end{aligned} \tag{20.34}$$

---

[1] We omit the details of the matching conditions, which are a little complicated.

which takes us to the exact answer

$$E_n = \hbar\omega\left(n + \frac{1}{2}\right) \tag{20.35}$$

Example: the $|x|$ potential

Let $V(x) = k|x|$, so the turning points are at $x = \pm E/k$. The WKB condition is

$$\left(n + \frac{1}{2}\right)\pi\hbar = \int_{-E/k}^{E/k} \mathrm{d}x\sqrt{2m(E - k|x|)} \tag{20.36}$$

Change variables to $x = (E/k)y$,

$$\begin{aligned}\left(n + \frac{1}{2}\right)\pi\hbar &= \int_{-E/k}^{E/k} \mathrm{d}x\sqrt{2m(E - k|x|)} \\ &= \frac{E}{k}\sqrt{2mE}\int_{-1}^{1} \mathrm{d}y\sqrt{1 - |y|}\end{aligned} \tag{20.37}$$

The integral is equal to 4/3, so the WKB result for the energy levels is

$$E_n = \left[\frac{3}{4\sqrt{2m}}\frac{k}{m}\left(n + \frac{1}{2}\right)\pi\hbar\right]^{2/3} \tag{20.38}$$

Example: the hydrogen atom

This is a three-dimensional problem involving a central potential, so the WKB approximation applies to the radial equation derived in section 11.5. With $R = u/r$ and $V(r) = -e^2/4\pi r$ the radial equation for an eigenstate of angular momentum takes the form

$$-\frac{\hbar^2}{2m}\frac{\mathrm{d}^2u}{\mathrm{d}r^2} = \left(E + \frac{e^2}{4\pi r} - \frac{\hbar^2}{2m}\frac{l(l+1)}{r^2}\right)u \tag{20.39}$$

Apart from the fact that the range of r is restricted to $r \geqslant 0$ this is much like a Schrödinger equation in one dimension, for a particle moving in an effective potential

$$V_{\mathrm{eff}}(r) = -\frac{e^2}{4\pi r} + \frac{\hbar^2}{2m}\frac{l(l+1)}{r^2} \tag{20.40}$$

which is sketched in figure 20.2. Turning points are, as always, at $E = V_{\mathrm{eff}}(r)$. There are two such points, a, b, which are the roots of the quadratic equation

$$Er^2 + \frac{e^2}{4\pi}r - \frac{\hbar^2}{2m}l(l+1) = 0 \tag{20.41}$$

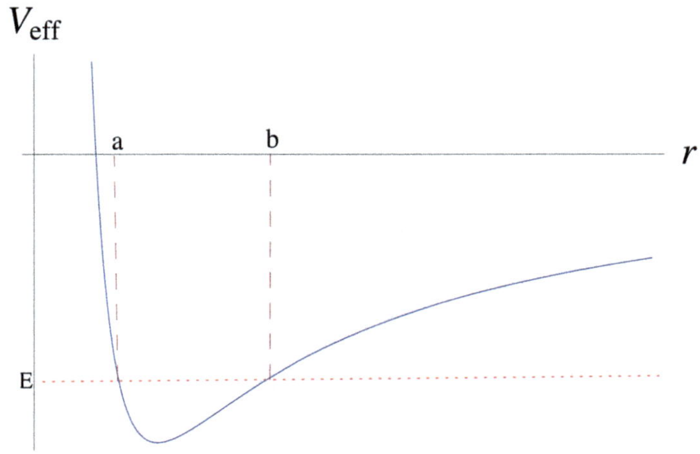

Figure 20.2. A sketch of the effective potential $V_{\text{eff}}(r)$ for the hydrogen atom, with classical turning points at $r = a$ and $r = b$.

with solutions

$$a = \frac{1}{2}\left(\alpha - \sqrt{\alpha^2 - 4\beta}\right)$$
$$b = \frac{1}{2}\left(\alpha + \sqrt{\alpha^2 - 4\beta}\right) \tag{20.42}$$

where

$$\alpha = \frac{e^2}{4\pi|E|}$$
$$\beta = \frac{\hbar^2}{2m|E|}l(l+1) \tag{20.43}$$

Applying the WKB method, the bound state condition is

$$\begin{aligned}\left(n - \frac{1}{2}\right)\pi\hbar &= \int_a^b \mathrm{d}r\sqrt{2m(E - V_{\text{eff}}(r)} \\ &= \sqrt{2m|E|}\int_a^b \mathrm{d}r\frac{1}{r}\sqrt{(r-a)(b-r)} \\ &= \sqrt{2m|E|}\,\frac{1}{2}\pi(a + b - 2\sqrt{ab})\end{aligned} \tag{20.44}$$

Inserting the expressions (20.42) and solving for the bound state energy $E < 0$, we find

$$E = -\left(\frac{e^2}{4\pi}\right)^2 \frac{m}{2\hbar^2}\frac{1}{\left(n - \frac{1}{2} + \sqrt{l(l+1)}\right)^2} \tag{20.45}$$

This is the correct answer for $n \gg \frac{1}{2}$ and $n \gg l$. It is not quite correct for small n, which is line with our expectations for the WKB method: it should work for highly excited states, which oscillate rapidly compared to the variation of the potential.

On the other hand, Bohr got the exact answer for all states, including the states at small n, and his quantization condition looks quite a lot like equation (20.32). But in fact the rules are different, not least because, in the case of the hydrogen atom, an electron in an S-wave state has zero angular momentum. Bohr's approach played a very important historical role in the development of quantum mechanics, but the fact that he got the right answer with the wrong physics must now be regarded as a very lucky coincidence.

20.2 Tunneling

In chapter 8 we saw that the probability that a particle of energy $\langle H \rangle = E$ will tunnel through a rectangular potential barrier of height greater than E is determined by solving for the energy eigenstate, which consists of a superposition of incident, reflected, and transmitted waves. The probability of tunneling is given by the transmission coefficient T. Using the WKB method we can rather easily estimate the tunneling probabilty through barriers of more complex shapes, if the WKB validity condition is fulfilled.

Let a, b be the classical turning points where $V(a) = V(b) = E$, and we assume that the WKB method applies outside two strips of width ϵ around those turning points as shown in figure 20.3. As in chapter 8, consider a particle approaching

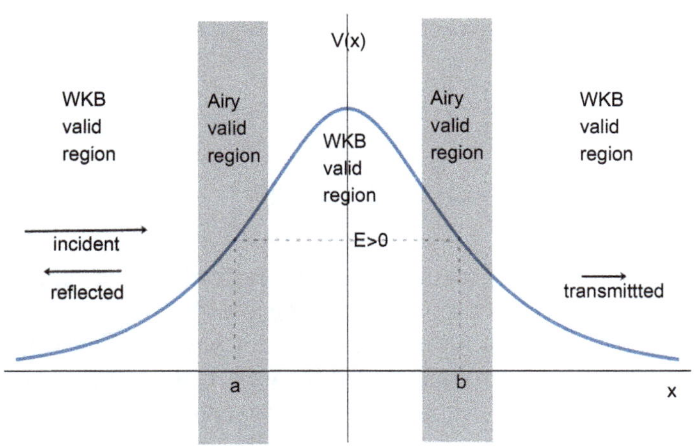

Figure 20.3. The WKB approximation can determine tunneling probabilities. In this case particles are incident on a repulsive potential from the left, and transmitted particles on the right are due to tunneling through the barrier. In a precise treatment, the matching of WKB and Airy functions is similar to the bound state case.

the potential barrier from the right. Then, away from the strips, the WKB approximation is

$$\begin{aligned}
\psi_{\rm III}(x) &= A\frac{1}{\sqrt{|p(x)|}}\exp\left[-\mathrm{i}\int_a^x p(x)/\hbar\right] + B\frac{1}{\sqrt{|p(x)|}}\exp\left[\mathrm{i}\int_a^x p(x)/\hbar\right] \\
\psi_{\rm II}(x) &= C\frac{1}{\sqrt{|p(x)|}}\exp\left[-\int_a^x |p(x)|/\hbar\right] + D\frac{1}{\sqrt{|p(x)|}}\exp\left[+\int_a^x |p(x)|/\hbar\right] \\
\psi_{\rm I}(x) &= E\frac{1}{\sqrt{|p(x)|}}\exp\left[\mathrm{i}\int_b^x p(x)/\hbar\right]
\end{aligned} \tag{20.46}$$

Far from the barrier the wave functions $\psi_{\rm III}$ and $\psi_{\rm I}$ describe free particle motion, and the reflection/transmission coefficients are the same as for the rectangular barrier, i.e. $T = |E/A|^2$, $R = |B/A|^2$.

To determine the constants A–E one has to again match the solution inside the strips, i.e. Airy functions, to the solution (20.46) outside. However, when the potential is very high (compared to the energy of the incident particle) or wide (compared to $\lambda(x)$), then the solution inside the classically forbidden region is mainly the exponentially decaying piece proportional to C. But because $\psi_{\rm II}$ has to be approximately equal to $\psi_{\rm III}$ in the neighborhood of turning point a, it must be that $A, B \sim C$, and because $\psi_{\rm II} \approx \psi_{\rm I}$ in the neighborhood of turning point b, it must also be true that

$$E \sim C\exp\left[-\int_a^b |p(x)|/\hbar\right] \tag{20.47}$$

where the '$\sim$' sign means that these quantities only differ by some numerical factor of order 1. Taking these approximate equalities together, we conclude that

$$\begin{aligned}
T &= \left|\frac{E}{A}\right|^2 \\
&\sim \exp\left[-2\int_a^b |p(x)|/\hbar\right]
\end{aligned} \tag{20.48}$$

Now we consider a different situation. Suppose, instead of a wavepacket approaching a potential barrier, we have a particle wavepacket bouncing back and forth in a potential well. Each time the wavepacket collides with the sides of the well, the particle has a probability T to tunnel out of the well. If the distance between turning points is L, and inside the well the particle wavepacket is traveling with some average group velocity v_g, then there are v_g/L collisions per second, and the escape probability per second is roughly $v_g T/L$. Of course, as the particle bounces back and forth inside the well, the wavepacket spreads out across the width of the well. Nevetheless, the expectation value of position continues to describe back-and-forth motion in the well, and while the picture of a well-localized wavepacket bouncing back and forth cannot be quite true, still the estimate $v_g T/L$ for escape probability per unit time is not so far off. In principle a more accurate approach would be to begin with a wave function that is localized within the well, and let it evolve in time, as shown in the animation of figure 20.4. As time goes on, the wave function will spread more and more outside the well, and the probability to find the particle outside the well will tend towards one as time goes on. It should be noted that a wave function which is well localized inside the well cannot be an eigenstate of energy. A wave function of that type would already be spread all over space.

Radioactive decay

There is an interesting application, due to George Gamow in 1928, of the WKB formula (20.48) to the problem of radioactive decay by α-particle emission. Gamow's model is that an α-particle inside a nucleus is trapped temporarily by the spherical 'square'-well potential, of depth $V(r) = -V_0$. Outside the nucleus the α-particle is subject to a repulsive Coulomb potential $V(r) = 2Ze^2/4\pi r$, since both the nucleus (atomic number Z) and the alpha particle (charge +2, mass m) are positively charged. Let us suppose that the α-particle has an energy $E > 0$ (of course with some uncertainty ΔE), and is in an S-wave state. The radial equation, with $R = u/r$ and $l = 0$ is

$$-\frac{\hbar^2}{2m}\frac{\mathrm{d}^2 u}{\mathrm{d}r^2} + V(r)u = Eu \tag{20.49}$$

which looks like a one-dimensional Schrödinger equation, except for the limitation that $r \geqslant 0$. The classical turning points are at the radius of the nucleus, $r = r_1$, and at $r = r_2$ where $V(r_2) = E$, i.e.

$$r_2 = \frac{2Ze^2}{4\pi E} \tag{20.50}$$

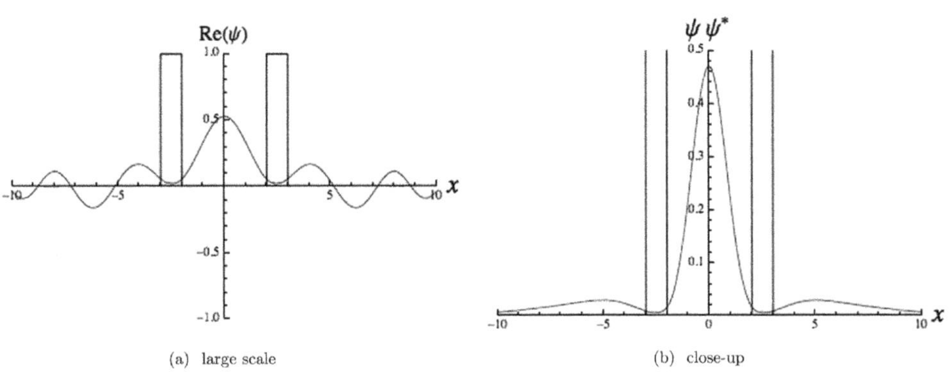

Figure 20.4. Tunneling in real time. The animation shows the evolution of (a) the real part of the wavefunction and (b) the probability density for a particle initially contained in a well. The height of the well is greater than the energy expectation value of the particle state, so classically the particle could never escape.

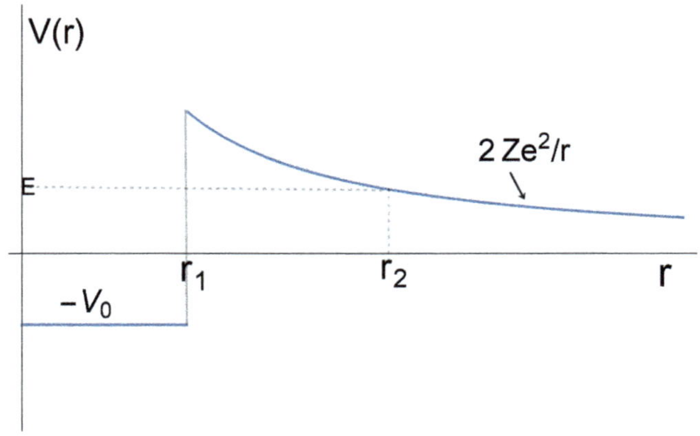

Figure 20.5. Hypothetical potential for the interaction of an alpha particle with the remaining nucleus of atomic number Z.

as shown in figure 20.5. Then

$$\begin{aligned} T &\sim \exp\left[-\frac{2}{\hbar}\int_{r_1}^{r_2} \mathrm{d}r\sqrt{2m\left(\frac{2Ze^2}{4\pi r} - E\right)}\right] \\ &= \exp\left[-2\frac{\sqrt{2mE}}{\hbar}\int_{r_1}^{r_2} \mathrm{d}r\sqrt{\frac{r_2}{r} - 1}\right] \\ &= \exp\left[-2\frac{\sqrt{2mE}}{\hbar}\left(r_2\cos^{-1}\sqrt{\frac{r_1}{r_2}} - \sqrt{r_1(r_2 - r_1)}\right)\right] \end{aligned} \tag{20.51}$$

Suppose that the number of nuclei at any given time is $N(t)$. Then, if the probability per unit time for an α-particle to escape the nucleus is $vT/2r_1$,

$$\frac{\mathrm{d}N}{\mathrm{d}t} = -\frac{v}{2r_1}TN \tag{20.52}$$

with solution

$$\begin{aligned} N(t) &= N(0)\exp\left[-\frac{vT}{2r_1}t\right] \\ &= N(0)\mathrm{e}^{-t/\tau} \end{aligned} \tag{20.53}$$

where $t_{\frac{1}{2}} = \tau \log 2$ is the half-life of the nucleus, and $\tau = 2r_1/(vT)$. Velocity v of the α-particle inside the nucleus is not known with any precision, however the logarithm of τ is

$$\log\tau = \frac{\sqrt{2mE}}{\hbar}\left(r_2\cos^{-1}\sqrt{\frac{r_1}{r_2}} - \sqrt{r_1(r_2 - r_1)}\right) + c \tag{20.54}$$

where c, which depends on $\log(v)$, has a relatively slow variation among different nuclei, while the rest of the parameters (E, m, r_2, r_1) are known for each nucleus which decays via α-emission. This formula turns out to be an excellent fit to the data for nuclear half-lives.

Problems

1. Use the **WKB** method to solve for the energy levels of a particle of mass m moving in the one-dimensional potential

$$V(x) = \frac{1}{2}kx^2 + \beta x \tag{20.55}$$

2. Consider a particle of mass m bound in a potential $V(x) = \lambda x^4$. Find the wavefunction deep in the classically forbidden regions, where $\lambda x^4 \gg E$.
3. Use the **WKB** method to determine the bound state energies of a particle of mass m moving in a potential well $V(x) = \lambda x^{2n}$, where n is any positive integer. Express your answer in terms of the constant

$$I_n = \int_{-1}^{1} \mathrm{d}x\sqrt{1 - x^{2n}} \tag{20.56}$$

4. Estimate the tunneling probability for a particle of mass m and energy $E < V_0$ approaching a potential

$$V(x) = \begin{cases} 0 & x < 0 \\ V_0 - kx & 0 < x < V_0/k \\ 0 & x > V_0/k \end{cases} \tag{20.57}$$

from the left.

5. Compute the tunneling probability as in the previous problem, but this time in a potential

$$V(x) = \begin{cases} 0 & x < -a \\ V_0 - \dfrac{1}{2}kx^2 & -a < x < a \\ 0 & x > a \end{cases} \tag{20.58}$$

where $x = \pm a$ are the classical turning points for a particle of mass m and energy E.

IOP Publishing

An Introduction to Quantum Theory

Jeff Greensite

Chapter 21

The variational method

With four parameters I can fit an elephant, and with five I can make him wiggle his trunk.

John von Neumann

Approximation methods have a limited range of validity. In quantum-mechanical problems it often happens that the potential is not among the handful for which the Schrödinger equation can be solved exactly, and moreover

- the potential is too strong for perturbative methods to be applicable;
- the physics of interest is far from the limit where semiclassical (i.e. WKB) methods can be trusted;
- there are so many degrees of freedom that it is impractical to solve the time-independent Schrödinger equation by numerical methods.

In cases like these, there is still one simple and powerful method which can be employed. Since there is no such thing as a free lunch, the method does have some drawbacks which will be discussed. But sometimes this technique, known as ***the variational method***, can produce an approximate answer to problems which defy any other method of solution.

Suppose we are interested in solving for the ground state of a Hamiltonian. Von Neumann's declaration that 'four parameters fits an elephant' means that for most data sets one can find an analytic function with a few adjustable parameters that can be made to fit the data. This also means that there are many simple functions, depending on a few parameters, that would closely approximate the low-lying eigenstates of any given Hamiltonian. The question is how to choose such a function, and how to adjust those few parameters to give the best possible approximation.

To pick a ***trial function*** which could approximate the ground state, if the parameters are adjusted correctly, there are some general guidelines. First there is the ***no-node theorem***: it can be shown that the ground state wave function should not pass through zero. So one might as well pick the trial function for the ground state to be real and positive. Secondly, if the Hamiltonian has some symmetry, e.g. parity or rotation symmetry, then the ground state usually has the same symmetry.[1] Finally, the ground state wave function is typically largest where the potential is smallest, so the trial function should also have this feature. Having chosen a trial function which stands a good chance of approximating the true ground state, the prescription of the variational method is to choose the values of the adjustable parameters so that the expectation value of the energy is minimized.

The justification for this procedure is that the true ground state $\phi_0(x)$ has the lowest energy expectation value of any state in Hilbert space. So if we have some normalized trial function $\psi(x; \{a_i\})$ that depends on a set of parameters $\{a_i\}$, the choice of parameters that minimizes

$$E_{\text{trial}} = \langle \psi | H | \psi \rangle \tag{21.1}$$

is, by this criterion, the set which brings the trial function as close as possible to the true ground state $\phi_0(x)$.

Let us first see that there is no state with a lower energy expectation value than the ground state. In general the eigenstates of a Hermitian operator span the Hilbert space, so any state can be written in the form

$$|\psi\rangle = \sum_n c_n |\phi_n\rangle \tag{21.2}$$

where $H\phi_n = E_n \phi_n$, and of course the sum should be replaced by an integral for unbound states. Then, using the orthogonality properties of energy eigenstates

$$\begin{aligned} \langle E \rangle &= \langle \psi | H | \psi \rangle \\ &= \sum_n E_n \, |c_n|^2 \\ &\geqslant E_0 \sum_n |c_n|^2 \\ &\geqslant E_0 \end{aligned} \tag{21.3}$$

where the last line follows from normalization, $\langle \psi | \psi \rangle = \sum_n |c_n|^2 = 1$. This demonstrates that the lowest energy expectation value is obtained for $\psi = \phi_0$. Now suppose that the trial state ψ differs from ϕ_0 by an $O(\epsilon)$ deviation, i.e.

$$\psi = \sqrt{1 - \epsilon^2}\,\phi_0 + \epsilon \phi' \tag{21.4}$$

[1] This may not be the case if the system has a very large number of degrees of freedom, in which case the ground state might not share the symmetry of the Hamiltonian. This is known as 'spontaneous symmetry breaking', and is of great importance in condensed matter and elementary particle physics. But the phenomenon is generally not relevant to systems with only a few degrees of freedom.

where $\epsilon \ll 1$, $\langle \phi' | \phi' \rangle = 1$, and

$$\epsilon \phi' = \sum_{n>0} c_n \phi_n \tag{21.5}$$

Then

$$\begin{aligned}\langle E \rangle &= \langle \psi | H | \psi \rangle \\ &= (1 - \epsilon^2) E_0 + \epsilon^2 \langle \phi' | H | \phi' \rangle \\ &= E_0 + O(\epsilon^2)\end{aligned} \tag{21.6}$$

This give us some idea of the potential power of the variational approach. An error of $O(\epsilon)$ in the approximation to the ground state wave function leads to an error of only $O(\epsilon^2)$ in the ground state energy.

Suppose the trial function is not initially normalized. It is easy to see that the expression

$$E_{\text{trial}}(\{a_i\}) = \frac{\langle \psi | H | \psi \rangle}{\langle \psi | \psi \rangle} \tag{21.7}$$

is identical to the energy expectation value (21.1) that would be obtained by first normalizing ψ, and then computing the expectation value. The parameters $\{a_i\}$ are to be chosen by minimizing this expectation, i.e.

$$\left(\frac{\partial}{\partial a_k} E_{\text{trial}}\left(\{a_i\} \right) \right)_{a_k = a_k^{\text{min}}} = 0 \tag{21.8}$$

must be satisfied for each of the parameters a_k, and then

$$\psi\left(x, \left\{a_i^{\text{min}}\right\}\right) \qquad \text{and} \qquad E_{\text{trial}}\left(\left\{a_i^{\text{min}}\right\}\right)$$

are the variational estimates for the ground state wave function and ground state energy, respectively

It is usually much easier to evaluate the integrals required to compute an expectation value than to solve a partial differential equation. This is the very great advantage of the variational method. The first drawback is that variational estimates are only as good as the choice of trial wave function, whose selection demands a certain amount of physical intuition. If the trial function is flexible enough to fit the true ground state reasonably well, then the variational method can produce accurate results. Unfortunately, there is no systematic method for estimating the error, so unless one can compare to experimental or numerical results, it is hard to tell whether the variational method is getting close to the right answer. The second limitation of the method is that it is usually only applicable to the lowest energy state of given (e.g. angular momentum or parity) quantum numbers.

Example: the anharmonic oscillator

The energy eigenstates and eigenvalues for systems with only one degree of freedom, usually a point particle moving in one dimension, are readily obtained numerically. In these cases the time-independent Schrödinger equation is an ordinary differential equation, and the eigenvalue problem is solved by some variant of the numerical method described in section 8.1. A presentation of methods for handling boundary/eigenvalue problems for ordinary differential equations can be found in any text on computational physics, and for most one-dimensional problems these numerical methods are preferred. However, one-dimensional problems still provide the simplest illustration of the variational method, and the results of that method can be compared to the numerical solution.

Let us then consider the potential of the ***anharmonic oscillator***

$$V(x) = \frac{1}{2}kx^2 + \lambda x^4 \tag{21.9}$$

For $\lambda \ll 1$ we can solve for the low-lying eigenstates and eigenvalues using time-independent perturbation theory.[2] But we are now interested in finding the ground state energy when $\lambda > 1$. Since the potential is invariant under reflections $x \to -x$, the trial wave function should have the same property, i.e. it should be an even parity state, and since the potential has a minimum at $x = 0$, the trial wave function should be maximum at $x = 0$. A trial wave function which has these properties, and which is real and positive everywhere, is e.g. a Gaussian wave function

$$\psi(x, a) = \mathrm{e}^{-ax^2} \tag{21.10}$$

where a is the variational parameter. Of course a wave function of this form, with the choice $a = \sqrt{km}/2\hbar$, is exact in the case of the harmonic oscillator, $\lambda = 0$. The question is how well it would approximate the true wave function at non-zero λ, for an optimal choice of the parameter a.

To make the comparison, let us choose units such that $\hbar = m = 1$, and also choose $k = 1$. The problem is to estimate the ground state energy E_0 and wave function ϕ_0 which solve the time-independent Schrödinger equation

$$\left\{-\frac{1}{2}\frac{\mathrm{d}^2}{\mathrm{d}x^2} + \frac{1}{2}x^2 + \lambda x^4\right\}\phi_0(x) = E_0\phi_0(x) \tag{21.11}$$

This problem was treated by time-independent perturbation theory in chapter 17, and the answer, to first order in λ, was given in equation (17.53) ($n = 0$):

$$E_{\text{pert}} = \frac{1}{2} + \frac{3}{4}\lambda \tag{21.12}$$

[2] Even for $\lambda \ll 1$, however, perturbation theory will break down for highly excited states, which extend substantially into regions where $\lambda x^4 > \lambda x^2$.

To apply the variational method, we first compute E_{trial} in equation (21.19) with the trial wave function (21.10), and obtain

$$E_{\text{trial}}(a) = \frac{a + a^3 + 3\lambda}{4a^2} \tag{21.13}$$

The condition for the minimum (21.8) is a cubic equation

$$a^3 - a - 6\lambda = 0 \tag{21.14}$$

and of course one must select the root which is real, and which minimizes $E_{\text{trial}}(a)$.

Figure 21.1(a) compares the minimum $E_{\text{trial}}(a)$, which is the variational estimate for the ground state energy, with the lowest-order perturbative value E_{pert} of equation (21.12), and the energy eigenvalue obtained by solving the Schrödinger equation numerically, at small values of λ. All methods agree for very small $\lambda < 0.01$, but already at $\lambda = 0.04$ the deviation of the numerical and variational values away from the perturbative value is quite significant. Figure 21.1(b) shows that the variational estimate for the ground state energy agrees very closely with the numerical solution, even for very large values of λ where perturbation theory has broken down completely. Finally, figure 21.1(c) shows the (normalized) ground state

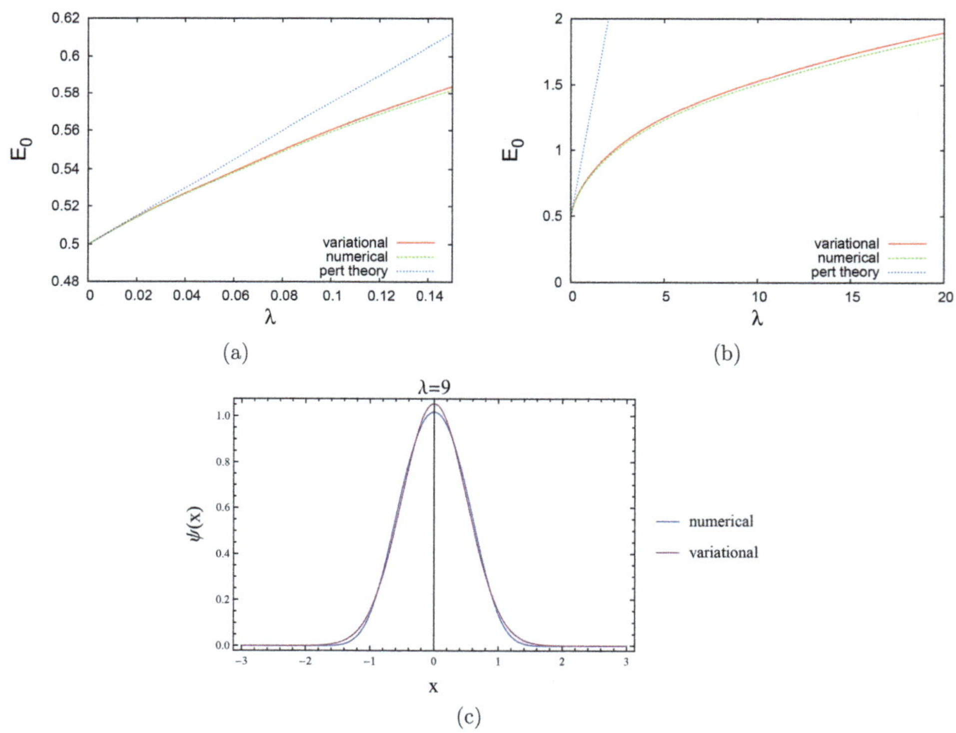

Figure 21.1. Comparison of perturbative and variational approximations to the ground state energy of the anharmonic oscillator, which can be accurately computed from a numerical solution of the time-independent Schrödinger equation. (a) Comparison for small λ; (b) results at larger λ; (c) numerical solution of the ground state wave function versus the variational estimate at $\lambda = 9$.

wave function computed by solving the Schrödinger equation numerically at $\lambda = 9$, with the (normalized) variational wave function at the optimal value of a. While these wave functions are not precisely the same, they are clearly quite close. It is the ability of the Gaussian wave function (21.10) to closely approximate the true ground state wave function that explains the success of the variational method in this example.

Example: the helium atom

Denote the positions of the two electrons in spherical coordinates as r_1, θ_1, ϕ_1 and r_2, θ_2, ϕ_2 respectively. The Schrödinger equation for the helium atom is

$$\left\{-\frac{\hbar^2}{2m}(\nabla_1^2 + \nabla_2^2) - \left(\frac{2e^2}{4\pi r_1} + \frac{2e^2}{4\pi r_2}\right) + \frac{e^2}{4\pi|\mathbf{r}_1 - \mathbf{r}_2|}\right\}\psi(\mathbf{r}_1, \mathbf{r}_2) = E\psi(\mathbf{r}_1, \mathbf{r}_2) \tag{21.15}$$

or

$$\{H_0(1) + H_0(2) + V_{\text{int}}(1, 2)\}\psi(\mathbf{r}_1, \mathbf{r}_2) = E\psi(\mathbf{r}_1, \mathbf{r}_2) \tag{21.16}$$

where

$$\begin{aligned} H_0(1) &= -\frac{\hbar^2}{2m}\nabla_1^2 - \frac{2e^2}{4\pi r_1} \\ H_0(2) &= -\frac{\hbar^2}{2m}\nabla_2^2 - \frac{2e^2}{4\pi r_2} \\ V_{\text{int}}(1, 2) &= \frac{e^2}{4\pi|\mathbf{r}_1 - \mathbf{r}_2|} \end{aligned} \tag{21.17}$$

If we could neglect the interaction term $V_{\text{int}}(1, 2)$ between the two electrons, then the Schrödinger equation would allow an exact solution by separation of the r_1, θ_1, ϕ_1 and r_2, θ_2, ϕ_2 variables. Separation of variables would turn the problem into two hydrogen atom problems, the only difference being that the charge of the nucleus is $+2e$ rather than $+e$, and the ground state solution would be a product of hydrogen atom solutions (12.36) for the spatial part. Since this spatial solution is symmetric under interchange of particle labels $1 \leftrightarrow 2$, the spin statistics theorem demands an antisymmetric spin wave function, i.e.

$$\Phi_{00} = \frac{1}{\sqrt{2}}\left(\chi_+^1\chi_-^2 - \chi_+^2\chi_-^1\right) \tag{21.18}$$

so that altogether

$$\begin{aligned} \psi(1, 2) &= \left(\frac{e^{-r_1/a_0}}{a_0^{3/2}\sqrt{\pi}}\right)\left(\frac{e^{-r_2/a_0}}{a_0^{3/2}\sqrt{\pi}}\right)\Phi_{00} \\ &= \psi(1)\psi(2)\Phi_{00} \end{aligned} \tag{21.19}$$

where the modified Bohr radius, appropriate to a nucleus of charge $+2e$, is

$$a_0 = \frac{4\pi\hbar^2}{2e^2 m} \tag{21.20}$$

But $V_{\text{int}}(1, 2)$ is not zero, so separation of variables does not work, nor is it very small compared to the Coulombic interactions $-2e^2/r_1$, $-2e^2/r_2$ with the nucleus, and therefore the validity of perturbation theory is a bit dubious. So this is a case where we might want to employ the variational approach.

A good trial wave function calls for some physical intuition, and the idea here is the following. If electron 1 were very far from the nucleus, then from the point of view of electron 1 the main effect of electron 2 would be to screen some of the positive charge of the nucleus, so that the effective positive charge is $+e$ rather than $+2e$. But since electron 1 is not so far away, electron 2 only partially screens the charge of the nucleus, down to some effective charge ρe, with $1 < \rho < 2$. The same can be said of the effect of electron 1 from the point of view of electron 2. With this in mind, we try a trial wave function of precisely the same form as (21.19), but this time with an adjustable Bohr radius

$$a_\rho = \frac{4\pi\hbar^2}{\rho e^2 m} \tag{21.21}$$

The dimensionless parameter ρ is the variational parameter. We see that the trial wave function (21.19) satisfies the usual requirements. It is real and positive everywhere, it shares the spherical symmetry of the potential, and its maximum is at $r_1 = r_2 = 0$, where the potential has its (infinitely negative) minimum.

Computation of the trial energy involves a few integrations:

$$\begin{aligned} E_{\text{trial}}(\rho) &= \langle\psi(1, 2)|H|\psi(1, 2)\rangle \\ &= \langle\psi(1)\psi(2)|H_0(1) + H_0(2) + V_{\text{int}}(1, 2)|\psi(1)\psi(2)\rangle\langle\Phi_{00}|\Phi_{00}\rangle \\ &= \langle\psi(1)|H_0(1)|\psi(1)\rangle\langle\psi(2)|\psi(2)\rangle + \langle\psi(2)|H_0(2)|\psi(2)\rangle\langle\psi(1)|\psi(1)\rangle \\ &\quad + \langle\psi(1)\psi(2)|V_{\text{int}}(1, 2)|\psi(1)\psi(2)\rangle \\ &= \langle\psi(1)|H_0(1)|\psi(1)\rangle + \langle\psi(2)|H_0(2)|\psi(2)\rangle \\ &\quad + \langle\psi(1)\psi(2)|V_{\text{int}}(1, 2)|\psi(1)\psi(2)\rangle \end{aligned} \tag{21.22}$$

where

$$\begin{aligned} \langle\psi(1)|H_0(1)|\psi(1)\rangle &= \langle\psi(2)|H_0(2)|\psi(2)\rangle \\ &= 4\pi\int_0^\infty \mathrm{d}r\; r^2\left(\frac{\mathrm{e}^{-r/a_\rho}}{a_\rho^{3/2}\sqrt{\pi}}\right)\left\{-\frac{\hbar^2}{2m}\frac{1}{r^2}\frac{\mathrm{d}}{\mathrm{d}r}r^2\frac{\mathrm{d}}{\mathrm{d}r} - \frac{2e^2}{4\pi r}\right\} \\ &\quad \times\left(\frac{\mathrm{e}^{-r/a_\rho}}{a_\rho^{3/2}\sqrt{\pi}}\right) \end{aligned} \tag{21.23}$$

and

$$\langle\psi(1)\psi(2)|V_{\mathrm{int}}(1,2)|\psi(1)\psi(2)\rangle = \int \mathrm{d}^3r_1 \int \mathrm{d}^3r_2 \psi^*(1)\psi^*(2)\frac{e^2}{4\pi|\mathbf{r}_1 - \mathbf{r}_2|}\psi(1)\psi(2) \tag{21.24}$$

The integrals in (21.23) are straightforward, and the result is

$$\langle\psi(1)|H(1)|\psi(1)\rangle + \langle\psi(2)|H(2)|\psi(2)\rangle = \frac{m}{\hbar^2}\left(\frac{e^2}{4\pi}\right)^2(\rho^2 - 4\rho) \tag{21.25}$$

As for $\langle V_{\mathrm{int}}\rangle$, we can adapt a useful result from electrostatics, which is that the potential of a thin spherical shell of charge $\mathrm{d}q$ of radius r_2, at the point $\mathbf{r}_1$, is given by

$$U(r_1, r_2) = \begin{cases} \dfrac{\mathrm{d}q}{4\pi r_1} & \text{if } r_1 > r_2 \\[2ex] \dfrac{\mathrm{d}q}{4\pi r_2} & \text{if } r_1 < r_2 \end{cases} \tag{21.26}$$

So let us write (21.24) in the form

$$\begin{aligned}\langle V_{\mathrm{int}}\rangle &= \int \mathrm{d}^3r_1\psi^*(1)\psi(1)e\int_0^\infty\left(\int \mathrm{d}\Omega_2\frac{e\psi^*(r_2)\psi(r_2)}{4\pi|\mathbf{r}_1 - \mathbf{r}_2|}\right)r_2^2\,\mathrm{d}r_2 \\ &= \int \mathrm{d}^3r_1\psi^*(1)\psi(1)e\int_0^\infty u(r_1, r_2)\mathrm{d}r_2\end{aligned} \tag{21.27}$$

We recognize that $u(r_1, r_2)\mathrm{d}r$ is the electric potential at the point $\mathbf{r}_1$ of a thin shell of radius r_2 and total charge

$$\mathrm{d}q = e\psi^*(r_2)\psi(r_2)4\pi r_2^2\mathrm{d}r_2 \tag{21.28}$$

Making use of (21.26), we have

$$\begin{aligned}\langle V_{\mathrm{int}}\rangle = \int_0^\infty \mathrm{d}r_1\;4\pi r_1^2\left(\frac{\mathrm{e}^{-r_1/a_p}}{a_p^{3/2}\sqrt{\pi}}\right)^2 \\ \times\left\{\int_0^{r_1}\mathrm{d}r_2\;4\pi r_2^2\frac{e^2}{4\pi r_1}\left(\frac{\mathrm{e}^{-r_2/a_p}}{a_p^{3/2}\sqrt{\pi}}\right)^2\right. \\ \left.+\int_{r_1}^\infty \mathrm{d}r_2\;4\pi r_2^2\frac{e^2}{4\pi r_2}\left(\frac{\mathrm{e}^{-r_2/a_p}}{a_p^{3/2}\sqrt{\pi}}\right)^2\right\}\end{aligned} \tag{21.29}$$

The integrations over angles no longer appear, and the integrals over r_2 yields a function depending only on r_1. The integration over r_1 is also standard, and the final answer is

$$\langle V_{\mathrm{int}}\rangle = \frac{m}{\hbar^2}\left(\frac{e^2}{4\pi}\right)^2\frac{5\rho}{8} \tag{21.30}$$

Adding (21.25) and (21.30) gives the trial energy

$$E_{\text{trial}}(\rho) = \frac{m}{\hbar^2}\left(\frac{e^2}{4\pi}\right)^2\left(\rho^2 - \frac{27}{8}\rho\right) \tag{21.31}$$

The condition for the minimum, $\mathrm{d}E_{\text{trial}}/\mathrm{d}\rho = 0$ has the solution $\rho = 27/16$. Substituting this value into $E_{\text{trial}}(\rho)$, and recalling that the ground state energy of the hydrogen atom is $-me^4/2(4\pi)^2\hbar^2 = -13.61$ eV gives the variational estimate for the ground state energy of helium

$$E_0 = -77.51 \text{ eV (variational)} \tag{21.32}$$

The ionization energy of helium in the ground state is the minimum energy required to remove both electrons, and the ground state energy is the ionization energy times -1. Experimentally, it is found that

$$E_0 = -78.88 \text{ eV (experimental)} \tag{21.33}$$

This is pretty good agreement. Of course, without the experimental value, we would have no way of knowing how close the variational estimate was to the true ground state energy.

Excited states and truncated bases

The variational method can be used to find the lowest energy state of given quantum numbers. For example, in the case of the anharmonic oscillator, the ground state must have the same symmetry as the potential, and is therefore even parity. The first excited state is odd parity. So if we make a trial wave function of odd parity, e.g.

$$\psi(x) = x\mathrm{e}^{-ax^2} \tag{21.34}$$

then this state is guaranteed to be orthogonal to the exact ground state. Choosing a to minimize the energy expectation value will give an estimate for the lowest energy state of odd parity, which is the first excited state. This example is left as an exercise. If, however, we would like to find the first excited state of even parity, that is a more difficult task, because it is not obvious how to choose a trial wave function which is orthogonal to the exact ground state. One could choose a trial wave function which was orthogonal to the variational estimate for the ground state, but this is piling an approximation on top of another uncontrolled approximation. Generally the variational method is only used for finding the lowest energy state among states of a given symmetry.

Another approach to calculating excited states is the use of a ***truncated basis*** for Hilbert space. If we pick any set of orthonormal eigenstates $\{\varphi_n\}$ spanning Hilbert space, then the energy eigenstates must be some linear combination

$$\phi_n = \sum_{m=0}^{\infty} a_{nm}\varphi_m \tag{21.35}$$

of the basis states. Now truncate the basis to the first N basis states, and compute the $N \times N$ matrix $[H]$ with matrix elements

$$H_{ij} = \langle \varphi_i | H | \varphi_j \rangle \tag{21.36}$$

Standard software programs will solve the eigenvalue problem

$$[H]\mathbf{a}^{(n)} = E_n \mathbf{a}^{(n)} \tag{21.37}$$

If we denote the components

$$\mathbf{a}^{(n)} = \begin{bmatrix} a_{n1} \\ a_{n2} \\ \vdots \\ a_{nN} \end{bmatrix} \tag{21.38}$$

then the eigenvalues $\{E_n\}$ of the $N \times N$ matrix $[H]$ are an estimate of the exact energy eigenvalues of the first N eigenstates, and the states

$$\phi_n(x) = \sum_{m=0}^{N} a_{nm}\varphi_m(x) \tag{21.39}$$

are an approximation to the first N eigenfunctions. In principle this approach should converge to the correct answers as N increases, at least for $n \ll N$.

Problems

1. Use the trial wave function

$$\begin{aligned} \phi(x) &= (a^2 - x^2)^2 \quad \text{for} \quad -a \leqslant x \leqslant a \\ &= 0 \qquad \text{for } |x| > a \end{aligned} \tag{21.40}$$

   where a is a variational parameter, to estimate the ground state energy of the harmonic oscillator. Compare to the exact answer.

2. Use the trial wave function

$$\phi(x) = x \exp[-|x|/a] \tag{21.41}$$

   to estimate the energy of the lowest excited state of odd parity for the potential $V(x) = \lambda x^4$. Choose units so that $\hbar^2/m = 1$, and also choose $k = 1$. Plot the odd parity energy versus λ.

3. Suppose $V(x) = V(-x)$ and you know the six lowest energy states of odd parity. Describe how you would use the variational method to find the seventh excited state of odd parity.

4. Computer exercise: this is an example of the truncated basis method. Let us consider the potential

$$V(x) = \begin{cases} \lambda \sin^2\left(\frac{\pi x}{L}\right) & 0 < x < L \\ \infty & \text{otherwise} \end{cases} \tag{21.42}$$

Again choose units $\hbar^2/m = 1$, and in these units let $L = 1$. Compute the Hamiltonian matrix elements

$$H_{mn} = \langle \phi_m | H | \phi_n \rangle, \qquad n, m = 1, 2, \ldots, N \tag{21.43}$$

of the $N \times N$ matrix $[H]$, where the ϕ_n are eigenstates of the infinite square well potential. Use a software package such as Mathematica, Matlab, Octave or Maple to find the eigenvalues of this matrix at $\lambda = 500$, starting with $N = 1$, and increasing to $N = 12$. Make a plot showing how the lowest three energy eigenvalues depend on the dimension N of the truncated basis.

IOP Publishing

An Introduction to Quantum Theory

Jeff Greensite

Chapter 22

Scattering theory

Consider an old-fashioned alarm clock of the pre-digital age, the sort powered by a spring that had to be wound at regular intervals. The mechanism of such clocks was hidden from view by a plastic or metal casing. Suppose one would try to deduce that mechanism from the collision of two identical clocks moving towards each other at very high velocities. After the clocks collide, springs and gears and bits of plastic and paint fly away from the collision point in every direction. All the information about the internal mechanism of the clocks is contained in the precise distribution of the scattered pieces of the two clocks. But deducing that internal mechanism from the distribution of those scattered pieces would be quite a challenge.

Nevertheless, almost everything we know about the structure of subatomic particles has been deduced, with some heroic effort, from collisions of subatomic particles. The colliding particles themselves may be simple (electrons and positrons), or complicated (protons and antiprotons), or very complicated (heavy ions such as lead nuclei). But the product of collisions at very high relativistic energies is almost always complicated. Figure 22.1 is an image of the particles produced in a collision of two gold nuclei at the Relativistic Heavy Ion Collider in Brookhaven National Laboratory. Literally thousands of particles of many different types are produced. Somehow modern physics can make sense out of such collisions, and in the case of heavy ion collisions can deduce the existence of a new state of matter, the quark gluon plasma, emerging only at temperatures of several trillion kelvins. Such extreme conditions were previously encountered an instant after the Big Bang.

At very high energies, which is the domain of relativistic quantum theory, energy is conserved as always, but the number of each type of particle is not. Particles can be created and destroyed in collision processes, and naturally this complicates matters in ways which are beyond the scope of this text. We will be concerned in this chapter with only the simplest scattering process in which a single particle, moving at some non-relativistic velocity, encounters a fixed static target, and interacts with that target via some time-independent potential. This is actually not a bad description of

doi:10.1088/978-0-7503-1167-0ch22

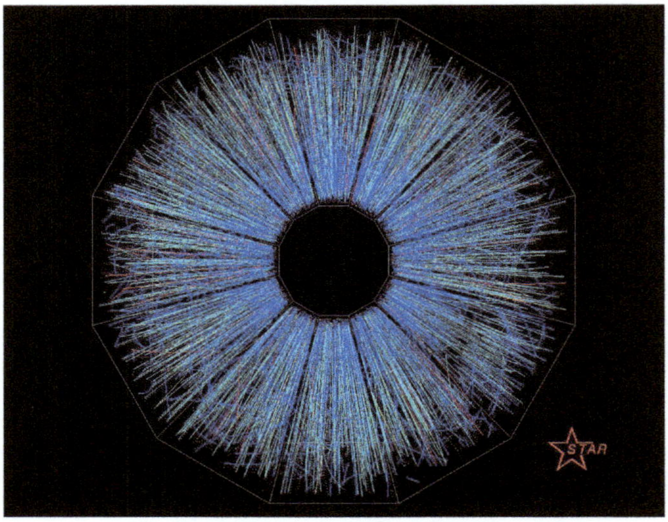

Figure 22.1. Tracks of particles produced by the collision of two gold ions, seen in the STAR detector at the Relativistic Heavy Ion Collider at Brookhaven National Laboratory in Long Island, New York. Image courtesy of Brookhaven National Laboratory.

the experimental situation relevant to atomic and low-energy nuclear physics. The tools we will develop to analyze this situation will provide a foundation for the more sophisticated analyses required in experimental nuclear and high-energy particle physics.

A starting point is that quantum mechanics describes the evolution of a wave function and therefore the scattering of waves, much as a wave in a pond would scatter from some fixed object like a rock. But an experimentalist does not record the arrival of waves, but rather the arrival of particles in a detector. The number of particles which arrives in a certain detector depends on details of the experiment such as the number of incoming particles per square centimeter per second, and the number of targets per square centimeter. These details are essential to a description of the experiment, but they are not essential to the physics. An observable which only depends on the physics of the collision, known as the *'differential cross section'*, will be described in the next section.

22.1 Scattering cross sections

Let us go back to classical physics, and imagine a beam of particles of mass m, all moving with some definite momentum $\mathbf{p}$ parallel to the z-axis. The beam has some intensity I, and we recall that intensity is defined as the amount of energy passing through a unit area perpendicular to the beam direction, per unit time. If N is the number of particles which pass through a small area ΔA perpendicular to the beam in time Δt, then

$$I = \frac{p^2}{2m}\frac{N}{\Delta A \Delta t} \tag{22.1}$$

Let us suppose that an impenetrable target is placed in front of the beam, and that any particle which strikes the target is scattered elastically (i.e. without losing any kinetic energy) in some direction. It is easy to see that the number of particles scattered per second is proportional to the cross-sectional area σ of the object, i.e.

$$\text{total no. scattered/s} = \sigma \frac{N}{\Delta A \Delta t} \tag{22.2}$$

Instead of reporting the outcome of the experiment in terms of the total number of particles scattered per second, which depends on experimental details (namely the beam intensity, or $N/(\Delta A \Delta t)$), it makes more sense to report the cross-sectional area σ, which is independent of those details.

If the target were a macroscopic object, and very large compared to the wavelength of visible light, we could think of its cross-sectional area σ as the area of the shadow cast by the object on a screen, if the object were illuminated by a beam of light along the z-axis. On the other hand, the concept of a cross section makes perfect sense even if the target is not an impenetrable object, but rather a region where the beam of particles encounters a scattering potential. We simply define (see figure 22.2)

The total scattering cross section σ is an (imaginary) area perpendicular to the beam such that the number of particles in the beam which cross this area per second equals the number of particles which are scattered per second.

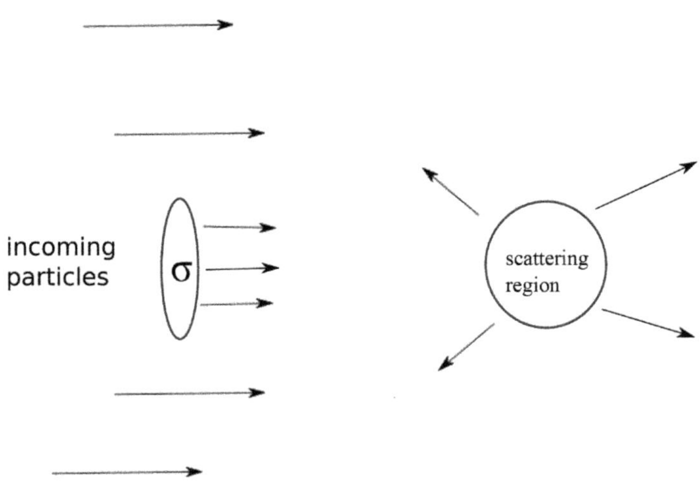

Figure 22.2. The number of incoming particles per second, crossing an imaginary area (the total cross section) σ perpendicular to the beam, equals the number of particles scattered per second by the scattering potential.

But a scattering experiment normally contains more information about the scattering potential (or about the shape and size of the impenetrable target), than just the total cross section. There is also information contained in the angular distribution of the scattered particles. In a spherical coordinate system, with the target at the origin of coordinates, let us consider a detector at coordinates (r, θ, ϕ) which spans a solid angle $\Delta\Omega = \sin\theta\Delta\theta\Delta\phi$. We recall that in a plane, $r\Delta\theta$ is the length of a small line a distance r from the origin, perpendicular to the radius vector, which subtends an arc $\Delta\theta$. Likewise, in three dimensions, $r^2\Delta\Omega$ is the area of a small surface a distance r from the origin, perpendicular to the radius vector, which subtends the angular intervals $\Delta\theta$, $\Delta\phi$. Then, in complete analogy to the total cross section, we may define $\Delta\sigma(\theta, \phi)$ to be an imaginary area perpendicular to the beam such that the number of particles in the beam which cross this area per second equals the number of particles per second which arrive at the detector. This gives us

$$\begin{aligned}\text{differential cross section} &= \frac{\mathrm{d}\sigma(\theta, \phi)}{\mathrm{d}\Omega} \\ &= \lim_{\Delta\Omega\to 0}\frac{\Delta\sigma}{\Delta\Omega}\end{aligned} \tag{22.3}$$

The task of the experimentalist is to report the differential cross section, as a function of scattering angles θ, ϕ. The task of theory is to either

- deduce the unknown interaction potential given the differential cross section; or
- predict the differential cross section given a hypothesis about the interaction potential.

22.2 Wave functions and the differential cross section

In a real scattering experiment, the state of the incoming particle is described by a wavepacket of some kind, moving towards the target. Part of the wavepacket will reach and overlap with the target (or a scattering potential of finite range), and this part will be scattered. For some short period, the incoming wavepacket will overlap with the scattered wave. As the uncertainty Δp of the momentum of the incoming particle becomes smaller, the duration of the overlap between the incoming wave-packet and scattered waves increases until, in the limit $\Delta p \to 0$, the wave function is an energy eigenstate, i.e. a stationary state, with the scattered and incoming waves overlapping at all times. We have already encountered this situation in chapter 8, which considered scattering in one dimension, and we have seen that the time-dependent scattering event can be represented by a time-independent energy eigenstate, where part of the eigenstate can be identified with the incoming wave-packet, and the rest, away from the potential well, corresponds to the scattered wave. From this stationary wave function we could work out the transmission and reflection coefficients.

Now we want to make the same sort of identification in three dimensions. It is assumed that the scattering potential has a finite range r_0; i.e. $V(r, \theta, \phi) \approx 0$ for

$r > r_0$. So for $r > r_0$, the wave function, up to some overall normalization constant, is a solution of the free particle Schrödinger equation, with

$$\psi = \psi_{\rm inc} + \psi_{\rm scatt} \tag{22.4}$$

The incident wave function $\psi_{\rm inc}$ is simply a plane wave in the z-direction,

$$\psi_{\rm inc} = e^{ikz} \tag{22.5}$$

where the momentum of the incoming particle is along the z-axis, $\mathbf{p} = p\mathbf{e}_z = \hbar k\mathbf{e}_z$. This is of course an energy eigenstate of the free particle Schrödinger equation, with energy $E = \hbar^2k^2/2m$. But $\psi_{\rm scatt}$ must also be a free particle energy eigenstate, and let us express it as a superposition of eigenstates of H, L^2, L_z:

$$\psi_{\rm scatt}(r, \theta, \phi) = \sum_{l=0}^{\infty}\sum_{m=-l}^{l} R_{kl}(r)Y_{lm}(\theta, \phi) \tag{22.6}$$

The function $R_{kl}(r)$ is a solution of the radial equation derived in section 11.5, with (at $r > r_0$) $V(r) = 0$. There are two types of solutions when $V(r) = 0$: the spherical Bessel functions $j_l(kr)$, and the spherical Neumann functions $n_l(kr)$, so in general R_{kl} is a superposition of the two:

$$R_{kl}(r) = aj_l(kr) + bn_l(kr) \tag{22.7}$$

Now if the potential were zero everywhere in space, then the expression for $R_{kl}(r)$ would have to hold all the way down to $r = 0$. However, at $r = 0$ the spherical Neumann functions are singular, and their derivatives are discontinuous. For that reason, the Neumann functions have to be excluded, and this is why they were ignored in the discussion of free particle propagation in chapter 11. But in the case of scattering it is *not* true that the potential is zero everywhere, and the free particle expression only holds for $r > r_0$. There will be some other wave function, different from spherical Bessel and Neumann functions, in the neighborhood of $r = 0$. For this reason there is no justification to exclude the spherical Neumann functions at $r > r_0$, and in fact we have to include them. The general solution, in the region where $V(r) = 0$ is

$$\psi_{\rm scatt}(r, \theta, \phi) = \sum_{l=0}^{\infty}\sum_{m=-l}^{l} \left(a_{lm}j_l(kr) + b_{lm}n_l(kr)\right)Y_{lm}(\theta, \phi) \qquad (r > r_0) \tag{22.8}$$

Particle detectors are placed at distances which are normally very much greater than the range of the potential, so let us consider the large distance limit $kr \gg 1$. In this limit

$$\begin{aligned} j_l(kr) &\to \frac{\sin(kr - l\pi/2)}{kr} \\ n_l(kr) &\to -\frac{\cos(kr - l\pi/2)}{kr} \end{aligned} \tag{22.9}$$

and therefore

$$\psi_{\text{scatt}}(r, \theta, \phi) = \sum_{l=0}^{\infty}\sum_{m=-l}^{l} \frac{a_{lm}\sin(kr - l\pi/2) - b_{lm}\cos(kr - l\pi/2)}{kr} Y_{lm}(\theta, \phi) \quad (kr \gg 1) \tag{22.10}$$

However, since

$$\begin{aligned} \sin(kr - l\pi/2) &= \frac{1}{2\mathrm{i}}(\mathrm{e}^{-\mathrm{i}l\pi/2}\mathrm{e}^{\mathrm{i}kr} - \mathrm{e}^{\mathrm{i}l\pi/2}\mathrm{e}^{-\mathrm{i}kr}) \\ \cos(kr - l\pi/2) &= \frac{1}{2}(\mathrm{e}^{-\mathrm{i}l\pi/2}\mathrm{e}^{\mathrm{i}kr} + \mathrm{e}^{\mathrm{i}l\pi/2}\mathrm{e}^{-\mathrm{i}kr}) \end{aligned} \tag{22.11}$$

we see that in general ψ_{scatt} contains waves which are traveling away from the scattering region, and these are proportional to $\mathrm{e}^{\mathrm{i}kr}$, but also waves which are traveling towards the scattering region, proportional to $\mathrm{e}^{-\mathrm{i}kr}$. As in one-dimensional scattering, section 8.3, the general energy eigenstate contains pieces which do not describe the scattering situation we are interested in, and these pieces can be consistently set to zero by a choice of constants. In section 8.3 we set the constant $D = 0$ on those grounds. For the same reason, we have to choose a particular combination of constants a_{lm}, b_{lm} to eliminate the ingoing waves $\mathrm{e}^{-\mathrm{i}kr}$. The choice has to be

$$\begin{aligned} \psi_{\text{scatt}}(r, \theta, \phi) &= \sum_{l=0}^{\infty}\sum_{m=-l}^{l} C_{lm}\big(j_l(kr) + \mathrm{i}n_l(kr)\big)Y_{lm}(\theta, \phi) \qquad (r > r_0) \\ &\approx \sum_{l=0}^{\infty}\sum_{m=-l}^{l} d_{lm}\frac{\mathrm{e}^{\mathrm{i}kr}}{r}Y_{lm}(\theta, \phi) \qquad (kr \gg 1) \\ &= f(\theta, \phi)\frac{\mathrm{e}^{\mathrm{i}kr}}{r} \end{aligned} \tag{22.12}$$

where $d_{lm} = (C_{lm}/k)\mathrm{e}^{-\mathrm{i}(l+1)\pi/2}$ and

$$f(\theta, \phi) \equiv \sum_{l=0}^{\infty}\sum_{m=-l}^{l} d_{lm}Y_{lm}(\theta, \phi) \tag{22.13}$$

Therefore, the superposition of the incoming and scattered waves is

$$\psi(r, \theta, \phi) = N\left(\mathrm{e}^{\mathrm{i}kz} + f(\theta, \phi)\frac{\mathrm{e}^{\mathrm{i}kr}}{r}\right) \quad (kr \gg 1) \tag{22.14}$$

where the overall normalization constant N will not be important here. We can simplify things a little further if we assume that nothing changes when we rotate the target around the z-axis, i.e. make a rotation around the azimuthal angle ϕ. This is normally the case in a collision where the target is an array of atoms or their nuclei. Rotations of the target will not normally change the collision cross sections. So we will assume that $V(r, \theta, \phi)$ is at least azimuthally symmetric. In that case ψ is also azimuthally symmetric, in which case

$$\psi(r, \theta) = N\left(e^{ikz} + f(\theta)\frac{e^{ipr/\hbar}}{r}\right) \quad (kr \gg 1) \tag{22.15}$$

This means that $d_{lm} = 0$ for $m \neq 0$ in (22.13) so that

$$f(\theta) = \sum_{l=0}^{\infty} A_l P_l(\cos(\theta)) \tag{22.16}$$

where $P_l(\cos(\theta)) = \sqrt{\frac{4\pi}{2l+1}}\, Y_{l0}(\theta, \phi)$ are known as Legendre polynomials. The function $f(\theta)$, which must have units of length, is called the ***scattering amplitude***. Now we will relate this scattering amplitude to the differential cross section.

Going back to equation (8.15), we know that the probability per unit time for a particle of momentum $\mathbf{p}$ to cross an area ΔA perpendicular to $\mathbf{p}$ is

$$\text{Prob/unit time} = \frac{p}{m}\psi^*\psi\Delta A \tag{22.17}$$

where $\psi^*\psi$ is evaluated at a point in the small area ΔA. By definition, the number of particles per unit time in the incoming beam crossing area $\Delta\sigma$ equals the number per unit time entering the detector, with entry area ΔS subtending a solid angle $\Delta\Omega$, a distance r from the target. Since the number/unit time is just proportional to the probability/unit time for a single particle, we must have

$$\begin{aligned} \frac{p}{m}\psi^*_{\text{inc}}\psi_{\text{inc}}\Delta\sigma(\theta, \phi) &= \frac{p}{m}\psi^*_{\text{scatt}}(r, \theta, \phi)\psi_{\text{scatt}}(r, \theta, \phi)\Delta S \\ \frac{p}{m}N^2\Delta\sigma &= \frac{p}{m}N^2\frac{f^*(\theta)f(\theta)}{r^2}r^2\Delta\Omega \end{aligned} \tag{22.18}$$

and therefore, in the limit $\Delta\Omega \to 0$,

$$\boxed{\frac{d\sigma}{\delta\Omega} = f^*(\theta)f(\theta)}$$

This equation is the starting point of scattering theory. It relates something that can be measured, namely the *differential cross section* $d\sigma/d\Omega$, to something which can be computed by solving the Schrödinger equation, i.e. the amplitude $f(\theta)$. The total cross section is just the integral over solid angle

$$\begin{aligned} \sigma &= \int d\Omega\ f^*(\theta)f(\theta) \\ &= 2\pi\int_0^{\pi} d\theta \sin\theta\ f^*(\theta)f(\theta) \end{aligned} \tag{22.19}$$

and, like the differential cross-section, it depends on the scattering amplitude $f(\theta)$. Scattering theory boils down to computing this amplitude, and for this purpose there are two general methods available: the method of partial waves, and the Born approximation.

22.3 The method of partial waves

For scattering in three dimensions the scattered wave far from the target can also be described entirely in terms of a set of phase shifts $\{\delta_l\}$, where $l = 0, 1, 2, \ldots$ is the angular momentum quantum number associated with operator L^2. To derive this fact, we begin with the expansion of the wave function $\psi(r, \theta)$ in the region $r > r_0$, where the potential vanishes, in terms of eigenstates of H, L^2, L_z, i.e.

$$\psi(r, \theta, \phi) = \sum_{l=0}^{\infty} \sum_{m=-l}^{l} B_{lm} R_{kl}(r) Y_{lm}(\theta, \phi) \tag{22.20}$$

where R_{kl} satisfies the free particle radial equation

$$\left(\frac{1}{r} \frac{\mathrm{d}^2}{\mathrm{d}r^2} r - \frac{l(l+1)}{r^2} + k^2 \right) R_{kl}(r) = 0 \tag{22.21}$$

and $E = p^2/2\,m = \hbar^2 k^2/2m$. This time, however, (22.20) represents the superposition of both the incoming *and* the scattered waves, in the region $r > r_0$. We have already noted that the general solution of this equation is

$$R_{kl}(r) = c_1 j_l(kr) + c_2 n_l(kr) \tag{22.22}$$

where $j_l(x)$ is a spherical Bessel function, and $n_l(x)$ is a spherical Neumann function. For $V(x) = 0$ everywhere (the case considered in section 11.5.1) we could dispense with the spherical Neumann functions, because they are singular and non-differentiable at the origin, which is forbidden in any finite potential. However, the free particle radial wave function is *not* valid all the way to the origin, because there is an interaction potential at distances $r < r_0$ which must be taken into account. So we have no reason and no right to neglect the Neumann functions.

Asymptotically, at $kr \gg 1$, the spherical Bessel and spherical Neumann functions tend to the limiting forms shown in equation (22.9). Then, writing $c_1 = A_l \cos \delta_l$, $c_2 = -A_l \sin \delta_l$, we have

$$\begin{aligned} R_{kl}(r) &= \frac{1}{kr} A_l \left\{ \cos \delta_l \sin\left(kr - \frac{\pi}{2} l\right) + \sin \delta_l \cos\left(kr - \frac{\pi}{2} l\right) \right\} \\ &= \frac{1}{kr} A_l \sin\left(kr - \frac{\pi}{2} l + \delta_l\right) \end{aligned} \tag{22.23}$$

Then we have the general free particle solution at $r \gg r_0$, $kr \gg 1$,

$$\psi(r, \theta, \phi) = \sum_{l=0}^{\infty} \sum_{m=-l}^{l} B_{lm} \frac{1}{kr} A_l \sin\left(kr - \frac{\pi}{2} l + \delta_l\right) Y_{lm}(\theta, \phi) \tag{22.24}$$

Now we again observe that if $\psi(r, \theta, \phi)$ has azimuthal symmetry, and is therefore independent of ϕ, it must be that only terms with quantum number $m = 0$ contribute to the sum, and therefore

$$\psi(r, \theta, \phi) = N\sum_l C_l \frac{1}{kr} \sin\left(kr - \frac{\pi}{2}l + \delta_l\right) P_l(\cos(\theta)) \tag{22.25}$$

So far, this is very general. Any free particle solution of the Schrödinger equation, with azimuthal symmetry, has this form. However, we are interested in solutions corresponding to scattering, so we should identify the wave function in (22.25) with the scattering wave function (22.15) in the region $r > r_0$. This means that

$$e^{ikz} + f(\theta)\frac{e^{ikr}}{r} = \sum_l C_l \frac{1}{kr} \sin\left(kr - \frac{\pi}{2}l + \delta_l\right) P_l(\cos(\theta)) \tag{22.26}$$

The next step is to expand the plane wave e^{ikz}, which is an eigenstate of H, p_z, in a superposition of eigenstates of H, L^2, L_z. It can be shown that

$$e^{ikz} = \sum_{l=0}^{\infty} (2l + 1) i^l j_l(kr) P_l(\cos(\theta)) \tag{22.27}$$

Inserting this identity into (22.26),

$$\begin{aligned} &\sum_{l=0}^{\infty} (2l + 1) i^l \frac{1}{kr} \sin\left(kr - \frac{\pi}{2}l\right) P_l(\cos(\theta)) + \frac{1}{r} f(\theta) e^{ikr} \\ &= \sum_l C_l \frac{1}{kr} \sin\left(kr - \frac{\pi}{2}l + \delta_l\right) P_l(\cos(\theta)) \end{aligned} \tag{22.28}$$

Next we express both sides in terms of ingoing e^{-ikr}/r and outgoing e^{ikr}/r spherical waves, using the identity $\sin x = (e^{ix} - e^{-ix})/2i$, and we equate separately the terms which are proportional to e^{ikr} and e^{-ikr}. This gives two equations

$$\sum_{l=0}^{\infty} (2l + 1) i^l e^{-il\pi/2} P_l(\cos(\theta)) + 2ikf(\theta) = \sum_l C_l e^{i(\delta_l - l\pi/2)} P_l(\cos(\theta)) \tag{22.29}$$

and

$$\sum_{l=0}^{\infty} (2l + 1) i^l e^{il\pi/2} P_l(\cos(\theta)) = \sum_l C_l e^{-i(\delta_l - l\pi/2)} P_l(\cos(\theta)) \tag{22.30}$$

Since the $P_l(\cos(\theta))$ are orthogonal polynomials, coefficients of P_l on the left- and right-hand sides of equation (22.30) must be equal to one another, and therefore

$$(2l + 1) i^l e^{il\pi/2} = C_l e^{-i(\delta_l - l\pi/2)} \tag{22.31}$$

which determines all the C_l in terms of the partial wave phase shifts δ_l

$$C_l = (2l + 1) i^l e^{i\delta_l} \tag{22.32}$$

Substituting this expression for C_l into (22.29), and solving for $f(\theta)$, we find

$$\begin{aligned} f(\theta) &= \frac{1}{2\mathrm{i}k}\sum_l (2l+1)\mathrm{i}^l(\mathrm{e}^{2\mathrm{i}\delta_l}-1)\mathrm{e}^{-\mathrm{i}l\pi/2}P_l(\cos(\theta)) \\ &= \frac{1}{k}\sum_{l=0}^{\infty}(2l+1)\,\mathrm{e}^{\mathrm{i}\delta_l}\sin(\delta_l)P_l(\cos(\theta)) \end{aligned} \tag{22.33}$$

and we see that the amplitude $f(\theta)$ is entirely determined by the partial wave phase shifts $\{\delta_l\}$.

The total cross section σ has an especially simple form in terms of δ_l. From (4.39) the differential cross section is

$$\frac{\mathrm{d}\sigma}{\mathrm{d}\Omega} = \frac{1}{k^2}\left|\sum_{l=0}^{\infty}(2l+1)\,\mathrm{e}^{\mathrm{i}\delta l}\sin(\delta_l)P_l(\cos(\theta))\right|^2 \tag{22.34}$$

Then

$$\begin{aligned} \sigma &= \int_0^{\pi}\mathrm{d}\theta\sin\theta\int_0^{2\pi}\mathrm{d}\phi\frac{\mathrm{d}\sigma}{\mathrm{d}\Omega} \\ &= 2\pi\frac{1}{k^2}\int\mathrm{d}\theta\sin\theta\sum_{l_1}\sum_{l_2}(2l_1+1)\mathrm{e}^{-\mathrm{i}\delta_{l_1}}\sin(\delta_{l_1}) \\ &\quad\times(2l_2+1)\mathrm{e}^{\mathrm{i}\delta_{l_2}}\sin(\delta_{l_2})P_{l_1}(\cos(\theta))P_{l_2}(\cos(\theta)) \end{aligned} \tag{22.35}$$

Using the orthogonality of Legendre polynomials

$$\int_0^{\pi}\mathrm{d}\theta\sin\theta P_{l_1}(\cos(\theta))P_{l_2}(\cos(\theta)) = \frac{2}{(2l_1+1)}\delta_{l_1l_2} \tag{22.36}$$

the expression for σ simplifies to

$$\sigma = \frac{4\pi}{k^2}\sum_{l=0}^{\infty}(2l+1)\sin^2(\delta_l) \tag{22.37}$$

22.3.1 How many partial waves do we need?

Since both the total cross section (22.37) and the differential cross section (22.34) involve an infinite sum over l, there is a question of how many partial waves need to be included in the sum before $\sin(\delta_l)$ becomes negligible. A very rough estimate is that we can ignore values $l > kr_0$. The argument is as follows. Consider, in classical physics, an incoming particle with parameter b, as shown in figure 22.3. The angular momentum of the particle is simply $L = pb$. This means that particles with angular momentum $L > pr_0$ will not be scattered, since the scattering potential has a finite

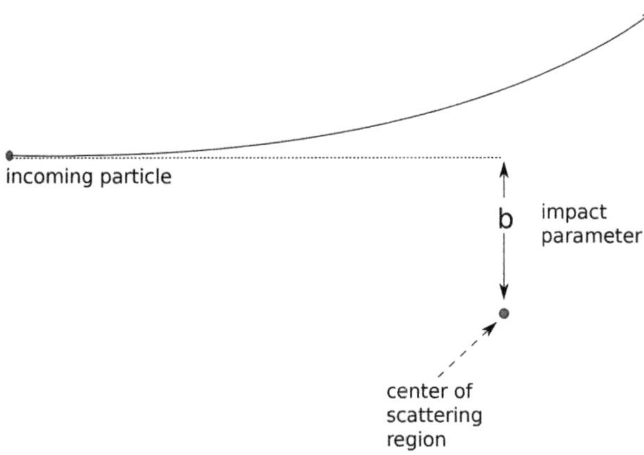

Figure 22.3. The impact parameter b.

range r_0. Quantum mechanically, $L = \sqrt{l(l+1)}\,\hbar$ or $L \approx l\hbar$ for large l. This suggests that scattering only occurs for partial waves with $l < pr_0/\hbar = kr_0$. As a practical matter, if partial wave phase shifts are difficult to compute, then the method of partial waves may only be useful at low energies, where just a few partial wave phase shifts (perhaps only one, at $l = 0$) need to be calculated.

22.3.2 Computing partial wave phase shifts

The idea is to solve the Schrödinger equation inside the scattering region, at $r < r_0$, and then match it to the free particle solution outside the scattering region, valid at $r > r_0$, at the interface r_0. We already know that in the exterior region, at $r \gg r_0$,

$$\begin{aligned}\psi^{\text{out}}(r, \theta) &= \sum_{l=0}^{\infty} C_l R_{kl}^{\text{out}}(r) P_l(\cos(\theta)) \\ &= \sum_{l=0}^{\infty} (2l+1) \mathrm{i}^l \mathrm{e}^{\mathrm{i}\delta_l} R_{kl}^{\text{out}}(r) P_l(\cos(\theta))\end{aligned} \tag{22.38}$$

where

$$R_{kl}^{\text{out}}(r) = \frac{\sin(kr - l\pi/2 + \delta_l)}{kr} \tag{22.39}$$

Equation (22.39) is just the asymptotic $kr \gg 1$ limit of the free particle solution of the radial equation,

$$R_{kl}^{\text{out}}(r) = \cos(\delta_l) j_l(kr) - \sin(\delta_l) n_l(kr) \tag{22.40}$$

and using this expression for $R_{kl}^{\text{out}}(r)$ in (22.38), the wave function ψ^{out} holds for all $r > r_0$. In the interior region, assuming a central potential $V(r)$ and azimuthal symmetry, the interior wave function must have the form

$$\psi^{\text{in}}(r,\theta) = \sum_{l=0}^{\infty} A_l R_{kl}^{\text{in}}(r) P_l(\cos(\theta)) \tag{22.41}$$

where $R_{kl}^{\text{in}}(r)$ is a solution of the radial equation at $r < r_0$, where the potential is non-zero.

Neither ψ^{in} nor ψ^{out} is normalized. This would not make sense anyway, since neither wave function is valid everywhere in space. On the other hand, we do know that a wave function and its first derivative should be everywhere continuous, and this means that ψ^{in}, ψ^{out}, and their first derivatives have to agree at $r = r_0$ for all θ. Then continuity of the wave function requires that, for each l,

$$A_l R_{kl}^{\text{in}}(r_0) = (2l+1) \mathrm{i}^l \mathrm{e}^{\mathrm{i}\delta_l} R_{kl}^{\text{out}}(r_0) \tag{22.42}$$

while continuity of the first derivative requires

$$A_l \left(\frac{\mathrm{d}R_{kl}^{\text{in}}}{\mathrm{d}r} \right)_{r=r_0} = (2l+1) \mathrm{i}^l \mathrm{e}^{\mathrm{i}\delta_l} \left(\frac{\mathrm{d}R_{kl}^{\text{out}}}{\mathrm{d}r} \right)_{r=r_0} \tag{22.43}$$

Dividing (22.43) by (22.42) results in the condition

$$\frac{1}{R_{kl}^{\text{in}}(r_0)} \left(\frac{\mathrm{d}R_{kl}^{\text{in}}}{\mathrm{d}r} \right)_{r=r_0} = \frac{1}{R_{kl}^{\text{out}}(r_0)} \left(\frac{\mathrm{d}R_{kl}^{\text{out}}}{\mathrm{d}r} \right)_{r=r_0} \tag{22.44}$$

or, equivalently, continuity of the derivative of the logarithm

$$\left(\frac{\mathrm{d}}{\mathrm{d}r} \log(R_{kl}^{\text{in}}(r)) \right)_{r=r_0} = \left(\frac{\mathrm{d}}{\mathrm{d}r} \log(R_{kl}^{\text{out}}(r)) \right)_{r=r_0} \tag{22.45}$$

where

$$R_{kl}^{\text{out}}(r) = \cos(\delta_l) j_l(kr) - \sin(\delta_l) n_l(kr) \tag{22.46}$$

Equations (22.45) and (22.46) determine the partial wave phase shifts δ_l, and from those the differential cross section. The catch is that it is necessary to solve the radial equation for $R_{kl}^{\text{in}}(r)$ in the scattering region, for as many l values as may be required. This might require a numerical solution, so if many partial wave phase shifts are required, the computational effort may be considerable. On the other hand, at low energies only a few l values (perhaps only $l = 0$) are needed.

22.3.3 Example: low energy scattering from a hard sphere

The potential of an impenetrable hard sphere of radius r_0 is much like that of an infinite square well:

$$V(r) = \begin{cases} \infty & r \leqslant r_0 \\ 0 & r > r_0 \end{cases} \tag{22.47}$$

As in the case of the infinite square well, the wave function vanishes in the region where the potential is infinite, so $\psi(r, \theta) = 0$ at $r < r_0$. We also know that infinite potentials are an exception to the rule that the first derivative of the wave function is continuous. The wave function itself must be continuous at $r = r_0$, but the first derivative is not. So the general method just outlined actually does not work in this case. Fortunately, the condition that the wave function itself is continuous at $r = r_0$ is sufficient, because it implies that

$$\begin{aligned} R_{kl}^{\mathrm{out}}(r_0) &= \cos(\delta_l)j_l(kr_0) - \sin(\delta_l)n_l(kr_0) \\ &= 0 \end{aligned} \tag{22.48}$$

or

$$\tan(\delta_l) = \frac{j_l(kr_0)}{n_l(kr_0)} \tag{22.49}$$

In the low energy limit

$$kr_0 \ll 1 \tag{22.50}$$

where the de Broglie wavelength of the incoming particle is much larger than the radius of the target, we can use the approximations

$$\begin{aligned} j_l(kr_0) &\propto (kr_0)^l \\ n_l(kr_0) &\propto \frac{1}{(kr_0)^{l+1}} \end{aligned} \tag{22.51}$$

and therefore

$$\tan(\delta_l) \propto (kr_0)^{2l+1} \tag{22.52}$$

This means that the leading contribution to the scattering amplitude at low energies comes from the $l = 0$ partial wave. Using the exact form of j_0, n_0 we have

$$\tan(\delta_0) = -\tan(kr_0) \tag{22.53}$$

so that for $kr_0 \ll 1$

$$\sigma \approx \frac{4\pi}{k^2}\sin^2(kr_0) \tag{22.54}$$

In the $k \to 0$ limit, using the small angle approximation of the sine function, this becomes

$$\sigma = 4\pi r_0^2 \tag{22.55}$$

Notice that at low energies the total cross section is four times the geometric cross section of a sphere, which of course is the area πr_0^2 of a circle of radius r_0. This reflects the fact that, as in optics, waves bend around objects whose dimensions are small compared to the wavelength. The geometric cross section is actually the answer for the total cross section in the opposite (high energy) limit $kr_0 \gg 1$ where the de Broglie wavelength λ of the incoming particle is much smaller than r_0. In that case the incoming wave, as in optics, would instead cast a shadow of area πa^2. But at high energies many partial wave phase shifts (up to $l_{\max} \approx kr_0$) must be included in the sum to get this result.

22.3.4 Example: low energy scattering from an attractive potential well

Instead of an impenetrable sphere, let us suppose that the scattering region is a spherical well of some finite depth

$$V(r) = \begin{cases} -V_0 & r \leqslant r_0 \\ 0 & r > r_0 \end{cases} \tag{22.56}$$

and this time the continuity equation (22.45) can be applied. We will again consider the low energy limit $kr_0 \ll 1$, and just compute the $l = 0$ partial wave phase shift. We recall that the radial equation

$$\left[\frac{1}{r}\frac{\mathrm{d}^2}{\mathrm{d}r^2}r - \frac{l(l+1)}{r^2} + k^2 - \frac{2mV(r)}{\hbar^2}\right]R_{kl} = 0 \tag{22.57}$$

(where $k = \sqrt{2mE}/\hbar$) is simplified by setting $u = R_{kl}/r$. For $l = 0$ the equation becomes

$$\frac{\mathrm{d}^2 u}{\mathrm{d}r^2} = -\frac{2m}{\hbar^2}(E - V(r))u \tag{22.58}$$

Let $u^{\mathrm{in}}(r)$ denote the interior solution at $r \leqslant r_0$, and $u^{\mathrm{out}}(r)$ denote the outer solution at $r \geqslant r_0$. Then, since $V(r) = -V_0$ at $r \leqslant r_0$, we have the general interior solution

$$u^{\mathrm{in}}(r) = A\sin(k'r) + B\cos(k'r) \quad \text{where} \quad k' = \sqrt{2m(E + V_0)}/\hbar \tag{22.59}$$

and therefore

$$R_{k0}^{\mathrm{in}}(r) = A\frac{\sin(k'r)}{r} + B\frac{\cos(k'r)}{r} \tag{22.60}$$

which is actually a superposition of spherical Bessel $j_0(k'r)$ and Neumann $n_0(k'r)$ functions. However, just as in the case of the free particle, the spherical Neumann function is singular at the origin, and its first derivative is discontinuous, which is impossible in a finite potential. Therefore we must set $B = 0$. There is no such restriction for the exterior solution, where we have

$$\begin{aligned} R_{k0}^{\mathrm{out}}(r) &= c_1 j_0(kr) + c_2 n_0(kr) \\ &= C_0\frac{\sin(kr + \delta_0)}{r} \end{aligned} \tag{22.61}$$

Now we impose the continuity equation (22.44)

$$\frac{1}{R^{\text{in}}_{k0}(r_0)}\left(\frac{\mathrm{d}R^{\text{in}}_{k0}(r)}{\mathrm{d}r}\right)_{r=r_0} = \frac{1}{R^{\text{out}}_{k0}(r_0)}\left(\frac{\mathrm{d}R^{\text{out}}_{k0}(r)}{\mathrm{d}r}\right)_{r=r_0} \tag{22.62}$$

which gives the condition

$$k' \cot(k'r_0) = k \cot(kr_0 + \delta_0) \tag{22.63}$$

If δ_0 would remain finite as $k \to 0$, then the left-hand side of this equation would be finite, while the right-hand side goes to zero. Therefore $\delta_0 \to 0$ in the $k \to 0$ limit, and in this limit we can use the small angle approximation of the cotangent $\cot(x) \approx 1/x$. Then

$$k' \cot(k'r_0) \approx \frac{k}{kr_0 + \delta_0} \tag{22.64}$$

and solving for δ_0

$$\delta_0 = kr_0\left(\frac{\tan(k'r_0)}{k'r_0} - 1\right) \tag{22.65}$$

Using the small angle approximation $\sin(x) \approx x$ we find for the total cross section

$$\begin{aligned}\sigma &\approx \frac{4\pi}{k^2}\delta_0^2 \\ &= 4\pi r_0^2\left(\frac{\tan(k'r_0)}{k'r_0} - 1\right)^2\end{aligned} \tag{22.66}$$

Ramsauer effect

An interesting special case of this formula occurs when the incoming energy $E = \hbar^2k^2/2m$ is such that

$$\tan(k'r_0) = k'r_0 \quad \Longrightarrow \quad \sigma = 0 \tag{22.67}$$

in which case the total cross section vanishes, assuming that the low energy limit applies. When this condition is satisfied, we can see from (22.63) that

$$\begin{aligned}\tan(kr_0 + \delta_0) &= \frac{k \tan(k'r_0)}{k'} \\ &= kr_0\end{aligned} \tag{22.68}$$

which implies in the low energy $kr_0 \ll 1$ limit that

$$\tan(\delta_0) \approx 0 \quad \Longrightarrow \quad \delta_0 = 0, \pi, 2\pi, \ldots \tag{22.69}$$

We have seen a very similar condition before, in chapter 8, for scattering by an attractive well in one dimension. In that case it was found that when precisely an integer number of de Broglie half-wavelengths fits inside the well, the reflection coefficient vanishes, and all particles are transmitted. In atomic physics this phenomenon is known as the Ramsauer effect, which is the fact that the scattering of electrons by atoms in a solid is minimized at certain definite electron energies. An atom is of course not a spherical well of finite depth, but the fact that an atom is electrically neutral means than it is seen by an incoming electron as a finite range potential

22.4 Resonance scattering and subatomic particles

The electron and the proton were the first subatomic particles to be discovered, and neither decays into other particles. Both of them are stable. Because these particles are charged, they leave tracks—essentially trails of ionized molecules—in detectors which can display such tracks, such as cloud chambers, bubble chambers, and drift chambers. All other charged particles are unstable, but some of them, e.g. muons, pions, kaons, Σ hyperons, all of which decay via the weak interaction, have lifetimes on the order of 10^{-10} seconds. For a particle traveling at nearly the speed of light, this lifetime is long enough for the particle to leave a track in a detector before decaying into other particles. On the other hand, beginning in the early 1960s, a host of other particles were discovered at particle accelerator facilities, and these particles have lifetimes on the order of 10^{-24} seconds, which is far too short to leave any detectable track. Nevertheless the existence, mass, and lifetimes of such particles have been determined from plots of scattering cross sections versus center-of-mass collision energy. An example of such a plot, for the process

$$\text{electron} + \text{positron} \longrightarrow \text{hadrons} \tag{22.70}$$

is shown in figure 22.4. The very short-lived particles indicated on this plot correspond to bumps and narrow peaks in the total scattering cross section. The masses of these unstable particles are given by energies where the bumps are highest, and the lifetimes of these particles are inversely proportional to the width of the bumps. Some of the peaks, e.g. for the J/ψ, are so narrow that their widths are not visible on the scale of this plot; such particles are comparatively long-lived. Others, such as the ρ, are fairly wide, meaning that these particles decay quite rapidly. But how do we know this? Why should peaks or bumps in a scattering cross section be associated with unstable particles, and what does the width of the peak have to do with the lifetime of the particle?

A full answer to those questions is beyond the scope of this text, but I will try to make the correspondence plausible with a simple example in non-relativistic quantum physics. Let us consider a particle incident on a δ-function scattering potential (a 'thin spherical shell')

$$V(r) = V_0\delta(r - r_0) \tag{22.71}$$

and we will calculate the contribution to the total cross section due to the $l = 0$ partial wave

$$\sigma_0 = \frac{4\pi}{k^2}\sin^2\delta_0 \tag{22.72}$$

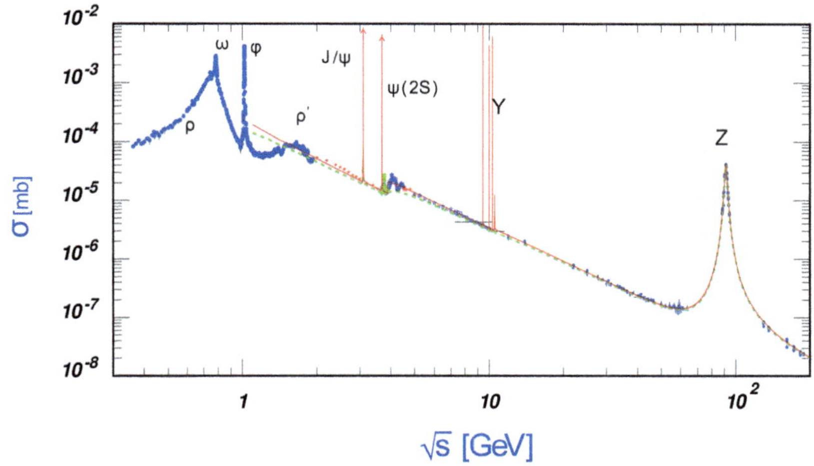

Figure 22.4. A plot of total cross section versus center of mass energy ($\sqrt{s}$) for electron–positron collisions. The broad peaks and narrow spikes indicate the existence of unstable subatomic particles, with masses corresponding to the center of mass energy at the peaks, and lifetimes inversely proportional to the width of the peaks. Image from Olive *et al* (Particle Data Group) 2014 *Chin. Phys.* C **38** 090001, courtesy of the Particle Data Group.

As before, this involves matching the interior radial wave function $R_{k0}^{\text{in}}(r < r_0)$ with the exterior solution $R_{k0}^{\text{out}}(r > r_0)$ at the boundary $r = r_0$. Now at $r < r_0$ the potential is zero, so the solution is just

$$R_{k0}^{\text{in}}(r) = Aj_0(kr) = A\frac{\sin(kr)}{kr} \tag{22.73}$$

Note that we have excluded the spherical Neumann function, which would be singular at $r = 0$. In the exterior region, however, we need to include the spherical Bessel and Neumann functions

$$\begin{aligned} R_{k0}^{\text{out}}(r) &= aj_0(kr) + bn_0(kr) \\ &= a\frac{\sin(kr)}{kr} - b\frac{\cos(kr)}{kr} \\ &= B\frac{\sin(kr + \delta_0)}{kr} \end{aligned} \tag{22.74}$$

Continuity of the wave function $R_{k0}^{\text{in}}(r_0) = R_{k0}^{\text{out}}(r_0)$ at $r = r_0$ gives us

$$A\sin(kr_0) = B\sin(kr_0 + \delta_0) \tag{22.75}$$

However, the continuity condition on the first derivative is a little trickier, because the derivative of a wave function is discontinuous at a delta function singularity. To get the condition correctly, we have to go back to the radial equation.

For $l = 0$ and $R = u/r$, and the potential (22.71), the radial equation reduces to

$$-\frac{\hbar^2}{2m}\frac{\mathrm{d}^2 u}{\mathrm{d}r^2} + V_0\delta(r - r_0)u(r) = E\mathrm{u}(r) \tag{22.76}$$

which we write in the form

$$\frac{\mathrm{d}}{\mathrm{d}r}\frac{\mathrm{d}u}{\mathrm{d}r} = -\frac{2\,mE}{\hbar^2}u(r) + \frac{2\,mV_0}{\hbar^2}\delta(r - r_0)u \tag{22.77}$$

Integrate both sides of the equation between $r = r_0 - \epsilon$ and $r = r_0 + \epsilon$

$$\begin{aligned}\int_{r_0-\epsilon}^{r=r_0+\epsilon} \mathrm{d}r\frac{\mathrm{d}}{\mathrm{d}r}\frac{\mathrm{d}u}{\mathrm{d}r} &= -\int_{r_0-\epsilon}^{r=r_0+\epsilon} \mathrm{d}r\frac{2\,mE}{\hbar^2}u(r) \\ &\quad + \int_{r_0-\epsilon}^{r=r_0+\epsilon} \mathrm{d}r\frac{2\,mV_0}{\hbar^2}\delta(r - r_0)u(r) \\ \left(\frac{\mathrm{d}u}{\mathrm{d}r}\right)_{r_0+\epsilon} - \left(\frac{\mathrm{d}u}{\mathrm{d}r}\right)_{r_0-\epsilon} &= O(\epsilon) + \frac{2\,mV_0}{\hbar^2}u(r_0)\end{aligned} \tag{22.78}$$

Then in the $\epsilon \to 0$ limit we have the condition for the discontinuity in slopes

$$kB\cos(kr_0 + \delta_0) - kA\cos(kr_0) = \frac{2\,mV_0}{\hbar^2}A\sin(kr_0) \tag{22.79}$$

which we rearrange a little bit,

$$A\left(\cos(kr_0) + \frac{\alpha}{kr_0}\sin(kr_0)\right) = B\cos(kr_0 + \delta_0) \tag{22.80}$$

where we have defined the dimensionless parameter

$$\alpha \equiv \frac{2\,mV_0r_0}{\hbar^2} \tag{22.81}$$

Now divide the left- and right-hand side of equation (22.80) by the left- and right-hand side of equation (22.75), and we find

$$\cot(kr_0) + \frac{\alpha}{kr_0} = \cot(kr_0 + \delta_0) \tag{22.82}$$

Then solve for δ_0:

$$\delta_0 = -kr_0 + \cot^{-1}\left[\cot(kr_0) + \frac{\alpha}{kr_0}\right] \tag{22.83}$$

We insert this solution into (22.72) to derive the $l = 0$ contribution to the total cross section, as a function of k.

The animation in figure 22.5 displays how the plot of σ_0/r_0^2 versus kr_0 changes as α increases (i.e. the barrier V_0 increases) from $\alpha = 0$ to $\alpha = 5$. In this range the shape of scattering cross section looks similar to hard sphere scattering, except that this cross section vanishes (as it must) in the $V_0 \to 0$ limit. But now look at the animation in figure 22.6, with α in the range $\alpha = 5$ to $\alpha = 50$. We see that, in this range, bumps develop in the σ_0 cross section, which become very narrow peaks, and eventually thin spikes, as α becomes large. If we look closely at the location of those spikes, it turns

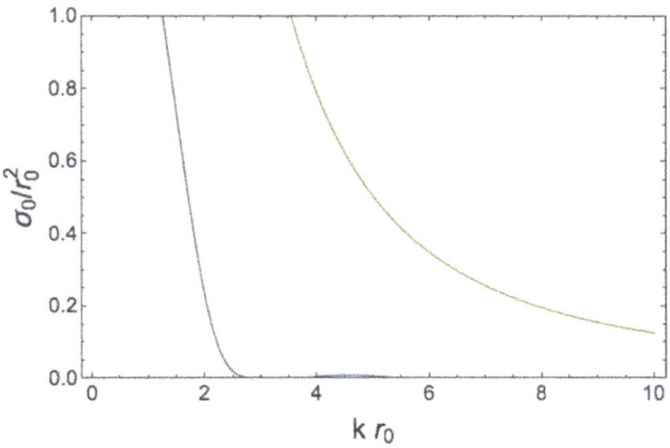

Figure 22.5. The $l = 0$ contribution to the total scattering cross section σ_0 from a thin shell potential of radius r_0, as a function of kr_0, which is proportional to incoming particle momentum. The animation shows how σ_0 versus kr_0 changes as the strength of the thin shell potential, described by the dimensionless parameter α, increases from $\alpha = 0$ to $\alpha = 5$.

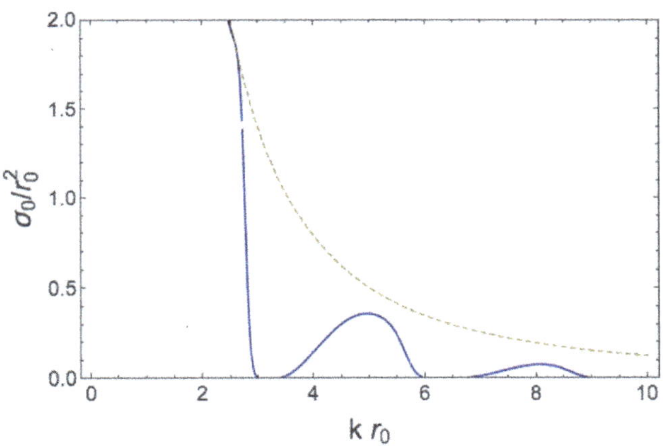

Figure 22.6. Same as figure 22.5, but for stronger potentials in the range $5 < \alpha < 50$. Note how 'spikes' appear in an otherwise smooth plot of scattering cross section, as the potential strength increases. The dashed line shown for comparison is $4\pi/k^2$, corresponding to $\sin(\delta_0) = \pm 1$.

out that they occur very near the values $kr_0 = n\pi$ where n is a positive integer, and in fact converge to exactly those values as $\alpha \to \infty$. This is very striking, and there must be some physics underlying those special locations.

Let us consider the limit V_0, $\alpha \to \infty$. In this limit, the discontinuity condition (22.80) for $l = 0$ states can only be satisfied if

$$A\frac{\sin(kr_0)}{kr_0} = 0 \tag{22.84}$$

which has the solutions

$$kr_0 = n\pi, \qquad n = 1, 2, 3, \ldots \tag{22.85}$$

By continuity, the external wave function $R^{\text{out}}(r)$ must vanish at $r = r_0$, and it is consistent for this wave function to vanish at all $r \geqslant r_0$. This is an energy eigenstate, and therefore stationary. If a particle is located inside the spherical shell, it will never get outside the spherical shell. This is therefore a bound state, and the bound state energies $E_n = \hbar^2 k_n^2$ for $l = 0$ angular momentum are determined by the condition $k_n r_0 = n\pi$. In other words, the positions of the spikes in the $l = 0$ scattering cross section precisely correspond to the bound state energies in the $V_0 \to \infty$ limit.

But what about finite V_0? In this case there is no bound state solution; an energy eigenstate must extend over all space, and the scattering cross section has finite width peaks, rather than spikes, at locations near but not exactly at $kr_0 = n\pi$. The animation in figure 22.7 focuses on the first peak in the interval $\alpha = 10$ to $\alpha = 100$.

It turns out that the height of this peak, and in fact all of the $l = 0$ other peaks, go in the $\alpha \to \infty$ limit to

$$\sigma_0 = \frac{4\pi}{k_n^2} \tag{22.86}$$

where k_n is the location of the nth peak. This of course implies that $\sin^2(\delta_0(E_R)) = 1$ where E_R is the energy where the peak occurs. This is true in general, and not just for this particular potential: a peak in any partial wave contribution σ_l to the total cross section corresponds to an energy at which $\sin^2(\delta_l) = 1$. From this fact, we can deduce the shape of the peak, independent of the details of the potential. We have

$$\begin{aligned} \sigma_l(E) &= \frac{4\pi}{k^2}(2l + 1)\sin^2(\delta_l) \\ &= \frac{4\pi}{k^2}(2l + 1)\frac{1}{1 + \cot^2(\delta_l(E))} \end{aligned} \tag{22.87}$$

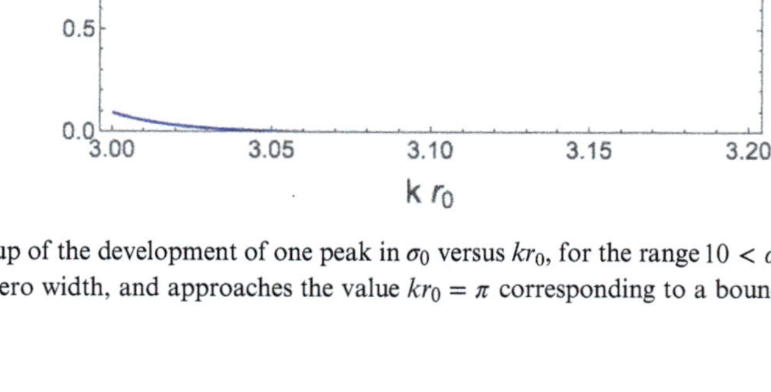

Figure 22.7. Closeup of the development of one peak in σ_0 versus kr_0, for the range $10 < \alpha < 100$. As $\alpha \to \infty$, the peak tends to zero width, and approaches the value $kr_0 = \pi$ corresponding to a bound state in an infinite spherical well.

Suppose that $\sin^2(\delta_l) \to 1$ at $E = E_R$. Then we can Taylor expand the cotangent for energies in the vicinity of $E = E_R$ as

$$\cot(\delta_l(E)) \approx \frac{2}{\Gamma}(E - E_R) \quad \text{where} \quad \frac{2}{\Gamma} = \left(\frac{\mathrm{d}}{\mathrm{d}E}\cot(\delta_l(E))\right)_{E=E_R} \tag{22.88}$$

which gives us the *Breit–Wigner formula* for the scattering cross section in the vicinity of a peak:

$$\sigma_l(E) = \frac{4\pi}{k^2}(2l+1)\frac{\frac{1}{4}\Gamma^2}{(E - E_R)^2 + \frac{1}{4}\Gamma^2} \tag{22.89}$$

Note that $\sigma_l(E)$ falls to half its peak value at $|E - E_R| = \frac{1}{2}\Gamma$, assuming k does not vary much in this region. In this sense, Γ is the width of the peak. Note also that the Breit–Wigner expression has exactly the same form as the amplitude of a forced, damped harmonic oscillator in the vicinity of the resonant frequency, with the correspondence being $E \to \omega$, where ω is the angular frequency of the driving force, $E_R \to \omega_R$ where ω_R is the resonant frequency, and Γ is proportional to a damping term. For this reason, peaks in the scattering amplitude are referred to as 'resonances'.

Let us return to our example. At finite V_0 there are no true bound states. Any particle placed in the region $r < r_0$ will eventually tunnel through the barrier. But the key word is 'eventually'. Suppose the barrier is very high. Then the normalized state

$$\psi = \begin{cases} \dfrac{1}{\sqrt{2\pi r_0}}\dfrac{\sin(n\pi r/r_0)}{r} & r < r_0 \\ 0 & r \geqslant r_0 \end{cases} \tag{22.90}$$

is almost an energy eigenstate, but not quite. It satisfies the time-independent Schrödinger equation at $r < r_0$ and at $r > r_0$, but not at $r = r_0$, where the slope discontinuity condition is slightly violated. This means that if a particle is placed in this state at time $t = 0$, it will slowly evolve, and the wave function will start to 'leak' into the region $r > r_0$. But any time-dependent state must have an energy uncertainty ΔE, and we know from the time-energy uncertainty relation that the minimum time for any observable to change by its uncertainty is $\tau = \hbar/\Delta E$. We may consider, for example, the expectation value of the radial position, and the uncertainty in the radial position in the state (22.90), which is easily calculated from the usual formulas to be

$$\begin{aligned} \langle r\rangle &= \frac{1}{2}r_0 \\ \Delta r &= r_0\left(\frac{1}{12} - \frac{1}{2n^2\pi^2}\right)^{\frac{1}{2}} \end{aligned} \tag{22.91}$$

As the wave function expands outside the sphere of radius $r = r_0$, the expectation value of the particle radial coordinate will increase. From the time-energy uncertainty relation, the time required for $\langle r \rangle$ to increase by Δr is at least $\hbar/\Delta E$. We know that if $V_0 = \infty$, the wave function is an energy eigenstate, and $\Delta E = 0$. If V_0 is finite but extremely large, then ΔE is non-zero but small, and τ, which is a rough measure of the time it takes for the wave function to expand appreciably beyond the potential shell, will be quite large. When this is the case, we can say that the state described by (22.90) is 'metastable', and τ is the approximate decay time, or 'lifetime', of the metastable state.

Now if metastable states of this kind are associated with peaks in the scattering amplitude, how broad should these peaks be or, to put it another way, which energies might be affected by the existence of a metastable state? It is reasonable that if a particle wavepacket, as it evolves in time, has no overlap whatever with a metastable state at any time, then the metastable state is irrelevant to particle scattering. But a particle wavepacket with a fairly definite energy can only have a substantial overlap with a metastable state if the energy of the incoming particle lies within the range of energies, of width ΔE, of the metastable state (otherwise the states are orthogonal). This means that if metastable states are somehow responsible for an enhancement of the scattering cross section, they can only appreciably affect particles whose energies lie in the range $E_R - \frac{1}{2}\Delta E < E < E_R + \frac{1}{2}\Delta E$. It follows that the energy uncertainty ΔE of the metastable state must be approximately the width of the peak, and therefore the width of the peak is approximately inverse to the lifetime τ of the metastable state. The precise statement, which will not be proved here, is that

$$\tau = \frac{\hbar}{\Gamma} \tag{22.92}$$

This brings us back to subatomic physics. The peaks in figure 22.4 correspond to unstable particles, or 'resonances' that are produced in electron–positron scattering, and the widths of those peaks, when fit to the Breit–Wigner form, provide us with the lifetimes of those particles. The center of mass energy at the maxima of these peaks gives us the masses (from $E = mc^2$) of these unstable particles, with the understanding that unstable particles cannot be said to have a definite mass, because of the energy uncertainty ΔE associated with their finite lifetimes.

22.5 The Born approximation

The second approach to computing differential cross sections is known as the Born approximation, and it amounts to a perturbative approach to computing energy eigenstates relevant to scattering. We assume again that the interaction potential is spherically symmetric, and write the Schrödinger equation in the form

$$(\nabla^2 + k^2)\psi = \frac{2m}{\hbar^2} V(r)\psi \tag{22.93}$$

where again $k = \sqrt{2mE}/\hbar$. Scattering solutions of the Schrödinger equation are also solutions of the integral equation

$$\psi(\mathbf{r}) = \mathrm{e}^{ikz} - \frac{m}{2\pi\hbar^2}\int \mathrm{d}^3r' \frac{\mathrm{e}^{ik|\mathbf{r}-\mathbf{r}'|}}{|\mathbf{r}-\mathbf{r}'|} V(r')\psi(\mathbf{r}') \tag{22.94}$$

The Born approximation is derived by solving (22.94) iteratively.

To prove that a solution of (22.94) is a solution of the Schrödinger equation, we consider

$$\psi(\mathbf{r}) = \mathrm{e}^{ikz} + \frac{2m}{\hbar^2}\int \mathrm{d}^3r' G(r-r')V(r')\psi(\mathbf{r}') \tag{22.95}$$

Operating on both sides of this equation with $\nabla^2 + k^2$, it is obvious that $\psi(\mathbf{r})$ is a solution of the Schrödinger equation providing that $G(r - r')$ is chosen such that

$$(\nabla^2 + k^2)G(r - r') = \delta^3(r - r') \tag{22.96}$$

In fact this equation has two linearly independent solutions

$$G(r - r') = -\frac{1}{4\pi}\frac{\mathrm{e}^{+\mathrm{i}k|r-r'|}}{|r-r'|} \quad \text{and} \quad -\frac{1}{4\pi}\frac{\mathrm{e}^{-\mathrm{i}k|r-r'|}}{|r-r'|} \tag{22.97}$$

so, once again, we have to choose the physically relevant solution. This is the first solution, proportional to exponent $\exp[+\mathrm{i}k|r - r'|]$, because this solution will correspond to a scattering state with waves propagating outwards, away from the scattering region.

The simplest way to check that this choice of $G(r - r')$ satisfies (22.96) is by direct substitution, using identities such as

$$\nabla|r - r'| = \frac{\mathbf{r}-\mathbf{r}'}{|r-r'|} \quad \text{and} \quad \nabla\frac{1}{|r-r'|} = -\frac{\mathbf{r}-\mathbf{r}'}{|r-r'|^3} \tag{22.98}$$

We find that everything cancels out except for one term,

$$(\nabla^2 + k^2)(-1)\frac{\mathrm{e}^{\mathrm{i}k|r-r'|}}{4\pi|r-r'|} = -\frac{\mathrm{e}^{\mathrm{i}k|r-r'|}}{4\pi}\nabla^2\frac{1}{|r-r'|} \tag{22.99}$$

and for this we can use the identity

$$\nabla^2\frac{1}{|r-r'|} = -4\pi\delta^3(\mathbf{r}-\mathbf{r}') \tag{22.100}$$

This last identity can be proven in various ways, but to see that it must be true we can refer back to electrostatics, and Gauss's Law $\nabla\cdot\mathbf{E} = \rho$, with ρ the charge density, and $\mathbf{E} = -\nabla\phi$ where ϕ is the electric potential. Then ϕ satisfies Poisson's equation $\nabla^2\phi = -\rho$, and for a point charge at r', where $\rho(r) = q\delta^3(\mathbf{r}-\mathbf{r}')$, we have the Coulomb potential

$$\phi(r) = \frac{q}{4\pi|r - r'|} \tag{22.101}$$

Given that $\phi(r)$ satisfies Poisson's equation (22.100) must be true, and from this we conclude that (22.96) is satisfied.

All solutions of the integral equation (22.94) are therefore solutions of the Schrödinger equation, but the converse is not true. The integral equation singles out a particular class of solutions, namely, those which become plane waves e^{ikz} at $V(r) = 0$, and which contain only radially outgoing scattered waves for non-zero potentials. Those are exactly the solutions we require to describe scattering.

The integral equation is solved by the methods of perturbation theory. We begin by writing the equation in the form

$$\psi(\mathbf{r}) = e^{i\mathbf{k}\cdot\mathbf{r}} + \lambda \int d^3r' G(r - r')W(r')\psi(\mathbf{r}') \tag{22.102}$$

where

$$\lambda W(r) \equiv \frac{2m}{\hbar^2} V(r) \tag{22.103}$$

and $\mathbf{k} = k\hat{\mathbf{e}}_z$ is a vector with magnitude k oriented in the z-direction. The constant λ might be some constant present in $V(r)$, e.g. the squared electric charge e^2, which controls the strength of the potential, or it might be just some parameter which we introduce for convenience, and set to $\lambda = 1$ at the end of the calculation. As in time-independent perturbation theory, we expand $\psi(\mathbf{r})$ in a power series

$$\psi = \psi^{(0)} + \lambda\psi^{(1)} + \lambda^2\psi^{(2)} + \lambda^3\psi^{(3)} + \cdots \tag{22.104}$$

and insert into (22.102). Equating like powers of λ on the left- and right-hand sides of (22.102), we see that

$$\begin{aligned} \psi^{(0)}(\mathbf{r}) &= e^{i\mathbf{k}\cdot\mathbf{r}} \\ \psi^{(n+1)}(\mathbf{r}) &= \int d^3r' G(r - r')W(r')\psi^{(n)}(\mathbf{r}') \end{aligned} \tag{22.105}$$

or, schematically,

$$\psi^{(n+1)} = \int GW\psi^{(n)} \tag{22.106}$$

This gives us a solution for ψ as an infinite power series in λ

$$\begin{aligned} \psi = \psi_0 &+ \lambda \int GW\psi_0 + \lambda^2 \int GW \int GW\psi_0 + \lambda^3 \int GW \int GW \int GW\psi_0 \\ &+ \lambda^4 \int GW \int GW \int GW \int GW\psi_0 + \cdots \end{aligned} \tag{22.107}$$

which is known as the ***Born series***. Written out in full, the $O(\lambda^n)$ term in the series is

$$\lambda^n \int d^3r_1 G(r - r_1) W(r_1) \int d^3r_2 G(r_1 - r_2) W(r_2) \int d^3r_3 G(r_2 - r_3) W(r_3)$$
$$\dots \int d^3r_{n-1} G(r_{n-2} - r_{n-1}) W(r_{n-1}) \int d^3r_n G(r_{n-1} - r_n) W(r_n) e^{i\mathbf{k}\cdot\mathbf{r}_n} \tag{22.108}$$

Now suppose that $V(r)$ is a weak potential, in the sense that each term in the Born series is smaller than the term before. Then we can approximate the solution to the integral equation (22.94) equation the way we approximate any power series, i.e. by throwing away all terms beyond some order. Throwing away all terms involving V would mean there is no scattering at all, but for a sufficiently weak potential it may be sufficient to truncate the series after first order,

$$\psi = \psi_0 + \lambda \int GW\psi_0 \tag{22.109}$$

or, in full,

$$\psi(\mathbf{r}) = e^{i\mathbf{k}\cdot\mathbf{r}} + \frac{2m}{\hbar^2} \int d^3r' G(r - r') V(r') e^{i\mathbf{k}\cdot\mathbf{r}'} \tag{22.110}$$

If more accuracy is desired, more terms in the Born series can be kept.

We now want to use (22.110) to work out the differential cross section, and this means identifying (22.110) with (22.15) in the region where (22.15) is valid, namely very far from the scattering region. Inside the integral of (22.110) only points where $r' < r_0$ contribute, where r_0 is the range of the potential, and we are interested in calculating $\psi(\mathbf{r})$ at points $r \gg r'$, because this is where we place the detector. Denote the angle between $\mathbf{r}$ and $\mathbf{r}'$ by θ'. This is an integration variable which varies with $\mathbf{r}'$ at fixed $\mathbf{r}$, and should not be confused with the angle θ between $\mathbf{r}$ and the z-axis. Then for $r \gg r_0$ we can approximate

$$\begin{aligned} |\mathbf{r} - \mathbf{r}'| &= \sqrt{r^2 + r'^2 - 2rr'\cos(\theta')} \\ &\approx r - \hat{\mathbf{e}}_r \cdot \mathbf{r}' \end{aligned} \tag{22.111}$$

Then

$$G(\mathbf{r} - \mathbf{r}') = -\frac{1}{4\pi} \frac{\exp[ik(r - \hat{\mathbf{e}}_r \cdot \mathbf{r}']}{r} \tag{22.112}$$

The reason that we keep the term $r - \hat{\mathbf{e}}_r \cdot \mathbf{r}'$ in the numerator but not the denominator is that even though $r \gg r'$, nevertheless the term $\exp[-ik\hat{\mathbf{e}}_r \cdot \mathbf{r}']$ in the numerator can vary wildly between -1 and 1 for kr' of $O(1)$ or greater, regardless of the value of kr. This is not true of the denominator.

Let us define

$$\mathbf{k}' = k\hat{\mathbf{e}}_r \tag{22.113}$$

Then for $r \gg r'$

$$\psi(\mathbf{r}) = e^{ikz} - \left\{ \frac{m}{2\pi\hbar^2} \int d^3r' e^{i(\mathbf{k}-\mathbf{k}')\cdot\mathbf{r}'} V(r') \right\} \frac{e^{ikr}}{r} \tag{22.114}$$

and by comparison with (22.15) we find

$$f(\theta) = -\frac{m}{2\pi\hbar^2}\int d^3r' e^{i(\mathbf{k}-\mathbf{k}')\cdot\mathbf{r}'} V(r') \tag{22.115}$$

This is the Born approximation for the scattering amplitude.

Assuming that $V(r)$ is spherically symmetric, the Born approximation can be simplified. Define

$$\mathbf{q} = \mathbf{k} - \mathbf{k}' \tag{22.116}$$

Since $\hbar\mathbf{k}$ is the momentum of the incoming particle, and $\hbar\mathbf{k}'$ is the momentum of the scattered particle moving in the $\hat{\mathbf{e}}_r$-direction, the vector $\hbar\mathbf{q}$ has the interpretation of a momentum transfer. Since the angle between $\mathbf{k}$ and $\mathbf{k}'$ is the scattering angle θ, and $|\mathbf{k}| = |\mathbf{k}'| = k$, we have

$$\begin{aligned} q^2 &= (\mathbf{k} - \mathbf{k}')\cdot(\mathbf{k} - \mathbf{k}') = 2k^2(1 - \cos(\theta)) \\ &= 4k^2 \sin^2\left(\frac{\theta}{2}\right) \\ q &= 2\,k \sin\left(\frac{\theta}{2}\right) \end{aligned} \tag{22.117}$$

Now transform to spherical coordinates (r', θ', ϕ') where we choose θ' to be the angle between $\mathbf{r}'$ and $\mathbf{q}$, and ϕ' the angle in a plane perpendicular to $\mathbf{q}$:

$$f(\theta) = -\frac{m}{2\pi\hbar^2}\int_0^\infty dr' r'^2 V(r') \int_0^\pi d\theta' \sin(\theta') \int_0^{2\pi} d\phi' e^{iqr'\cos(\theta')} \tag{22.118}$$

The integral over ϕ' is trivial, and the integral over θ' becomes trivial after a change of variables $u = \cos(\theta')$. The final result is

$$\boxed{\begin{aligned} f(\theta) &= -\frac{2\,m}{\hbar^2 q}\int_0^\infty dr \;\; r \sin(qr) V(r) \\ q &= 2\,k \sin\left(\frac{\theta}{2}\right) \end{aligned}}$$

which is the Born approximation for the scattering of a particle in a central potential.

Example: the shielded Coulomb potential

An example of great historical interest is the scattering of particles from the shielded Coulomb potential, also known as the 'Yukawa potential'

$$V(r) = -V_0 \frac{e^{-ar}}{r} \tag{22.119}$$

This is the electrostatic potential of an atom, as seen by an incident charged particle. An atom is electrically neutral, so of course the potential falls rapidly to zero beyond the atomic radius (of order $1/a$). Close to the positively charged nucleus, on the other hand, the shielding effect of the electrons is absent, and the electrostatic potential is simply the Coulomb potential $-V_0/r$. The shielded Coulomb potential interpolates between these limits.

The Born approximation for the scattering amplitude of this potential is

$$\begin{aligned} f(\theta) &= \frac{2m}{\hbar^2 q} V_0 \int_0^\infty \mathrm{d}r \mathrm{e}^{-ar} \sin(qr) \\ &= \frac{2m}{\hbar^2 q} \frac{V_0}{2\mathrm{i}} \int_0^\infty \mathrm{d}r [\mathrm{e}^{-(a-\mathrm{i}q)r} - \mathrm{e}^{-(a+\mathrm{i}q)r}] \\ &= \frac{2m}{\hbar^2 q} \frac{V_0}{2\mathrm{i}} \left[\frac{1}{a - \mathrm{i}q} - \frac{1}{a + \mathrm{i}q} \right] \\ &= \frac{2mV_0}{\hbar^2} \frac{1}{a^2 + q^2} \end{aligned} \tag{22.120}$$

with differential cross section

$$\frac{\mathrm{d}\sigma}{\mathrm{d}\Omega} = \left(\frac{2mV_0}{\hbar^2} \right)^2 \left(\frac{1}{a^2 + q^2} \right)^2 \tag{22.121}$$

Now consider this cross section in the limit $q \gg a$, which is relevant to the scattering of energetic particles at large angles (large momentum transfer = large q). This is also the relevant limit for $a \to 0$, which describes particles scattering from the unshielded Coulomb potential. In either limit

$$\begin{aligned} \frac{\mathrm{d}\sigma}{\mathrm{d}\Omega} &= \left(\frac{2mV_0}{\hbar^2 q^2} \right)^2 \\ &= \left(\frac{2mV_0}{\hbar^2 k^2 4 \sin^2(\theta/2)} \right)^2 \\ &= \left(\frac{V_0}{4E} \right)^2 \frac{1}{\sin^4(\theta/2)} \end{aligned} \tag{22.122}$$

where we have used $E = \hbar^2 k^2/2m$.

In 1911 Ernest Rutherford carried out his famous experiment, in which a beam of alpha particles, generated by the decay of cesium atoms, is directed onto a sheet of gold foil, and is scattered by the gold atoms. The differential cross section was then determined experimentally, and at large angles this cross section was found to agree with the formula above. Rutherford correctly concluded that the scattering of alpha particles at large angles was due to scattering from a Coulomb potential due to a fixed, positively charged source, namely the atomic nucleus. This led to the planetary picture of an atom, in which light electrons circle the heavy nucleus. The planetary picture is not entirely correct, but, more importantly, the tools that we have just used to derive the differential cross section were not available to Rutherford in 1911, since Schrödinger did not publish his famous equation until 1926. What Rutherford was using was the expression for the differential cross section of a Coulomb potential, derived from *classical* mechanics. In that formulation, the angle of scattering just depends on the impact parameter b (see figure 22.3), and wide angle scattering is associated with small b. That meant, for Rutherford, that an alpha particle scattered at wide angles had come close to the nucleus, and had probed the Coulomb potential due to a positive electric charge which was unshielded, at these short distances, by the negatively charged electrons.

The reason that the Coulombic cross section derived from classical mechanics can agree with the quantum result is apparent in (22.122): Planck's constant is absent, and we know that quantum mechanics should converge to classical mechanics in the $\hbar \to 0$ limit. So the Coulombic cross section must agree with the classical result even without taking this limit, and that is a historically fortunate coincidence. For most potentials, $\hbar$ will definitely appear in the expression for the differential cross section; its absence here is a special property of the $1/r$ potential. Because of this coincidence, Rutherford was able to deduce a correct result, namely the existence of the atomic nucleus, from a theoretical framework which does not actually apply at the atomic level.[1]

Example: the thin shell

The Born approximation for scattering from the δ-function 'thin shell' potential of equation (22.71) does not require a long calculation. The scattering amplitude is simply

$$f(\theta) = -\frac{2\,m}{\hbar^2 q} r_0 \sin(qr_0) V_0 \tag{22.123}$$

and the differential cross section is the square of this expression. But let us first consider the low energy limit, kr_0, $qr_0 \ll 1$. In this limit, where $\sin(qr_0) \approx qr_0$, the momentum transfer cancels out, so that

$$f(\theta) = -\frac{2\,mV_0 r_0^2}{\hbar^2} \tag{22.124}$$

and

$$\sigma = \int \mathrm{d}\Omega\, |f(\theta)|^2 = \frac{16\pi m^2}{\hbar^4} V_0^2 r_0^4 \tag{22.125}$$

Let us compare this expression with our previous partial wave result. At low energies only S-wave ($l = 0$) scattering contributes, and in equation (22.83) for δ_0 we use the fact that for $x \ll 1$

[1] For a derivation of the Coulomb scattering cross section in classical physics, see e.g.[1].

$$\cot(x) \approx \frac{1}{x} \quad \text{and} \quad \cot^{-1}\left(\frac{1}{x}\right) \approx x \tag{22.126}$$

Then for thin shell scattering equation (22.83) becomes

$$\delta_0 \approx -kr_0 + \frac{kr_0}{1+\alpha} \tag{22.127}$$

If we make the further assumption that the potential is weak, i.e. $\alpha \ll 1$, and use the fact that $1/(1+x) \approx 1 - x$ for $x \ll 1$, then $\delta_0 \approx -kr_0\alpha$ and

$$\begin{aligned} \sigma &\approx \frac{4\pi}{k^2}\delta_0^2 = 4\pi\alpha^2 r_0^2 \\ &= 4\pi\left(\frac{2\,mV_0r_0}{\hbar^2}\right)^2 r_0^2 \end{aligned} \tag{22.128}$$

in perfect agreement with the Born result (22.125) at $kr_0 \to 0$.

The partial wave result is exact for the $l = 0$ contribution, but the Born approximation is not. The latter is only the first term of a power series expansion, and this expansion will diverge for sufficiently strong scattering potentials, which in this thin-shell example is controlled by the dimensionless parameter α. On the other hand, the Born term contains implicitly *all* the partial waves, while in the partial wave approach one has to truncate the sum over all partial waves at some maximum l_{max}. If one is interested in the total cross section, rather than just a specific partial wave contribution, then a fixed truncation will break down for $kr_0 \gg l_{\text{max}}$. So both the partial wave approach and the Born approximation have their limitations, and it is interesting to contrast those limitations in the thin shell example.

To see the breakdown of the Born approximation at strong potentials, we begin by isolating the $l = 0$ contribution in the Born scattering amplitude. From (22.33), and using the orthogonality of the Legendre polynomials with $P_0(\cos(\theta)) = 1$, we have

$$|\sin(\delta_0)| = \left|\frac{k}{2}\int_0^\pi \mathrm{d}\theta \;\sin\theta f(\theta)\right| \tag{22.129}$$

Then the $l = 0$ contribution to the total scattering amplitude, from (22.37), is

$$\sigma_0^{\text{Born}} = \pi\left|\int_0^\pi \mathrm{d}\theta \;\sin\theta\, f^{\text{Born}}(\theta)\right|^2 \tag{22.130}$$

where $f^{\text{Born}}(\theta)$ for the thin shell potential is given in (22.123), and the integral can be evaluated numerically.

The Born approximation to σ_0/r_0^2 is compared to the exact result, calculated by the method of partial waves, over a range of momenta $2 < kr_0 \leqslant 10$ in figure 22.8. Figure 22.8(a) shows the comparison for a weak potential, at $\alpha = 0.1$, where the Born approximation should be valid, and in fact we see that the Born result and the exact result are indistinguishable. However, this agreement cannot last. Inspection of equations (22.81) and (22.123) shows that $\sigma_0^{\text{Born}} \propto \alpha^2$, so while the amplitude of

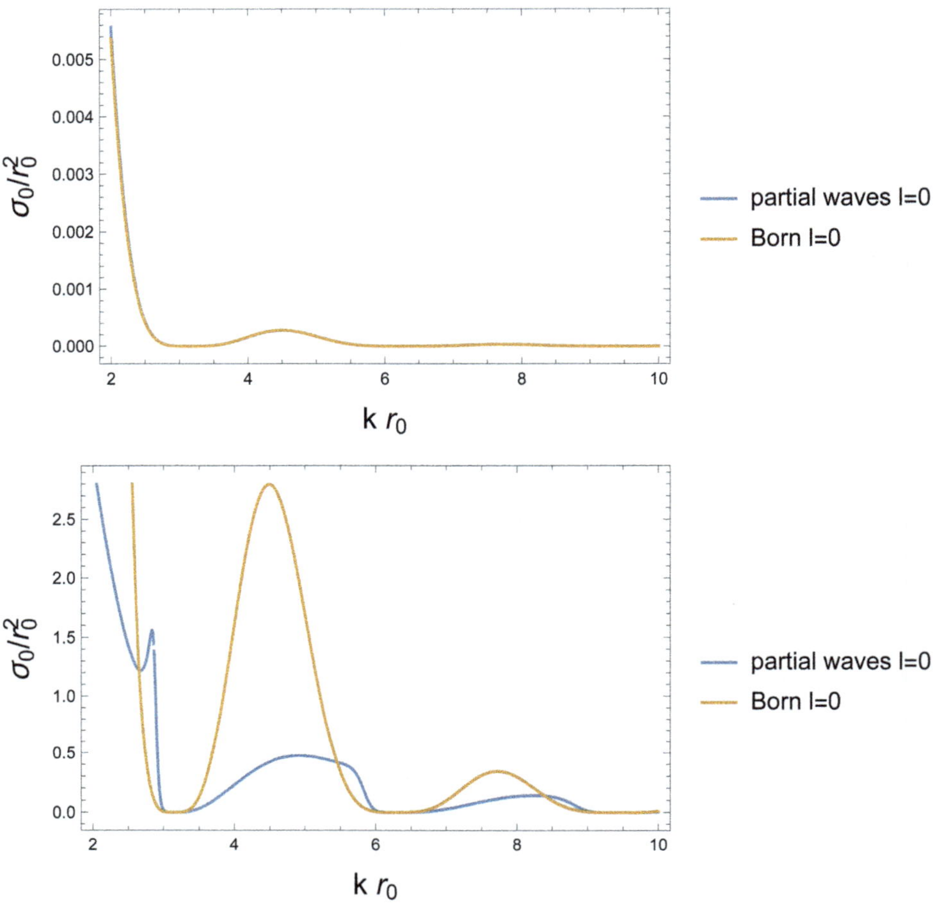

Figure 22.8. Advantage of the method of partial waves. The Born approximation agrees with the exact $l = 0$ contribution for weak scattering potentials, but not for strong potentials. The comparison is shown in (a) for a weak thin-shell potential at $\alpha = 0.1$, and in (b) at a strong thin-shell potential at $\alpha = 10$, where the exact $l = 0$ contribution is computed by the method of partial waves. The Born approximation is seen to break down completely at a strong potential.

the curve increases with increasing α, the overall shape does not. In contrast, we already know that the exact result develops spikes near $kr_0 = n\pi$, and this is beginning to appear at $\alpha = 10$, shown in figure 22.8(b). In this second plot, it is clear that the Born approximation has completely broken down, due to the non-convergence of the Born series at this large value of α.

On the other hand, the partial wave method also has its limitations at high momentum, where it may be necessary to include many terms in the series to get an accurate answer for the total cross section. Going back to $\alpha = 0.1$, where the Born approximation can be trusted, we compare σ_0 from the partial wave analysis with the total cross section σ^{Born}

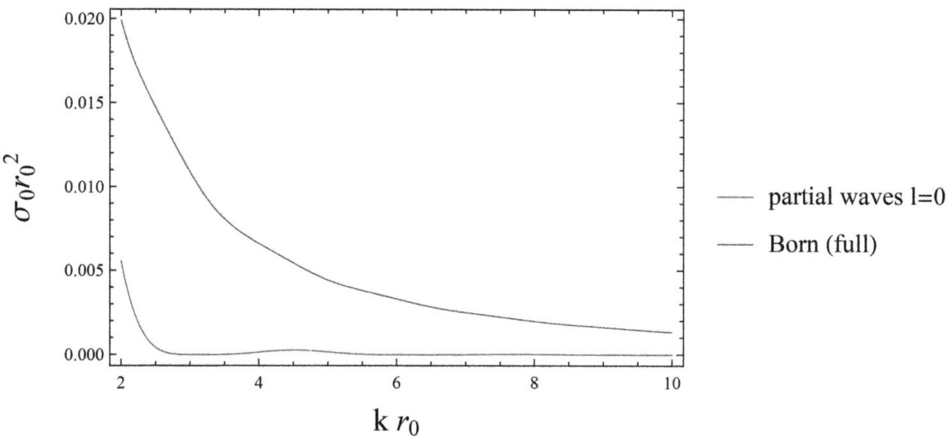

Figure 22.9. Advantage of the Born approximation. The Born result implicitly contains all partial wave contributions, albeit as an approximation, while a truncated partial wave expansion inevitably fails at large momentum. The figure compares the $l = 0$ partial wave contribution to the cross section at $kr_0 \geqslant 2$, with the full cross section computed in the Born approximation. The calculation is for a weak ($\alpha = 0.1$) thin shell potential where the Born approximation can be trusted.

$$\sigma^{\text{Born}} = 2\pi\alpha^2 r_0^2 \int_0^{\pi} d\theta \ \sin\theta \left(\frac{\sin[2kr_0 \sin(\theta/2)]}{2kr_0 \sin(\theta/2)} \right)^2 \tag{22.131}$$

The comparison is shown in figure 22.9, at $kr_0 > 2$, and we see that the $l = 0$ contribution is very far from the total cross section. However, in particle physics we are often interested in partial wave contributions at definite l, since those are associated with resonances of definite spin.

In relativistic collisions described by quantum field theory, scattering amplitudes are computed in a generalization of the Born series known as the Feynman diagrammatic expansion. For the interaction of quarks via the strong nuclear force this expansion is divergent, just as the Born series diverges when applied to a sufficiently strong thin shell potential. Non-perturbative methods based on computer simulations have been developed to circumvent this problem, but a full understanding of the strong nuclear force, in particular the property of quark confinement, does not yet exist.

Problems

1. Compute the differential cross section of a spherically symmetric potential

$$V(r) = \begin{cases} V_0 & r \leqslant R \\ 0 & r > R \end{cases} \tag{22.132}$$

and a delta-function potential

$$V(x, y, z) = \lambda\delta(x)\delta(y)\delta(z) \tag{22.133}$$

in the Born approximation. Show that these potentials give the same cross section, for a certain choice of λ, in the limit that $R \to 0$ and $V_0 \to \infty$ with $R^3 V_0$ held fixed.

2. Consider a spherically symmetric potential which falls off at large distances as $1/r^p$. Show from the Born approximation that the total cross section is finite if and only if $p > 2$.
3. Obtain the differential cross section for the potential

$$V(r) = \frac{c}{r^2} \tag{22.134}$$

via the Born approximation.

4. Find the differential cross section for a particle of mass m and energy E scattered from a Gaussian potential

$$V(r) = V_0 \mathrm{e}^{-r^2/a^2} \tag{22.135}$$

5. For the potential

$$V(r) = \begin{cases} \infty & r < R_0 \\ -U_0 & R_0 < r < R_1 \\ 0 & r > R_1 \end{cases} \tag{22.136}$$

it is found that the $l = 0$ partial wave phase shift is $\delta_0 = 0$ for particles of energy $E = U_0$. From this information, compute R_0 as a function of U_0 and R_1.

6. A beam of particles is scattered by heavy nuclear targets, and it is found that only the partial wave phase shifts of the S and P waves (i.e. δ_0 and δ_1) are non-zero.
   a) Show that the differential cross-section can be expressed as

$$\frac{\mathrm{d}\sigma}{\mathrm{d}\Omega} = a + b\cos(\theta) + c\cos^2(\theta) \tag{22.137}$$

   and give the values of the constants a, b, c in terms of the two partial wave phase shifts.

   b) Let $\delta_0 = 37°$, and $\delta_1 = 15°$. Suppose the beam intensity (number of incident particles per second per unit area) is 10^{16} particles $\mathrm{s}^{-1}\ \mathrm{cm}^{-2}$, and the beam is incident on a thin film of material containing 10^{14} targets per cm^2. If the cross-sectional area of the beam is $0.03\ \mathrm{cm}^2$, calculate the total number of particles scattered per second, as a function of $k = \sqrt{2mE}/\hbar$. What is this number if the particles are electrons, and the incoming energy is 5 keV?

Reference

[1] Goldstein H, Poole C and Safko J 2014 *Classical Mechanics* (Pearson Education Limited, Essex, England)

Part V

Advanced topics

IOP Publishing

An Introduction to Quantum Theory

Jeff Greensite

Chapter 23

Bras and kets, and canonical quantization

There is a certain loose analogy between bra-ket notation in quantum mechanics, and div-grad-curl notation in electromagnetism. Neither is absolutely necessary. It is certainly possible to formulate Maxwell's equations without the use of div-grad-curl, and Maxwell himself did so. And there are many books on quantum theory which forgo bra-ket notation entirely. But div-grad-curl notation, in addition to its many other advantages, allows us to formulate Maxwell's equations independent of any particular choice of coordinate system. In principle there is nothing sacred about, e.g., Cartesian coordinates, and it seems unnatural that something as fundamental as the laws of electromagnetism should require, for their formulation, a particular choice of coordinate system such as Cartesian. Div-grad-curl allows us to avoid making such a choice when formulating Maxwell's equations. A coordinate system is only introduced when it is finally needed for calculations.

Bra-ket notation is not as important to quantum mechanics as div-grad-curl is in electromagnetism. Nevertheless, like div-grad-curl, bra-ket notation frees us from a specific choice of reference frame, which in this case means a specific choice of basis vectors in Hilbert space. We have seen that a physical state is regarded as a vector, denoted by the ket symbol $|\psi\rangle$, in an infinite-dimensional linear vector space known as Hilbert space. The components of $|\psi\rangle$ are specified by the wave function $\psi(x)$, where x is to be thought of as a continuous index, analogous to the discrete index $i = 1, 2, 3$ which labels the components v_i of a vector $\mathbf{v}$. But the components of any vector depend on the choice of basis vectors, such as the familiar $\hat{\mathbf{i}}$, $\hat{\mathbf{j}}$, $\hat{\mathbf{k}}$ unit vectors which span three-dimensional space. Change the basis, and the components of all vectors will also change. A wave function $\psi(x)$ can be thought of as the components of $|\psi\rangle$ in one particular (albeit very useful) basis in Hilbert space, consisting of eigenstates of the position operator. For this reason $\psi(x)$ is known as the ***position-space wave function***. But there are an infinite number of possible bases in Hilbert space, and sometimes a different basis, consisting, e.g., of eigenstates of the momentum operator, is preferable, with the ***momentum-space wave function***

doi:10.1088/978-0-7503-1167-0ch23

corresponding to $|\psi\rangle$ denoted $\psi(p)$. In any event, since there is nothing sacred about any choice of basis in Hilbert space, it ought to be possible to formulate the laws of quantum mechanics in a manner which is independent of a choice of basis, just as Maxwell's equations can be expressed independent of a choice of coordinates. For this purpose, bra-ket notation is indispensable.

Apart from notation, there is another direction for generalization. We have so far focused on the quantum mechanics of a point particle moving in some potential. Is there a general scheme for quantizing an arbitrary mechanical system, e.g. a system of gears, levers, and springs? Or an electrical LC circuit? This question will be addressed in the last section of this chapter.

23.1 Outer products and unitary matrices

We begin with a review of how the components of vectors and matrices in a finite-dimensional vector space are related in different bases, which will bring us to the concept of unitary matrices which transform one basis to another.

A set of orthogonal unit vectors $\{|e^{(i)}\rangle,\ i = 1, 2, \ldots, D\}$ in a D-dimensional vector space can be regarded as a set of basis vectors, or simply a 'basis', in the sense that any vector $|v\rangle$ can be represented as a sum

$$|v\rangle = \sum_{n=1}^{D} v_n|e^{(n)}\rangle \tag{23.1}$$

For the corresponding bra vector,

$$\langle v| = \sum_{n=1}^{D} v_n^*\langle e^{(n)}| \tag{23.2}$$

The fact that a basis is orthonormal means, by definition, that

$$\langle e^{(i)}|e^{(j)}\rangle = \delta_{ij} \tag{23.3}$$

and from this it follows that the components v_i of vector $|v\rangle$ in the basis $\{|e^{(i)}\rangle\}$ are given by

$$\begin{aligned} v_i &= \langle e^{(i)}|v\rangle \\ v_i^* &= \langle v|e^{(i)}\rangle \end{aligned} \tag{23.4}$$

In particular, the jth vector component $e_j^{(i)}$ of the basis vector $|e^{(i)}\rangle$ is simply δ_{ij}. A matrix M is a linear operator acting on vectors $|v\rangle$. The components of this operator, i.e. the matrix elements of the matrix, are also given in terms of inner products:

$$\begin{aligned} M_{ij} &= \sum_{nm} \delta_{in}M_{nm}\delta_{mi} \\ &= \sum_{nm} e_n^{(i)}M_{nm}e_m^{(j)} \\ &= \langle e^{(i)}|M|e^{(j)}\rangle \end{aligned} \tag{23.5}$$

We are quite familiar with the concept of an inner product of two vectors

$$\langle u|v\rangle = \sum_n u_n^* v_n \tag{23.6}$$

But it is also useful to define a so-called ***outer product***[1]

$$Q = |u\rangle\langle v| \tag{23.7}$$

which takes the pair of vectors into a linear operator (i.e. a matrix) whose matrix elements are $Q_{ij} = u_i v_j^*$. To see the connection, starting from the general definition of matrix components (23.5),

$$\begin{aligned} Q_{ij} &= \langle e^{(i)}|Q|e^{(j)}\rangle \\ &= \langle e^{(i)}|(|u\rangle\langle v|)e^{(j)}\rangle \\ &= \langle e^{(i)}|u\rangle\langle v|e^{(j)}\rangle \\ &= u_i v_j^* \end{aligned} \tag{23.8}$$

We can then express *any* linear operator as a sum of outer products

$$M = \sum_{nm} M_{nm}|e^{(n)}\rangle\langle e^{(m)}| \tag{23.9}$$

because it is easy to see that this expression, combined with (23.5) and the orthonormality condition (23.3), gives $M_{ij} = \langle e^{(i)}|M|e^{(j)}\rangle$.

A very useful operator is the identity operator $\mathbb{1}$, which has the property that $\mathbb{1}|v\rangle = |v\rangle$ for any vector $|v\rangle$. It has the outer product bra-ket representation

$$\mathbb{1} = \sum_i |e^{(i)}\rangle\langle e^{(i)}| \tag{23.10}$$

which is sometimes called the ***resolution of the identity***. This representation is proved as follows:

$$\begin{aligned} \mathbb{1}|v\rangle &= \sum_i |e^{(i)}\rangle\langle e^{(i)}|\sum_j v_j|e^{(j)}\rangle \\ &= \sum_i v_i|e^{(i)}\rangle \\ &= |v\rangle \end{aligned} \tag{23.11}$$

It is also easy to see, using (23.5), that $\mathbb{1}_{ij} = \delta_{ij}$, meaning that $\mathbb{1}$ corresponds to the unit matrix.

Although any matrix can be expressed in the form (23.9), Hermitian matrices have a special bra-ket representation in terms of their eigenvalues and eigenstates. Let L be a $D \times D$ Hermitian matrix, with corresponding eigenvalue equation

[1] Sometimes written $Q = \mathbf{u} \otimes \mathbf{v}$.

$$L|\phi^{(n)}\rangle = \lambda_n|\phi^{(n)}\rangle \tag{23.12}$$

Then L can be expressed in bra-ket form as

$$L = \sum_n \lambda_n|\phi^{(n)}\rangle\langle\phi^{(n)}| \tag{23.13}$$

with matrix elements

$$\begin{aligned} L_{ij} &= \sum_n \lambda_n\langle e^{(i)}|\phi^{(n)}\rangle\langle|\phi^{(n)}|e^{(j)}\rangle \\ &= \sum_n \lambda_n\phi_i^{(n)}\phi_j^{(n)*} \end{aligned} \tag{23.14}$$

If the basis happens to be the eigenstates of L, i.e. $\{|e^{(n)}\rangle\} = \{|\phi^{(n)}\rangle\}$, then in that basis L is diagonal, with the diagonal matrix elements equal to the eigenvalues:

$$\begin{aligned} L_{ij} &= \langle\phi^{(i)}|\left(\sum_n \lambda_n|\phi^{(n)}\rangle\langle|\phi^{(n)}|\right)|\phi^{(j)}\rangle \\ &= \lambda_i\delta_{ij} \end{aligned} \tag{23.15}$$

This is why solving the eigenvalue problem for a Hermitian operator (or matrix) is often called ***diagonalizing*** the operator. It amounts to finding a basis in which the matrix is diagonal.

This is all a bit abstract, so let us consider a two-dimensional vector space with basis vectors

$$\begin{aligned} |e^{(1)}\rangle &\leftrightarrow \begin{bmatrix}1\\0\end{bmatrix} \\ |e^{(2)}\rangle &\leftrightarrow \begin{bmatrix}0\\1\end{bmatrix} \end{aligned} \tag{23.16}$$

Then, using the notation [...] to denote the matrix representation, the resolution of the identity is

$$\begin{aligned} \mathbb{1} &= |e^{(1)}\rangle\langle e^{(1)}| + |e^{(2)}\rangle\langle e^{(2)}| \\ [\mathbb{1}] &= \begin{bmatrix}1&0\\0&0\end{bmatrix} + \begin{bmatrix}0&0\\0&1\end{bmatrix} \\ &= \begin{bmatrix}1&0\\0&1\end{bmatrix} \end{aligned} \tag{23.17}$$

which is the unit matrix. Next let us consider a Hermitian matrix

$$L = \begin{bmatrix}0&1\\1&0\end{bmatrix} \tag{23.18}$$

which is simply the Pauli matrix σ_1. We have already solved the eigenvalue problem $H\phi^n = \lambda_n\phi^{(n)}$ and found the solutions

$$\lambda_1 = 1,\ \phi^{(1)} = \frac{1}{\sqrt{2}}\begin{bmatrix}1\\1\end{bmatrix} \quad \text{and} \quad \lambda_2 = -1,\ \phi^{(2)} = \frac{1}{\sqrt{2}}\begin{bmatrix}1\\-1\end{bmatrix} \tag{23.19}$$

Then

$$\begin{aligned} L &= +1|\phi^{(1)}\rangle\langle\phi^{(1)}| + (-1)|\phi^{(2)}\rangle\langle\phi^{(2)}| \\ [L] &= \frac{1}{2}\begin{bmatrix}1 & 1\\1 & 1\end{bmatrix} - \frac{1}{2}\begin{bmatrix}1 & -1\\-1 & 1\end{bmatrix} \\ [L] &= \begin{bmatrix}0 & 1\\1 & 0\end{bmatrix} \end{aligned} \tag{23.20}$$

which confirms that the bra-ket representation (23.13) does in fact reproduce $[L] = \sigma_1$. Note also the resolution of the identity in this basis:

$$\begin{aligned} \mathbb{1} &= |\phi^{(1)}\rangle\langle\phi^{(1)}| + |\phi^{(2)}\rangle\langle\phi^{(2)}| \\ [\mathbb{1}] &= \frac{1}{2}\begin{bmatrix}1 & 1\\1 & 1\end{bmatrix} + \frac{1}{2}\begin{bmatrix}1 & -1\\-1 & 1\end{bmatrix} \\ &= \begin{bmatrix}1 & 0\\0 & 1\end{bmatrix} \end{aligned}$$

Unitary matrices and change of basis

Since any set of D orthonormal vectors can serve as a basis, let us consider another set $\{|e'^{(i)}\rangle,\ i = 1, 2, \ldots, D\}$, and ask how the components v_i, M_{ij} of vectors and matrices change in going to the new basis. What is needed are the vector components of the old basis vectors in the new basis, i.e.

$$U_{mn} = \langle e'^{(m)}|e^{(n)}\rangle \tag{23.21}$$

Likewise

$$U^{\dagger}_{mn} = \langle e^{(m)}|e'^{(n)}\rangle \tag{23.22}$$

gives the vector components of the new basis vectors in the old basis. As an example in $D = 2$ dimensions, let the new basis vectors be

$$\begin{aligned} |e'^{(1)}\rangle &\leftrightarrow \frac{1}{\sqrt{2}}\begin{bmatrix}1\\i\end{bmatrix} \\ |e'^{(2)}\rangle &\leftrightarrow \frac{1}{\sqrt{2}}\begin{bmatrix}1\\-i\end{bmatrix} \end{aligned} \tag{23.23}$$

where, from the right-hand side, we read off the vector components of the new basis vectors in the old basis. Taking the indicated inner products in (23.21) and (23.22), we have

$$U = \frac{1}{\sqrt{2}}\begin{bmatrix} 1 & 1 \\ \mathrm{i} & -\mathrm{i} \end{bmatrix}$$
$$U^\dagger = \frac{1}{\sqrt{2}}\begin{bmatrix} 1 & -\mathrm{i} \\ 1 & \mathrm{i} \end{bmatrix} \tag{23.24}$$

With the U, $U^\dagger$ matrices, we can convert vector components and matrix elements in the old basis to the corresponding quantities in the new basis. Starting with

$$|v\rangle = \sum_n v_n'|e'^{(n)}\rangle = \sum_n v_n|e^{(n)}\rangle \tag{23.25}$$

we multiply both sides by the bra $\langle e'^{(m)}|$, which gives

$$\begin{aligned} v_m' &= \langle e'^{(m)}|v\rangle \\ &= \sum_n \langle e'^{(m)}|e^{(n)}\rangle v_n \\ &= \sum_n U_{mn} v_n \end{aligned} \tag{23.26}$$

Let M_{ij}' be the components of matrix M in the new basis. They are also related to the matrix elements of M in the old basis via the U matrix:

$$\begin{aligned} M_{ij}' &\equiv \langle e'^{(i)}|M|e'^{(j)}\rangle \\ &= \langle e'^{(i)}|\left(\sum_{mn} M_{mn}|e^{(m)}\rangle\langle e^{(n)}|\right)|e'^{(j)}\rangle \\ &= \sum_{mn}\langle e'^{(i)}|e^{(m)}\rangle M_{mn}\langle e^{(n)}|e'^{(j)}\rangle \\ &= \sum_{mn} U_{im} M_{mn} U_{nj}^\dagger \end{aligned} \tag{23.27}$$

The matrix U_{mn} has the following property, which makes it a ***unitary matrix***, namely,

$$UU^\dagger = U^\dagger U = \mathbb{1} \tag{23.28}$$

To see this, we just consider the matrix product

$$\begin{aligned} \sum_j U_{ik} U_{kj}^\dagger &= \sum_k \langle e^{(i)}|e'^{(k)}\rangle\langle e'^{(k)}|e^{(j)}\rangle \\ &= \langle e^{(i)}|\left(\sum_k |e'^{(k)}\rangle\langle e'^{(k)}|\right)|e^{(j)}\rangle \\ &= \langle e^{(i)}|\mathbb{1}||e^{(j)}\rangle \\ &= \delta_{ik} \end{aligned} \tag{23.29}$$

which means that $UU^\dagger$ is the unit matrix. This property can be verified explicitly in the example (23.24), by multiplying the U, $U^\dagger$ matrices shown.

Unitarity of the matrix U insures that inner products $\langle u|v\rangle$ and $\langle u|M|v\rangle$, when expressed in terms of components, are the same whether the components are in the $\{|e^{(i)}\rangle\}$ or $\{|e^{(i)}\rangle\}$ basis. For example,

$$\begin{aligned}\langle u|v\rangle &= \sum_i u_i^* v_i \\ &= \sum_{ij} u_i^* \delta_{ij} v_j \\ &= \sum_{ijk} u_i^* U_{ik}^\dagger U_{kj} v_j \\ &= \sum_k u_k'^* v_k'\end{aligned} \tag{23.30}$$

We leave the basis independence of $\langle u|M|v\rangle$ as a short exercise.

The eigenvalue problem for unitary matrices has the usual form

$$U|\phi^{(n)}\rangle = c_n|\phi^{(n)}\rangle \tag{23.31}$$

so that a unitary operator can be put in the bra-ket form

$$U = \sum_n c_n |\phi^{(n)}\rangle\langle|\phi^{(n)}| \tag{23.32}$$

and the matrix U_{nm} is diagonal in the basis of eigenstates $\{|\phi^{(n)}\rangle\}$. But then the unitarity condition $UU^\dagger = \mathbb{1}$ requires that

$$c_n = \mathrm{e}^{\mathrm{i}\theta_n} \tag{23.33}$$

Now define the Hermitian operator

$$H = \sum_n \theta_n |\phi^{(n)}\rangle\langle\phi^{(n)}| \tag{23.34}$$

It follows that

$$U = \mathrm{e}^{\mathrm{i}H} \tag{23.35}$$

because the matrices

$$\begin{aligned}U_{ij} &= \langle\phi^{(i)}|\left(\sum_n \mathrm{e}^{\mathrm{i}\theta_n}|\phi^{(n)}\rangle\langle|\phi^{(n)}|\right)|\phi^{(j)}\rangle \\ &= \mathrm{e}^{\mathrm{i}\theta_j}\delta_{ij}\end{aligned} \tag{23.36}$$

and

$$\begin{aligned}[\mathrm{e}^{\mathrm{i}H}]_{ij} &= \langle\phi^{(i)}|\mathrm{e}^{\mathrm{i}H}|\phi^{(j)}\rangle \\ &= \langle\phi^{(i)}|\mathrm{e}^{\mathrm{i}\theta_j}||\phi^{(j)}\rangle \\ &= \mathrm{e}^{\mathrm{i}\theta_j}\delta_{ij}\end{aligned} \tag{23.37}$$

have the same matrix elements. The conclusion is that a unitary matrix can be expressed as the exponential of iH, where H is a Hermitian matrix.

In this section we have used the terms 'linear operator' and 'matrix' interchangeably. In fact, there is a difference. A linear operator maps vectors into other vectors, while a matrix is some array of numbers. A linear operator can be represented as a matrix only after a choice of basis; the components of the matrix depend on the basis. Likewise, a vector can only be represented as a column of numbers after a choice of basis; these too depend on the basis. One of the advantages of the bra-ket outer product representation of operators is that this representation of an operator, as opposed to a representation by matrix components, remains true in any basis.

23.2 Position- and momentum-space wave functions

Any complete set of states, which are the eigenstates of a complete set of operators as described in chapter 10, can be regarded as a basis in Hilbert space. For a spinless point particle moving in one dimension, the position operator X and momentum operator P are each a complete set of operators, and the eigenstates of each operator constitute two different bases for the Hilbert space of the point particle. The defining equations are

$$\begin{aligned} X|x_0\rangle &= x_0|x_0\rangle \\ P|p_0\rangle &= p_0|p_0\rangle \\ [X, P] &= i\hbar\mathbb{1} \end{aligned} \tag{23.38}$$

and the X and P operators can be formally expressed in the outer product form

$$\begin{aligned} X &= \int dx\ x|x\rangle\langle x| \\ P &= \int dp\ p|p\rangle\langle p| \end{aligned} \tag{23.39}$$

The components of a ket vector $|\psi\rangle$ in the basis $\{|x_0\rangle, -\infty < x_0 < \infty\}$

$$\psi(x) = \langle x|\psi\rangle \tag{23.40}$$

are referred to as the position-space wave function. Likewise, the components of $|\psi\rangle$ in the basis $\{|p_0\rangle, -\infty < p_0 < \infty\}$

$$\psi(p) = \langle p|\psi\rangle \tag{23.41}$$

are referred to as the momentum-space wave function. We denote operators X, P that act on bras and kets by capital letters, reserving the notation $\tilde{x}$, $\tilde{p}$ for operators which act on either position- or momentum-space wave functions.

Many results can now be taken over directly from the last section, with the understanding that the space is infinite dimensional, and indices run over a continuous range. That said, we can write expansions in the two bases,

$$\begin{aligned} |\psi\rangle &= \int dx\ \psi(x)|x\rangle \\ |\psi\rangle &= \int dp\ \psi(p)|p\rangle \end{aligned} \tag{23.42}$$

orthonormality of basis vectors,

$$\begin{aligned}\langle x_1|x_2\rangle &= \delta(x_2 - x_1)\\ \langle p_1|p_2\rangle &= \delta(p_2 - p_1)\end{aligned} \tag{23.43}$$

the identity operator

$$\begin{aligned}\mathbb{1} &= \int \mathrm{d}x \ \ |x\rangle\langle x|\\ &= \int \mathrm{d}p \ \ |p\rangle\langle p|\end{aligned} \tag{23.44}$$

matrix elements

$$\begin{aligned}\langle x_1|X|x_2\rangle &= x_2\delta(x_2 - x_1)\\ \langle p_1|P|p_2\rangle &= p_2\delta(p_2 - p_1)\end{aligned} \tag{23.45}$$

and the unitary matrix elements which go from the x to the p basis

$$\begin{aligned}U_{px} &= \langle p|x\rangle\\ U^{\dagger}_{xp} &= \langle x|p\rangle\end{aligned} \tag{23.46}$$

Unitarity of the U_{xp} matrix is easily checked using (23.44):

$$\begin{aligned}\int \mathrm{d}x U_{p_1x}U^{\dagger}_{xp_2} &= \langle p_1|\Big(\int \mathrm{d}x|x\rangle\langle x|\Big)|p_2\rangle\\ &= \langle p_1|p_2\rangle\\ &= \delta(p_1 - p_2)\end{aligned} \tag{23.47}$$

The problem is to work out what the unitary matrix elements $\langle p|x\rangle$ actually are, since this is what takes us from one basis to another, and also to go from the defining relations (23.38) to the operators $\tilde{x}$, $\tilde{p}$ which act on position- and momentum-space wave functions. Here the commutation relation $[X, P] = \mathrm{i}\hbar\mathbb{1}$ is the key.

In earlier chapters of the book we have represented the action of an operator Q on a state $|\psi\rangle$ by the notation $\tilde{Q}\psi(x)$. Let us see what this amounts to in bra-ket notation, and use the commutator relations to rederive old results for $\tilde{x}$, $\tilde{p}$ in position space, as well as derive new results for these operators in momentum space. First recall that in a finite-dimensional vector space, the ith component of the vector $M|v\rangle$ is

$$(M|v\rangle)_i = \langle e^{(i)}|M|v\rangle \tag{23.48}$$

In a similar way, for quantum mechanics in one space dimension, the interpretation of $\tilde{Q}\psi(x)$ is that it is the x-component of the ket vector $Q|\psi\rangle$ in Hilbert space, i.e.

$$\tilde{Q}\psi(x) = (Q|\psi\rangle)_x = \langle x|Q|\psi\rangle \tag{23.49}$$

It is easy to recover, from this expression, the action of the $\tilde{x}$ operator in position space

$$\begin{aligned}\tilde{x}\psi(x) &= \langle x|X|\psi\rangle \\ &= x\langle x|\psi\rangle \\ &= x\psi(x)\end{aligned} \tag{23.50}$$

where we have used the Hermiticity of X: $\langle x|X = x\langle x|$. As we already knew, the position operator operates on a position-space wave function $\psi(x)$ by multiplying the wave function by the function $f(x) = x$. To find $\tilde{p}$ in position space, we need to use the commutator relation, plus Hermiticity of X. We have

$$\langle x_1|[X, P]|x_2\rangle = (x_2 - x_1)\langle x_1|P|x_2\rangle \tag{23.51}$$

and also

$$\begin{aligned}\langle x_1|[X, P]|x_2\rangle &= \mathrm{i}\hbar\langle x_1|\mathbb{1}|x_2\rangle \\ &= \mathrm{i}\hbar\delta(x_2 - x_1)\end{aligned} \tag{23.52}$$

Then

$$(x_1 - x_2)\langle x_1|P|x_2\rangle = \mathrm{i}\hbar\delta(x_2 - x_1) \tag{23.53}$$

This equation is solved by choosing the matrix element $\langle x_1|P|x_2\rangle$ to be

$$\langle x_1|P|x_2\rangle = -\mathrm{i}\hbar\frac{\partial}{\partial x_1}\delta(x_1 - x_2) \tag{23.54}$$

Then

$$\begin{aligned}\langle x|P|\psi\rangle &= \langle x|P|\int \mathrm{d}x'\psi(x)|x'\rangle \\ &= \int \mathrm{d}x'\langle x|P|x'\rangle\psi(x') \\ &= -\mathrm{i}\hbar\frac{\mathrm{d}}{\mathrm{d}x}\psi(x') \\ &= \tilde{p}\psi(x)\end{aligned} \tag{23.55}$$

In this way we recover the usual result that $\tilde{p} = -\mathrm{i}\hbar\frac{\mathrm{d}}{\mathrm{d}x}$.

Next, we want to determine the action of $\tilde{x}, \tilde{p}$ on momentum-space wave functions, which is something that has not be done up to now. The easy case, in momentum space, is $\tilde{p}$:

$$\begin{aligned}\tilde{p}\psi(p) &= \langle p|P|\psi\rangle \\ &= p\langle p|\psi\rangle \\ &= p\psi(p)\end{aligned} \tag{23.56}$$

So for momentum space wave functions $\tilde{p}$ acts by multiplying the wave function by the function $f(p) = p$. On the other hand

$$\langle p_1|[X, P]|p_2\rangle = (p_2 - p_1)\langle p_1|X|p_2\rangle \tag{23.57}$$

and

$$\begin{aligned}\langle p_1|[X, P]|p_2\rangle &= \mathrm{i}\hbar\langle p_1|\mathbb{1}|p_2\rangle \\ &= \mathrm{i}\hbar\delta(p_2 - p_1)\end{aligned} \tag{23.58}$$

so

$$(p_2 - p_1)\langle p_1|X|p_2\rangle = \mathrm{i}\hbar\delta(p_2 - p_1) \tag{23.59}$$

This last equation is solved by

$$\langle p_1|X|p_2\rangle = \mathrm{i}\hbar\frac{\partial}{\partial p_1}\delta(p_2 - p_1) \tag{23.60}$$

Then

$$\begin{aligned}\langle p|X|\psi\rangle &= \langle p|X|\int \mathrm{d}p'\psi(p')|p'\rangle \\ &= \int \mathrm{d}p'\langle p|X|p'\rangle\psi(p') \\ &= \mathrm{i}\hbar\frac{d}{\mathrm{d}p}\psi(p) \\ &= \tilde{x}\psi(p)\end{aligned} \tag{23.61}$$

To summarize:

Position and momentum space representations

$$\text{Position space}\begin{cases}\psi(x) = \langle x|\psi\rangle \\ \widetilde{Q}\psi(x) = \langle x|Q|\psi\rangle \\ \tilde{x}\psi(x) = x\psi(x) \\ \tilde{p}\psi(x) = -\mathrm{i}\hbar\frac{d}{\mathrm{d}x}\psi(x)\end{cases}$$

$$\text{Momentum space}\begin{cases}\psi(p) = \langle p|\psi\rangle \\ \widetilde{Q}\psi(p) = \langle p|Q|\psi\rangle \\ \tilde{x}\psi(p) = \mathrm{i}\hbar\frac{d}{\mathrm{d}p}\psi(p) \\ \tilde{p}\psi(p) = p\psi(p)\end{cases}$$

To go from position-space wave functions to momentum-space wave functions, we still need the components of the unitary matrix $U_{px} = \langle p|x\rangle$. To obtain those, we begin with equation (23.55), with $|\psi\rangle = |p\rangle$:

$$\begin{aligned} \langle x|P|p\rangle &= \tilde{p}\psi(x) \qquad \text{where} \qquad \psi(x) = \langle x|p\rangle \\ p\langle x|p\rangle &= -\mathrm{i}\hbar\frac{d}{\mathrm{d}x}\langle x|p\rangle \end{aligned} \tag{23.62}$$

The second line is an eigenvalue equation, with the normalized solution

$$\langle x|p\rangle = \frac{1}{\sqrt{2\pi\hbar}}\mathrm{e}^{\mathrm{i}px/\hbar} \tag{23.63}$$

This expression can also be interpreted as the wave function of a momentum eigenstate in position space. Taking the complex conjugate of both sides

$$\langle p|x\rangle = \frac{1}{\sqrt{2\pi\hbar}}\mathrm{e}^{-\mathrm{i}px/\hbar} \tag{23.64}$$

which can likewise be interpreted as the wave function of a position eigenstate in momentum space. In the notation of chapter 7, we would write

$$\begin{aligned} \phi^{p}_{p_0}(x) = \langle x|p_0\rangle &= \frac{1}{\sqrt{2\pi\hbar}}\mathrm{e}^{\mathrm{i}p_0x/\hbar} \\ \phi^{x}_{x_0}(p) = \langle p|x_0\rangle &= \frac{1}{\sqrt{2\pi\hbar}}\mathrm{e}^{-\mathrm{i}px_0/\hbar} \end{aligned} \tag{23.65}$$

Now we can simply take over the result (23.26) to convert a position-space wave function to a momentum-space wave function:

$$\begin{aligned} \psi(p) &= \int \mathrm{d}x \;\; U_{px}\psi(x) \\ &= \int \mathrm{d}x \;\; \langle p|x\rangle\psi(x) \\ &= \int \mathrm{d}x \;\; \psi(x)\frac{1}{\sqrt{2\pi\hbar}}\mathrm{e}^{-\mathrm{i}px/\hbar} \end{aligned} \tag{23.66}$$

In other words, a wave function in momentum space is simply the Fourier transform of the corresponding wave function in position space, and we recognize from chapter 6 that the probability density for momentum is just the squared modulus of the momentum-space wave function

$$P(p) = \psi^*(p)\psi(p) \tag{23.67}$$

23.2.1 The harmonic oscillator basis

Of course the $\{|x\rangle\}$ and $\{|p\rangle\}$ bases, which consist of eigenstates of the X and P operators respectively, are not the only possible bases in Hilbert space. As was already mentioned, any complete set of states, which are the eigenstates of some complete set of operators, can serve as a basis for the Hilbert space. For a spinless point particle, the eigenstates $|\varphi_n\rangle$ of the harmonic oscillator Hamiltonian H^{ho} can be used as a basis, and this basis is perfectly well defined whether or not the point particle is actually moving in a harmonic oscillator potential (although that would be the most useful application).

From the orthonormality of eigenstates of a Hermitian operator, the wave function of the eigenstate $|\varphi_n\rangle$ is

$$\varphi_n(m) = \langle \varphi_m | \varphi_n \rangle = \delta_{mn} \qquad m = 0, 1, 2, \ldots \tag{23.68}$$

In other words, the wave function of eigenstates of the harmonic oscillator, in the harmonic oscillator representation, can be represented as infinite-dimensional column vectors

$$\varphi_0 = \begin{bmatrix} 1 \\ 0 \\ 0 \\ \vdots \end{bmatrix} \qquad \varphi_1 = \begin{bmatrix} 0 \\ 1 \\ 0 \\ \vdots \end{bmatrix} \qquad \varphi_2 = \begin{bmatrix} 0 \\ 0 \\ 1 \\ \vdots \end{bmatrix} \cdots \tag{23.69}$$

In fact, because $n = 0, 1, 2, \ldots$ is a discrete index, all 'eigenfunctions' $\psi(n) = \langle \varphi_n | \psi \rangle$ can be thought of as the components of a column vector. In particular, an eigenfunction of position $|x\rangle$, in the harmonic oscillator basis, is

$$\begin{aligned} \psi_x(n) &= \langle \varphi_n | x \rangle \\ &= \langle x | \varphi_n \rangle^* \\ &= \varphi_n(x) \end{aligned} \tag{23.70}$$

So in this basis an eigenstate of position has the form of a column vector

$$\psi_x = \begin{bmatrix} \varphi_0(x) \\ \varphi_1(x) \\ \varphi_2(x) \\ \vdots \end{bmatrix} \tag{23.71}$$

Then the orthogonality of position eigenstates implies

$$\begin{aligned} \delta(x - y) &= \langle x | y \rangle \\ &= \left[\varphi_0^*(x), \varphi_1^*(x), \varphi_2^*(x), \ldots\right] \begin{bmatrix} \varphi_0(y) \\ \varphi_1(y) \\ \varphi_2(y) \\ \vdots \end{bmatrix} \end{aligned} \tag{23.72}$$

Taking the inner product of the row and column vectors, we obtain the identity

$$\sum_{n=0}^{\infty} \varphi_n(y)\varphi_n^*(x) = \delta(x - y) \tag{23.73}$$

In bra-ket notation this is simply the usual bra-ket expression for the identity operator ('resolution of the identity'), this time using harmonic oscillator basis states

$$\sum_n |\varphi_n\rangle\langle\varphi_n| = \mathbb{1} \tag{23.74}$$

Multiplying both sides of this equation by $|x\rangle$ on the left and $\langle y|$ on the right gives (23.73).

Next we turn to the matrix elements of operators. Taking the inner product of the eigenvalue equation $H^{\text{ho}}|\varphi_n\rangle = E_n|\varphi_n\rangle$ with the bra $\langle\varphi_m|$ gives

$$H^{\text{ho}}_{mn} \equiv \langle\varphi_m|H^{\text{ho}}|\varphi_n\rangle = E_n\delta_{mn} \tag{23.75}$$

i.e. H^{ho}_{mn} is a diagonal matrix

$$H^{\text{ho}} = \begin{bmatrix} E_1 & 0 & 0 & \ldots \\ 0 & E_2 & 0 & \ldots \\ 0 & 0 & E_3 & \ldots \\ \vdots & \vdots & \vdots & \vdots\vdots\vdots \\ \vdots & \vdots & \vdots & \vdots\vdots\vdots \end{bmatrix} \tag{23.76}$$

This is another example of the fact that the matrix elements of a Hermitian operator, in the basis of Hilbert space formed by its eigenstates, are always diagonal.

The matrix representations X_{mn} and P_{mn} are easily obtained using the raising and lowering operator relations

$$\begin{aligned} X &= \sqrt{\frac{\hbar}{2m\omega}}(a^\dagger + a) \\ P &= \mathrm{i}\sqrt{\frac{m\omega\hbar}{2}}(a^\dagger - a) \\ a^\dagger|\varphi_n\rangle &= \sqrt{n+1}\,|\varphi_{n+1}\rangle \\ a|\varphi_n\rangle &= \sqrt{n}\,|\varphi_{n-1}\rangle \end{aligned} \tag{23.77}$$

Then

$$\begin{aligned} X_{mn} &= \sqrt{\frac{\hbar}{2m\omega}}[\langle\varphi_m|a^\dagger|\varphi_n\rangle + \langle\varphi_m|a|\varphi_m\rangle] \\ &= \sqrt{\frac{\hbar}{2m\omega}}[\sqrt{n+1}\,\delta_{m,n+1} + \sqrt{n}\,\delta_{m,n-1}] \end{aligned} \tag{23.78}$$

and likewise

$$P_{mn} = \mathrm{i}\sqrt{\frac{m\omega\hbar}{2}}[\sqrt{n+1}\,\delta_{m,n+1} - \sqrt{n}\,\delta_{m,n-1}] \tag{23.79}$$

or

$$X = \sqrt{\frac{\hbar}{2m\omega}}\begin{bmatrix} 0 & 1 & 0 & . & . & . \\ 1 & 0 & \sqrt{2} & . & . & . \\ 0 & \sqrt{2} & 0 & \sqrt{3} & . & . \\ 0 & 0 & \sqrt{3} & 0 & \sqrt{4} & . \\ . & . & . & . & . & . \end{bmatrix}$$
$$P = \sqrt{\frac{m\omega\hbar}{2}}\begin{bmatrix} 0 & -\mathrm{i} & 0 & . & . & . \\ \mathrm{i} & 0 & -\mathrm{i}\sqrt{2} & . & . & . \\ 0 & \mathrm{i}\sqrt{2} & 0 & -\mathrm{i}\sqrt{3} & . & . \\ 0 & 0 & \mathrm{i}\sqrt{3} & 0 & -\mathrm{i}\sqrt{4} & . \\ . & . & . & . & . & . \end{bmatrix} \tag{23.80}$$

Notice that the matrix representations of H^{ho}, X, and P, are all Hermitian.

23.2.2 Landau levels again

In chapter 10 we considered the case of a particle of charge q and mass m, restricted to motion in the x–y plane, interacting with a constant magnetic field $\mathbf{B} = B\mathbf{e}_z$ oriented in the z-direction. The Hamiltonian operator, unlike the classical equations of motion, depends on the vector potential $\mathbf{A}$ rather than the $\mathbf{B} = \nabla \times \mathbf{A}$ field directly, and $\mathbf{B}$ does not determine $\mathbf{A}$ uniquely. A gauge transformation $\mathbf{A} \rightarrow \mathbf{A} + \nabla\chi$ does not change $\mathbf{B}$, nor does it change any physical observable (all such observables, in both classical and quantum physics, are gauge-invariant). Still, we must make some choice for $\mathbf{A}$ to solve the time-dependent Schrödinger equation. In chapter 10, the choice was

$$A_x = 0, \quad A_y = Bx, \quad A_z = 0 \tag{23.81}$$

Suppose we now apply a gauge transformation to obtain a different, yet physically equivalent, vector potential

$$A_x = -\frac{1}{2}By, \quad A_y = \frac{1}{2}Bx, \quad A_z = 0 \tag{23.82}$$

Now the Hamiltonian operator has the form

$$H = \frac{1}{2m}\left\{\left(P_x + \frac{1}{2}qBY\right)^2 + \left(P_y - \frac{1}{2}qBX\right)^2\right\} \tag{23.83}$$

The easy way to solve the corresponding time-independent Schrödinger equation is to define the four operators

$$\begin{aligned} P_1 &= P_y - \frac{1}{2}qBX, \qquad Q_1 = \frac{1}{qB}\left(P_x + \frac{1}{2}qBY\right) \\ P_2 &= \frac{1}{qB}\left(P_x - \frac{1}{2}qBY\right), \qquad Q_2 = P_y + \frac{1}{2}qBX \end{aligned} \tag{23.84}$$

A little algebra shows that

$$H = \frac{1}{2m}P_1^2 + \frac{1}{2}m\omega^2 Q_1^2 \quad \text{where} \quad \omega = \frac{qB}{m} \tag{23.85}$$

and that

$$\begin{aligned} [H, P_2] &= [H, Q_2] = 0 \\ [Q_1, P_1] &= [Q_2, P_2] = i\hbar\mathbb{1} \\ [Q_1, Q_2] &= [P_1, P_2] = 0 \end{aligned} \tag{23.86}$$

Landau level eigenstates in chapter 10 were expressed in a position-space basis

$$\varphi_{kn}(x, y) = \langle x, y|\varphi_{kn}\rangle \tag{23.87}$$

But for the Hamiltonian in the form (23.85), it is more convenient to express the wave function, and all the operators, in the basis of Q_1, Q_2 eigenstates

$$\begin{aligned} Q_1|q_1, q_2\rangle &= q_1|q_1, q_2\rangle, \qquad Q_2|q_1, q_2\rangle = q_2|q_1, q_2\rangle \\ \varphi_{kn}(q_1, q_2) &= \langle q_1, q_2|\varphi_{kn}\rangle \end{aligned} \tag{23.88}$$

and from the commutation relations it follows that

$$\begin{aligned} \langle q_1, q_2|P_1|\psi\rangle &= \tilde{p}_1\psi(q_1, q_2) = -i\hbar\frac{\partial}{\partial q_1}\psi(q_1, q_2) \\ \langle q_1, q_2|P_2|\psi\rangle &= \tilde{p}_2\psi(q_1, q_2) = -i\hbar\frac{\partial}{\partial q_2}\psi(q_1, q_2) \end{aligned} \tag{23.89}$$

Since H and P_2 commute, they share a common set of eigenstates, and in the $|q_1, q_2\rangle$ basis the eigenvalue equations are

$$\begin{aligned} \left\{-\frac{\hbar^2}{2m}\frac{\partial^2}{\partial q_1^2} + \frac{1}{2}m\omega^2\right\}\varphi_{kn}(q_1, q_2) &= E_{kn}\varphi_{kn}(q_1, q_2) \\ -i\hbar\frac{\partial}{\partial q_2}\varphi_{kn}(q_1, q_2) &= \hbar k\varphi_{kn}(q_1, q_2) \end{aligned} \tag{23.90}$$

The solutions are

$$\phi_{kn}(q_1, q_2) = \frac{1}{\sqrt{2\pi\hbar}} e^{ikq_2}\phi_n(q_1)$$
$$E_{kn} = E_n = \hbar\omega\left(n + \frac{1}{2}\right) \tag{23.91}$$
$$n = 0, 1, 2, \ldots, \quad \infty < k < \infty$$

where the ϕ_n are the usual harmonic oscillator wave functions. The Landau level energies and degeneracies are (as they must be) the same as found in chapter 10, with the degeneracy a consequence of the fact that H commutes with Q_2 and P_2, which do not commute with each other. But the point is that it is useful in this case to express the wave functions in the basis spanned by eigenstates of Q_1, Q_2, rather than in the usual position-space basis spanned by eigenstates of X, Y.

Subtleties

Almost everything that can be said about vectors and matrices in finite dimensions translates directly to infinite-dimensional Hilbert space, e.g. the fact that the eigenvalues of a Hermitian operator are real, and that eigenstates corresponding to different eigenvalues are orthogonal. Sometimes, however, there is a surprise, and here is an example.

The ***trace*** of a matrix A is the sum of diagonal elements of the matrix, i.e.

$$\mathrm{Tr}A \equiv \sum_{n=1}^{D} A_{nn} \tag{23.92}$$

and it is easy to show that if A and B are two $D \times D$ matrices, then

$$\mathrm{Tr}(AB) = \mathrm{Tr}(BA) \tag{23.93}$$

This immediately implies that the trace of a commutator vanishes:

$$\mathrm{Tr}[A, B] = 0 \tag{23.94}$$

On the other hand, in quantum mechanics $[X, P] = i\hbar\mathbb{1}$, which means that if X and P are represented by (infinite-dimensional) matrices, e.g. in the harmonic oscillator basis, then

$$\mathrm{Tr}[X, P] = i\hbar\mathrm{Tr}[\mathbb{1}] = i\infty \tag{23.95}$$

How can this be? We will leave the answer to an exercise at the end of the chapter.

23.3 The angular momentum basis

The norm of a wave function in spherical coordinates involves an integration over solid angle

$$1 = \int dr r^2 \int d\Omega\ \psi^*(r, \theta, \phi)\psi(r, \theta, \phi) \tag{23.96}$$

where

$$\int d\Omega \equiv \int_0^\pi d\theta \sin\theta \int_0^{2\pi} d\phi \tag{23.97}$$

Now consider the wave function for a particle moving on the surface of a sphere, with $r = 1$. Wave functions of that sort are only functions of the angular variables, and the norm is

$$1 = \int d\Omega \;\psi^*(\theta, \phi)\psi(r, \theta, \phi) \tag{23.98}$$

As usual, we may think of $\psi(\theta, \phi)$ as the (θ, ϕ)-component of the ket vector $|\psi\rangle$, which lives in the Hilbert space of all square-integrable functions $f(\theta, \phi)$, where in this case the integration is over solid angle. The set of all ket vectors $|\theta, \phi\rangle$ are a basis in this space, and

$$\begin{aligned} \psi(\theta, \phi) &= \langle\theta, \phi|\psi\rangle \\ |\psi\rangle &= \int d\Omega \;\psi(\theta, \phi)|\theta, \phi\rangle \end{aligned} \tag{23.99}$$

If we multiply the left- and right-hand sides of the second line of this equation by the bra vector $\langle\theta', \phi'|$, then consistency with the first line requires

$$\langle\theta', \phi'|\theta, \phi\rangle = \frac{1}{\sin\theta}\delta(\theta - \theta')\delta(\phi - \phi') \tag{23.100}$$

It is worth noting that the right-hand side is the angular part of the Dirac delta function in spherical coordinates

$$\begin{aligned} \delta^3(\mathbf{r} - \mathbf{r}') &= \delta(x - x')\delta(y - y')\delta(z - z') \\ &= \frac{1}{r^2}\delta(r - r')\frac{1}{\sin\theta}\delta(\theta - \theta')\delta(\phi - \phi') \end{aligned} \tag{23.101}$$

We are interested in working out the matrix elements of angular momentum operators in the position basis $\{|\theta, \phi\rangle\}$. In chapter 11 we found a representation for the angular momentum operators $\tilde{L}_x$, $\tilde{L}_y$, $\tilde{L}_z$, where in particular

$$\tilde{L}_z\phi(\theta, \phi) = -\mathrm{i}\hbar\frac{\partial}{\partial\phi}\psi(\theta, \phi) \tag{23.102}$$

In bra-ket language, this equation gives us the (θ, ϕ)-component of the ket vector $|\psi\rangle$,

$$\tilde{L}_z\phi(\theta, \phi) = \langle\theta, \phi|L_z|\psi\rangle \tag{23.103}$$

Since the unit operator is

$$\mathbb{1} = \int d\Omega \;|\theta, \phi\rangle\langle\theta, \phi| \tag{23.104}$$

we must have

$$
\begin{aligned}
\tilde{L}_z\phi(\theta,\phi) &= \langle\theta,\phi|L_z\psi\rangle \\
&= \int \mathrm{d}\Omega' \ \langle\theta,\phi|L_z|\theta',\phi'\rangle\langle\theta',\phi'|\psi\rangle \\
&= \int \mathrm{d}\Omega' \ \langle\theta,\phi|L_z|\theta',\phi'\rangle\psi(\theta',\phi') \\
&= -\mathrm{i}\hbar\frac{\partial}{\partial\phi}\psi(\theta,\phi)
\end{aligned}
\tag{23.105}
$$

From the third and fourth lines we deduce that matrix element of L_z in the position basis has to be

$$
\langle\theta,\phi|L_z|\theta',\phi'\rangle = -\mathrm{i}\hbar\frac{\partial}{\partial\phi}\left(\frac{1}{\sin\theta}\delta(\theta-\theta')\delta(\phi-\phi')\right) \tag{23.106}
$$

By the same arguments

$$
\langle\theta,\phi|L_x|\theta',\phi'\rangle = \mathrm{i}\hbar\left(\sin\phi\frac{\partial}{\partial\theta} + \cot\theta\cos\phi\frac{\partial}{\partial\phi}\right)\left(\frac{1}{\sin\theta}\delta(\theta-\theta')\delta(\phi-\phi')\right)
$$

$$
\left\langle\theta,\phi|L_y|\theta',\phi'\right\rangle = \mathrm{i}\hbar\left(-\cos\phi\frac{\partial}{\partial\theta} + \cot\theta\sin\phi\frac{\partial}{\partial\phi}\right)\left(\frac{1}{\sin\theta}\delta(\theta-\theta')\delta(\phi-\phi')\right)
$$

(23.107)

However, the set of position eigenstates $\{|\theta,\phi\rangle\}$ is not the only basis in the Hilbert space of square-integrable functions $f(\theta,\phi)$ defined on the unit sphere. An alternative basis is the set of eigenstates of the operators L^2, L_z, denoted $|lm\rangle$. These satisfy the eigenvalue equations

$$
\begin{aligned}
L^2|lm\rangle &= l(l+1)\hbar^2|lm\rangle \\
L_z|lm\rangle &= m\hbar|lm\rangle
\end{aligned}
\tag{23.108}
$$

with orthogonality relations

$$
\langle l'm'|l'm'\rangle = \delta_{l'l}\delta_{m'm} \tag{23.109}
$$

The action of the angular momentum operators L_+, L_- on angular momentum eigenstates $\{|lm\rangle\}$ was worked out in chapter 11, and using $L_x = \frac{1}{2}(L_+ + L_-)$, $L_y = \frac{1}{2i}(L_+ - L_-)$ we have the matrix elements of angular momentum operators in the angular momentum basis

$$
\begin{aligned}
\langle l'm'|L_x|lm\rangle &= \frac{1}{2}\hbar\sqrt{(l-m)(l+m+1)}\,\delta_{l'l}\delta_{m',\,m+1} \\
&\quad + \frac{1}{2}\hbar\sqrt{(l+m)(l-m+1)}\,\delta_{l'l}\delta_{m',\,m-1} \\
\left\langle l'm'\middle|L_y\middle|lm\right\rangle &= \frac{1}{2\mathrm{i}}\hbar\sqrt{(l-m)(l+m+1)}\,\delta_{l'l}\delta_{m',\,m+1} \\
&\quad - \frac{1}{2\mathrm{i}}\hbar\sqrt{(l+m)(l-m+1)}\,\delta_{l'l}\delta_{m',\,m-1} \\
\langle l'm'|L_z|lm\rangle &= m\hbar\delta_{l'l}\delta_{m'm} \\
\langle l'm'|L^2|lm\rangle &= l(l+1)\hbar^2\delta_{l'l}\delta_{mm'}
\end{aligned}
\tag{23.110}
$$

Here once again we see that the Hermitian operators L^2 and L_z, in a basis defined by their eigenstates $\{|lm\rangle\}$, are represented by diagonal matrices.

The θ, ϕ-components of the ket vectors $|lm\rangle$, which can be viewed as the wave functions of angular momentum eigenstates in position space, are the spherical harmonics

$$Y_{lm}(\theta, \phi) = \langle \theta, \phi|lm\rangle \tag{23.111}$$

The identity operator can be represented by position-space bra-ket vectors, as in equation (23.104), or by bra-ket vectors in the angular momentum basis

$$\mathbb{1} = \sum_{l=0}^{\infty}\sum_{m=-l}^{l} |lm\rangle\langle lm| \tag{23.112}$$

These lead to two identities for the spherical harmonics. The first is obtained by multiplying both sides of (23.104) on the left by the bra $\langle lm|$ and on the right by the ket $|l'm'\rangle$. The result is

$$\int \mathrm{d}\Omega\ \ Y^*_{lm}(\theta, \phi)\,Y_{l'm'}(\theta, \phi) = \delta_{ll'}\delta_{mm'} \tag{23.113}$$

which is just the orthogonality relation for spherical harmonics. The second result is obtained, in a similar way, by multiplying both sides of (23.112) on the left by the bra $\langle\theta, \phi|$ and on the right by the ket $|\theta', \phi'\rangle$, which gives the identity

$$\sum_{l=0}^{\infty}\sum_{m=-l}^{l} Y_{lm}(\theta, \phi)\,Y^*_{lm}(\theta', \phi') = \frac{1}{\sin\theta}\delta(\theta-\theta')\delta(\phi-\phi') \tag{23.114}$$

which is known as the ***closure relation*** for spherical harmonics.

From the fact that spherical harmonics have the form (23.111), and from the orthogonality relations (23.100), (23.109), it can be seen that the spherical harmonics transform basis vectors in position space to basis vectors in the angular momentum basis, and vice versa, i.e.

$$|lm\rangle = \int \mathrm{d}\Omega \; Y_{lm}(\theta, \phi)|\theta, \phi\rangle$$
$$|\theta, \phi\rangle = \sum_{l=0}^{\infty} \sum_{m=-l}^{l} Y_{lm}^{*}(\theta, \phi)|lm\rangle \tag{23.115}$$

The spherical harmonics are therefore analogous, in some ways, to the unitary matrices discussed in section 23.1, although in this case we are mixing discrete indices l, m with continuous indices θ, ϕ.

It was shown in chapter 11 that the commutation relations for angular momenta allow for half-integer values of l, m, but that these could not be realized in terms of spherical harmonics. Nevertheless, the spin angular momentum quantum numbers of some elementary particles, such as electrons and quarks, do take on half-integer values. In general, the position basis $\{|\theta, \phi\rangle\}$ can be used to represent states of orbital angular momentum, but that basis is not able to represent states of internal spin angular momentum. For this purpose we have to represent physical states as column vectors, and employ a spin angular momentum basis $\{|ss_z\rangle\}$ with half-integral quantum number s. Spin angular momentum operators are then represented as matrices in this basis. That was the approach taken in chapter 13, which we need not repeat here.

23.4 Density matrices, pure states, and mixed states

The most that can ever be known about a system is its physical state, represented by some unit vector in Hilbert space. But it is often the case that we do not have complete information. The most familiar example comes from thermodynamics, where we normally do not know the microstate s of a macroscopic object in thermal equilibrium, but we do know the probability that the object is in any given microstate, which is given by the Boltzmann distribution

$$P(s) = \frac{1}{Z} e^{-E(s)/kT} \tag{23.116}$$

In quantum theory there is an elegant way to represent situations where we have only incomplete, probabilistic knowledge of the state of the system.[2]

Let us begin by associating, to each physical state $|\psi\rangle$, an operator

$$\rho = |\psi\rangle\langle\psi| \tag{23.117}$$

with a rule for computing expectation values

$$\begin{aligned}\langle Q\rangle &= \mathrm{Tr}\, Q\rho \\ &= \sum_n \langle \phi_n|Q\rho|\phi_n\rangle\end{aligned} \tag{23.118}$$

[2] This should not be confused with incomplete knowledge about the value of observables, which is of course probabilistic even if we know the state with certainty.

where the $\{\phi_n\}$ is *any* complete set of states spanning the Hilbert space. It is easy to see that this rule gives the usual expression for expectation values

$$\begin{aligned}\langle Q\rangle &= \sum_n \langle \phi_n|Q|\psi\rangle\langle\psi|\phi_n\rangle \\ &= \sum_n \langle\psi|\phi_n\rangle\langle\phi_n|Q|\psi\rangle \\ &= \langle\psi|\left(\sum_n |\phi_n\rangle\langle\phi_n|\right)Q|\psi\rangle\end{aligned} \tag{23.119}$$

$$=\langle\psi|Q|\psi\rangle \tag{23.120}$$

where we have used the resolution of the identity (23.74). Likewise, the probability that a measurement of Q will result in the eigenvalue q_n, where $Q|q_n\rangle = q_n|q_n\rangle$, is

$$\begin{aligned}P(q_n) &= \langle q_n|\rho|q_n\rangle \\ &= \langle q_n|\psi\rangle\langle\psi|q_n\rangle\end{aligned} \tag{23.121}$$

and this is again the usual rule.

So far nothing is new. Now, however, consider a Hermitian operator ρ, not necessarily of the form (23.117), with a trace equal to unity

$$\begin{aligned}\mathrm{Tr}\rho &= \sum_n \langle\phi_n|\rho|\phi_n\rangle \\ &= 1\end{aligned} \tag{23.122}$$

Consider $\langle Q\rangle$ defined according to (23.118). What physical meaning might this expression have in general? To see that, we note that any Hermitian operator can be diagonalized in terms of its eigenstates, so in this case let $\rho|\omega_n\rangle = p_n|\omega_n\rangle$ be the relevant eigenvalue equation, and therefore

$$\rho = \sum_n p_n|\omega_n\rangle\langle\omega_n| \tag{23.123}$$

Then, taking the trace in the basis $\{\omega_n\}$, we have

$$\langle Q\rangle = \sum_n p_n\langle\omega_n|Q|\omega_n\rangle \tag{23.124}$$

This expression has a simple interpretation. The operator ρ is known in this context as a ***density operator***, and when expressed as a matrix, with components $\rho_{ij} = \langle\phi_i|\rho|\phi_j\rangle$ in some basis, it is referred to as a ***density matrix***. The density operator represents a situation in which the state of the system is not known with certainty. Rather, it is known that the system is in one of the states ω_n with corresponding probabilities p_n. The expectation value of any observable Q in state

ω_n is of course $\langle\omega_n|Q|\omega_n\rangle$. Then the expected value of observable Q is simply the sum of expectation values in each of the states ω_n, weighted by their probabilities p_n. The condition that the trace of ρ is unity is the condition that the sum of probabilities adds up to one:

$$\begin{aligned}\mathrm{Tr}\rho &= \sum_m \langle\omega_m|\rho|\omega_m\rangle \\ &= \sum_m \langle\omega_m|\left(\sum_n p_n|\omega_n\rangle\langle\omega_n|\right)|\omega_m\rangle \\ &= \sum_m p_m \\ &= 1\end{aligned} \tag{23.125}$$

In particular, the situation described by the Boltzmann probability distribution is represented by the density operator

$$\rho = \frac{1}{Z}\mathrm{e}^{-\tilde{H}/kT} \tag{23.126}$$

where the partition function is the trace

$$Z = \mathrm{Tr}\mathrm{e}^{-\tilde{H}/kT} \tag{23.127}$$

If we evaluate the trace in a complete set of energy eigenstates $|s\rangle$, then it is obvious that Z is equivalent to the usual expression for the partition function in statistical mechanics,

$$Z = \sum_s \mathrm{e}^{-E(s)/kT} \tag{23.128}$$

although of course the trace is independent of the basis chosen.

When the state of a system is known with certainty, so that ρ can be put in the form (23.117), the system is said to be in a ***pure state***. Otherwise, the system is said to be in a 'mixed state', or ***mixture***. When a system is in a pure state, the corresponding density operator has the property

$$\rho = \rho\rho \tag{23.129}$$

as is clear from (23.117). If this is not the case, then the system is in a mixed state. A good example is provided by the spin states of a spin $\frac{1}{2}$ particle, with the Hilbert space spanned by the two eigenstates $\chi_\pm$ of the spin angular momentum operator $S_z = \frac{1}{2}\sigma_z$. The most general possible density matrix has the form

$$\rho = \frac{1}{2}(\mathbb{1} + \vec{P}\cdot\vec{\sigma}) \tag{23.130}$$

where the x-, y-, z-components of $\vec{\sigma}$ are the Pauli matrices σ_x, σ_y, σ_z. This is a matrix representation of the ρ operator, whose components are evaluated in the basis $\chi_\pm$ of eigenstates of σ_z.[3] Since the Pauli matrices are traceless, and $\mathbb{1}$ is a 2×2 unit matrix, it is obvious that $\mathrm{Tr}\rho = 1$. Moreover, from the fact that $\sigma_i\sigma_j = -\sigma_j\sigma_i$ if $i \neq j$, it is easy to verify that ρ represents a pure state, meaning that the condition (23.129) is satisfied, if and only if $\vec{P} \cdot \vec{P} = 1$, and it is a mixture if $|P| < 1$. The expectation value of any component of spin angular momentum is

$$\begin{aligned}\langle S_i\rangle &= \mathrm{Tr}\left[\frac{\hbar}{2}\sigma_i\rho\right] \\ &= \frac{\hbar}{2}P_i\end{aligned} \tag{23.131}$$

The vector $\vec{P}$ is known as the ***polarization***. When $|P| = 1$, the spin of the particle is up in the $\vec{P}$-direction. When $|P| = 0$ it means that nothing is known about the spin; it is equally likely to be spin up or spin down measured in any direction. In this case, the spin is said to be unpolarized.

Under Schrödinger evolution, a pure state evolves into another pure state, i.e.

$$\rho(t_0) = |\psi(t_0)\rangle\langle\psi(t_0)| \longrightarrow \rho(t_1) = |\psi(t_1)\rangle\langle\psi(t_1)| \tag{23.132}$$

where $\psi(t_1) = \mathrm{e}^{-\mathrm{i}\tilde{H}(t_1-t_0)/\hbar}\psi(t_0)$, and the property $\rho = \rho\rho$ is preserved. Likewise a mixture evolves into another mixture

$$\rho(t_0) = \sum_n p_n|\omega_n(t_0)\rangle\langle\omega_n(t_0)| \longrightarrow \rho(t_1) = \sum_n p_n|\omega_n(t_1)\rangle\langle\omega_n(t_1)| \tag{23.133}$$

where

$$|\omega_n(t_1)\rangle = \mathrm{e}^{-\mathrm{i}\tilde{H}(t_1-t_0)/\hbar}|\omega_n(t_0)\rangle \tag{23.134}$$

A mixture *cannot* evolve into a pure state, or vice versa. The eigenstates of ρ change with time, but the set of probabilities $\{p_n\}$ does *not* change under Schrödinger evolution. Ignorance does not evolve into knowledge, nor does knowledge evolve into ignorance. At least, that is the case in the quantum theory. This fact has important consequences, when we turn to the problem of measurement in chapter 25.

23.5 Poisson brackets and canonical quantization

We are familiar with the Schrödinger equation for a system of point particles. But it is natural to ask if there are general rules for going from classical to quantum mechanics for any mechanical system, no matter how complicated. In fact there are such rules, they were discovered by Dirac, and they are based on the concept of Poisson brackets in classical Hamiltonian mechanics. An alternative (but equivalent) quantization scheme, based on the classical action, is known as path integral quantization. That scheme is discussed in a later chapter.

[3] For the corresponding density operator we replace the Pauli matrices by Pauli operators, whose matrix representation depends on the basis (see problem 2 at the end of the chapter).

We suppose that the classical Hamiltonian $H(q, p)$ is known in terms of a set $\{q^i, p_i\}$ of canonical coordinates and momenta. Let $A(q, p)$, $B(q, p)$ be any two observables which are functions of those canonical variables. The ***Poisson bracket*** of A and B is defined to be

$$[A, B]_{PB} = \sum_a \left[\frac{\partial A}{\partial q^a}\frac{\partial B}{\partial p_a} - \frac{\partial A}{\partial p_a}\frac{\partial B}{\partial q^a} \right] \tag{23.135}$$

Note, in particular, that

$$[q^a, q^b]_{PB} = 0 \qquad [p_a, p_b]_{PB} = 0 \qquad [q^a, p_b]_{PB} = \delta_{ab} \tag{23.136}$$

An important feature is that any equation expressed in terms of Poisson brackets holds true for every possible choice of canonical conjugate variables $\{q^a, p_a\}$. For example, for a spinless point particle, $\{x, y, z, p_x, p_y, p_z\}$ is one choice of canonical coordinates and momenta, but one could also go to spherical coordinates, in which $\{r, \theta, \phi, p_r, p_\theta, p_\phi\}$ is an alternative set of canonical coordinates and momenta. Suppose $\{q^a, p_a\}$ and $\{q'^a, p'_a\}$ are two sets of canonical conjugate variables, and of course it is possible to express one set as functions of the other, i.e. $q'^a = q'^a(\{q^c, p_c\})$, $p'^a = p'^a(\{q^c, p_c\})$. Observables A, B can therefore be expressed as functions of either set. The statement is that the Poisson bracket in the primed variables equals the Poisson bracket in the unprimed variables

$$\sum_a \left[\frac{\partial A}{\partial q^a}\frac{\partial B}{\partial p_a} - \frac{\partial A}{\partial p_a}\frac{\partial B}{\partial q^a} \right] = \sum_a \left[\frac{\partial A}{\partial q'^a}\frac{\partial B}{\partial p'_a} - \frac{\partial A}{\partial p'_a}\frac{\partial B}{\partial q'^a} \right] \tag{23.137}$$

The proof can found in, e.g., Goldstein [1]. So Poisson brackets are, in a sense, coordinate independent.

Canonical quantization is the quantization scheme devised by Dirac, which is based on reinterpreting the Poisson bracket. Starting from a Hamiltonian system with canonical conjugate variables $\{q^a, p_a\}$, quantization is the procedure which identifies the coordinates and momenta as operators $\{Q^a, P_a\}$, and Poisson brackets are replaced by operator commutators according to the rule

$$[A, B]_{PB} \to \frac{1}{i\hbar}[A, B] \tag{23.138}$$

Making this identification for the Poisson brackets (23.136) brings us again to the basic operator commutator equations[4]

$$\begin{aligned} [Q^a, Q^b] &= 0 \\ [P_a, P_b] &= 0 \\ [Q^a, P_b] &= i\hbar\delta_{ab}\mathbb{1} \end{aligned} \tag{23.139}$$

[4] A word of caution, however. It is not always true that if $[A, B]_{PB} = C(q, p)$, then $[A, B] = i\hbar C(Q, P)$. It is only assumed that this identity holds for the commutators of canonical coordinates and momenta, and that assumption is the basis of canonical quantization.

Now in classical physics,

$$\begin{aligned}
[A, H]_{PB} &= \sum_a \left[\frac{\partial A}{\partial q^a}\frac{\partial H}{\partial p_a} - \frac{\partial A}{\partial p_a}\frac{\partial H}{\partial q^a} \right] \\
&= \sum_a \left[\frac{\partial A}{\partial q^a}\frac{\mathrm{d}q^a}{\mathrm{d}t} + \frac{\partial A}{\partial p_a}\frac{\mathrm{d}p_a}{\mathrm{d}t} \right] \\
&= \frac{\mathrm{d}A}{\mathrm{d}t}
\end{aligned} \tag{23.140}$$

where we have used Hamilton's equations in the second line. In quantum mechanics (in the so-called ***Schrödinger picture***) it is physical states, rather than operators, which have time dependence, so this Poisson bracket equation is interpreted in terms of expectation values

$$\frac{\mathrm{d}}{\mathrm{d}t}\langle\psi(t)|A|\psi(t)\rangle = \frac{\mathrm{i}}{\hbar}\langle\psi(t)|[A, H]|\psi(t)\rangle \tag{23.141}$$

which has been seen before (equation (7.134)) in an earlier chapter. In fact this equation has a simple, albeit rather formal, solution for the time dependence of a state $|\psi(t)\rangle$:

$$|\psi(t)\rangle = \mathrm{e}^{\mathrm{i}Ht/\hbar}|\psi(0)\rangle \tag{23.142}$$

Taking the time derivative of both sides takes us to the Schrödinger equation for ket vectors

$$\mathrm{i}\hbar\frac{d}{\mathrm{d}t}|\psi(t)\rangle = H[Q, P]|\psi(t)\rangle \tag{23.143}$$

We note in passing that there is an alternative scheme known as the ***Heisenberg picture*** which puts the time dependence in the operators, according to

$$\frac{\mathrm{d}A}{\mathrm{d}t} = \frac{\mathrm{i}}{\hbar}[A, H] \tag{23.144}$$

while the ket vectors stay fixed. This equation has the formal solution

$$A(t) = \mathrm{e}^{\mathrm{i}Ht/\hbar}A(0)\mathrm{e}^{-\mathrm{i}Ht/\hbar} \tag{23.145}$$

Of course the Schrödinger and Heisenberg pictures are equivalent, as can be seen by computing the expectation value of A at time t, starting from the Heisenberg picture:

$$\begin{aligned}
\langle A(t)\rangle &= \langle\psi(0)|A(t)|\psi(0)\rangle \\
&= \langle\psi(0)|\mathrm{e}^{\mathrm{i}Ht/\hbar}A(0)\mathrm{e}^{-\mathrm{i}Ht/\hbar}|\psi(0)\rangle \\
&= \langle\psi(t)|A(0)|\psi(t)\rangle
\end{aligned} \tag{23.146}$$

where the last line is the expectation value of A in the Schrödinger picture.

Denote eigenstates of the Q^a operators by $|q\rangle$ and eigenstates of the conjugate momentum operators P_a by $|p\rangle$. We get the Schrödinger equation for position-space wave functions by multiplying both sides of (23.143) by the bra $\langle q|$

$$\begin{aligned} i\hbar\frac{d}{dt}\langle q|\psi\rangle &= \langle q|H[Q, P]|\psi\rangle \\ i\hbar\frac{d}{dt}\psi(q) &= H[\tilde{q}, \tilde{p}]\psi(q) \end{aligned} \tag{23.147}$$

where, following the discussion of position- and momentum-space operators in the previous section,

$$\tilde{q}^a\psi(q) = q^a\psi(q), \qquad \tilde{p}_a\psi(q) = -i\hbar\frac{\partial}{\partial q^a}\psi(q) \tag{23.148}$$

For a point particle moving in one dimension, (23.147) is just the usual Schrödinger equation

$$i\hbar\frac{d}{dt}\psi(x) = \left\{\frac{1}{2m}\left(-i\hbar\frac{\partial}{\partial x}\right)^2 + V(x)\right\}\psi(x) \tag{23.149}$$

On the other hand, multiplying both sides of (23.143) by the bra $\langle p|$, we find the Schrödinger equation in a less familiar form, that describes the time evolution of momentum-space wave functions

$$\begin{aligned} i\hbar\frac{d}{dt}\langle p|\psi\rangle &= \langle p|H[Q, P]|\psi\rangle \\ i\hbar\frac{d}{dt}\psi(p) &= H[\tilde{q}, \tilde{p}]\psi(p) \end{aligned} \tag{23.150}$$

where this time

$$\tilde{q}^a\psi(p) = i\hbar\frac{\partial}{\partial p_a}\psi(p), \qquad \tilde{p}_a\psi(p) = p_a\psi(p) \tag{23.151}$$

For the point particle in one dimension, the Schrödinger equation is then

$$i\hbar\frac{d}{dt}\psi(p) = \left\{\frac{p^2}{2m} + V\left(i\hbar\frac{\partial}{\partial p}\right)\right\}\psi(p) \tag{23.152}$$

The is usually not the most practical version of the Schrödinger equation, since the potential operator $V(\tilde{x})$ is now difficult to handle. The exception is for the harmonic oscillator, where the momentum-space version of the Schrödinger equation is exactly soluble (see problem 6 below).

Problems

1. Verify the outer product representation of the matrix

$$[L] = \begin{bmatrix} 2 & 1 - i \\ 1 + i & 3 \end{bmatrix} \tag{23.153}$$

in terms of its eigenstates and eigenvalues (i.e. repeat steps (23.18)–(23.20) for this matrix).

2. Express the Pauli operators as matrices in the basis spanned by the eigenstates of σ_y.

3. Show that the expression

$$\langle u|M|v\rangle = \sum_{ij} u_i^* M_{ij} v_j \tag{23.154}$$

is basis independent; i.e. components depend on the basis, but $\langle u|M|v\rangle$ does not.

4. Resolve the trace paradox discussed at the end of section 23.2. Do this by truncating the infinite matrices X, P in equation (23.80) to $D \times D$ dimensions, and explicitly compute the commutator $[X, P]$ of the truncated matrices. How is the commutator relation $[X, P] = i\hbar\mathbb{1}$ obtained in the $D \to \infty$ limit?

5. Let Q be the charge stored in a capacitor in a simple LC circuit. The energy stored in the capacitor is $\frac{1}{2}Q^2/C$, and the energy stored in the inductor is $\frac{1}{2}L(\mathrm{d}Q/\mathrm{d}t)^2$. From this information, construct the Lagrangian and the Hamiltonian. Apply canonical quantization, and obtain the Schrödinger equation for a quantum LC circuit in both the Q-basis and the 'momentum' basis. Compute, in the ground state of the system, the uncertainty in the charge stored in the capacitor, and the uncertainty in the current $I = \mathrm{d}Q/\mathrm{d}t$.

6. The energy eigenstates of the quantized harmonic oscillator were obtained, in position space, in equation (9.53) of chapter 9. Using a similar approach, find the harmonic oscillator eigenstates of the Schrödinger equation in momentum space.

7. A particle of mass m is moving in one dimension along the x-axis. Assuming the corresponding Hamiltonian has a discrete set of energy eigenvalues E_n and corresponding eigenstates $|\phi_n\rangle$, prove the identity

$$\sum_j (E_j - E_k)\left|\langle\phi_j|x|\phi_k\rangle\right|^2 = \frac{\hbar^2}{2m}$$

and check this identity in the case of the harmonic oscillator.

Reference

[1] Goldstein H, Poole C and Safko J 2002 *Classical Mechanics* (Reading, MA: Addison Wesley)

IOP Publishing

An Introduction to Quantum Theory

Jeff Greensite

Chapter 24

Entangled states

No man is an Iland, intire of itselfe; every man is a peece of the Continent, a part of the maine; if a Clod bee washed away by the Sea, Europe is the lesse, as well as if a Promontorie were...

John Donne

Suppose we have a baseball in Philadelphia, and a football in Detroit, and we want to specify the physical state of the combined baseball–football system. In classical physics, this is done by first specifying the state of the baseball (its mass, composition, location, velocity ...), and then specifying, in a similar way, the state of the football. There are no baseball+football states that *cannot* be described in this way. If we decide to change the physical state of the football (by giving it a kick, say) then of course this will not very much affect the state of the far away baseball.

The point here is that the physical states of composite systems in classical physics are ***separable***, in the sense that the description of the state of the whole system can always be reduced to a specification of the state of the parts. Suppose, instead of sports equipment, we consider the state of a system of two equal mass particles, which is represented by a point in 12-dimensional phase space $\{\mathbf{x}_1, \mathbf{x}_2, \mathbf{p}_1, \mathbf{p}_2\}$. Obviously, this state completely determines, and is determined by, the physical state $\{\mathbf{x}_1, \mathbf{p}_1\}$ of particle 1, and the physical state $\{\mathbf{x}_2, \mathbf{p}_2\}$ of particle 2. Moreover, if particle 1 is not in interaction with particle 2, then measurement of its location and/or momentum will not in any way affect the position or momentum of particle 2.

Physical states of composite systems, which have the same 'common sense' property of separability, certainly exist in quantum mechanics. For example, suppose particle 1 is in state ψ_A and particle 2 is in state ψ_B. Then the state of the combined 1–2 system is determined to be

$$\psi(x_1, x_2) = \psi_A(x_1)\psi_B(x_2) \tag{24.1}$$

doi:10.1088/978-0-7503-1167-0ch24

Likewise, if particle 1 is in state ψ_C, and particle 2 is in state ψ_D, the combined system is in state

$$\psi(x_1, x_2) = \psi_C(x_1)\psi_D(x_2) \tag{24.2}$$

But what do we make of the (normalized) superposition

$$\psi(x_1, x_2) = a\psi_A(x_1)\psi_B(x_2) + b\psi_C(x_1)\psi_D(x_2) \tag{24.3}$$

According to the principles of quantum mechanics, physical states are normalized vectors in Hilbert space, and the sum of two vectors is another vector. If that vector is normalized, it is a physical state too. So the wave function (24.3) is, from this point of view, entirely kosher: it is a physically allowed state of the two-particle system. But then if we ask: what is the state of particle 1, when the two-particle system is in state (24.3), we discover that we cannot answer the question! Particle 1 is not in state ψ_A, nor is it in state ψ_C; neither is it in a linear combination such as $c\psi_A(x_1) + d\psi_C(x_1)$. The fact is, the state of particle 1 has become entangled with that of particle 2; it is impossible to specify the state of either particle separately, even if the two particles are non-interacting, and very far apart.

The fact that the quantum states of composite systems can be inseparable or, to use the standard term, ***entangled*** in this way was first noticed by Einstein, Podolsky, and Rosen (EPR) in 1933. The consequences of this entanglement are truly mind-bending, even by the generous mind-bending standards of quantum mechanics.

24.1 The EPR 'paradox'

Einstein, Podolsky, and Rosen were of the opinion that quantum mechanics is in some way incomplete, because it leads to (what they considered to be) a paradox in certain circumstances. To illustrate their point, we consider the following experiment (figure 24.1). Every time a button is pushed, two spin 1/2 particles are ejected, in opposite directions, each particle eventually passing through a detector which measures the spin of the particle along a given axis. The detectors can be rotated,

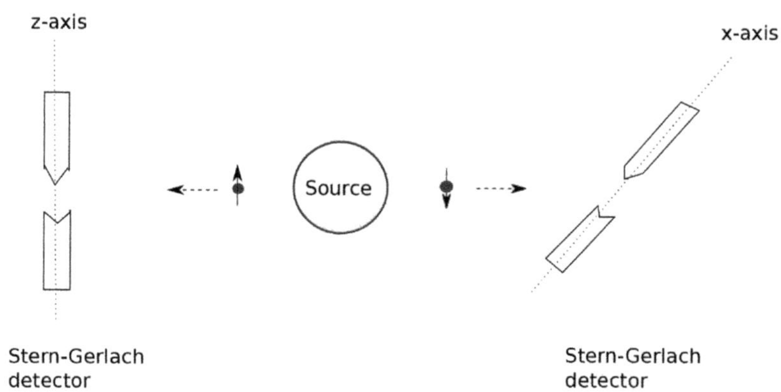

Figure 24.1. Setup illustrating the EPR 'paradox' concerning entangled states. Two spin 1/2 particles are emitted by the source, and their spins along any axis (e.g. the z- and x-axes) are measured by a Stern–Gerlach apparatus. Does a measurement carried out on one particle affect the state of the other?

so that the spin along any axis, in particular along the x- and z-axes, can be measured. After many trials, the following result is reported:

A. Both detectors are set to measure spin along the z-axis. Whenever the spin of the particle on the left is found to be spin up, and this occurs 50% of the time, the spin of the particle on the right is found to be spin down. Likewise, whenever the particle on the left is found to be spin down, which is the other 50% of the time, the particle on the right is found to be spin up.

This means that by measuring the z-component of the spin of the particle on the left, we can determine the z-component of the spin of the particle on the right, without actually *interacting* with the particle on the right. Suppose, in a particular run, that the particle which moves to the left (which we will now call 'L') is found to be spin-up. Since the measurement on the left (goes the EPR argument) could not possibly affect the spin of the other particle ('R') on the right, and yet we *know* that R is spin down, it must be that R was spin down even *before* we made the measurement of L. Therefore, the spin state of the particles before entering the detectors must have been

$$\psi_1 = \alpha_z^L \beta_z^R \tag{24.4}$$

where α_z, β_z refer to spin up and down states, respectively, along the z-axis. For future reference, let us recall from chapter 13 that

$$\alpha_z = \begin{bmatrix} 1 \\ 0 \end{bmatrix} \qquad \beta_z = \begin{bmatrix} 0 \\ 1 \end{bmatrix} \tag{24.5}$$

and

$$\alpha_x = \frac{1}{\sqrt{2}} \begin{bmatrix} 1 \\ 1 \end{bmatrix} \qquad \beta_x = \frac{1}{\sqrt{2}} \begin{bmatrix} 1 \\ -1 \end{bmatrix} \tag{24.6}$$

Now, by the same reasoning, if the particle on the left were found to be spin down, then the two-particle state would have been be

$$\psi_2 = \beta_z^L \alpha_z^R \tag{24.7}$$

even before the measurement was made. According to (A), in half the runs the particles are in state ψ_1, and in the other half they are in state ψ_2.

This reasoning seems pretty solid, but now a troubling discovery is made. After many more runs, it is found that:

B. When both detectors are set to measure spin in the x-direction, the two particles are *never* found to each have spin up, or each have spin down.

The problem is that, if the particles were initially in state ψ_1, then the probability that a measurement would find both spins up in the x-direction is

$$|\langle\alpha_x^L\alpha_x^R|\psi_1\rangle|^2 = \left|\left\langle\alpha_x^L\middle|\alpha_z^L\right\rangle\left\langle\alpha_x^R\middle|\beta_z^R\right\rangle\right|^2 = \frac{1}{4} \tag{24.8}$$

Likewise, if the particles were initially in state ψ_2, the probability to find both spins up in the x-direction would be

$$|\langle\alpha_x^L\alpha_x^R|\psi_2\rangle|^2 = \left|\left\langle\alpha_x^L|\beta_z^L\right\rangle\langle\alpha_x^R|\alpha_z^R\rangle\right|^2 = \frac{1}{4} \tag{24.9}$$

Then if the particles are half the time in state ψ_1, and half the time in state ψ_2, the total probability for finding both particles with spin up along the x-axis is simply the average of the two probabilities, i.e.

$$P(\text{up up})_{x\text{-axis}} = \frac{1}{2}\left|\left\langle\alpha_x^L\alpha_x^R\middle|\psi_1\right\rangle\right|^2 + \frac{1}{2}\left|\left\langle\alpha_x^L\alpha_x^R\middle|\psi_2\right\rangle\right|^2 = \frac{1}{4} \tag{24.10}$$

in total contradiction to the reported fact (B)! Conclusion: no experiment should report both results (A) *and* (B).

And yet, as EPR pointed out, there are quantum-mechanical states which lead to precisely these results. Suppose that the two particles, before reaching the detectors, are in a spin-0 state

$$\psi = \frac{1}{\sqrt{2}}(\psi_1 - \psi_2) = \frac{1}{\sqrt{2}}\left(\alpha_z^L\beta_z^R - \beta_z^L\alpha_z^R\right) \tag{24.11}$$

Then the probability to find L with spin up and R with spin down along the z-axis is

$$\begin{aligned} P(\text{up down})_{z\text{-axis}} &= \left|\left\langle\alpha_z^L\beta_z^R|\psi\right\rangle\right|^2 \\ &= \frac{1}{2}\left|\left\langle\alpha_z^L\beta_z^R|\psi_1\right\rangle - \left\langle\alpha_z^L\beta_z^R|\psi_2\right\rangle\right|^2 \\ &= \frac{1}{2}(1-0)^2 \\ &= \frac{1}{2} \end{aligned} \tag{24.12}$$

Likewise, the probability to find L with spin down and R with spin up in the direction is

$$
\begin{aligned}
P(\text{down up})_{z\text{-axis}} &= \left|\left\langle \beta_z^L \alpha_z^R | \psi \right\rangle\right|^2 \\
&= \frac{1}{2}\left|\left\langle \beta_z^L \alpha_z^R | \psi_1 \right\rangle - \left\langle \beta_z^L \alpha_z^R | \psi_2 \right\rangle\right|^2 \\
&= \frac{1}{2}(0 - 1)^2 \\
&= \frac{1}{2}
\end{aligned}
\tag{24.13}
$$

These two probabilities are exactly what is reported in (A). Then we compute the probability to find both L and R with spin up along the x-axis

$$
\begin{aligned}
P(\text{up up})_{z\text{-axis}} &= \left|\left\langle \alpha_x^L \alpha_x^R \middle| \psi \right\rangle\right|^2 \\
&= \frac{1}{2}\left|\left\langle \alpha_x^L \alpha_x^R \middle| \psi_1 \right\rangle - \left\langle \alpha_x^L \alpha_x^R \middle| \psi_2 \right\rangle\right|^2 \\
&= \frac{1}{2}\left|\left\langle \alpha_x^L \middle| \alpha_z^L \right\rangle\left\langle \alpha_x^R \middle| \beta_z^R \right\rangle - \left\langle \alpha_x^L \middle| \beta_z^L \right\rangle\left\langle \alpha_x^R \middle| \alpha_z^R \right\rangle\right|^2 \\
&= \frac{1}{2}\left(\frac{1}{2} - \frac{1}{2}\right)^2 \\
&= 0
\end{aligned}
\tag{24.14}
$$

which is the same as result (B).

Now if both (A) and (B) are true, it means that before any measurement takes place, R is neither in state β_z^R (as in state ψ_1), nor in state α_z^R (as in ψ_2), because this would contradict (B). So, suppose we measure the spin of L along the z-axis, and it is found to be in spin up. At that point, we know that R is now spin down. But we have already concluded that before the measurement took place, R was neither spin up nor spin down. This means that, although R was not in a spin-down state initially, it must have 'jumped' into a spin-down state after L was found in spin up, *even if, at the time of measurement, L and R were very far apart*! But how could a measurement of L, which perhaps is in Philadelphia, influence the state of R, which is maybe in Detroit? This is the EPR 'paradox', which led Einstein, Podolsky, and Rosen to conclude that the quantum-mechanical description of Nature is somehow incomplete.

24.2 Bell's theorem

Quantum mechanics might be wrong, but so far there is no evidence of that. So in what way could it be incomplete? One possibility is that, contrary to everything said so far in this book, the physical state of a particle is *not* completely specified by its wave function. Perhaps the particle also has some other, 'hidden' degrees of freedom, which determine with certainty, rather than probabilistically, the outcome of any measurement of the particle. The probabilistic nature of quantum mechanics, according to this view, is simply due to our ignorance of the actual values of these 'hidden variables'.

If the hidden variable idea were true, it seems that we might be able to wriggle out of the EPR paradox. Denote the hidden variable by μ, and suppose the following values imply the following results of spin measurements along the z- and x-axes, respectively:

$$\begin{array}{llll} \mu = 1 & \text{leads to} & \text{up}_z & \text{or up}_x \\ \mu = 2 & \text{"} & \text{up}_z & \text{" down}_x \\ \mu = 3 & \text{"} & \text{down}_z & \text{" up}_x \\ \mu = 4 & \text{"} & \text{down}_z & \text{" down}_x \end{array} \tag{24.15}$$

Then we could explain (A) and (B), for example, by saying that the two particles come out of the apparatus only with the hidden variable combinations

$$(\mu^L, \mu^R) = (1, 4) \quad \text{or} \quad (4, 1) \quad \text{or} \quad (2, 3) \quad \text{or} \quad (3, 2) \tag{24.16}$$

Of course this is not yet a theory, only an idea for a theory, but the idea itself was squelched by a celebrated theorem proved by John Bell in 1964, which essentially says the following:

Bell's theorem

No local hidden variable theory can agree with all of the predictions of quantum mechanics.

The restriction to a local theory means that theory does not contain any non-local 'action-at-a-distance' interactions or, equivalently, any effects which propagate from one place to another at greater than the speed of light. This means that a measurement of L cannot affect the hidden variables of R; at least, not before the time required for light to travel from the left detector to the particle on the right.

In order to demonstrate Bell's theorem, we generalize the two-electron experiment slightly. There is no restriction to measuring the component of spin angular momentum in only the x- or z-directions, spin can be measured in any direction, so let us imagine that Alice, on the left, measures the component of spin in the **a**-direction, and Bob, on the right, measures the component of spin in the **b**-direction. It is assumed that the two electrons moving to the left and right are in an antisymmetric spin 0 state

$$\psi = \frac{1}{\sqrt{2}}\left(\alpha_z^L \beta_z^R - \beta_z^L \alpha_z^R\right) \tag{24.17}$$

Spin up along the **a**- or **b**-directions is assigned the value +1, and spin down is assigned the value −1. If Alice and Bob take the product of these two values in each experiment, and then average their results for the product over many identical experiments, then their average should approach the expectation value

$$P(\mathbf{a}, \mathbf{b}) = \langle\psi|(\mathbf{a}\cdot\sigma^L)(\mathbf{b}\cdot\sigma^R)|\psi\rangle \tag{24.18}$$

where the Pauli matrices $\sigma^{L,\,R}$ act on the spin state of the particle moving to the left or right, respectively, and

$$\begin{aligned} \mathbf{a}\cdot\sigma^L &= a_x\sigma_x^L + a_y\sigma_y^L + a_z\sigma_z^L \\ &= \begin{bmatrix} a_z & a_x - \mathrm{i}a_y \\ a_x + \mathrm{i}a_y & -a_z \end{bmatrix}^L \\ \mathbf{b}\cdot\sigma^R &= b_x\sigma_x^R + b_y\sigma_y^R + b_z\sigma_z^R \\ &= \begin{bmatrix} b_z & b_x - \mathrm{i}b_y \\ b_x + \mathrm{i}b_y & -b_z \end{bmatrix}^R \end{aligned} \tag{24.19}$$

Then

$$\begin{aligned} P(\mathbf{a}, \mathbf{b}) = \frac{1}{2}([10]^L[01]^R - [01]^L[10]^R)&\begin{bmatrix} a_z & a_x - \mathrm{i}a_y \\ a_x + \mathrm{i}a_y & -a_z \end{bmatrix}^L \\ \times &\begin{bmatrix} b_z & b_x - \mathrm{i}b_y \\ b_x + \mathrm{i}b_y & -b_z \end{bmatrix}^R \\ \times &\left(\begin{bmatrix}1\\0\end{bmatrix}^L\begin{bmatrix}0\\1\end{bmatrix}^R - \begin{bmatrix}0\\1\end{bmatrix}^L\begin{bmatrix}1\\0\end{bmatrix}^R\right) \end{aligned} \tag{24.20}$$

Grouping together the L vectors and matrices, and the R vectors and matrices, and carrying out the required matrix multiplications and inner products, the result is simply

$$P(\mathbf{a}, \mathbf{b}) = -\mathbf{a}\cdot\mathbf{b} \tag{24.21}$$

Bell's theorem says that the above result cannot be reproduced, for arbitrary $\mathbf{a}$, $\mathbf{b}$ by any local hidden variables theory, and the proof is fairly simple. We assume that there is some hidden variable (or set of variables) μ that determines the result of any experiment, for any choice of directions $\mathbf{a}$, $\mathbf{b}$. This means that there is some function $L(\mathbf{a}, \mu) = \pm 1$ such that when $L(\mathbf{a}, \mu) = +1$ the particle on the left will be found by Alice to have spin up in the $\mathbf{a}$-direction, and when $L(\mathbf{a}, \mu) = -1$ the particle on the left will be found to have spin down in that direction. There must exist a similar function, $R(\mathbf{b}, \mu) = \pm 1$ for the particle moving to the right. Neither Alice nor Bob is aware of the value of μ, which varies from one experiment to the next, but we can assume that on average the number N of experiments with μ in some finite range $\Delta\mu$ is given by a distribution $N = \rho(\mu)\Delta\mu$. If there are such hidden variables, then

$$P(\mathbf{a}, \mathbf{b}) = \int d\mu \ \rho(\mu)L(\mathbf{a}, \mu)R(\mathbf{b}, \mu) \tag{24.22}$$

Now according to (24.21), $P(\mathbf{a}, \mathbf{a}) = -1$, i.e. the spins of the two electrons are always anti-aligned, if Alice and Bob both measure the spin component in the same direction. If the hidden variable idea is right, it implies that

$$R(\mathbf{a}, \mu) = -L(\mathbf{a}, \mu) \tag{24.23}$$

and therefore

$$P(\mathbf{a}, \mathbf{b}) = -\int d\mu \ \rho(\mu)L(\mathbf{a}, \mu)L(\mathbf{b}, \mu) \tag{24.24}$$

Now let $\mathbf{c}$ be any other unit vector. Clearly

$$\begin{aligned} P(\mathbf{a}, \mathbf{b}) - P(\mathbf{a}, \mathbf{c}) &= -\int d\mu \ \rho(\mu)[L(\mathbf{a}, \mu)L(\mathbf{b}, \mu) - L(\mathbf{a}, \mu)L(\mathbf{c}, \mu)] \\ &= -\int d\mu \ \rho(\mu)[1 - L(\mathbf{b}, \mu)L(\mathbf{c}, \mu)]L(\mathbf{a}, \mu)L(\mathbf{b}, \mu) \qquad (24.25) \\ &= -\int d\mu \ \rho(\mu)[1 + L(\mathbf{b}, \mu)R(\mathbf{c}, \mu)]L(\mathbf{a}, \mu)L(\mathbf{b}, \mu) \end{aligned}$$

where in the second line we have used the fact that $L^2(\mathbf{b}, \mu) = 1$. Now take the absolute value of both sides

$$|P(\mathbf{a}, \mathbf{b}) - P(\mathbf{a}, \mathbf{c})| = \left| \int d\mu \ \rho(\mu)[1 + L(\mathbf{b}, \mu)R(\mathbf{c}, \mu)]L(\mathbf{a}, \mu)L(\mathbf{b}, \mu) \right| \tag{24.26}$$

and note that part of the integrand is positive semi-definite, i.e.

$$\rho(\mu)[1 + L(\mathbf{b}, \mu)R(\mathbf{c}, \mu)] \geqslant 0 \tag{24.27}$$

while the other factor, $L(\mathbf{a}, \mu)L(\mathbf{b}, \mu)$, can only be ± 1. It follows that dropping the $L(\mathbf{a}, \mu)L(\mathbf{b}, \mu)$ factor can only increase the magnitude of the integral, and therefore

$$|P(\mathbf{a}, \mathbf{b}) - P(\mathbf{a}, \mathbf{c})| \leqslant \int d\mu \ \rho(\mu)[1 + L(\mathbf{b}, \mu)R(\mathbf{c}, \mu)] \tag{24.28}$$

So if there are hidden variables which determine, in any experiment, the results obtained by Alice and Bob, then those results must obey ***Bell's inequality***

$$|P(\mathbf{a}, \mathbf{b}) - P(\mathbf{a}, \mathbf{c})| \leqslant 1 + P(\mathbf{b}, \mathbf{c}) \tag{24.29}$$

The punchline is that the inequality (24.29) is inconsistent with the quantum-mechanical prediction (24.21). For example, let $\mathbf{a}$ be the unit vector in the x-direction,

b the unit vector in the y-direction, and **c** a unit vector which lies in the x–y-plane at an angle of 45 degrees as measured from the x-axis. Then inserting the values determined by (24.21):

$$\begin{aligned} |P(\mathbf{a}, \mathbf{b}) - P(\mathbf{a}, \mathbf{c})| &= 0.707 \\ 1 + P(\mathbf{b}, \mathbf{c}) &= 0.293 \end{aligned} \tag{24.30}$$

which is in clear violation of the inequality (24.29) required by a local hidden variable theory. Locality enters in the assumption that $L(\mathbf{a}, \mu)$ and $R(\mathbf{b}, \mu)$ are independent. Bob's results do not depend on whatever **b**-direction Alice chooses in a given experimental run, meaning that whatever adjustments Alice makes to her apparatus will not affect the particle on its way to Bob.

This brings us to an experimental question: maybe quantum mechanics is wrong! Perhaps when an appropriate experiment is carried out, it will be found that the quantum-mechanical prediction (24.21) is wrong, and the hidden variable inequality (24.29) is right? The relevant experimental tests, in which the two particles were two entangled photons produced in a 'cascade' transition (two sequential transitions) in Calcium atoms, were performed by Aspect and his collaborators in the 1980s [1]. Various refinements of the Aspect experiments have been made since then, and in all cases the quantum-mechanical predictions have been confirmed. It seems that the mysterious non-local behavior of quantum theory, in which a measurement of particle 1 apparently causes the distant particle 2 to jump into a state of definite spin, cannot be cheaply explained by a local hidden variables theory.

24.3 Entangled states and causality

Is it possible to use entanglement to send signals from Alice to Bob faster than the speed of light? If a measurement by Alice instantaneously affects the state of the electron heading towards Bob, then it seems like this might be a possibility. But if one could send messages faster than the speed of light, then we know from special relativity that it would also be possible to send messages backwards in time. And that would raise all of the famous paradoxes associated with time travel. The fact that the past determines the future, but the future cannot in any way change the past, is known as ***causality***. Most physicists believe in it.

For better or worse, entanglement respects causality; Alice cannot send a message to Bob by either performing, or not performing, some measurement on her part of the entangled state. It is not necessary that the entangled objects are electrons, so let us suppose that Alice measures some observable A of system 1, and Bob measures some other observable B of system 2, and denote the respective eigenstates by ket vectors $\{|a_m(1)\rangle, |b_n(2)\rangle\}$, where

$$\tilde{A}|a_m(1)\rangle = a_m|a_m(1)\rangle, \qquad \tilde{B}|b_m(2)\rangle = b_m|b_m(2)\rangle \tag{24.31}$$

In other words, $|a_m(1)\rangle$ represents an eigenstate of observable A in system 1 with eigenvalue a_m, and similarly for $|b_m(2)\rangle$. Now let us suppose that $\{|\phi_m(1)\rangle, |\varphi_n(2)\rangle\}$ is some complete set of states for systems 1 and 2, so that any state $\psi(1, 2)$ of the combined system can be expanded as a sum

$$\psi(1, 2) = \sum_{mn} c_{mn}\phi_m(1)\varphi_n(2) \tag{24.32}$$

First imagine that Alice does not carry out any measurement; she just lets object 1 go whizzing past. Then the expectation value for Bob is

$$\begin{aligned}\langle B\rangle &= \langle\psi(1, 2)|B|\psi(1, 2)\rangle \\ &= \sum_{m'n'}\sum_{m,n} c^*_{m'n'}c_{mn}\big\langle\phi_{m'}(1)|\phi_m(1)\big\rangle\langle\varphi_{n'}(2)|B|\varphi_n(2)\rangle \\ &= \sum_{nn'm} c^*_{mn'}c_{mn}\langle\varphi_{n'}(2)|B|\varphi_n(2)\rangle\end{aligned} \tag{24.33}$$

Next suppose Alice does measure observable A, but does not have the time or inclination to communicate the result to Bob. We want to see if the fact that Alice makes a measurement, or the particular observable she measures, makes any difference to the expectation value of B as measured by Bob. Begin from the joint probability that Alice finds $A = a_i$, and Bob finds $B = b_j$, which according to the standard rule is

$$\begin{aligned}\text{Prob}(a_i, b_j) &= \left|\big\langle a_i(1)b_j(2)|\psi(1, 2)\big\rangle\right|^2 \\ &= \left|\sum_{mn} c_{mn}\big\langle a_i(1)|\phi_m(1)\big\rangle\big\langle b_j(2)|\varphi_n(2)\big\rangle\right|^2\end{aligned} \tag{24.34}$$

Then Bob's expectation value for B, not knowing Alice's result, is (according to the usual rules of probability)

$$\langle B\rangle = \sum_j b_j \sum_i \text{Prob}(a_i, b_j) \tag{24.35}$$

Substituting (24.34) on the right-hand side,

$$\begin{aligned}\langle B\rangle &= \sum_{ij} b_j \left|\sum_{mn} c_{mn}\big\langle a_i(1)|\phi_m(1)\big\rangle\big\langle b_j(2)|\varphi_n(2)\big\rangle\right|^2 \\ &= \sum_{ij} b_j \sum_{m'n'}\sum_{mn} c^*_{m'n'}c_{mn}\big\langle\phi_{m'}(1)|a_i(1)\big\rangle\big\langle a_i(1)|\phi_m(1)\big\rangle\big\langle\varphi_{n'}(2)|b_j(2)\big\rangle\big\langle b_j(2)|\varphi_n(2)\big\rangle \\ &= \sum_{nn'm} c^*_{mn'}c_{mn}\langle\varphi_{n'}(2)|B|\varphi_n(2)\rangle\end{aligned} \tag{24.36}$$

which is exactly the same as if Alice had not made any measurement.[1] Note also that $\langle B \rangle$ does not depend on *which* observable A Alice chooses to measure.

The conclusion is that there is nothing Alice can do, using an entangled state or a large set of entangled states, which would communicate some message to Bob. If she wants to communicate, she will have to use one of the more conventional methods, which are limited by the speed of light.

24.4 Quantum teleportation

If faster-than-light communication by entanglement is ruled out, what about another idea from science fiction, namely teleportation? Surprisingly, entangled states can actually be used for this purpose, although not (yet) for objects of macroscopic size.

To explain this feat, it is important to understand what we mean by the term 'teleportation'. The idea was popularized in the television/movie series Star Trek, in which Captain Kirk and members of his starship crew would routinely step into the teleporter room, and 'beam down' to the surface of a planet, or perhaps to another starship. The captain and his team would disappear from one place, and reappear in another. The first question to ask is this: if Captain Kirk can be made to appear on the planet, was it necessary for him to disappear from the starship? Could we not leave the original captain intact, and end up with two identical Captain Kirks, both of whom could then be engaged in galactic exploration? In other words, could the teleporter room be used to 'clone' Captain Kirk?

Specifically, suppose we have a system (perhaps Captain Kirk, or maybe we could start with just a spin $\frac{1}{2}$ particle) in state $|\psi\rangle_A$ in region A, and another system (maybe the chemicals required to assemble another Captain Kirk, or perhaps just another spin $\frac{1}{2}$ particle) in a 'readiness' state $|0\rangle$ in a region B, distant from region A. The state of the system in region A is not necessarily known, at least not completely. Without this knowledge, is it possible to transfer the state ψ from A to B, without changing the state of the system at A? In other words, can the transformation

$$|\psi\rangle_A|0\rangle_B \Longrightarrow |\psi\rangle_A|\psi\rangle_B \tag{24.37}$$

be made to occur, by any physical process?

The no-cloning theorem

The answer is no, as shown in [2]. The argument is simple. We imagine that the cloning will work for an object in any state ψ. In particular, it will work for $\psi = \psi_1$ or $\psi = \psi_2$:

[1] There are a few missing steps in going from the second line to the third line of (24.36). These are reserved for a problem at the end of the chapter.

$$
\begin{aligned}
|\psi_1\rangle_A |0\rangle_B &\Longrightarrow |\psi_1\rangle_A |\psi_1\rangle_B \\
|\psi_2\rangle_A |0\rangle_B &\Longrightarrow |\psi_2\rangle_A |\psi_2\rangle_B
\end{aligned}
\tag{24.38}
$$

But now suppose $\psi = \psi'$ where $\psi' = a\psi_1 + b\psi_2$. Because of the linearity of the Schrödinger equation, we must have

$$
|\psi'\rangle_A |0\rangle_B \Longrightarrow a\,|\psi_1\rangle_A |\psi_1\rangle_B + b\,|\psi_2\rangle_A |\psi_2\rangle_B \tag{24.39}
$$

But this is *not* the same as cloning, which would require

$$
\begin{aligned}
|\psi'\rangle_A |0\rangle_B &\Longrightarrow |\psi'\rangle_A |\psi'\rangle_B \\
&= \left(a\,|\psi_1\rangle_A + b\,|\psi_2\rangle_A\right)\left(a\,|\psi_1\rangle_B + b\,|\psi_2\rangle_B\right) \\
&\neq a\,|\psi_1\rangle_A |\psi_1\rangle_B + b\,|\psi_2\rangle_A |\psi_2\rangle_B
\end{aligned}
\tag{24.40}
$$

The conclusion is that the cloning of arbitrary quantum states is not possible.

However, teleportation is not the same thing as cloning. In teleportation, the information about quantum state ψ disappears from region A, and reappears in region B, without any transport of matter. We imagine that there is some object, initially in some (possibly unknown) state ψ in region A, and some other object in a state of readiness in region B. After the teleportation has occurred, the object in region B is in state ψ, while the object in region A is no longer in that state. It is not the material object which disappears in A and reappears in B, rather it is the quantum state ψ which has been transferred from one object to another. This is the meaning of the term 'quantum teleportation', and it turns out that such a miracle is possible, and has actually been carried out for microscopic objects in laboratory experiments.

Teleportation of spin states

Imagine that Alice, in lab A, has a spin $\frac{1}{2}$ particle in some spin state ψ, and wishes to 'teleport' (i.e. transfer) that state to another particle in Bob's lab, lab B. Alice and Bob can communicate, say by cell phone, but cannot or will not transfer the particle of interest from one lab to another.

Teleportation of the spin state will require three spin $\frac{1}{2}$ particles, whose identities are denoted by a superscript 1, 2, or 3. Particle 1 in lab A is in some (possibly unknown) spin state

$$
\psi^1 = a\alpha^1 + b\beta^1 \tag{24.41}
$$

where $\alpha = \alpha_z$ is the spin up state and $\beta = \beta_z$ is the spin down state along the z-axis, and we have dropped the z subscript which will be redundant here. This is the state to be teleported. It is arranged beforehand that particles 2 and 3 are in an entangled state

$$
\frac{1}{\sqrt{2}}\left(\alpha^2\beta^3 + \beta^2\alpha^3\right) \tag{24.42}
$$

and, before the teleportation occurs, particle 2 has been transported to lab A, and particle 3 to lab B. The total state, at the beginning of the teleportation session, is

$$\begin{aligned}\Psi &= \frac{1}{\sqrt{2}}(a\alpha^1 + b\beta^1)(\alpha^2\beta^3 + \beta^2\alpha^3) \\ &= \frac{1}{\sqrt{2}}(a\alpha^1\alpha^2\beta^3 + a\alpha^1\beta^2\alpha^3 + b\beta^1\alpha^2\beta^3 + b\beta^1\beta^2\alpha^3)\end{aligned} \tag{24.43}$$

To begin the session, Alice measures some observable Q of particles 1 and 2, whose corresponding Hermitian operator $\tilde{Q}$ has the following eigenstates

$$\begin{aligned}\Phi_1^{12} &= \frac{1}{\sqrt{2}}(\alpha^1\alpha^2 + \beta^1\beta^2) \\ \Phi_2^{12} &= \frac{1}{\sqrt{2}}(\alpha^1\alpha^2 - \beta^1\beta^2) \\ \Phi_3^{12} &= \frac{1}{\sqrt{2}}(\alpha^1\beta^2 + \beta^1\alpha^2) \\ \Phi_4^{12} &= \frac{1}{\sqrt{2}}(\alpha^1\beta^2 - \beta^1\alpha^2)\end{aligned} \tag{24.44}$$

with corresponding eigenvalues q_1, q_2, q_3, q_4. The eigenstates of $S_z^1 S_z^2$ can also be expressed in terms of the eigenstates of $\tilde{Q}$:

$$\begin{aligned}\alpha^1\alpha^2 &= \frac{1}{\sqrt{2}}(\Phi_1^{12} + \Phi_2^{12}) \\ \alpha^1\beta^2 &= \frac{1}{\sqrt{2}}(\Phi_3^{12} + \Phi_4^{12}) \\ \beta^1\alpha^2 &= \frac{1}{\sqrt{2}}(\Phi_3^{12} - \Phi_4^{12}) \\ \beta^1\beta^2 &= \frac{1}{\sqrt{2}}(\Phi_1^{12} - \Phi_2^{12})\end{aligned} \tag{24.45}$$

Inserting these expressions into (24.43) and rearranging terms, we can express the initial Ψ as

$$\begin{aligned}\Psi = \frac{1}{2}\{&\Phi_1(a\beta^3 + b\alpha^3) + \Phi_2(a\beta^3 - b\alpha^3) \\ &+ \Phi_3(a\alpha^3 + b\beta^3) + \Phi_4(a\alpha^3 - b\beta^3)\}\end{aligned} \tag{24.46}$$

Now Alice makes a measurement of Q, after which particles 1 and 2 are left (with equal probability) in one of the eigenstates of $\tilde{Q}$, and, because particles 2 and 3 were entangled, the total state of the three particles is in one of the four states

$$\begin{aligned}\Psi_1 &= \Phi_1(a\beta^3 + b\alpha^3)\\ \Psi_2 &= \Phi_2(a\beta^3 - b\alpha^3)\\ \Psi_3 &= \Phi_3(a\alpha^3 + b\beta^3)\\ \Psi_4 &= \Phi_4(a\alpha^3 - b\beta^3)\end{aligned} \tag{24.47}$$

At this point Alice communicates her result by conventional means (e.g. cell phone) to Bob, together with the following instructions, depending on the measured value of Q:

1. If $Q = q_1$, rotate the spin state of particle 3 by 180° around the x-axis.
2. If $Q = q_2$, rotate the spin state of particle 3 by 180° around the y-axis.
3. If $Q = q_3$, do nothing.
4. If $Q = q_4$, rotate the spin state of particle 3 by 180° around the z-axis.

After this operation is carried out by Bob, particle 3 is left in the state ψ. The spin state of particle 1 has been 'teleported' from Alice to Bob.

But how does Bob rotate the spin state of particle 3 around some axis? This can be accomplished if, e.g., particle 3 has a magnetic moment aligned with its spin, and we turn on a magnetic field, directed along the x-, y-, or z-axes for a short period. Let us recall the discussion of electron precession in section 13.2.1, where the electron is subjected to an external magnetic field, and the Hamiltonian is effectively

$$H = \frac{1}{2}\hbar\Omega\sigma_z \tag{24.48}$$

If the electron starts out in the spin state ψ_0 at time $t = 0$, then the solution of the Schrödinger equation gives us the spin state at any later time

$$\begin{aligned}\psi(t) &= e^{-iHt/\hbar}\psi(0)\\ &= \left(\cos\left(\frac{1}{2}\Omega\sigma_z t\right) - i\sin\left(\frac{1}{2}\Omega\sigma_z t\right)\right)\psi(0)\\ &= \left(\cos\left(\frac{1}{2}\Omega t\right) - i\sigma_z\sin\left(\frac{1}{2}\hbar\Omega t\right)\right)\psi(0)\end{aligned} \tag{24.49}$$

where, in going from the second to the last line, we have used the Taylor series expansions of the sine and cosine, and the fact that $(\sigma_z)^n = I_2$ for even n, and $(\sigma_z)^n = \sigma_z$ for odd n. What we found was that after a time $t = \pi/\Omega$, an electron whose initial spin was initially polarized along the $+x$-axis now has its spin polarized in the $-x$-direction, and likewise if the initial spin was in the $+y$-direction, it is left with spin polarized in the $-y$-direction. In other words, the spin state has been rotated counterclockwise around the z-axis by 180 degrees. We see that for $t = \pi/\Omega$,

$$e^{iHt/\hbar} = -i\sigma_z \tag{24.50}$$

So a rotation by 180 degrees around the z-axis amounts to the operation

$$\psi \rightarrow -\mathrm{i}\sigma_z\psi \tag{24.51}$$

and it can be implemented by turning on a magnetic field for the specified amount of time. Obviously, rotations around the x- and y-axes can be implemented in a similar way, and correspond to operations

$$\psi \rightarrow -\mathrm{i}\sigma_x\psi \quad \text{and} \quad \psi \rightarrow -\mathrm{i}\sigma_y\psi \tag{24.52}$$

respectively.

Now let us begin with the case that Alice found observable $Q = q_3$, and Bob does nothing. We see from (24.47) that particle 3 is already in the state $\psi = a\alpha^3 + b\beta^3$, and the spin state has been teleported. Suppose instead Alice found $Q = q_4$. Then, following the instructions (4), Bob applies a rotation by 180° around the z-axis to particle 3, and ends up with

$$\begin{aligned} -\mathrm{i}\sigma_z\left(a\alpha_3 - b\beta_3\right) &= -\mathrm{i}(a\alpha^3 + b\beta^3) \\ &= -\mathrm{i}\psi \end{aligned} \tag{24.53}$$

Now states ψ and $\mathrm{e}^{\mathrm{i}\delta}\psi$ are the same physical state, so after the rotation the spin state of particle 3 is the same as the original spin state of particle 1. Again the spin state has been teleported.

If Alice found $Q = q_1$, then Bob is instructed to rotate the spin state of particle 3 around the x-axis, in which case particle 3 ends in the state

$$\begin{aligned} -\mathrm{i}\sigma_x(a\beta^3 + b\alpha^3) &= -\mathrm{i}\begin{bmatrix} 0 & 1 \\ 1 & 0 \end{bmatrix}\begin{bmatrix} b \\ a \end{bmatrix} \\ &= -\mathrm{i}\begin{bmatrix} a \\ b \end{bmatrix} \\ &= -\mathrm{i}(a\alpha^3 + b\beta^3) \end{aligned} \tag{24.54}$$

Here too, the spin state has been teleported. Finally, for the case that Alice found $Q = q_2$, Bob rotates the spin state of particle 3 around the y-axis, to reach the state

$$\begin{aligned} -\mathrm{i}\sigma_y(a\beta^3 - b\alpha^3) &= -\mathrm{i}\begin{bmatrix} 0 & -\mathrm{i} \\ \mathrm{i} & 0 \end{bmatrix}\begin{bmatrix} -b \\ a \end{bmatrix} \\ &= -\begin{bmatrix} a \\ b \end{bmatrix} \\ &= -(a\alpha^3 + b\beta^3) \end{aligned} \tag{24.55}$$

and in this last case also, the spin has been teleported. Note that although the spin state has been transferred from particle 1 to a distant particle 3, there is no question here of faster-than-light communication, since an essential element is communicating the result of the Q measurement from Alice to Bob by conventional means. Also there is no cloning. In the process of teleporting the state ψ, the state of particle 1 has necessarily been altered, and the combined quantum state of particles 1 and 2 is one of the four states Φ_{1-4}, none of which carries any information about the original state ψ of particle 1.

It is hard to imagine that a procedure like this will be applicable to teleportation of macroscopic objects anytime soon. It is nonetheless amazing that teleportation is a possibility, even for microscopic objects. The interested reader may consult e.g. [3], for technical details on how quantum teleportation has actually been carried out in the laboratory.

24.5 Quantum computation

In a computer's random access memory (RAM), information is stored in the form of binary numbers, and each number consists of a set of binary digits, which can take values of either 0 or 1. Every digit is stored in an object that can be in one of two states, corresponding to 0 or 1, and these objects are known as ***bits***.[2] Computer calculations are based on sequences of simple operations (known as AND, OR, XOR, NOT) applied to these bits, and, if we interpret 1 and 0 as 'true' and 'false' respectively, then these operations correspond to operations in symbolic logic. For example, let A and B represent statements which are either true or false. Then A. AND.B is true only if A and B are both true, and false otherwise. Likewise, A.OR.B is true if either A, or B, or both A and B are true, and false only if both A and B are both false. The binary representation of AND, OR, and XOR (pronounced 'excluded or'), is shown in table 24.1 below. In computer science, these operations are known as ***logic gates***. Computer science deals with how to combine such operations to perform, e.g. arithmetic calculations, while electrical engineering is concerned with the realization of logic gates using transistors. Both of these topics are beyond the scope of this book, but it may be interesting to give one simple example from computer science.

Table 24.1. Truth tables for logic gates.

A	B	A.AND.B	A.OR.B	A.XOR.B
0	0	0	0	0
0	1	0	1	1
1	0	0	1	1
1	1	1	1	0

In ordinary arithmetic, when we add two numbers digit by digit, the result is in general two digits, referred to as a 'sum' and a 'carry' digit. Consider, for example, 87 + 65. Adding 7 + 5 gives a sum digit of 2 and a carry digit of 1, which is carried to

[2] Bits are organized into 'words', which typically contain 64 bits each, but details such as these will not concern us here.

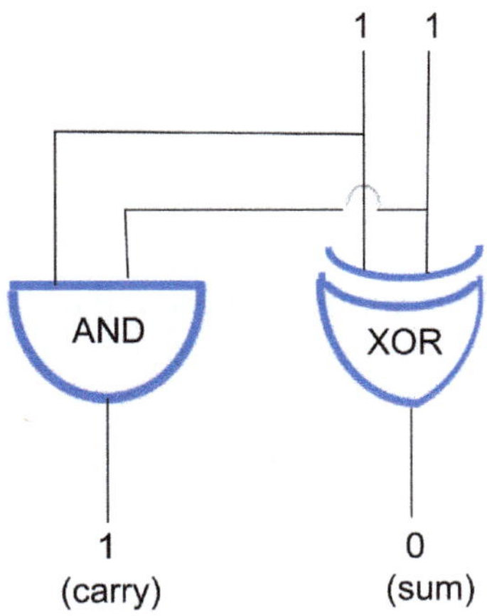

Figure 24.2. Classical computation. The half-adder consists of two connected logic gates operating on two single digit binary numbers. The case of 1 + 1 = 10 (binary arithmetic) is shown.

the next column. Then adding the carry digit 1 to 8 + 6 gives a sum digit of 5 and a carry digit of 1, carried to the third column. Result: 152. Now consider adding two binary digits in binary arithmetic. The sum digit corresponds to the XOR operation, and the carry digit corresponds to the AND operation. The combination of logic gates which takes binary digits A and B as input, and outputs the sum and carry digits, is shown in figure 24.2.

For our purposes a computer takes an N-digit binary number $I = i_1 i_2 i_3 \dots i_N$ ranging from 0 to $2^N - 1$ as input, operates on it with some sequence of logic gates determined by the computer program, and delivers as an output a binary number which is some function $F(I)$ of the input. A ***quantum computer*** carries out logical operations, closely related to those just mentioned, on quantum bits, known as ***qubits***.[3] A qubit can be in the quantum state $|0\rangle$ or $|1\rangle$, just as an ordinary bit can take on values 0 or 1, but the qubit can also be in any superposition

$$a|0\rangle + b|1\rangle \tag{24.56}$$

of those two states. A set of N qubits can be in the quantum state $|I\rangle = |i_1 i_2 i_3 \dots i_N\rangle$, but it can also be in any superposition of those states. An integer state

$$|I\rangle = |i_1 i_2 i_3 \dots i_N\rangle = |i_1\rangle|i_2\rangle \dots |i_N\rangle \tag{24.57}$$

is not an entangled state, because it can be written as a product of states of each of the qubits. However, a superposition of $|I\rangle$ states is in general entangled.

[3] In quantum computing, a logical operation evolves an input state of qubits into an output state via Schrödinger evolution, and therefore a quantum logic gate must (unlike the AND and XOR operations) have the same number of input and output bits.

In a quantum computer the initial state evolves according to the Schrödinger equation, and we suppose that it is possible to write a program such that an initial state $|I\rangle$ evolves into a state

$$|J(I)\rangle = |j_1 j_2 \ldots j_N\rangle \tag{24.58}$$

where the binary number $J(I)$ is some function of the binary number I and contains, perhaps in a subset of digits, the answer to some computation. Then if we would start instead with input state

$$|\Psi_{\text{in}}\rangle = \frac{1}{\sqrt{2^N}} \sum_{I=0}^{2^N-1} |I\rangle \tag{24.59}$$

the linearity of the Schrödinger equation implies that the output state is a superposition

$$|\Psi_{\text{out}}\rangle = \frac{1}{\sqrt{2^N}} \sum_{I=0}^{2^N-1} |J(I)\rangle \tag{24.60}$$

of all possible answers. In a sense, the quantum computer has carried out 2^N computations in parallel. But it is not so easy to take advantage of this tremendous parallelization. Ultimately the output must be read by a human being, a single measurement results in only one of the possible answers, and it is not obvious which input integer I corresponds to the J which is observed. In order to take advantage of the tremendous potential for parallel processing which quantum mechanics suggests is possible, some degree of ingenuity is required.

It has been shown that a quantum computer can, in principle, solve problems which are simply impossible to solve on an ordinary computer. The classic example is Shor's algorithm [4] for finding the prime number factors of very large integers. While it is trivial to find the factors of two-digit numbers, e.g. $15 = 3 \times 5$, the difficulty, i.e. the computation time required to find all the factors, increases exponentially as the number of digits increases. The standard RSA encryption algorithm, used e.g. in online shopping to protect credit card data, is based on the fact that at present it is impossible to factorize a very large number. Shor demonstrated that a quantum computer could break such encryption. However, Shor's algorithm, and most other quantum computation algorithms, are quite complex. To get something of the flavor of these methods, we will look at what is probably the simplest example of a quantum computing algorithm, invented by Deutsch and Jozsa [5]. To introduce their idea, let us first consider a question which was posed by the father of computer science, Alan Turing.[4]

Anyone who has written a lot of computer programs has encountered programs which never end, at least not until someone aborts the program. An example is an

[4] For an interesting biography of Turing and his contributions, see Hodges [6].

infinite loop, caused (in the simplest case) by some old-fashioned code of this kind, embedded somewhere in the program:

```
    .
    .
10 X = 2 + 2
   PRINT('2+2 = ',X)
   GO TO 10
    .
    .
```

This is just a program error, and such errors are usually easy to find and fix. But the problem is not always so easy to spot. A program might be designed to end only when a certain condition is satisfied, but it might not be obvious whether that condition would ever be satisfied. Turing asked whether it was possible to design a computer which could take, as input, an *arbitrary* program, and produce, as output, the answer to whether or not that program (if it were run) would ever end. Turing showed that the answer is no.[5] The idea we want to extract from this is that we may want to consider computer programs whose input is itself a computer program.

Deutsch and Jozsa consider a very special set of imaginable programs, namely, programs which, when presented with an N-digit binary number I, will output a single binary digit $F(I)$, with the following strong condition: the programs are designed to be either 'constant' or 'balanced' in the sense that

$$\begin{aligned} &\text{either} \quad F(I) = 0 \quad \text{all integers} && \text{“constant”} \\ &\text{or} \quad F(I) = \begin{cases} 0 & \text{half of the integers} \\ 1 & \text{half of the integers} \end{cases} && \text{“balanced”} \end{aligned} \tag{24.61}$$

For most programs one could determine whether they were constant or balanced just by inspection of the computer program. But if we are considering all possible such programs, some of which might be horribly complicated, inspection is not guaranteed to work. An obvious and simple alternative is just to run a given program for each of the binary integers, one by one, starting with $I = 0$. We might be lucky and discover that $F(I) = 1$ for some low value of I, in which case the program is balanced. But if $F(I)$ is constant, or if $F(I)$ is balanced and we are very unlucky, then it would be necessary to run the program on half of the 2^N possible integers to know the answer for sure. If, say, $N = 100$, this approach is completely impractical, even for the fastest existing machines.

What Deutsch and Jozsa showed is that an appropriately designed quantum computer can determine whether $F(I)$ is constant or balanced by running a certain program (implementing the $F(I)$ program as a subroutine), only once. Two

[5] See Penrose [7] for an exposition of the proof.

operations are required. The first is called *FT*. It is assumed that there is some Hamiltonian such that, when the computer evolves under this Hamiltonian for a certain period of time, beginning at an integer state $|I\rangle$, the following state evolution takes place:

$$FT: \quad |I\rangle \Longrightarrow \frac{1}{\sqrt{2^N}} \sum_{J=0}^{2^N-1} \left(\prod_{n=1}^{N} (-1)^{i_n j_n} \right) |J\rangle \tag{24.62}$$

It is left as an exercise to show that

- *FT* preserves the norm of any state, and the orthogonality of any two orthogonal states, as required by any time-evolution according to the Schrödinger equation.
- When the *FT* operation is performed twice, it leaves any state invariant, i.e.

$$(FT)\,(FT)|\psi\rangle = |\psi\rangle \tag{24.63}$$

The N qubits that store an integer I will be referred to as the register bits. The second operation is called the 'Oracle', and this is where the program that determines $F(I)$ enters. The computer memory contains, in addition to the N qubits comprising an integer I, one additional qubit (we will just call it the 'extra' qubit), and the Oracle operation does the following:

$$\text{Oracle:} \quad |I\rangle|j\rangle \Longrightarrow |I\rangle|j.\ \text{XOR}.\ F(I)\rangle \tag{24.64}$$

The function $F(I)$ is determined to be constant or balanced in four steps.

Step 0: Initialize

The register bits are initially in the state

$$|I\rangle = |00\ldots0\rangle \tag{24.65}$$

and the extra qubit is in the superposition

$$|\text{extra}\rangle = \frac{1}{\sqrt{2}}(|0\rangle - |1\rangle) \tag{24.66}$$

Step 1

Perform the *FT* operation on the initial state, resulting in the final register+extra qubit state

$$|\Psi_0\rangle = \frac{1}{\sqrt{2^N}} \sum_{I=0}^{2^N-1} |I\rangle \times \frac{1}{\sqrt{2}}(|0\rangle - |1\rangle) \tag{24.67}$$

Step 2

Carry out the Oracle operation. The qubits are left in the state

$$|\Psi_1\rangle = \frac{1}{\sqrt{2^N}} \sum_{I=0}^{2^N-1} (-1)^{F(I)}|I\rangle \times \frac{1}{\sqrt{2}}(|0\rangle - |1\rangle) \tag{24.68}$$

This is verified by trying out the four possible combinations

$$\begin{aligned} F(I)=0: &\quad |I\rangle|0\rangle \Longrightarrow |I\rangle|0\rangle \\ F(I)=1: &\quad |I\rangle|0\rangle \Longrightarrow |I\rangle|1\rangle \\ F(I)=0: &\quad |I\rangle|1\rangle \Longrightarrow |I\rangle|1\rangle \\ F(I)=1: &\quad |I\rangle|1\rangle \Longrightarrow |I\rangle|0\rangle \end{aligned} \tag{24.69}$$

and applying the results to the superposition $|I\rangle(|0\rangle - |1\rangle)$. Note that the register qubits are in an entangled state. But the register is *not* entangled with the extra qubit.

Suppose that $F(I)$ is constant. In that case $\Psi_1 = \Psi_0$. Conversely, suppose $F(I)$ is balanced. In that case it is easy to see that $\langle\Psi_1|\Psi_0\rangle = 0$.

Step 3

Apply the FT operation again to the register qubits, to obtain

$$|\Psi_2\rangle = |\psi\rangle \times \frac{1}{\sqrt{2}}(|0\rangle - |1\rangle) \tag{24.70}$$

If $F(I)$ is constant so that $\Psi_1 = \Psi_0$, then, because two applications of FT is the identity operation, the register is left in the original state $|\psi\rangle = |000\ldots0\rangle$. Conversely, if $F(I)$ is balanced so that Ψ_1 is orthogonal to Ψ_0, then, since FT preserves orthogonality, $\langle\psi|000\ldots0\rangle = 0$.

Step 4

Make a measurement of the register bits. The observable has, as its eigenstates, the non-entangled integer states $\{|I\rangle\}$. If $F(I)$ is constant, then this measurement will obtain, with certainty, the integer $I = 0$. If $F(I)$ is balanced, the probability of obtaining $I = 0$ is zero, since $|\psi\rangle$ is orthogonal to $|000\ldots0\rangle$. We conclude that if the measurement finds $I = 0$, then $F(I)$ is constant, and if the integer is anything different from zero, then $F(I)$ is balanced.

It is obvious that this example is a little contrived. Determining whether some function, already known to be either constant or balanced, is in fact constant or not, does not seem like a particularly useful application of such sophisticated technology. But the example does illustrate some of the tricks which are employed by much more complex and useful algorithms, and the potential power of quantum computing. Technological development of qubits, and logic gates that act on qubits, is currently under rapid development. But it is likely that the first application of qubits will not be to quantum computing per se, but rather to the simulation of certain quantum many-body systems, whose dynamics are not easily soluble by either analytic or standard numerical methods.

24.6 For whom the bell tolls

Entangled states are the norm, not the exception, in quantum mechanics. Generally speaking, when any two systems come into interaction, the resulting state of the combined system will be entangled. In this technical sense of quantum inseparability, the poet was right: no man is an island. We all live in entangled states.

As we have seen, the existence of entangled states means that a measurement of one system can cause the second system, perhaps very distant from the first, to jump into one or another quantum state. It is time to return to the EPR criticism: *Is this situation not paradoxical? Does it not imply that something must be wrong with quantum theory?*

Now first of all, as pointed out by Niels Bohr in his response to EPR, this situation is not really a paradox. The non-locality pointed out by EPR is certainly very, very strange. But quantum non-locality is not actually a paradox in the sense of being a logical contradiction.

Well then, does this non-local behavior not violate the theory of special relativity? According to relativity theory, after all, no signal can propagate faster than the speed of light, so how can the state of particle 2 change instantaneously, in the laboratory reference frame, upon measurement of particle 1? Here we must be careful: the relativistic prohibition is against *information* propagating faster than the speed of light. If such propagation were possible, it would also be possible to send messages backwards in time, which would raise many other (authentic!) paradoxes. However, we have just seen that the observer at detector 2 can never, by simply observing one of the particles, conclude anything about the settings of detector 1. Non-locality is not a violation of relativity in the sense of information transfer.

At this point it is worth noting that non-locality is not really unique to entangled states; some kind of non-local effect occurs in almost any measurement. Consider, for example, a single particle, whose wavepacket $\psi(x)$ is distributed uniformly in a large, dark room, filled with particle detectors. At a given time $t = 0$ the detectors are suddenly all switched on and, just as suddenly, the position of the particle is measured. This means that the wavepacket of the particle, which was spread all over the room at $t < 0$, suddenly collapses to a narrow Gaussian in the region of one of the detectors at $t = 0$. This is known as the 'collapse of the wave function', and it also seems to be a form of non-local behavior. If we think of the particle wavepacket as some sort of fluid, by what mechanism does it instantaneously collapse?

Throughout this text I have urged you to avoid thinking of the wave function as representing some sort of material fluid. There are just too many difficulties inherent in that kind of picture. Rather, one should think of Hilbert space, rather than ordinary space, as being the arena of dynamics, where the state (vector) of a system evolves continuously, according to the Schrödinger equation, by rotation. This change of arena alleviates a lot of the problems associated with non-locality. On the other hand, I also said that a measurement causes the state to 'jump', probabilistically, into one of a number of eigenstates of the quantity being measured.

It is time to face up to the real problem, the hard problem: by what process does a measurement cause a state to suddenly jump, probabilistically, into one of a set of orthogonal states? Can this behavior itself be explained quantum mechanically? This *very* puzzling issue has not been settled to this day. It is known as the ***problem of measurement,*** and it is central to the question of what quantum mechanics really means.

Problems

1. Show that (24.20) leads to (24.21). Then:
   a) Suppose that instead of the two spin 1/2 particles being ejected in an antisymmetric $S = 0$ state, the particles are coming out in an $S = 1$, $S_z = 0$ state. What is $P(\mathbf{a}, \mathbf{b})$ in this case?
   b) Now suppose that the particles are ejected 50% of the time in the $\alpha_z^L \beta_z^R$ state, and 50% of the time in the $\beta_z^L \alpha_z^R$ state. In this case, calculate $P(\mathbf{a}, \mathbf{b})$. (This problem anticipates the distinction between pure and mixed states, to be discussed in the next chapter.)

2. Fill in the missing steps in equation (24.36). The outer product notation of chapter 23 greatly simplifies this task. Otherwise, you will first need to show that if $\tilde{Q}$ is any operator (including the identity operator) with eigenvalue equation $\tilde{Q}|q_n\rangle = q_n|q_n\rangle$, then

$$\langle\psi_1|\tilde{Q}|\psi_2\rangle = \sum_n q_n\langle\psi_1|q_n\rangle\langle q_n|\psi_2\rangle \tag{24.71}$$

   where ψ_1, ψ_2 are any two states.

3. Each time a button is pushed, two spin 1/2 particles emerge, one moving towards Bob (R), and one towards Alice (L). The spin state is unknown, but always the same:

$$\psi = a\alpha_z^L\alpha_z^R + b\alpha_z^L\beta_z^R + c\beta_z^L\alpha_z^R + d\beta_z^L\beta_z^R \tag{24.72}$$

   Bob and Alice can both make spin up/down measurements on the R and L particles, respectively, along the x-, y-, z-axes, and they can communicate their results. Explain how Bob and Alice together can determine the unknown state (i.e. the constants a, b, c, d) from observed probabilities. Since ψ and $e^{i\delta}\psi$ are physically equivalent, you are free to choose a to be a real, positive number. The other numbers are in general complex, and you need to determine both the real and imaginary parts.

4. Show that FT is its own inverse. In other words, show that

$$(FT)(FT)|I\rangle = |I\rangle \tag{24.73}$$

   Using the same logic, show that if

$$|\Psi_I\rangle = FT|I\rangle \quad \text{and} \quad |\Psi_J\rangle = FT|J\rangle \tag{24.74}$$

   then

$$\langle\Psi_I|\Psi_J\rangle = \delta_{IJ} \tag{24.75}$$

   Explain why this implies that the norm of *any* state is preserved by the FT transformation, and that if any two states are orthogonal, they are still orthogonal after each is exposed to the FT transformation.

References

[1] Aspect A, Grangier P and Roger G 1982 Experimental realization of Einstein–Podolsky–Rosen–Bohm gedanken experiment: a new violation of Bell's inequalities *Phys. Rev. Lett.* **49** 91–7

[2] Wootters W K and Zurek W H 1982 A single quantum cannot be cloned *Nature* **299** 802–3

[3] Riebe M *et al* 2004 Deterministic quantum teleportation with atoms *Nature* **429** 734–7

[4] Shor P W 1997 Polynomial time algorithms for prime factorization and discrete logarithms on a quantum computer *SIAM J. Sci. Stat. Comput.* **26** 1484

[5] Deutsch D and Jozsa R 1992 Rapid solutions of problems by quantum computation *Proc. R. Soc. Lond.* A **439** 553

[6] Hodges A 2014 *Alan Turing: The Enigma* (Princeton, NJ: Princeton University Press)

[7] Penrose R 1994 *Shadows of the Mind: A Search For the Missing Science of Consciousness* 1st edn (New York: Oxford University Press)

IOP Publishing

An Introduction to Quantum Theory

Jeff Greensite

Chapter 25

The interpretations of quantum mechanics

Do not keep saying to yourself, if you can possibly avoid it, 'But how can it [quantum mechanics] *be like that?' because you will get 'down the drain' into a blind alley from which nobody has yet escaped. Nobody knows how it can be like that.*

Richard Feynman

More than 90 years have passed since the invention of quantum mechanics. It is a little disquieting that in all that time, despite its triumphant application to atomic, nuclear, sub-nuclear, and condensed matter physics, no general agreement has been reached on what quantum theory really *means*. This could indicate one of two things. First, it may be that quantum theory is wrong in some unexpected way, and this fact will someday be discovered by experiment. The other possibility is that the problem to be discussed in this chapter is not really a scientific question, which can ultimately be settled by either calculation or experiment, but is more a philosophical issue, in which arguments for or against various views are destined to be endlessly refined, but no general consensus will ever be reached. In either case, the interpretation of quantum mechanics is a deep and interesting subject. In this chapter I will try to give a brief outline of some of the contending views, although I do not pretend, in this short space, to do justice to any of them.

25.1 Mixtures and pure states

What does a measurement *do*?

We have asked that question before, back in chapter 7. The answer is: if the detector is designed to measure some observable A, it will leave the measured object, at least for an instant, in a zero-uncertainty state $\Delta A = 0$ of that observable, i.e. in an eigenstate of the operator A, and record the eigenvalue found.

If A is a complete set of observables, then any state ψ can be expressed as a superposition of eigenstates of $\tilde{A}$

doi:10.1088/978-0-7503-1167-0ch25

$$\psi = \sum_n a_n \phi_n \quad , \qquad \tilde{A}\phi_n = \alpha_n \phi_n \tag{25.1}$$

Now what happens if A is measured by some apparatus, but we are not yet aware of the result? Then we know that the system is in one of the states ϕ_n but we do not know which one. We do know the probabilities, however. This state of partial ignorance is often referred to as a *mixed state*:[1]

$$M = \{\text{in one of the states } \phi_n \text{ with probability } |a_n|^2\} \tag{25.2}$$

Now the question is: can we distinguish between the state ψ, which is usually called a *pure state*—it means that we know the physical state of the system—and a mixed state? Put another way, could we determine by further measurements, and without checking the measurement apparatus, whether a measurement of observable A was actually carried out? If we immediately measure A again, the mixture and the pure state have the same expectation value:

$$\begin{aligned} \langle A\rangle_{\text{pure}} &= \sum_m \sum_n a_m^* a_n \langle \phi_m | A | \phi_n \rangle \\ &= \sum_n |a_n|^2 \alpha_n \\ \langle A\rangle_{\text{mix}} &= \sum_n \alpha_n \text{Prob}(\alpha_n) \\ &= \sum_n \alpha_n \, |a_n|^2 \end{aligned} \tag{25.3}$$

where, in the second line, we have used the orthogonality of eigenstates of A. On the other hand, suppose a measurement is made of some observable B which does not commute with A

$$\begin{aligned} \langle B\rangle_{\text{pure}} &= \sum_m \sum_n a_m^* a_n \langle \phi_m | B | \phi_n \rangle \\ \langle B\rangle_{\text{mix}} &= \sum_n |a_n|^2 \langle \phi_n | B | \phi_n \rangle \end{aligned} \tag{25.4}$$

In general these expectation values are *not* the same:

$$\langle B\rangle_{\text{pure}} - \langle B\rangle_{\text{mix}} = \sum_{m\neq n} a_m^* a_n \langle \phi_m | B | \phi_n \rangle \tag{25.5}$$

The expression on the right-hand side is sometimes called the *interference term*. It means that the pure state ψ and the mixture M are, in principle, experimentally distinguishable. It is therefore possible to tell whether a measurement has been made, without checking the measurement apparatus (e.g. to be sure it was turned on).

[1] Mixed states were introduced, using the density operator formalism, in section 23.4, and this is the best way to treat mixed and pure states on an equal footing. However, in order to keep this chapter independent of chapter 23, I will avoid the density operator formalism here.

It is, however, necessary to measure some observable which does not commute with whatever observable is measured by the apparatus.

Now this conclusion may seem a little at odds with the conclusions of section 24.3, where we concluded that Bob could not tell whether or not Alice had made a measurement of one object (moving to the left) which was entangled with another object, moving to the right. If Alice made a measurement, then surely from Bob's point of view the system is in a mixed state. So from the previous discussion one might think that he could distinguish between a mixed and pure state, and therefore deduce that Alice had, or had not, made a measurement. The fallacy lies in the fact that to distinguish between a pure and mixed state, Bob will need to make measurements on the whole system, and not just the half in his lab.

Let us examine this case more closely. Can we tell the difference between an entangled pure state

$$\psi(1, 2) = \sum_n a_n \phi_n(1) \varphi_n(2) \tag{25.6}$$

where

$$\tilde{A}\phi_n(1) = \alpha_n \phi_n(1), \qquad \tilde{B}\varphi_n(2) = \beta_n \varphi_n(2) \tag{25.7}$$

and a mixed state

$$M = \{\text{in one of the states } \phi_n(1)\varphi_n(2) \text{ with probability } |a_n|^2\}? \tag{25.8}$$

The answer is yes, but only if we make measurements on both systems 1 and 2, and measure observables which do not commute with A and B. So suppose we measure C on system 1, and D on system 2. Then we find for a pure state

$$\langle CD\rangle_{\text{pure}} = \sum_{m,n} a_m^* a_n \langle \phi_m(1)|C|\phi_n(1)\rangle \langle \varphi_m(2)|D|\varphi_n(2)\rangle \tag{25.9}$$

whereas for a mixed state

$$\langle CD\rangle_{\text{mix}} = \sum_n |a_n|^2 \langle \phi_n(1)|C|\phi_n(1)\rangle \langle \varphi_n(2)|D|\varphi_n(2)\rangle \tag{25.10}$$

Again, the difference is the interference term

$$\langle CD\rangle_{\text{pure}} - \langle CD|\rangle_{\text{mix}} = \sum_{m\neq n} a_m^* a_n \langle \phi_m(1)|C|\phi_n(1)\rangle \langle \varphi_m(2)|D|\varphi_n(2)\rangle \tag{25.11}$$

Note that this interference term vanishes if either C or D is the identity (meaning that no measurement has been performed on one of the systems), or if all the $\phi_n(1)$ are eigenstates of C, or if all the $\varphi_n(2)$ are eigenstates of D. The interference term is non-zero when $[A, C] \neq 0$, and $[B, D] \neq 0$, and we make measurements on both systems.

From this discussion we extract the following conclusions:

1. In principle, it is always possible to distinguish experimentally between a mixture and a pure state;

2. To distinguish between pure and mixed states of composite systems, one must make measurements on *each* of the components.

Thus, to the question, 'what does a measurement *do*?', our tentative response is: 'it leaves the measured system in a mixed state'. And this is why the distinction between pure states and mixed states is useful.

25.2 The problem of measurement

The problem of measurement is the problem that no closed system, evolving according to the Schrödinger equation, can ever make a transition from a pure state to a mixed state. A measurement, as we have described it, can therefore never occur if *all* relevant objects (the detector as well as the detected object) obey the Schrödinger equation.

In order to explain how a detector works, one would begin by analyzing the interaction between the detector and the particle, and describe the evolution of the particle–detector system in terms of some (probably very complicated) Hamiltonian, involving the degrees of freedom of both the particle and the detector. But to understand why there is a problem connected with the measurement process, one does not need to know anything about this complicated Hamiltonian. It is enough to simply assume that an appropriate Hamiltonian exists such that, if the particle starts out in an eigenstate of the relevant observable, then the corresponding eigenvalue is somehow registered by the detector. Once again, take the measurement of electron spin as an example. The detector is designed to measure S_z. This means that if the initial state of the particle–detector system is

$$|\Psi_0\rangle = |\text{up}\rangle|\text{ready}\rangle \tag{25.12}$$

where $|\text{up}\rangle$ means the electron is in a spin-up state, and $|\text{ready}\rangle$ means that the detector has been switched on, then the interaction between the electron and detector leaves the combined system in the final state

$$|\Psi_f\rangle = |\text{up}\rangle|\text{red}\rangle \tag{25.13}$$

where $|\text{red}\rangle$ indicates that red light on the detector is on. It is assumed that the transition from the initial to final state is adequately explained by the Schrödinger equation, and thus we write, schematically,

$$|\text{up}\rangle|\text{ready}\rangle \longrightarrow |\text{up}\rangle|\text{red}\rangle \tag{25.14}$$

Similarly, if the particle starts in a spin-down state,

$$|\text{down}\rangle|\text{ready}\rangle \longrightarrow |\text{down}\rangle|\text{green}\rangle \tag{25.15}$$

Now for the crucial point. The Schrödinger equation is a *linear* equation. This means that if

$$\psi_a(x, t) \quad \text{and} \quad \psi_b(x, t) \tag{25.16}$$

are both solutions of the Schrödinger equation, and if the initial state is

$$\Psi_0 = A\psi_a(x, 0) + B\psi_b(x, 0) \tag{25.17}$$

then the corresponding final state (at time $t = T$, say) is

$$\Psi_f = A\psi_a(x, T) + B\psi_b(x, T) \tag{25.18}$$

Suppose, then, that the electron entering the detector is in the spin state

$$|\psi\rangle = A|\text{up}\rangle + B|\text{down}\rangle \tag{25.19}$$

so that the initial particle–detector state is

$$\begin{aligned} |\Psi_0\rangle &= |\psi\rangle|\text{ready}\rangle \\ &= A|\text{up}\rangle|\text{ready}\rangle + B|\text{down}\rangle|\text{ready}\rangle \end{aligned} \tag{25.20}$$

Then, from the linearity of the Schrödinger equation, we can conclude that the particle–detector system must end up in the entangled final state

$$|\Psi\rangle = A|\text{up}\rangle|\text{red}\rangle + B|\text{down}\rangle|\text{green}\rangle \tag{25.21}$$

This is a pure state, not a mixture! A mixture, for the particle–detector system would be

$$M = \begin{cases} \text{either in state } |\text{up}\rangle|\text{red}\rangle, & \text{with probability } |A|^2 \\ \text{or else in state } |\text{down}\rangle|\text{green}\rangle, & \text{with probability } |B|^2 \end{cases} \tag{25.22}$$

We have seen that pure states can always be distinguished, in principle, from mixtures. Then we are forced to conclude, after the measurement process has supposedly taken place, that the light on the detector is neither flashing red, nor green, but is somehow in a superposition of the two possible states!

When we add a human observer to the system, the quantum-mechanical assault on common sense only gets worse. Denote the initial state of the observer by the ket '$|\text{waiting}\rangle$'. When the observer sees a red light flash, her state changes to a new state, denoted $|\text{I saw red}\rangle$; likewise, if the green light flashes, the observer is left in the state $|\text{I saw green}\rangle$. The Hamiltonian describing the interaction between particle, detector, and observer is assumed to lead to

$$|\text{up}\rangle|\text{ready}\rangle|\text{waiting}\rangle \longrightarrow |\text{up}\rangle|\text{red}\rangle|\text{I saw red}\rangle \tag{25.23}$$

Similarly, if the particle starts in a spin-down state,

$$|\text{down}\rangle|\text{ready}\rangle|\text{waiting}\rangle \longrightarrow |\text{down}\rangle|\text{green}\rangle|\text{I saw green}\rangle \tag{25.24}$$

Then, if the instead the particle starts off in the superposition of spin states $|\psi\rangle$ of equation (25.19), so the initial state of the combined system is

$$\begin{aligned} |\Psi_0\rangle &= |\psi\rangle|\text{ready}\rangle|\text{waiting}\rangle \\ &= A|\text{up}\rangle|\text{ready}\rangle|\text{waiting}\rangle + B|\text{down}\rangle|\text{ready}\rangle|\text{waiting}\rangle \end{aligned} \tag{25.25}$$

Therefore, by the linearity of the Schrödinger equation, we are forced to conclude that the particle–detector–observer system ends up in the entangled final state

$$|\Psi_f\rangle = A|\text{up}\rangle|\text{red}\rangle|\text{I saw red}\rangle + B|\text{down}\rangle|\text{green}\rangle|\text{I saw green}\rangle \tag{25.26}$$

The observer has also ended up in a quantum superposition, in which she is neither in the $|\text{I saw red}\rangle$ state, nor in the $|\text{I saw green}\rangle$ state. Instead, her physical state has become entangled with the state of the detector and the state of the electron. And this strange conclusion is deduced simply from the assumption that a particle in a spin up (down) state always leads the observer to an 'I saw red (green)' state, and from the linearity of the Schrödinger equation.

It is time to stand back and reconsider. A simple line of reasoning, founded on the well-established principles of quantum mechanics, has led us to a nonsensical result. It is not just that the conclusion is counterintuitive; by now you should be used to that. Rather, the problem is that the apparent conclusion is just *wrong*: human beings never find themselves in superpositions of this kind. There must, somewhere, be an error in our reasoning. But where?

25.3 A sample of proposals

Here is a partial list of suggested answers to the previous question. They are each understood to be an 'interpretation' of quantum mechanics.

1. Collapse by consciousness

The simplest answer was given by von Neumann and Wigner, who urge us to follow the chain of events into the brain of the observer. The detector is in a superposition of 'red light/green light' states, and it emits photons in a superposition of the corresponding frequencies. The photons reach the retina of the observer, and certain neurons are left in a superposition of excited/un-excited states. The message from the retina travels to the cerebral cortex; very large collections of neurons are now in quantum superpositions, and the brain remains in such a superposition until, at some point, a *sensation* occurs. At the instant of conscious perception, the observer, the detector, and even the particle, jump into one or the other of the 'up/down, red/green' states.

What von Neumann and Wigner were suggesting is that human consciousness causes the wave function to 'collapse', with a certain probability, into one or another of the possible neural states; the collapse of the wave function occurs due to the awareness of the observer. It follows, since the Schrödinger equation by itself can never cause a wave function to collapse, that the mental function described as 'awareness' or 'consciousness' cannot be described by the Schrödinger equation; it is a process which follows different rules.

The notion that there is something special about consciousness, and that it cannot be explained by the dynamical laws that govern all other physical processes, is anathema to most scientists. It is reminiscent of vitalism; a theory which held that one must look beyond the usual laws of physics and chemistry to explain the processes which occur in living organisms. This theory was, of course, long ago discredited by spectacular advances in molecular biology.

Still, the von Neumann–Wigner idea should not be rejected out of hand. Philosophers have argued for centuries about the so-called mind/body problem, and the emergence of awareness from physical processes in the brain is not something which is well understood. So here we have two mysteries: the collapse of the wave function, and the existence of consciousness. Perhaps the two are somehow related. But for now, this is only speculation.

2. The Copenhagen interpretation

Niels Bohr's view (also known as 'the Copenhagen interpretation') was that the wave function which represents the physical state of a microscopic object does not refer to the state of the object in isolation from the observer. Rather, the wave function is a compact representation of the observer's *information* about the observables of a given object, and merely describes the possible outcome of a series of measurements. Put another way, the wave function does not refer to 'physical reality' per se, in the absence of an observer; it serves only to predict and correlate measurements. If there is no observer, then no meaning can be attached to a quantum-mechanical wave function. From this point of view the 'collapse' of the wave function does not describe a new physical process; it is just a change in the information available to the observer, obtained by measurement.

The main criticism that can be leveled at Bohr's interpretation is that it becomes meaningless to speak of the physical state of an object in the absence of an observer. How, then, can we describe quantum-mechanical processes that may have occurred in Nature prior to the evolution of human beings, or events which, for one reason or another, may have escaped human scrutiny? In classical physics, every object in the Universe has a physical state, a definite position and momentum, regardless of whether or not human beings are around to measure that state. Not so for quantum mechanics, at least in the Bohr interpretation. It is impossible, in this view, to imagine that any object is in any definite quantum state, without at least an implicit reference to an observer. Yet an active research area, at the border between astrophysics and quantum physics, is the application of quantum theory to early Universe cosmology. It does not seem clear how to make sense of that application in the context of the Copenhagen interpretation.

The fact that human observers are somehow an essential feature of the quantum-mechanical description of Nature is therefore a very troubling aspect of Bohr's view. In general, the Copenhagen interpretation has something of the flavor of logical positivism, a philosophy which holds that the purpose of science is to correlate measurements, rather than to describe some 'objective' reality. 'Objective reality', in the positivist view, is a meaningless concept, and science should be formulated entirely in terms of quantities which can be measured directly. This view, in the hands of Mach and others, had considerable influence on physics in the early 20th century, and certainly the Copenhagen interpretation show traces of this influence.

It is a little distressing to think that science is not about Nature, but only about correlating measurements. Still, the consistency of the Copenhagen interpretation, and its ability to evade puzzles connected with the apparent non-local 'collapse' of entangled wave functions, should not be underestimated. To the extent that there is an 'official' interpretation of quantum theory, it would probably be the Copenhagen view.

3. The many-Universe interpretation

This view originated in a 1957 paper by Everett [1], and there exist many variations of the general idea. We consider a detector which measures some observable A of an object, where the eigenvalues and eigenstates of $\tilde{A}$ are denoted α_n and ϕ_n respectively. The detector has a pointer which points to the integer n when the eigenvalue α_n is detected. Things are set up so that if the initial state of the Universe is

$$|\text{Universe}\rangle_{\text{initial}} = |\phi_n\rangle|\text{ready}\rangle|\text{rest of Universe}\rangle \tag{25.27}$$

Then immediately after the measurement, the state is

$$|\text{Universe}\rangle_{\text{final}} = |\phi_n\rangle|\text{pointer} \to n\rangle|\text{rest of Universe}\rangle \tag{25.28}$$

Now suppose that the initial state is

$$|\text{Universe}\rangle_{\text{initial}} = \left(\sum_n a_n|\phi_n\rangle\right)|\text{ready}\rangle|\text{rest of Universe}\rangle \tag{25.29}$$

Then the final state is an entangled state

$$|\text{Universe}\rangle_{\text{final}} = \left(\sum_n a_n|\phi_n\rangle|\text{pointer} \to n\rangle\right)|\text{rest of Universe}\rangle \tag{25.30}$$

In the most common version of the many-Universe scenario, the measurement causes the Universe to split into a huge number of universes, and in each of these universes the pointer position takes on one definite value; these macroscopic pointer positions are never in a superposition. The connection to the Born interpretation is that the 'measure' of the number of universes in which the pointer is in the n-position is given by $|a_n|^2$. If the experiment is repeated, then each 'branch' of the Universe will split again.

Now suppose that each time the experiment is carried out, the pointer position is recorded. Then the idea is that if the experiment is carried out N times, where N is a huge number, then in the overwhelming majority of branches of the Universe, the detector record will indicated that pointer position n was recorded $N|a_n|^2$ times, in agreement with the predictions of quantum mechanics.

Of course this whole idea can be extended away from pointer positions, and in general the statement is that in any branch of the Universe, macroscopic observables take on definite values, and are never in superpositions. The standard criticism of the many-Universe interpretation is that it places a special emphasis on pointer positions, or more generally macroscopic variables, as being observables which should be definite. It is not clear what gives these observables such a special status, and the border between macroscopic and microscopic observables is not entirely well-defined. Still, this interpretation (like the others listed here) has fervent supporters.[2]

4. Non-local hidden variables: the Bohm–de Broglie interpretation

In chapter 3, which presented and discussed the consequences of electron diffraction in the simplified two-slit example, I was very definite that this phenomenon simply rules out the possibility that an electron can follow a particular trajectory in ordinary three-dimensional space, as its physical state evolves in time. Supporters of the interpretation advanced by Bohm and de Broglie (see [2]) would take vehement exception to this view, because in their picture an electron actually *does* follow a definite trajectory, and in particular goes through one slit or another in the double-slit example, although we can never know exactly what that trajectory is. But in order to make this idea work, it is necessary to invoke non-local interactions. For how else could the path of an electron, passing through slit A, be affected by the presence of the other slit B? More generally, since the Bohm–de Broglie idea is essentially a version of hidden variables, non-locality is essential to avoid Bell's theorem.

The idea is that the quantum-mechanical wave function, which obeys the usual Schrödinger equation, is a 'guiding wave' which, together with an initial condition, determines the particle trajectory. The wave function knows about both slits, in the double-slit example, and passes this information on to the electron. The framework is as follows. The position of the particle $x(t)$ is definite at every time t (for simplicity we will deal with a single point particle in one space dimension; the generalization to many particles in three dimensions is straightforward). The motion of the particle is determined from its initial position x_0 at some time $t = 0$, and from the equations

$$\begin{aligned} \frac{\mathrm{d}x}{\mathrm{d}t} &= v(x, t) \\ v(x, t) &= \frac{\hbar}{m}\mathrm{Im}\left[\frac{1}{\psi(x, t)}\frac{\mathrm{d}\psi(x, t)}{\mathrm{d}x}\right] \end{aligned} \tag{25.31}$$

In the Bohm–de Broglie setup the initial position of the particle is not known exactly, but there is some probability density for the initial position which is related to the initial wave function in the usual way introduced by Born, i.e. $P(x, 0) = \psi^*(x, 0)\psi(x, 0)$. What can then be shown is that if the subsequent trajectory of the particle is given by (25.31), then the probability density for the particle position at any later time is *also* given by the Born expression, $P(x, t) = \psi^*(x, t)\psi(x, t)$, where $\psi(x, t)$ is obtained by solving the Schrödinger equation, starting from the initial wave function $\psi(x, 0)$ at $t = 0$. The ignorance implied by a probability density does not mean (in contrast to standard quantum mechanics) that the particle position is indefinite, it only means that our knowledge of particle position is imperfect.

[2] For a discussion of some of the subtleties involved, see http://plato.stanford.edu/entries/qm-manyworlds/.

The Bohm–de Broglie idea seems to work well enough for non-relativistic quantum mechanics. It has difficulties, however, with relativity. This is to be expected, since non-local interactions usually call for introducing some special reference frame. For this reason, in fact, locality requires that interactions are mediated by fields, which themselves have dynamics and must be quantized. Particles are not fundamental in quantum field theory. As we will see in chapter 27, they are simply quantum excitations of an underlying field, and there is nothing fundamental about their position or even their number, which can change in time. The Bohm–de Broglie framework has difficulties in coping with relativity, spin statistics, and quantum field theory. Although these difficulties have been (and are being) addressed by various people, I am not aware of a fully satisfactory resolution.

5. Decoherence

We have seen that mixtures and pure states are in principle distinguishable, but it is necessary to make measurements on all parts of an entangled pure state to make the distinction. However, it has been pointed out that the quantum state of a macroscopic object such as a detector is very rapidly entangled not only with the quantum state of a microscopic object, such as the spin state of an electron, but also with the environment. For example, a macroscopic object is constantly radiating photons through thermal emission and reflection, and its quantum state is entangled with those photons. It is essentially impossible in practice to track down and make measurements on every one of those emitted photons, or every part of the environment that the object has interacted with. So in practice, a very short while after a detector has made a measurement, it is simply impossible to tell that the detector is in a pure, entangled state, rather than a mixed state. This is what is meant by 'decoherence'.

But if we believe that the detector is in one or another definite state corresponding to a definite value of the measured observable (e.g. red light on or green light on), the decoherence idea does not really explain how that definite state emerges. It only says that we cannot in practice distinguish between a mixture and an entangled pure state, when a macroscopic object is part of the entangled system. Opinions differ on whether that fact really solves the problem of measurement. For an advocate's presentation, see Zurek [3].

6. Spontaneous collapse

We should not discard the possibility that quantum mechanics is actually wrong, or at least incomplete. It could be that the collapse of the wave function requires some new rule, which is open to experimental investigation.

The first proposal of this kind was due to Ghirardi, Rimini, and Weber [4]. Their idea was that the wave function of a particle $\psi(x)$ is, every so often, multiplied by a highly localized Gaussian factor

$$\psi(x) \rightarrow \psi'(x) = N \exp\left[(x - x_0)^2/2d^2\right]\psi(x) \qquad (25.32)$$

where N is some appropriate normalization factor. Hence the particle wave function collapses, every so often, to a highly localized state around some position x_0. There are two stochastic elements. First, the collapse happens at random times, with some average frequency. Secondly, the point x_0 in the Gaussian factor has a probability distribution $P(x_0) = \psi^*(x_0)\psi(x_0)$, so the collapse tends to happen where the modulus of the wave function is largest. It is suggested that for a single particle the collapse happens with a frequency on the order of once every few hundred million years, so this collapse is not readily observed at the microscopic level. However, a macroscopic system has on the order of Avogadro's number of particles, and the collapse of a single particle wave function in this system localizes the entire system. Hence a macroscopic system is localized in perhaps a millionth of a second or less. The main critique is that the rule for spontaneous collapse is a bit ad hoc, and at present has no experimental support. In addition, a process of this sort will violate conservation of energy. Nevertheless, it is an idea which is *in principle* testable, although perhaps not with current technology.

A variant of the spontaneous collapse idea has been proposed by Penrose. His suggestion is that superpositions of positional states of a massive object are associated with some gravitational self-energy E_G, and that a spontaneous collapse of the wave function to a well-localized state will occur in a time of the order of $\hbar/E_G$. See Penrose [5] for a presentation of this idea.

7. Other

I have listed the most well-known ideas, but this list is by no means exhaustive, and the interested reader may care to investigate other suggestions. A general reference is the book by Jammer [6].

8. 'Shut up and calculate'

'Shut up and calculate' is not an interpretation, it is an attitude. I include it here because it is the attitude of most working physicists with respect to the interpretations listed above. The fact is, we know how to use quantum mechanics to investigate the behavior of solids, atoms, nuclei, elementary particles, the dynamics of the early Universe,… really any phenomenon which is accessible to experiment or observation. The rules are clear, and all we need is the Schrödinger equation (or its relativistic field or string-theory generalizations), and the Born rule for probabilities. Philosophical issues (like what does quantum mechanics really mean) are in this view a distraction if there is no experimental path of investigation. And for all but the spontaneous collapse idea there is no such path; the other interpretations have been tailored so as to not disagree with any of the standard rules. Although philosophers are attracted to deep issues which cannot be decided experimentally, most scientists prefer to work on questions which can be decided, at least in principle, by some experiment or observation.

It is hard, nonetheless, to avoid thinking about such a profound mystery, and if any of the interpretations listed above satisfy the reader, well and good. Otherwise I think that Richard Feynman, whose wise advice was quoted at the beginning of this chapter, should have the last word. At least for now.

References

[1] Everett H 1957 Relative state formulation of quantum mechanics *Rev. Mod. Phys.* **29** 454–62

[2] Bohm D and Hiley B 1993 *The Undivided Universe: An Ontological Interpretation of Quantum Theory* (London: Routledge)

[3] Zurek W H 2003 Decoherence, einselection, and the quantum origins of the classical *Rev. Mod. Phys.* **75** 715–75

[4] Ghirardi G C, Rimini A and Weber T 1986 A unified dynamics for micro and macro systems *Phys. Rev.* D **D34** 470

[5] Penrose R 2014 On the gravitization of quantum mechanics 1: quantum state reduction *Found. Phys.* **44** 557–75

[6] Jammer M 1974 *The Philosophy of Quantum Mechanics: The Interpretations of Quantum Mechanics in Historical Perspective* (New York: Wiley)

IOP Publishing

An Introduction to Quantum Theory

Jeff Greensite

Chapter 26

The Feynman path integral

One feels as Cavalieri must have felt calculating the volume of a pyramid before the invention of calculus.

Richard Feynman

In classical physics the Euler–Lagrange equations are derived from the condition that the action $S[x(t)]$ should be stationary. These second-order equations are equivalent to the first-order Hamilton equations of motion, which are obtained by taking appropriate derivatives of the Hamiltonian function $H[q, p]$. Now the Hamiltonian is a quantity that we have encountered frequently in these pages. But what about the action? It would be surprising if something so fundamental to classical physics had no part at all to play in quantum theory. In fact, the action has the central role in an alternative approach to quantum mechanics known as the 'path-integral formulation'.

Let's start with the concept of a ***propagator***. Given the wave function $\psi(x, t)$ at time t, the wave function at any later time $t + T$ can be expressed as

$$\psi(x, t + T) = \int dy\ G_T(x, y)\psi(y, t) \tag{26.1}$$

where $G_T(x, y)$ is known as the 'propagator', and of course it depends on the Hamiltonian of theory. In fact, given a time-independent Hamiltonian with eigenstates

$$H\phi_n(x) = E_n\phi_n(x) \tag{26.2}$$

it is easy to see that

$$G_T(x, y) = \sum_n \phi_n(x)\phi_n^*(y)e^{-iE_nT/\hbar} \tag{26.3}$$

Richard Feynman, in 1948, discovered a very beautiful expression for the propagator in terms of an integral over all paths that the particle can follow, from point y at time t, to point x at time $t + T$, with an integrand

$$e^{iS[x(t)]/\hbar} \tag{26.4}$$

As we will see, his expression can be taken as a new way of quantizing a mechanical system, equivalent to the 'canonical' approach based on exchanging observables for operators.

To see how a sum over paths comes in, let us suppose that we knew the propagator, for *any* Hamiltonian, when the time difference $T = \epsilon$ is tiny, i.e.

$$\psi(x, t + \epsilon) = \int \mathrm{d}y\ G_\epsilon(x, y)\psi(y, t) \tag{26.5}$$

In that case, we could get the propagator for large time differences by using the propagator successively, i.e.

$$\begin{aligned} \psi(x, t + 2\epsilon) &= \int \mathrm{d}y\mathrm{d}x_1\ G_\epsilon(x, x_1)G_\epsilon(x_1, y)\psi(y, t) \\ \psi(x, t + 3\epsilon) &= \int \mathrm{d}y\mathrm{d}x_2\mathrm{d}x_1\ G_\epsilon(x, x_2)G_\epsilon(x_2, x_1)G_\epsilon(x_1, y)\psi(y, t) \end{aligned} \tag{26.6}$$

and by induction

$$\psi(x, t + N\epsilon) = \int \mathrm{d}y \int \left(\prod_{i=1}^{N-1} \mathrm{d}x_n \right) G_\epsilon(x, x_{N-1})G_\epsilon(x_{N-1}, x_{N-2}) \ldots G_\epsilon(x_1, y)\psi(y, t) \tag{26.7}$$

So the propagator, for the time difference $T = N\epsilon$, has the form

$$G_T(x, y) = \int \left(\prod_{i=1}^{N-1} \mathrm{d}x_n \right) G_\epsilon(x, x_{N-1})G_\epsilon(x_{N-1}, x_{N-2}) \ldots G_\epsilon(x_1, y) \tag{26.8}$$

Now a set of points

$$\begin{array}{lll} y & \text{at time} & t \\ x_1 & \text{at time} & t + \epsilon \\ x_2 & \text{at time} & t + 2\epsilon \\ \ldots & \ldots\ldots\ldots & \ldots\ldots \\ x_{N-1} & \text{at time} & t + (N - 1)\epsilon \\ x & \text{at time} & t + T \end{array} \tag{26.9}$$

connected by straight lines, as shown in figure 26.1, is a path between point (y, t) and point $(x, t + T)$. The integral in equation (26.8) is therefore a sum over all paths between these two points, which consist of N straight-line segments of equal time duration $\epsilon = T/N$. As $N \to \infty$ any continuous path between (y, t) and $(x, t + T)$ can be approximated to arbitrary precision.

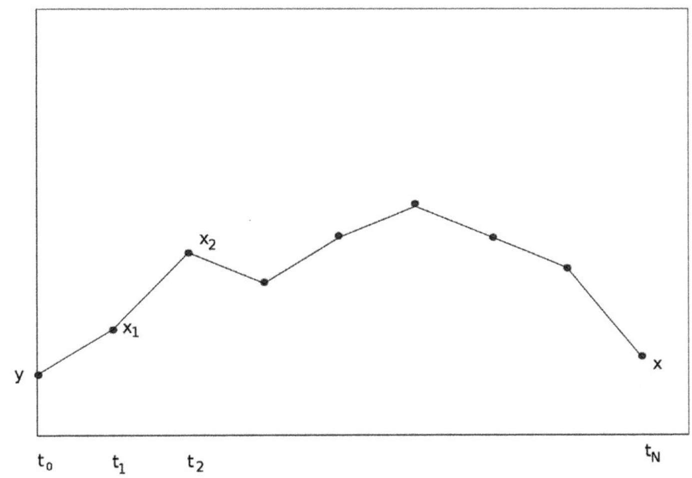

Figure 26.1. A sample path in the sum over all (discretized) paths, where $t_n = t + n\epsilon$.

It is clear, then, that a propagator can be represented by a sum over paths. The problem is to find an expression for $G_\epsilon(x, y)$, which does not involve first solving the eigenvalue equation (26.2) and plugging the result into (26.3). Fortunately, if ϵ is very small, we can deduce $G_\epsilon(x, y)$ from the time-dependent Schrödinger equation. First, approximate the time derivative with a finite difference

$$i\hbar\frac{\psi(x, t+\epsilon) - \psi(x, t)}{\epsilon} \approx H\psi(x, t) \tag{26.10}$$

so that

$$\begin{aligned}\psi(x, t+\epsilon) &= \left(1 + \frac{\epsilon}{i\hbar}H\right)\psi(x, t) + O(\epsilon^2) \\ &= \left(1 - i\frac{\epsilon}{\hbar}V + \frac{i\epsilon\hbar}{2m}\frac{d^2}{dx^2}\right)\psi(x, t) + O(\epsilon^2) \\ &= e^{-i\epsilon V(x)/\hbar}\left(\psi(x, t) + \frac{i\epsilon\hbar}{2m}\psi''(x, t)\right) + O(\epsilon^2)\end{aligned} \tag{26.11}$$

Now let us write $y = x + \eta$, and expand $\psi(y, t)$ in equation (26.5) in a Taylor series

$$\begin{aligned}\psi(x, t+\epsilon) &= \int dy\; G_\epsilon(x, y)\psi(y, t) \\ &= \int d\eta\; G_\epsilon(x, x+\eta)\left[\psi(x) + \psi'(x)\eta + \frac{1}{2}\psi''(x)\eta^2 + \cdots\right]\end{aligned} \tag{26.12}$$

Comparing (26.11) and (26.12) we see that, to first order in ϵ, the propagator $G_\epsilon(x, x+\eta)$ must satisfy

$$\int d\eta \ G_\epsilon(x, x+\eta) = e^{-i\epsilon V(x)/\hbar}$$
$$\int d\eta \ G_\epsilon(x, x+\eta)\eta = 0 \tag{26.13}$$
$$\int d\eta \ G_\epsilon(x, x+\eta)\eta^2 = \frac{i\epsilon\hbar}{m}$$

it is also clear from (26.5) that, in the $\epsilon \to 0$ limit, the propagator must become a delta function, i.e.

$$\lim_{\epsilon\to 0} G_\epsilon(x, x+\eta) = \delta(\eta) \tag{26.14}$$

Therefore, $G_\epsilon(x, x+\eta)$ is a delta sequence. We have seen two examples of delta sequences in chapter 4; one involved the integral of e^{ikx} over a finite range, the other was a Gaussian. The Gaussian delta sequence, together with the first of the conditions in equation (26.13), motivates trying, as an ansatz

$$G_\epsilon(x, x+\eta) = e^{-i\epsilon V(x)/\hbar}\sqrt{\frac{A}{\pi}}\, e^{-A\eta^2} \tag{26.15}$$

which will be a delta sequence if $A \to \infty$ as $\epsilon \to 0$. It is easy to check that this ansatz satisfies the first and second conditions in (26.13). The constant A is determined by the third condition

$$\begin{aligned}\frac{i\epsilon\hbar}{m} &= \sqrt{\frac{A}{\pi}}\int d\eta\ \eta^2 e^{-A\eta^2}\\ &= \frac{1}{2A}\end{aligned} \tag{26.16}$$

Solving for A, and recalling $\eta = x - y$, we then have to first order in ϵ,

$$G_\epsilon(x, y) = \left(\frac{m}{2i\pi\epsilon\hbar}\right)^{1/2} \exp\left[i\frac{\epsilon}{\hbar}\left\{\frac{1}{2}m\left(\frac{x-y}{\epsilon}\right)^2 - V(x)\right\}\right] \tag{26.17}$$

The expression above for $G_\epsilon(x, y)$ is not exact; there are $O(\epsilon^2)$ corrections. If we insert this expression into equation (26.8), the result for $G_T(x, y)$ will not be exact either: the product of T/ϵ terms will give an overall error of order ϵ. However, we can obtain the *exact* result for $G_T(x, y)$ by taking the $N \to \infty$ ($\epsilon = \frac{T}{N} \to 0$) limit

$$\begin{aligned}G_T(x, y) &= \lim_{N\to\infty}\int\left(\prod_{i=1}^{N-1} dx_n\right) G_\epsilon(x, x_{N-1})G_\epsilon(x_{N-1}, x_{N-2})\ldots G_\epsilon(x_1, y)\\ &= \lim_{N\to\infty}\int\left(\prod_{i=1}^{N-1} dx_n\right)\left(\frac{m}{2i\pi\epsilon\hbar}\right)^{N/2}\exp\left[\frac{i}{\hbar}\sum_{n=1}^{N}\epsilon\left(\frac{1}{2}m\frac{(x_n - x_{n-1})^2}{\epsilon^2} - V(x_n)\right)\right]\end{aligned} \tag{26.18}$$

where we have defined $x_0 \equiv y$ and $x_N \equiv x$. We now introduce the notation for the integral over all possible paths

$$\int Dx(t) \equiv \lim_{N\to\infty} \int \left(\prod_{i=1}^{N-1} \mathrm{d}x_i\right)\left(\frac{m}{2\mathrm{i}\pi\epsilon\hbar}\right)^{N-1/2} \tag{26.19}$$

and note that in the $\epsilon \to 0$ limit

$$\lim_{N\to\infty} \sum_{n=1}^{N} \epsilon\left(\frac{1}{2}m\frac{(x_n - x_{n-1})^2}{\epsilon^2} - V(x_n)\right) = \int_t^{t+T} \mathrm{d}t\left(\frac{1}{2}m\dot{x}^2 - V(x)\right) = S[x(t)] \tag{26.20}$$

where $S[x(t)]$ is the action functional defined back in chapter 1. At last we have arrived at the ***Feynman path integral***

$$G_T(x, y) = \int Dx(t)\mathrm{e}^{\mathrm{i}S[x(t)]/\hbar} \tag{26.21}$$

In words, this equation says that the amplitude for a particle to propagate from point y at time t to point x at time $t + T$ is given by the integral over all possible paths running between those two points, with each path weighted by the factor $\exp[\mathrm{i}S[x(t)]/\hbar]$, where $S[x(t)]$ is the action of the given path. Of course, the *precise* meaning of $\int Dx(t)$, the sum over paths, must be given in terms of a limit, namely, the limit of a multiple integral, as the number of integrals goes to infinity. But recall that the *precise* meaning of an ordinary integral is also in terms of a limit: the limit of a sum over very many points, as the number of points tends to infinity.

26.1 The free particle propagator

One might wonder if the path-integral formulation is useful: is it really possible to carry out an infinite multiple integral? The answer is yes, if the integrand is Gaussian. *Any* Gaussian integral can be done, no matter whether it is a single integral, a multiple integral, or a path integral.

As an example, we will work out the propagator for a free particle by two methods: first, by using equation (26.3), and, second, from the Feynman path integral. For a free particle

$$\phi_p(x) = \frac{1}{\sqrt{2\pi\hbar}}\mathrm{e}^{\mathrm{i}px/\hbar}$$
$$E_p = \frac{p^2}{2m} \tag{26.22}$$

so equation (26.3) becomes

$$G_T(x, y) = \int \frac{\mathrm{d}p}{2\pi\hbar} \exp\left[-\frac{\mathrm{i}T}{2m\hbar}p^2 + \frac{\mathrm{i}(x-y)}{\hbar}p\right] \tag{26.23}$$

Using the Gaussian integral formula

$$\int \mathrm{d}p\, \mathrm{e}^{-ap^2+bp} = \sqrt{\frac{\pi}{a}}\,\mathrm{e}^{b^2/4a} \tag{26.24}$$

we find

$$\begin{aligned} G_T(x, y) &= \frac{1}{2\pi\hbar}\left(\frac{\pi}{\mathrm{i}T/(2m\hbar)}\right)^{1/2} \exp\left[-\frac{(x-y)^2/\hbar^2}{4\mathrm{i}T/(2m\hbar)}\right] \\ &= \left(\frac{m}{2\mathrm{i}\pi\hbar T}\right)^{1/2} \exp\left[\mathrm{i}\frac{m(x-y)^2}{2\hbar T}\right] \end{aligned} \tag{26.25}$$

Now let us try to compute the same quantity by the path-integral method. Start from the expression (26.18) for the propagator, with $V(x) = 0$ for a free particle:

$$G_T(x, y) = \lim_{N\to\infty} B^{-N} \int \mathrm{d}x_{N-1}\mathrm{d}x_{N-2}\ldots\mathrm{d}x_1 \exp\left[-a\sum_{n=1}^{N}(x_n - x_{n-1})^2\right] \tag{26.26}$$

where

$$a = \frac{m}{2\mathrm{i}\epsilon\hbar}, \qquad B = \left(\frac{2\pi\mathrm{i}\epsilon\hbar}{m}\right)^{1/2} \tag{26.27}$$

This expression has the form of a multi-dimensional Gaussian integral, and Gaussian integrals, even multi-dimensional Gaussian integrals, are generally easy to compute. There are several different ways to tackle this one. Probably the easiest is to use this identity:

$$\mathrm{e}^{-ax^2} = \frac{1}{\sqrt{4\pi a}}\int_{-\infty}^{\infty} \mathrm{d}q \exp\left[-\frac{q^2}{4a} + \mathrm{i}qx\right] \tag{26.28}$$

Then we have

$$\begin{aligned} G_T(x, y) &= \lim_{N\to\infty} B^{-N}\left(\frac{1}{4\pi a}\right)^{N/2} \int \mathrm{d}x_{N-1}\mathrm{d}x_{N-2}\ldots\mathrm{d}x_1\mathrm{d}q_N\mathrm{d}q_{N-1}\ldots\mathrm{d}q_1 \\ &\quad\times \exp\left[-\frac{1}{4a}\sum_{n=1}^{N} q_n^2 + \mathrm{i}\sum_{n=1}^{N} q_n(x_n - x_{n-1})\right] \\ &= \lim_{N\to\infty} B^{-N}\left(\frac{1}{4\pi a}\right)^{N/2} \int \mathrm{d}x_{N-1}\mathrm{d}x_{N-2}\ldots\mathrm{d}x_1\mathrm{d}q_N\mathrm{d}q_{N-1}\ldots\mathrm{d}q_1 \\ &\quad\times \exp\left[-\frac{1}{4a}\sum_{n=1}^{N} q_n^2 + \mathrm{i}\sum_{n=1}^{N-1}(q_n - q_{n+1})x_n + \mathrm{i}(q_N x_N - q_1 x_0)\right)\right] \end{aligned} \tag{26.29}$$

Now interchange the order of integrations and do all the x-integrals first, which gives a product of δ-functions, and then integrate over all q_1 up to q_{N-1}

$$
\begin{aligned}
G_T(x, y) &= \lim_{N\to\infty} B^{-N}\left(\frac{1}{4\pi a}\right)^{N/2} \int \mathrm{d}q_N \mathrm{d}q_{N-1} \dots \mathrm{d}q_1 \\
&\quad \times \exp\left[-\frac{1}{4a}\sum_{n=1}^{N} q_n^2 + \mathrm{i}(q_N x_N - q_1 x_0)\right] \\
&\quad \times (2\pi)^{N-1}\delta(q_N - q_{N-1})\delta(q_{N-1} - q_{N-2}) \dots \delta(q_2 - q_1) \\
&= \lim_{N\to\infty} B^{-N}\left(\frac{1}{4\pi a}\right)^{N/2} (2\pi)^{N-1} \int \mathrm{d}q_N \exp\left[-\frac{N}{4a}q_N^2 + \mathrm{i}q_N(x_N - x_0)\right]
\end{aligned} \tag{26.30}
$$

Do the last Gaussian integral,

$$
G_T(x, y) = \lim_{N\to\infty} B^{-N}\left(\frac{1}{4\pi a}\right)^{N/2} (2\pi)^{N-1}\sqrt{\frac{4\pi a}{N}}\, \mathrm{e}^{-a(x_N - x_0)^2/N} \tag{26.31}
$$

and finally insert the expressions in (26.27) for the constants a, B, and the definitions $x \equiv x_N$, $y \equiv x_0$. The remarkable fact is that ϵ and N are absorbed into a finite product $T = \epsilon N$, and all other powers of N cancel out:

$$
\begin{aligned}
G_T(x, y) &= \lim_{N\to\infty} \left(\frac{2\pi \mathrm{i}\epsilon\hbar}{m}\right)^{-N/2}\left(\frac{4\pi m}{2\mathrm{i}\epsilon\hbar}\right)^{-N/2}(2\pi)^{N-1}\left(\frac{4\pi m}{2\mathrm{i}N\epsilon\hbar}\right)^{1/2}\exp\left[\mathrm{i}\frac{m}{2N\epsilon\hbar}(x - y)^2\right] \\
&= \left(\frac{m}{2\pi \mathrm{i}T\hbar}\right)\exp\left[\mathrm{i}\frac{m}{2T\hbar}(x - y)^2\right]
\end{aligned} \tag{26.32}
$$

which is in perfect agreement with the expression obtained using equation (26.3).

26.2 Stationary phase and the functional derivative

A function f is a kind of input/output device. Input is a number x, output is another number $f(x)$. A generalization of the concept of function is the ***functional***, which is also an input/output device. Input is a function $f(x)$, output is a number $F[f(x)]$. An example of a functional is the action $S[x(t)]$. Input is trajectory $x(t)$ between times t_1 and t_2. Output is a number, the action of the trajectory

$$
S[x(t)] = \int_{t_1}^{t_2} \mathrm{d}t\left\{\frac{1}{2}m\dot{x}^2 - V(x(t))\right\} \tag{26.33}
$$

An ordinary integral sums the values of a function over all points in a certain range. A *functional integral* sums the values of a functional over all functions in a certain range. The Feynman path integral

$$
\int Dx(t)\mathrm{e}^{\mathrm{i}S[x(t)]/\hbar} \tag{26.34}
$$

is a functional integral.

If there is such a thing as integration of a functional, one expects also a corresponding concept of differentiation, which would express how fast the functional changes when its argument (the function), changes in a certain way. The *functional derivative* is defined, in analogy to ordinary derivatives, in the following way:

$$\frac{\delta F}{\delta f(x)} \equiv \lim_{\epsilon \to 0} \frac{F[f(x') + \epsilon\delta(x - x')] - F[f(x')]}{\epsilon} \tag{26.35}$$

This quantity is essentially a measure of how fast the functional changes when its argument (the function) is changed in the region of point $x' = x$. Most of the usual lore about ordinary derivatives, e.g. the chain rule, holds true for functional derivatives as well. A particular, very important case is when the functional depends only on the value of the input function at one particular point, e.g. $F(f) = f(y)$. In this case

$$\begin{aligned} \frac{\delta}{\delta f(x)} f(y) &= \lim_{\epsilon \to 0} \frac{(f(y) + \epsilon\delta(y - x)) - f(y)}{\epsilon} \\ &= \delta(x - y) \end{aligned} \tag{26.36}$$

This example generalizes in a straightforward way

$$\begin{aligned} \frac{\delta}{\delta f(x)} f^n(y) &= \lim_{\epsilon \to 0} \frac{(f(y) + \epsilon\delta(y - x))^n - f^n(y)}{\epsilon} \\ &= n f^{n-1}(y)\delta(x - y) \\ \frac{\delta}{\delta f(x)} R(f(y)) &= \lim_{\epsilon \to 0} \frac{R(f(y) + \epsilon\delta(y - x)) - R(f(y))}{\epsilon} \\ &= \frac{\partial R(f(y))}{\partial f(y)} \delta(x - y) \end{aligned} \tag{26.37}$$

Functional differentiation is the easiest and quickest way to derive the Euler–Lagrange equations of motion, from the condition

$$\frac{\delta S}{\delta x(t)} = 0 \tag{26.38}$$

that the action should be stationary with respect to an infinitesimal change of path. This works as follows:

$$\begin{aligned} 0 &= \frac{\delta S}{\delta x(t)} \\ &= \frac{\delta}{\delta x(t)} \int \mathrm{d}t' \left\{ \frac{1}{2} m\dot{x}^2 - V(x(t')) \right\} \\ &= \int \mathrm{d}t' \left\{ \frac{1}{2} m \frac{\delta}{\delta x(t)} \dot{x}^2 - \frac{\delta}{\delta x(t)} V(x(t')) \right\} \end{aligned} \tag{26.39}$$

Now, using the definition of the functional derivative and the properties of delta functions, we have

$$
\begin{aligned}
m\frac{\delta}{\delta x(t)}\dot{x}^2 &= \lim_{\epsilon\to 0}\frac{[\partial_{t'}(x(t') + \epsilon\delta(t-t'))]^2 - (\partial_{t'}x(t'))^2}{\epsilon} \\
&= 2\dot{x}(t')\partial_{t'}\delta(t-t') \\
&= -\, 2\partial_{t'}^2 x(t')\delta(t-t')
\end{aligned} \tag{26.40}
$$

Then

$$
\begin{aligned}
0 &= \int \mathrm{d}t'\left\{-m\partial_{t'}^2 x(t')\delta(t-t') - \frac{\partial V(x(t'))}{\partial x(t')}\delta(t-t')\right\} \\
&= -\, m\partial_t^2 x - \frac{\partial V}{\partial x}
\end{aligned} \tag{26.41}
$$

which is, of course, Newton's law of motion $F = ma$.

To obtain a little more practice with functional derivatives, let us find the equations of motion for a wave on a string. Denote the wave function of the string (which in this case just means the displacement of the string at point x and time t) by $\phi(x, t)$. The action of the string is known to be

$$
S[\phi] = \mu\int \mathrm{d}t'\int \mathrm{d}x'\left\{\frac{1}{2}(\partial_{t'}\phi(x', t'))^2 - \frac{v^2}{2}(\partial_{x'}\phi(x', t'))^2\right\} \tag{26.42}
$$

where μ is the string mass per unit length. Then the equation of motion for the string is found from the condition that the action is stationary with respect to small variations in $\phi(x, t)$, i.e.

$$
\begin{aligned}
0 &= \frac{\delta}{\delta\phi(x, t)}S[\phi(x', t')] \\
&= \lim_{\epsilon\to 0}\frac{S[\phi(x', t') + \epsilon\delta(x-x')\delta(t-t')] - S[\phi(x', t')]}{\epsilon} \\
&= \mu\int \mathrm{d}t'\int \mathrm{d}x'\left\{\frac{1}{2}\frac{\delta}{\delta\phi(x, t)}(\partial_{t'}\phi(x', t'))^2 - \frac{v^2}{2}\frac{\delta}{\delta\phi(x, t)}(\partial_{x'}\phi(x', t'))^2\right\} \\
&= \mu\int \mathrm{d}t'\mathrm{d}x'\left\{\partial_{t'}\phi(x', t')\partial_{t'}\frac{\delta\phi(x', t')}{\delta\phi(x, t)} - v^2\partial_{x'}\phi(x', t')\partial_{x'}\frac{\delta\phi(x', t')}{\delta\phi(x, t)}\right\}
\end{aligned} \tag{26.43}
$$

and using

$$
\frac{\delta\phi(x', t')}{\delta\phi(x, t)} = \delta(x-x')\delta(t-t') \tag{26.44}
$$

and the property

$$f(x)\partial_x\delta(x-y) = -[\partial_x f(x)]\delta(x-y) \tag{26.45}$$

we obtain

$$0 = \mu\int dx'dt'\Big\{-\partial_{t'}^2\phi(x',t') + v^2\partial_{x'}^2\phi(x',t')\Big\}\delta(x-x')\delta(t-t') \tag{26.46}$$

which finally leads us to the classical equation of motion, namely, the wave equation

$$\frac{\partial^2\phi}{\partial t^2} - v^2\frac{\partial^2\phi}{\partial x^2} = 0 \tag{26.47}$$

It is now easy to understand the relationship between the Feynman path integral and classical physics. Let us ask the question of which paths, out of all possible paths between two points (x_1, t_1) and (x_2, t_2), give the largest contribution to the path integral

$$\int Dx(t)e^{iS[x(t)]/\hbar} \tag{26.48}$$

Now the contribution of each path to the integral has the same magnitude, since $e^{iS/\hbar}$ is just a complex number of modulus 1. However, if $S \gg \hbar$, then for most paths a small variation $\delta x(t)$ in the path will cause an enormous change in the phase $S/\hbar$ of the integrand. Therefore, the contributions from nearby paths oscillate wildly in phase, and tend to cancel each other out. The exception is for paths in the vicinity of a path where the phase $S/\hbar$ is stationary. For paths in that vicinity, the contributions to the functional integral have nearly the same phase, and hence sum up constructively. But the path where the phase is stationary is the path $x_{cl}(t')$ such that

$$\left(\frac{\delta S}{\delta x(t)}\right)_{x=x_{cl}} = 0 \tag{26.49}$$

As we have seen, this is just the condition that the path $x_{cl}(t)$ is a solution of the classical equations of motion. Therefore, for $S \gg \hbar$, *the path integral is dominated by paths in the immediate vicinity of the classical trajectory* $x_{cl}(t)$. In the limit $\hbar \to 0$, *only* the classical path (and paths infinitesimally close to the classical path) contributes. The 'semiclassical' or WKB approximation to the Feynman path integral is, in fact, the approximation of evaluating the integral in the neighborhood single configuration $x_{cl}(t)$, i.e.

$$\begin{aligned} G_T(x,y) &= \int Dx(t)e^{iS[x(t)]/\hbar} \\ &\approx \text{prefactor} \times e^{iS[x_{cl}(t)]/\hbar} \end{aligned} \tag{26.50}$$

where the prefactor is a slowly varying term which comes from integrating over small variations δx around $x_{cl}(t)$.

The free particle

It is interesting to apply the semiclassical approximation to the very simplest case, namely the free particle. In that case the classical path between points (x_0, t_0) and (x_1, t_1) is a straight line, with

$$x_{cl}(t) = x_0 + \frac{x_1 - x_0}{t_1 - t_0}(t - t_0) \tag{26.51}$$

and corresponding action

$$\begin{aligned} S[x_{cl}(t)] &= \int_{t_0}^{t_1} \mathrm{d}t \, \frac{1}{2} m \dot{x}_{cl}^2 \\ &= \frac{1}{2} m \frac{(x_1 - x_0)^2}{t_1 - t_0} \end{aligned} \tag{26.52}$$

The approximation to the propagator that we obtain from including *only* the classical path is

$$G_{t_1 - t_0}(x_1, x_0) \approx \exp\left[\mathrm{i} \frac{m(x_1 - x_0)^2}{2\hbar(t_1 - t_0)} \right] \tag{26.53}$$

Apart from the prefactor, which is proportional to $1/(t_1 - t_0)$ and which varies very slowly in time compared to the rapidly varying exponential factor, this is precisely the free particle propagator (26.3).

26.3 The Aharonov–Bohm effect

It was noted in chapter 19 that the Hamiltonian for a particle of mass m and charge q, in the presence of an electromagnetic field, is

$$H = \frac{1}{2m}\left(\mathbf{p} - \frac{q}{c}\mathbf{A}\right) \cdot \left(\mathbf{p} - \frac{q}{c}\mathbf{A}\right) + q\phi \tag{26.54}$$

Hamilton's equations are equivalent to $\mathbf{F} = m\mathbf{a}$, supplemented with the Lorentz force law for $\mathbf{F}$, and that force involves only the $\mathbf{E}$ and $\mathbf{B}$ fields. Classically, there is no force on the particle in any region where $\mathbf{E} = \mathbf{B} = 0$. On the other hand, it is possible to have $\mathbf{B} = \nabla \times \mathbf{A} = 0$ in some region, and yet have a non-zero vector potential $\mathbf{A}$ in that region. A simple example is a solenoid. Suppose this solenoid is a long cylinder, oriented along the z-axis, of radius R. Then in polar coordinates z, r, θ the B-field is non-zero inside the solenoid, at $r < R$, and vanishes outside, at $r > R$. Classically, a particle following a trajectory which does not intersect the solenoid cannot possibly be affected by the magnetic field inside the solenoid.

On the other hand, the vector potential $\mathbf{A}$ outside the solenoid is necessarily non-zero. From Stoke's Law, $\mathbf{B} = \nabla \times \mathbf{A}$ implies that, for any integration contour which encircles the solenoid,

$$\begin{aligned} \oint \mathrm{d}x^i A_i &= \int \mathrm{d}\mathbf{S} \cdot \mathbf{B} \\ &= \Phi_B \end{aligned} \tag{26.55}$$

where $\Phi_B = \pi R^2 B$ is the magnetic flux inside the solenoid. We can, for example, take the contour to be a circle of radius $r > R$, and choose, in polar coordinates, A_θ to be constant at fixed r, with all other vector components set to zero.[1] Then

$$A_\theta(r) = \frac{\Phi_B}{2\pi r} \quad (r > R) \tag{26.56}$$

clearly satisfies equation (26.55), and also has the property that $\mathbf{B} = \nabla \times \mathbf{A} = 0$ at $r > R$, where Φ_B is a constant. The question is this: since the Hamiltonian depends on $\mathbf{A}$ rather than $\mathbf{B}$, and quantum dynamics depends on the Hamiltonian operator rather than the derivatives of the Hamiltonian, is it possible that in quantum mechanics the vector potential can affect the dynamics of a charged particle, even if the magnetic field vanishes in the region where the wave function is non-zero?

To investigate this possibility, we will return to the two-slit setup discussed back in chapter 3.

26.3.1 Two-slit interference revisited

Imagine that we have a spherical wavepacket $\psi(\mathbf{x}, t)$ describing the state of an electron at time $t = t_0$, which is concentrated in the neighborhood of point S in front of a double slit barrier, as seen in figure 26.2 (ignore the solenoid for now). Let the energy expectation value be E, with $p = \sqrt{2mE}$, and the packet travels at some group velocity $v_g = p/m$. If L_0 is the distance from S to either slit A or slit B, then the maximum of the packet will arrive at the two slits at time $t_1 = t_0 + L_0/v_g$. We are interested in the amplitude of the wavefunction on the screen at point P, but it is necessary to allow some time for wavepackets to travel, again with group velocity v_g,

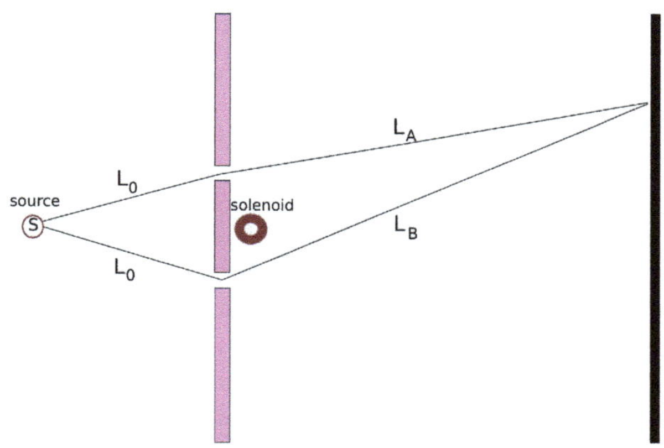

Figure 26.2. Relevant paths for the semiclassical approximation to the path integral for double slit interference, in the presence of a solenoid containing magnetic flux Φ_B.

[1] This is possible because of gauge invariance. One may add to $\mathbf{A}$ the gradient of any function ∇f without affecting the magnetic field. From this it can be shown that in the region outside the solenoid we can choose $A_\theta(z, r, \theta) = A_\theta(r)$ and $A_r = A_z = 0$.

from slit A and slit B to point P. Let L_A, L_B be the distance to point P, from slits A and B respectively, and let

$$\begin{aligned} L_{\rm av} &= \frac{1}{2}(L_A + L_B) \\ \Delta L &= L_B - L_A \end{aligned} \tag{26.57}$$

If wavepackets traveling from slit A and slit B are both traveling with the same group velocity v_g, then the maxima cannot reach point P at exactly the same time. The best we can do is to choose $t_2 = t_1 + L_{\rm av}/v_g$, which is a little while after the maximum of the wavepacket from A has reached P, and a little while before the maximum of the wavepacket from B has reached P.

Let us use these choices of t_1, t_2 to make a rough approximation of the relevant path integral

$$\begin{aligned} \psi(P, t_2) &= \int {\rm d}x G_{t_2-t_1}(P, x)\psi(x, t_0) \\ &= \int Dx(t_0 \le t < t_2)\exp\left[\frac{\rm i}{\hbar}\int_{t_0}^{t_2} {\rm d}t\, L\right]\psi(x, t_0) \end{aligned} \tag{26.58}$$

where $x(t_2) = P$. The first approximation is that we restrict to paths $x(t)$ which pass through either slit A or slit B at time $t_1 = t_0 + L_0/v_g$. Then

$$\begin{aligned} \psi(P, t_2) &\approx G_{t_2-t_1}(P, A)\int {\rm d}x G_{t_1-t_0}(A, x)\psi(x, t_0) \\ &+ G_{t_2-t_1}(P, B)\int {\rm d}x G_{t_1-t_0}(B, x)\psi(x, t_0) \end{aligned} \tag{26.59}$$

Define

$$\begin{aligned} \psi_A &= \int {\rm d}x G_{t_1-t_0}(A, x)\psi(x, t_0) \\ \psi_B &= \int {\rm d}x G_{t_1-t_0}(B, x)\psi(x, t_0) \end{aligned} \tag{26.60}$$

We assume that, from the symmetry of the setup, $\psi_A = \psi_B = \psi_{slit}$. Let $x_A(t)$ be a path from slit A at time t_1 to point P at time t_2, with a similar definition for $x_B(t)$. Then

$$\begin{aligned} \psi(P, t_2) &= (G_{t_2-t_1}(P, A) + G_{t_2-t_1}(P, B))\psi_{\rm slit} \\ &= \left\{\int Dx_A(t_1 < t < t_2)\exp\left[\frac{\rm i}{\hbar}\int_{t_1}^{t_2} {\rm d}t\, L(x_A(t), \dot{x}_A(t)\right]\right. \\ &+ \left.\int Dx_B(t_1 < t < t_2)\exp\left[\frac{\rm i}{\hbar}\int_{t_1}^{t_2} {\rm d}t\, L(x_B(t), \dot{x}_B(t))\right]\right\}\psi_{\rm slit} \end{aligned} \tag{26.61}$$

The next approximation is to replace the integral over all paths from A to P by one classical path, and likewise for paths from B to P (see figure 26.2). The result is

$$\begin{aligned}
\psi(P, t_2) &\approx \left\{\exp\left[\mathrm{i}\frac{mL_A^2}{2\hbar(t_2 - t_1)}\right] + \exp\left[\mathrm{i}\frac{mL_B^2}{2\hbar(t_2 - t_1)}\right]\right\}\psi_{\mathrm{slit}} \\
&\approx \psi_{\mathrm{slit}}\exp\left[\mathrm{i}\frac{mL_{\mathrm{av}}^2}{2\hbar(t_2 - t_1)}\right] \\
&\quad\times\left\{\exp\left[-\mathrm{i}\frac{mL_{\mathrm{av}}}{2\hbar(t_2 - t_1)}\Delta L\right] + \exp\left[\mathrm{i}\frac{mL_{\mathrm{av}}}{2\hbar(t_2 - t_1)}\Delta L\right]\right\} \\
&= 2\psi_{\mathrm{slit}}\exp\left[\mathrm{i}\frac{mL_{\mathrm{av}}^2}{2\hbar(t_2 - t_1)}\right]\cos\left[\frac{1}{2\hbar}\frac{mL_{\mathrm{av}}}{t_2 - t_1}\Delta L\right]
\end{aligned} \tag{26.62}$$

where we have used (26.57) and dropped terms of order $(\Delta L)^2$. Let d be the distance between slits A and B, and θ the angle (assumed to be approximately the same) from each slit to point P, so that $\Delta L = d\sin\theta$. Using also

$$\frac{L_{\mathrm{av}}}{t_2 - t_1} = v_g = \frac{p}{m} \tag{26.63}$$

we have

$$|\psi(P, t_2)|^2 \propto \cos^2\left[\frac{pd\sin\theta}{2\hbar}\right] \tag{26.64}$$

which is exactly the same expression that was deduced back in chapter 3.

26.3.2 The effect of the solenoid

Now we consider how the interference pattern changes when a solenoid is inserted just behind the barrier, as shown in figure 26.2. Things can be arranged so that the wavepackets emerging from slits A and B have almost no overlap with the interior of the solenoid, which is equivalent to the statement that electron paths which pass through the solenoid have almost no effect on the relevant Green's function. Of course the straight-line paths contributing to the semiclassical calculation have no overlap whatever with the solenoid.

Despite the fact that the electron never passes through a region where $\mathbf{B} \neq 0$, the interference pattern is nevertheless shifted by the **A**-field exterior to the solenoid. This is the Aharonov–Bohm effect. To understand how it arises in the path-integral formulation, we need to understand how the **A**-field appears in the action. It is not hard to show that the action which leads to Hamiltonian (26.54) (with scalar potential $\phi = 0$ and charge $q = -e$) is

$$\begin{aligned}
S &= \int \mathrm{d}t\, \frac{1}{2}m\dot{\mathbf{x}}^2 - \frac{e}{c}\int \mathrm{d}t\frac{\mathrm{d}x^i}{\mathrm{d}t}A_i \\
&= \int \mathrm{d}t\, \frac{1}{2}m\dot{\mathbf{x}}^2 - \frac{e}{c}\int_C \mathrm{d}x^i A_i
\end{aligned} \tag{26.65}$$

where the second integration is a line integral along the curve $\mathbf{x}(t)$ denoted by the subscript C. Since the B-field is zero there is no force exerted on the particles away from the slits, and the classical paths from S to P are unchanged. Nevertheless, since $\mathbf{A} \neq 0$ along the upper and lower paths, the classical action *is* changed, but by different amounts. Let

$$
\begin{aligned}
\Delta S_{\rm up} &= -\frac{e}{c}\int_{\rm up} \mathrm{d}x^i A_i \\
\Delta S_{\rm low} &= -\frac{e}{c}\int_{\rm low} \mathrm{d}x^i A_i \\
\Delta S_{\rm av} &= \frac{1}{2}(\Delta S_{\rm low} + \Delta S_{\rm up}) \\
\delta S &= \Delta S_{\rm low} - \Delta S_{\rm up}
\end{aligned}
\tag{26.66}
$$

where up and low refer to line integrals over the upper path and lower path respectively. Taking these contributions along each classical straight-line path into account, the semi-classical path integral estimate becomes

$$
\begin{aligned}
\psi(P, t_2) &\approx \left\{\exp\left[\mathrm{i}\left(\frac{mL_A^2}{2\hbar(t_2-t_1)} + \frac{\Delta S_{\rm up}}{\hbar}\right)\right] + \exp\left[\mathrm{i}\left(\frac{mL_B^2}{2\hbar(t_2-t_1)} + \frac{\Delta S_{\rm low}}{\hbar}\right)\right]\right\}\psi_{\rm slit} \\
&\approx \psi_{\rm slit}\exp\left[\mathrm{i}\frac{mL_{\rm av}^2}{2\hbar(t_2-t_1)}\right]\mathrm{e}^{\mathrm{i}\Delta S_{\rm av}/\hbar} \\
&\quad\times\left\{\exp\left[-\mathrm{i}\frac{mL_{\rm av}}{2\hbar(t_2-t_1)}\Delta L\right]\mathrm{e}^{-\mathrm{i}\frac{1}{2}\delta S/\hbar} + \exp\left[\mathrm{i}\frac{mL_{\rm av}}{2\hbar(t_2-t_1)}\Delta L\right]\mathrm{e}^{\mathrm{i}\frac{1}{2}\delta S/\hbar}\right\} \\
&= 2\psi_{\rm slit}\exp\left[\mathrm{i}\frac{mL_{\rm av}^2}{2\hbar(t_2-t_1)}\right]\mathrm{e}^{\mathrm{i}\Delta S_{\rm av}/\hbar}\cos\left[\frac{1}{2\hbar}\left(\frac{mL_{\rm av}}{t_2-t_1}\Delta L + \delta S\right)\right] \\
&= 2\psi_{\rm slit}\exp\left[\mathrm{i}\frac{mL_{\rm av}^2}{2\hbar(t_2-t_1)}\right]\mathrm{e}^{\mathrm{i}\Delta S_{\rm av}/\hbar}\cos\left[\frac{pd\sin\theta + \delta S}{2\hbar}\right]
\end{aligned}
\tag{26.67}
$$

Now observe that the difference δS can be expressed as the integral over a closed contour, consisting in figure 26.2 of the lower path from S to P, and the upper path P to S, encircling the solenoid. Referring back to (26.55)

$$
\begin{aligned}
\delta S &= -\frac{e}{c}\left(\int_{\rm low}\mathrm{d}x^i A_i - \int_{\rm up}\mathrm{d}x^i A_i\right) \\
&= -\frac{e}{c}\oint \mathrm{d}x^i A_i \\
&= -\frac{e}{c}\Phi_B
\end{aligned}
\tag{26.68}
$$

The probability density at the screen will therefore be proportional to a factor

$$|\psi(P, t_2)|^2 \propto \cos^2\left[\frac{p d \sin\theta - \frac{e}{c}\Phi_B}{2\hbar}\right] \tag{26.69}$$

and we see that the interference pattern has been shifted by an amount proportional to the magnetic flux in the solenoid. This happens despite the fact that the relevant electron paths in the path integral avoid the interior of the solenoid, which is the only region where $\mathbf{B} \neq 0$. The effect has been observed experimentally in various ways (see Tonomura [1]).

26.4 Operators from path integrals

Given a trajectory, $x(t)$, one can always define a momentum, $m\dot{x}(t)$. Then a natural way to define a momentum operator acting on a wave function at time t_f is in terms of the path integral

$$\tilde{p}\psi(x_f, t_f) \equiv \int \mathrm{d}y \int Dx(t)\; m\dot{x}(t_f)\mathrm{e}^{\mathrm{i}S/\hbar}\psi(y, t_0) \tag{26.70}$$

where the paths run from (x_0, t_0) to (x_f, t_f), and $\psi(x, t_0)$ is the wave function at any earlier time $t_0 < t_f$. Let us take $t = t_f$ and $t_0 = t - \epsilon$, and then go to the limit $\epsilon \to 0$. In that case

$$\begin{aligned}
\tilde{p}\psi(x, t) &\equiv \lim_{\epsilon\to 0} \int \mathrm{d}y\; m\frac{(x-y)}{\epsilon} G_\epsilon(x, y)\psi(y, t-\epsilon) \\
&= \lim_{\epsilon\to 0} \int \mathrm{d}y\; m\frac{(x-y)}{\epsilon}\left(\frac{m}{2\pi\mathrm{i}\epsilon\hbar}\right)^{1/2} \exp\left[\mathrm{i}\frac{m}{2\epsilon\hbar}(x-y)^2 - \mathrm{i}\frac{\epsilon}{\hbar}V(x)\right] \\
&\quad \times \psi(y, t-\epsilon) \\
&= \lim_{\epsilon\to 0} \int \mathrm{d}y\; m\frac{(x-y)}{\epsilon}\left(\frac{m}{2\pi\mathrm{i}\epsilon\hbar}\right)^{1/2} \exp\left[\mathrm{i}\frac{m}{2\epsilon\hbar}(x-y)^2\right]\psi(y, t-\epsilon)
\end{aligned} \tag{26.71}$$

where we have dropped the term in the exponent proportional to $V(x)$, since that term is negligible in the $\epsilon \to 0$ limit. Now change variables to $z = y - x$. Then

$$\begin{aligned}
\tilde{p}\psi(x, t) &= \lim_{\epsilon\to 0}\left(\frac{m}{2\pi\mathrm{i}\epsilon\hbar}\right)^{1/2}\frac{m}{\epsilon}\int \mathrm{d}z\;(-z)\exp\left[\mathrm{i}\frac{m}{2\epsilon\hbar}z^2\right]\psi(x+z, t-\epsilon) \\
&= \lim_{\epsilon\to 0}\left(\frac{m}{2\pi\mathrm{i}\epsilon\hbar}\right)^{1/2}\frac{m}{\epsilon}\int \mathrm{d}z\;(-z)\exp\left[\mathrm{i}\frac{m}{2\epsilon\hbar}z^2\right] \\
&\quad \times \left[\psi(x, t-\epsilon) + \psi'(x, t-\epsilon)z + O(z^2)\right] \\
&= -\lim_{\epsilon\to 0}\left(\frac{m}{2\pi\mathrm{i}\epsilon\hbar}\right)^{1/2}\frac{m}{\epsilon}\frac{\mathrm{d}\psi}{\mathrm{d}x}\int \mathrm{d}z\; z^2 \exp\left[\mathrm{i}\frac{m}{2\epsilon\hbar}z^2\right]
\end{aligned} \tag{26.72}$$

The term proportional to z drops out in the integration, because z is an odd function whereas the exponent is even. Also, terms of order z^3 and higher disappear in the $\epsilon \to 0$ limit. Performing the Gaussian integral and taking the $\epsilon \to 0$ limit we finally obtain

$$\begin{aligned} \tilde{p}\psi(x, t) &= -\lim_{\epsilon \to 0} \left(\frac{m}{2\pi i \epsilon \hbar} \right)^{1/2} \frac{m}{\epsilon} \frac{d\psi}{dx} \left(-\frac{\pi}{im/(2\epsilon\hbar)} \right)^{1/2} \frac{i\epsilon\hbar}{m} \\ &= -i\hbar \frac{d\psi}{dx} \end{aligned} \tag{26.73}$$

which is the same rule that we had deduced previously, in chapter 5, from Ehrenfest's principle.

26.5 Path integration as quantization

Having derived the path integral from the Schrödinger equation, one can of course go in the other direction, i.e. derive the Schrödinger equation starting from the concept of an integration over paths. We have seen that path integrals with Gaussian integrands can be evaluated exactly; integrands with non-Gaussian terms can often be evaluated approximately by a perturbation technique. We have also seen that path-integrals lead to the usual form of the momentum operator. Logically, the path-integral approach is an alternative to canonical quantization based on commutators; either method can be used to quantize a classical theory.

Why then, have we spent the last 25 chapters following the Schrödinger equation approach? Why not begin with the path integral, and solve bound-state and scattering problems by that method? In fact, such an approach to teaching non-relativistic quantum mechanics can be and has been followed, by Feynman himself. The results are enshrined in a textbook by Feynman and Hibbs [2]. However, no other elementary textbook, and (to my knowledge) not many instructors, have followed Feynman's example. The reason is simply that, in non-relativistic quantum mechanics, the path integral is a rather cumbersome procedure for solving problems, as compared to the Schrödinger equation.

In *relativistic* quantum field theory, however, the situation is different: for very many problems it is the path-integral technique which is easier to use, and better adapted than operator methods to the thorny technicalities that are encountered. As a bonus, the path-integral formulation is naturally suited to various non-perturbative approximation methods, such as the lattice Monte Carlo procedure, in cases where perturbation theory cannot be applied. Finally, there are interesting and deep connections between quantum field theory, based on the Feynman path integral, and statistical mechanics, based on the analysis of a partition function. But this is a long story, which can be found in textbooks devoted to these subjects.

Problems

1. Feynman, in his original derivation of the free particle propagator via the path integral, carried out the integrals in equation (26.26) by first integrating over x_1, keeping x_2 fixed, then integrating over x_2, keeping x_3 fixed, and so on. Apply this method to compute $G_T(x - y)$ in equation (26.26).
2. Use equation (26.50) to calculate the propagator $G_T(x - y)$ for the harmonic oscillator up to an undetermined prefactor. (Hint: The classical solution between times t_0 and t_1 can be expressed as $A \sin(\omega(t - t_0)) + B \sin(\omega(t_1 - t)$, where A and B are determined by the endpoints.)
3. Use the same method as in section 24.3 to find the operator corresponding to $\tilde{p}^2$. Note that if you just replace

$$m\frac{(x - y)}{\epsilon} \quad \text{by} \quad m^2\frac{(x - y)^2}{\epsilon^2} \tag{26.74}$$

something goes wrong! Can you figure out how to fix it?

References

[1] Peshkin M and Tonomura A 1989 *The Aharonov-Bohm Effect (Lecture Notes in Economics and Mathematical Systems)* (Springer)
[2] Feynman R and Hibbs A 1965 *Quantum Mechanics and Path Integrals* (McGraw-Hill)

IOP Publishing

Jeff Greensite

Chapter 27

A glimpse of quantum field theory

27.1 The quantization of sound

Strike a solid at any point, and you generate a sound wave traveling through the solid. At the microscopic level, a solid is a regular arrangement of atoms, and a sound wave is simply a particular vibration of those atoms. But the motion of atoms should be treated quantum mechanically. What, then, is the quantum mechanics of sound?

As usual, simplify. Our model will be a one-dimensional solid with periodic boundary conditions. This is similar to the Kronig–Penney model, except the atoms are allowed to vibrate. There are N atoms, each of mass m, the coordinate of the nth atom is x_n with index n running from 0 to $N-1$, and its classical equilibrium position is x_{n0}. Periodic boundary conditions mean that $x_{n+N} = x_n$. Denote the displacement of the nth atom from its equilibrium position by

$$q_n = x_n - x_{n0} \tag{27.1}$$

Assuming only nearest-neighbor atoms interact, we can write the potential as a sum of two-body potentials

$$\begin{aligned} V_{\text{total}} &= \sum_{n=0}^{N-1} V(x_{n+1} - x_n) \\ &= \sum_{n=0}^{N-1} V(q_{n+1} - q_n + x_{n+1,0} - x_{n0}) \end{aligned} \tag{27.2}$$

By assumption, the two body potentials are minimized when the distance between atoms is the same as the distance $a = x_{n+1,0} - x_{n0}$ between equilibrium positions. Then we expand the two body potentials in a Taylor series around the minima

$$V = \sum_{n=0}^{N-1} \left\{ V(a) + V'(a)(q_{n+1} - q_n) + \frac{1}{2} V''(a)(q_{n+1} - q_n)^2 + \cdots \right\} \tag{27.3}$$

doi:10.1088/978-0-7503-1167-0ch27

The constant term $V(a)$ is physically irrelevant, and can be set to zero for convenience. At the minimum, $V'(a) = 0$. So for small displacements, dropping terms of order $(q_{n+1} - q_n)^3$ and higher (and keeping in mind the periodicity $q_{N+1} = q_1$),

$$V_{\text{total}} = \sum_{n=0}^{N-1} \frac{1}{2} K (q_{n+1} - q_n)^2 \tag{27.4}$$

where $K = V''(a)$. The Lagrangian (kinetic − potential energy) is

$$L = \sum_{n=0}^{N-1} \left(\frac{1}{2} m \dot{q}_n^2 - \frac{1}{2} K (q_{n+1} - q_n)^2 \right) \tag{27.5}$$

By the standard method outlined in chapter 1 we derive the Hamiltonian

$$H = \sum_{n=0}^{N-1} \left(\frac{p_n^2}{2m} + \frac{1}{2} K (q_{n+1} - q_n)^2 \right) \tag{27.6}$$

and the Dirac quantization rule tells us that the q_n and p_n become operators, with commutation rule

$$\left[\tilde{q}_i, \tilde{p}_j \right] = \mathrm{i}\hbar \delta_{ij} \longrightarrow \tilde{p}_n = -\mathrm{i}\hbar \frac{\partial}{\partial q_n} \tag{27.7}$$

The N-body Schrödinger equation for the solid is therefore

$$\sum_{n=0}^{N-1} \left[-\frac{\hbar^2}{2m} \frac{\partial^2}{\partial q_n^2} + \frac{1}{2} K (q_{n+1} - q_n)^2 \right] \Psi\left[\{q_i\} \right] = E \Psi\left[\{q_i\} \right] \tag{27.8}$$

The system has a ground state, denoted Ψ_0. Quantized sound waves can only correspond to excited states of the system.

The Schrödinger equation above is a partial differential equation in $N \sim 10^{23}$ variables. The only chance of solving it is by the method of separation of variables. Introduce the discrete inverse Fourier transform

$$\begin{aligned} q_n &= \frac{1}{\sqrt{N}} \sum_{k=0}^{N-1} Q_k \exp[2\pi \mathrm{i} nk/N] \\ p_n &= \frac{1}{\sqrt{N}} \sum_{k=0}^{N-1} P_k \exp[2\pi \mathrm{i} nk/N] \end{aligned} \tag{27.9}$$

which automatically incorporates the periodicity $q_{N+n} = q_N$. Using the identity

$$\sum_{n=0}^{N-1} \exp\left[\mathrm{i} \frac{2\pi(k + k')}{N} n \right] = N \delta_{k+k', N} \tag{27.10}$$

we can write the discrete Fourier transform

$$
\begin{aligned}
Q_k &= \frac{1}{\sqrt{N}} \sum_{n=0}^{N-1} q_n \exp[-2\pi i nk/N] \\
P_k &= \frac{1}{\sqrt{N}} \sum_{n=0}^{N-1} p_n \exp[-2\pi i nk/N]
\end{aligned}
\tag{27.11}
$$

By taking the complex conjugate of these inverse transforms, and using the fact that q_n and p_n are real numbers, we see that

$$
Q_k^* = Q_{N-k} \qquad P_k^* = P_{N-k} \tag{27.12}
$$

The wonderful thing about this transformation is that the Hamiltonian, written in terms of Q_k, P_k, is separable, i.e.

$$
\begin{aligned}
\sum_n q_n^2 &= \frac{1}{N} \sum_n \sum_{k_1} \sum_{k_2} Q_{k_1} Q_{k_2} \exp[2\pi i(k_1 + k_2)n/N] \\
&= \sum_k Q_k Q_{N-k} \\
\sum_n p_n^2 &= \frac{1}{N} \sum_n \sum_{k_1} \sum_{k_2} P_{k_1} P_{k_2} \exp[2\pi i(k_1 + k_2)n/N] \\
&= \sum_k P_k P_{N-k} \\
\sum_n q_{n+1} q_n &= \frac{1}{N} \sum_n \sum_{k_1} \sum_{k_2} Q_{k_1} Q_{k_2} \exp[2\pi i(k_1 + k_2)n/N] e^{i2\pi k_1/N} \\
&= \sum_k Q_k Q_{N-k} e^{i2\pi k/N} \\
&= \sum_k Q_k Q_{N-k} \cos[2\pi k/N]
\end{aligned}
\tag{27.13}
$$

Put it all together,

$$
H = \sum_k \left\{ \frac{1}{2m} P_k P_{N-k} + K\left[1 - \cos\left(\frac{2\pi k}{N}\right)\right] Q_k Q_{N-k} \right\} \tag{27.14}
$$

We still need to know the commutation relations between P and Q:

$$
\begin{aligned}
[Q_k, P_{k'}] &= \frac{1}{N} \sum_n \sum_{n'} [q_n, p_{n'}] \exp\left[-i\frac{2\pi}{N}(kn + k'n')\right] \\
&= i\hbar \delta_{k+k', N}
\end{aligned}
\tag{27.15}
$$

Thus,

$$P_k = -\mathrm{i}\hbar \frac{\partial}{\partial Q_{N-k}} \tag{27.16}$$

In this way the Hamiltonian of the one-dimensional solid has been rewritten as a sum of harmonic oscillator Hamiltonians

$$H = \sum_k \left[\frac{1}{2m} P_k P_{N-k} + \frac{1}{2} \sigma_k Q_k Q_{N-k} \right] \tag{27.17}$$

where

$$\frac{1}{2}\sigma_k \equiv K\left[1 - \cos\left(\frac{2\pi k}{N}\right)\right] \tag{27.18}$$

As in the ordinary harmonic-oscillator problem, we solve $H\Psi = E\Psi$ by introducing raising/lowering operators

$$\begin{aligned}
a_k &= \sqrt{\frac{m\omega_k}{2\hbar}}\left[Q_k + \frac{\mathrm{i}}{m\omega_k} P_k\right] \\
a_k^\dagger &= \sqrt{\frac{m\omega_k}{2\hbar}}\left[Q_{N-k} - \frac{\mathrm{i}}{m\omega_k} P_{N-k}\right] \\
Q_k &= \sqrt{\frac{\hbar}{2m\omega_k}}(a_k + a_{N-k}^\dagger) \\
P_k &= \frac{1}{2\mathrm{i}}\sqrt{2\hbar m\omega_k}(a_k - a_{N-k}^\dagger)
\end{aligned} \tag{27.19}$$

where

$$\omega_k = \omega_{N-k} = \sqrt{\frac{\sigma_k}{m}} \tag{27.20}$$

This is for $k > 0$ since $\sigma_0 = 0$. In terms of these operators, one finds that

$$H = \frac{P_0^2}{2m} + \sum_{k=1}^{N-1} \hbar\omega_k\left(a_k^\dagger a_k + \frac{1}{2}\right) \tag{27.21}$$

where

$$[a_k, a_{k'}^\dagger] = \delta_{k,k'} \tag{27.22}$$

The problem has been reduced to solving the dynamics of a set of uncoupled harmonic oscillators; one oscillator for each wavenumber k in the range $1 \le k \le N - 1$. The ground state is an eigenstate of operator P_0 with eigenvalue zero, whose physical interpretation we leave as an exercise (Problem 1).

All lowering operators annihilate the harmonic oscillator ground state, so we require that, for all wavenumbers $k > 0$

$$a_k \Psi_0[Q] = 0 \tag{27.23}$$

or

$$\hbar \frac{\partial}{\partial Q_{N-k}} \Psi_0 = -m\omega_k Q_k \psi_0 \tag{27.24}$$

This equation can be solved by inspection, and we have, for the ground state of the solid,

$$\begin{aligned} \Psi_0 &= N \exp\left[-\frac{m}{2\hbar} \sum_{k>0} \omega_k Q_k Q_{N-k} \right] \\ &= N \exp\left[-\frac{\sqrt{m}}{2\hbar} \sum_{k>0} \left\{ 2K\left(1 - \cos\left(\frac{2\pi k}{N}\right)\right)\right\}^{1/2} Q_k Q_k^* \right] \end{aligned} \tag{27.25}$$

with ground-state (or 'zero-point') energy

$$E_0 = \sum_{k>0} \frac{1}{2} \hbar\omega_k \tag{27.26}$$

In classical physics there are no sound vibrations at all in a solid at zero temperature; each atom is at rest in its equilibrium position $x_n = x_{n0}$. Quantum mechanically, of course, the atoms cannot be in such a position eigenstate, because, due to the uncertainty principle, this would imply an infinite kinetic energy for each atom. Thus, even at zero temperature, the atoms have a finite kinetic/potential energy of vibration, which is why E_0, the vibrational energy at zero temperature, is sometimes called the 'zero-point' energy

To obtain the excited states, it is sufficient to observe that

$$[H, a_k^\dagger] = \hbar\omega_k a_k^\dagger \quad \Longrightarrow \quad H a_k^\dagger = a_k^\dagger (H + \hbar\omega_k) \tag{27.27}$$

Repeated use of this commutator shows us that

$$H \Psi_{n_a n_b \ldots n_p} = E_{n_a n_b \ldots n_p} \Psi_{n_a n_b \ldots n_p} \tag{27.28}$$

where

$$\Psi_{n_a n_b \ldots n_p} = \left(a_{k_a}^\dagger\right)^{n_a} \left(a_{k_b}^\dagger\right)^{n_b} \cdots \left(a_{k_p}^\dagger\right)^{n_p} \Psi_0 \tag{27.29}$$

are the wave functions of the excited energy eigenstates, and

$$E_{n_a n_b \ldots n_p} = n_a \hbar\omega_a + n_b \hbar\omega_b + \cdots + n_p \hbar\omega_p + E_0 \tag{27.30}$$

Interpretation: a glance at (27.9) shows that Q_k is essentially the amplitude of a harmonic wave of wavenumber k, propagating through the solid. Suppose we try to excite such waves in the solid, by striking it at some given frequency. In classical physics the corresponding sound waves could have any energy at all, depending on how vigorously we strike the solid. But quantum mechanically, we have just discovered that the energy of such sound waves is quantized in units of $\hbar\omega_k$; i.e. the energy that can be added to the solid, by exciting vibrations of wavenumber k, can only be one of the possible values

$$\Delta E = n\hbar\omega_k \tag{27.31}$$

Where have we seen this equation before? It is, of course, Planck's condition for the energy in a radiation field, where $\omega_k = 2\pi f$ is the angular frequency of the radiation. Einstein's interpretation of Planck's formula, back in 1905, was that the energy of the radiation field of a given frequency is subdivided in the form of *photons*, each of which propagates through space like a particle of energy hf. By quantizing the oscillations of a solid, we have found the same phenomenon. The energy of sound vibrations of a solid, of a given wavenumber, is carried in the form of particle-like excitations known as ***phonons***, which propagate through the solid and carry energy equal to $\hbar\omega_k$. For this reason, we drop the 'raising/lowering' operator terminology, and refer to $a_k^\dagger$ and a_k as ***creation/destruction*** operators, respectively, because they create/destroy individual phonons in the solid.

Of course, a phonon is not 'really' a particle, in the sense that one could trap such an object, extract it from the solid, and (say) look at it under a microscope. What is 'really' in the solid are the atoms which compose it. Phonons, even though they carry energy, and even though they scatter (like particles) off impurities in the solid and off one another, are merely convenient labels for the quantized vibrational states of the underlying solid.

What to say, then, about photons? Are they, or are they not, elementary particles?

27.2 The quantization of light

Quantum mechanics began with Planck's discovery that the energy of an electromagnetic field of frequency f is always an integer multiple of hf. Then came Einstein's identification of those units as particles, 'photons'. Then came the Compton effect, the Bohr atom, de Broglie waves, the Schrödinger equation,... and then came a lot, some of which we have touched on in this book. Now it is time to return to the beginning. *Why* is the energy of a radiation field quantized in units of hf? What happens if we apply the principles of quantum mechanics, which were developed to describe the motion of electrons and other particles, to the dynamics of the electromagnetic field?

First of all, what are the degrees of freedom of the system we are quantizing? An electron only has a few: the position degrees of freedom (x, y, z), and the spin state. The electromagnetic field, on the other hand, has an electric field **E** and a magnetic field **B** defined at every point in space; the number of degrees of freedom must be infinite! But how many independent degrees of freedom are there, exactly, per point?

A first guess might be that there are six degrees of freedom per point, namely the three components of the electric field, and the three components of the magnetic field. But this cannot be right. First of all, not all of the components of **E** and **B** are independent (think of the no-monopole equation $\nabla \cdot \mathbf{B} = 0$). Secondly, Maxwell's equations are first-order differential equations, which means that they must be Hamiltonian, rather than Euler–Lagrange, equations of motion. This implies that some of the E, B fields are actually canonical 'momenta', rather than canonical 'coordinates'.

To figure out the actual number of generalized 'coordinates' there are per point for the electromagnetic field, let us start with Maxwell's equations in Heaviside–Lorentz units:

$$\begin{aligned} \nabla \cdot \mathbf{E} &= \rho && \text{Gauss's law} \\ \nabla \cdot \mathbf{B} &= 0 && \text{no magnetic monopoles} \\ \frac{1}{c}\frac{\partial \mathbf{B}}{\partial t} + \nabla \times \mathbf{E} &= 0 && \text{Maxwell–Faraday equation} \\ \frac{1}{c}\frac{\partial \mathbf{E}}{\partial t} - \nabla \times \mathbf{B} &= -\frac{1}{c}\mathbf{j} && \text{Ampere–Maxwell equation} \end{aligned} \tag{27.32}$$

The first thing to notice is that we can solve the equations without sources, i.e. the no-monopole equation and the Maxwell–Faraday equation, by simply expressing **E**, **B** in terms of a vector potential **A** and scalar potential ϕ:

$$\begin{aligned} \mathbf{E} &= -\frac{1}{c}\frac{\partial \mathbf{A}}{\partial t} - \nabla\phi \\ \mathbf{B} &= \nabla \times \mathbf{A} \end{aligned} \tag{27.33}$$

Now we just have Gauss's law and the Ampere–Maxwell equation, which are equations for ϕ and **A**. To keep things as simple as possible, we will imagine that there are no charged particles and no currents, $\rho = \mathbf{j} = 0$, and we will go a little further with units and also set $c = 1$. In high-energy physics it is common to set $\hbar = 1$ as well, but we will not go quite that far here. Our remaining Maxwell equations are

$$\begin{aligned} \nabla^2\phi + \frac{\partial}{\partial t}\nabla \cdot \mathbf{A} &= 0 && \text{Gauss's law} \\ -\frac{\partial^2 A}{\partial^2 t} - \frac{\partial}{\partial t}\nabla\phi - \nabla \times \nabla \times \mathbf{A} &= 0 && \text{Ampere–Maxwell equation} \end{aligned} \tag{27.34}$$

Now it looks like we are down to four degrees of freedom. But notice that these equations have a special symmetry known as 'gauge invariance', which stems from the fact that **E**, **B** are unchanged if we transform the vector and scalar potentials in this way:

$$\begin{aligned} \mathbf{A} &\rightarrow \mathbf{A} - \nabla\theta \\ \phi &\rightarrow \phi + \frac{\partial\theta}{\partial t} \end{aligned} \tag{27.35}$$

where $\theta(x, t)$ is any function. We can always find a function such that the transformed **A**-field satisfies the Coulomb gauge condition

$$\nabla \cdot \mathbf{A} = 0 \tag{27.36}$$

But with this condition, Gauss's law becomes $\nabla^2\phi = 0$, and for electromagnetic radiation it is completely consistent to set $\phi = 0$. So with the above choice of Coulomb gauge, and in the absence of sources, Maxwell's equations boil down to one equation for the vector potential

$$\frac{\partial^2 \mathbf{A}}{\partial^2 t} - \nabla^2 \mathbf{A} = 0 \tag{27.37}$$

Now it seems that we have the degrees of freedom down to three per point, but this is still too many. The Coulomb gauge condition actually reduces the number to two degrees of freedom per point. This is related to the fact that an electromagnetic plane wave has only two possible polarizations, which are perpendicular to the direction of propagation. To make the final reduction, we need to introduce the inverse Fourier transform

$$A_i(x) = \int \frac{\mathrm{d}^3k}{(2\pi)^{3/2}}\, A_i(k)\mathrm{e}^{\mathrm{i}\mathbf{k}\cdot\mathbf{x}} \tag{27.38}$$

Usually we divide d^3k by $(2\pi)^3$, but in this case it will be convenient to divide by the fractional power.[1] Then, taking the divergence of both sides of this equation, we see that $\nabla \cdot \mathbf{A}(x) = 0$ is equivalent to

$$\mathbf{k} \cdot \mathbf{A}(k) = 0 \tag{27.39}$$

which says that $\mathbf{A}(k)$ is a vector which is perpendicular to the wavevector $\mathbf{k}$. Let us then introduce, for each wavevector $\mathbf{k}$, two unit vectors $\boldsymbol{\epsilon}(k, \lambda)$, $\lambda = 1, 2$ (known as polarization vectors), which are orthogonal to $\mathbf{k}$ and to each other

$$\begin{aligned} \mathbf{k} \cdot \boldsymbol{\epsilon}(k, \lambda) &= 0 \\ \boldsymbol{\epsilon}(k, \lambda') \cdot \boldsymbol{\epsilon}(k, \lambda) &= \delta_{\lambda\lambda'} \end{aligned} \tag{27.40}$$

Then the Coulomb gauge condition is satisfied by setting

$$\mathbf{A}(k) = \sum_{\lambda=1,2} \boldsymbol{\epsilon}(k, \lambda) A(k, \lambda) \tag{27.41}$$

So for each $\mathbf{k}$ there are just two degrees of freedom, $A(k, 1)$, $A(k, 2)$, related to the Fourier transform $\mathbf{A}(k)$ of the vector potential by

$$A(k, \lambda) = \boldsymbol{\epsilon}(k, \lambda) \cdot \mathbf{A}(k) \tag{27.42}$$

We will not actually need an explicit formula for the polarization vectors, but it is usual to choose them as corresponding to left and right circular polarization, known

[1] Other conventions are possible. This one is chosen to avoid having extra factors in the commutation relations for creation/annihilation operators, equation (27.54).

as 'helicities'. For **k** directed along the positive z-axis, $\epsilon(k, 1) = (1, -\mathrm{i}, 0)$, $\epsilon(k, 1) = (1, \mathrm{i}, 0)$, and we recall that for complex vectors the inner product involves a complex conjugate, i.e. $\mathbf{u} \cdot \mathbf{v} = \sum_j u_j^* v_j$. We also note that because $\mathbf{A}(x)$ is real, and the Fourier transform in this case is

$$\mathbf{A}(k) = \int \frac{\mathrm{d}^3x}{(2\pi)^{3/2}} \, \mathbf{A}(x)\mathrm{e}^{-\mathrm{i}\mathbf{k}\cdot\mathbf{x}} \tag{27.43}$$

it follows, taking the complex conjugate of both sides, that $\mathbf{A}^*(k) = \mathbf{A}(-k)$. This implies that $\mathbf{A}^*(k, \lambda) = \mathbf{A}(-k, \lambda)$ and $\boldsymbol{\epsilon}^*(k, \lambda) = \boldsymbol{\epsilon}(-k, \lambda)$. One last fact we will need about polarization vectors is that

$$\sum_{\lambda=1,2} \epsilon_i^*(k, \lambda)\epsilon_j(k, \lambda) = \delta_{ij} - \frac{k_i k_j}{k^2} \tag{27.44}$$

which follows from (27.40).

In terms of the independent degrees of freedom, the equation of motion becomes

$$\begin{aligned} 0 &= \frac{\partial^2 A_i(x)}{\partial t^2} - \nabla^2 A_i(x) \\ &= \int \frac{\mathrm{d}^3k}{(2\pi)^{3/2}} \mathrm{e}^{-\mathrm{i}\mathbf{k}\cdot\mathbf{x}} \sum_{\lambda=1,2} \epsilon_i(k, \lambda)\left[\partial_t^2 A(k, \lambda) + k^2 A(k, \lambda)\right] \end{aligned} \tag{27.45}$$

or

$$\partial_t^2 A(k, \lambda) = -k^2 A(k, \lambda) \tag{27.46}$$

for every $\mathbf{k}, \lambda$. But this looks exactly like a harmonic oscillator equation $\partial_t^2 x = -\omega^2 x$! It is easy, then, to guess the Hamiltonian which gives this equation of motion:

$$H = \frac{1}{2} \int \mathrm{d}^3k \, \sum_\lambda \left[P(k, \lambda)P(-k, \lambda) + k^2 A(k, \lambda)A(-k, \lambda)\right] \tag{27.47}$$

By analogy to (27.15) it is natural to choose $A(k, \lambda)$, $P(-k, \lambda)$ as conjugate variables, with commutation relations

$$[A(k, \lambda), P(k', \lambda')] = \mathrm{i}\hbar\delta_{\lambda\lambda'}\delta^3(\mathbf{k} + \mathbf{k}') \longrightarrow P(\mathbf{k}, \lambda) = -\mathrm{i}\hbar \frac{\delta}{\delta A(-\mathbf{k}, \lambda)} \tag{27.48}$$

With this choice, the commutator $[A_i(x), P_j(y)]$ involves $\delta^3(\mathbf{x} - \mathbf{y})$ rather than $\delta^3(\mathbf{x} + \mathbf{y})$ (see problem 3). Classically, the Hamiltonian equations of motion are then

$$\begin{aligned} \partial_t A(k, \lambda) &= \frac{\delta H}{\delta P(-k, \lambda)} = P(k, \lambda) \\ \partial_t P(k, \lambda) &= -\frac{\delta H}{\delta A(-k, \lambda)} = -k^2 A(k, \lambda) \end{aligned} \tag{27.49}$$

and the second-order equation of motion (27.46) follows immediately. Now that we know that $P(k, \lambda) = \partial_t A(k, \lambda)$ (which implies that the transverse **E** field has the role of canonical 'momentum' variables in the Hamiltonian formulation, conjugate to the transverse A-field), we can express H in terms of the original **E**, **B** fields:

$$\begin{aligned}
H &= \frac{1}{2}\int \mathrm{d}^3k \sum_\lambda \big[\partial_t A(k,\lambda)\partial_t A(-k,\lambda) + k^2 A(k,\lambda)A(-k,\lambda)\big] \\
&= \frac{1}{2}\int \mathrm{d}^3k \sum_\lambda \epsilon_i(k,\lambda)\epsilon_j^*(k,\lambda)\big[\partial_t A_i(k)\partial_t A_j(-k) + k^2 A_i(k)A_j(-k)\big] \\
&= \frac{1}{2}\int \mathrm{d}^3k \left(\delta_{ij} - \frac{k_i k_j}{k^2}\right)\big[\partial_t A_i(k)\partial_t A_j(-k) + k^2 A_i(k)A_j(-k)\big] \\
&= \frac{1}{2}\int \mathrm{d}^3k \big[\partial_t A_i(k)\partial_t A_i(-k) + k^2 A_i(k)A_i(-k)\big] \\
&= \frac{1}{2}\int \mathrm{d}^3k \int \frac{\mathrm{d}^3x}{(2\pi)^{3/2}} \int \frac{\mathrm{d}^3y}{(2\pi)^{3/2}}\Big[\partial_t \mathbf{A}(x)\cdot\partial_t\mathbf{A}(y) \\
&\quad + (\nabla\times\mathbf{A}_i(x))\cdot(\nabla\times\mathbf{A}_i(y))\Big]\mathrm{e}^{\mathrm{i}\mathbf{k}\cdot(\mathbf{x}-\mathbf{y})} \\
&= \frac{1}{2}\int \mathrm{d}^3x[\mathbf{E}^2 + \mathbf{B}^2]
\end{aligned} \tag{27.50}$$

The last line is just the usual expression for the energy of an electromagnetic field, in Heaviside–Lorentz units with $c = 1$, in the absence of charge and current sources.

With the operator identification (27.48) the Hamiltonian operator $\tilde{H}$ is

$$\tilde{H} = \frac{1}{2}\int \mathrm{d}^3k \sum_{\lambda=1,2}\left(-\hbar^2\frac{\delta^2}{\delta A(k,\lambda)\delta A(-k,\lambda)} + k^2 A(k,\lambda)A(-k,\lambda)\right) \tag{27.51}$$

Since H is just a collection of independent harmonic oscillators, we follow the previous treatment of sound waves and introduce raising and lowering (or 'creation and annihilation') operators

$$\begin{aligned}
a(k,\lambda) &= \frac{1}{\sqrt{2\hbar}}\left(\sqrt{|\mathbf{k}|}\,A(k,\lambda) + \frac{\mathrm{i}}{\sqrt{|\mathbf{k}|}}P(k,\lambda)\right) \\
a^\dagger(k,\lambda) &= \frac{1}{\sqrt{2\hbar}}\left(\sqrt{|\mathbf{k}|}\,A(-k,\lambda) - \frac{\mathrm{i}}{\sqrt{|\mathbf{k}|}}P(-k,\lambda)\right)
\end{aligned} \tag{27.52}$$

so that

$$\mathbf{A}(x) = \int \frac{\mathrm{d}^3k}{(2\pi)^{3/2}}\sqrt{\frac{\hbar}{2|\mathbf{k}|}}\sum_{\lambda=1}^{2}\left(\boldsymbol{\epsilon}(k,\lambda)a(k,\lambda)\mathrm{e}^{\mathrm{i}\mathbf{k}\cdot\mathbf{x}} + \boldsymbol{\epsilon}^*(k,\lambda)a^\dagger(k,\lambda)\mathrm{e}^{-\mathrm{i}\mathbf{k}\cdot\mathbf{x}}\right) \tag{27.53}$$

It is straightforward to verify that

1. The creation/annihilation operators have the standard raising/lowering commutation relations

$$[a(k, \lambda), a^\dagger(k', \lambda')] = \delta_{\lambda\lambda'}\delta^3(k - k') \tag{27.54}$$

2. The Hamiltonian has the form

$$\begin{aligned} H &= \frac{1}{2}\int d^3k \;\; \hbar|\mathbf{k}|\sum_\lambda \left[a^\dagger(k, \lambda)a(k, \lambda) + a(k, \lambda)a^\dagger(k, \lambda)\right] \\ &= \int d^3k \;\; \hbar|\mathbf{k}|\sum_\lambda \left[a^\dagger(k, \lambda)a(k, \lambda) + \frac{1}{2}\delta^3(0)\right] \end{aligned} \tag{27.55}$$

3. The creation operator and the Hamiltonian satisfy the commutation relation

$$[H, a^\dagger(k, \lambda)] = \hbar|\mathbf{k}|a^\dagger(k, \lambda) \tag{27.56}$$

4. The destruction operators all annihilate the ground state

$$a(k, \lambda)\Psi_0 = 0 \qquad \text{for all } k, \lambda \tag{27.57}$$

from which it follows that the ground state has the form

$$\begin{aligned} \Psi_0[A] &= \exp\left[-\frac{1}{2\hbar}\int d^3k \sum_{\lambda=1,2} |\mathbf{k}|A(k, \lambda)A(-k, \lambda)\right] \\ &= \exp\left[-\frac{1}{2\hbar}\int d^3k \frac{(\mathbf{k}\times\mathbf{A}(k))\cdot(\mathbf{k}\times\mathbf{A}_i(-k))}{|\mathbf{k}|}\right] \\ &= \exp\left[-\frac{1}{4\pi\hbar}\int d^3x d^3y \frac{\mathbf{B}(x)\cdot\mathbf{B}(y)}{|\mathbf{x}-\mathbf{y}|^2}\right] \end{aligned} \tag{27.58}$$

The ground state $\Psi_0[A]$ is also called the 'vacuum' state because, as we are about to see, it represents a state of no photons. But the ground state is not a vacuum in the sense of 'nothing'. In fact there is always a probability to find non-zero electric and magnetic fields in the vacuum, far away from any charge or current sources, and in the absence of any propagating electromagnetic radiation. This is completely analogous to the case of an ordinary harmonic oscillator. Classically, the lowest energy state of the harmonic oscillator is to have the mass sitting at the minimum of the potential, say at $x = 0$. In the same way, the lowest energy state of the electromagnetic field is when there is no electromagnetic field at all, i.e. $\mathbf{A}(x) = \mathbf{E}(x) = \mathbf{B}(x) = 0$. But in quantum mechanics, there is always some finite probability to find the mass at some $x \neq 0$ away from the minimum of the potential. Likewise for the electromagnetic field. Even away from all charges, sources, and propagating electromagnetic waves, there is always a finite probability in the vacuum state to detect a non-zero electromagnetic field.

Making use of items (2) and (3) above we can immediately find all of the excited energy eigenstates, which are obtained by simply operating on the ground state Ψ_0 with the raising (or 'creation') operators

$$\Psi_{(n_a,\,\lambda_a),\,(n_b,\,\lambda_b)\ldots(n_p,\,\lambda_p)}[A] = (a^\dagger(k_a,\,\lambda))^{n_a}(a^\dagger(k_b,\,\lambda))^{n_b}\ldots(a^\dagger(k_p,\,\lambda))^{n_p}\Psi_0[A] \tag{27.59}$$

which have the corresponding energy eigenvalues (restoring, for clarity, the factor c of the speed of light)

$$E_{n_a n_b \ldots n_p} = n_a\hbar c|k_a| + n_b\hbar c|k_b| + \cdots + n_p\hbar c|k_p| + E_0 \tag{27.60}$$

where

$$E_0 = 2\hbar\delta^3(0)\int \mathrm{d}^3k\ \frac{1}{2}|\mathbf{k}|c \tag{27.61}$$

is the (infinite) ground state energy. The factor of 2 in (27.61) comes from summing over two polarizations λ. The ground state energy itself is not directly measurable[2], but what we can measure is the energy of a state, call it ΔE, above the ground state energy

$$\begin{aligned}\Delta E &= n_a\hbar c|k_a| + n_b\hbar c|k_b| + \cdots + n_p\hbar c|k_p| \\ &= n_a hf_a + n_b hf_b + \cdots + n_p hf_p\end{aligned} \tag{27.62}$$

where f is the frequency of radiation with a wavevector of modulus $|k|$.

In classical physics, any amount of energy can be stored in an electromagnetic field of a given frequency f. The energy of the field is proportional to the square of the amplitude, and the amplitude is proportional to the strength of the source. What we learn from quantizing the electromagnetic field is that, as in an ordinary harmonic oscillator, the energy (above the ground state) that can be stored at any given frequency comes in multiples of hf, i.e.

$$\Delta E_f = nhf \tag{27.63}$$

This rule, that the energy of an electromagnetic field at each frequency is quantized, brings us full circle: it explains black-body radiation, it explains the photoelectric effect, it gives us the formulae of Planck and Einstein. We learn that the energy of electromagnetic radiation at a given frequency comes in integer units, and each of these units can be interpreted as a particle. The photon. But we have now learned quite a bit more: we have learned what a photon *is*. It is an excited vibrational quantum state of the underlying electromagnetic field, much as phonons in a solid

[2] Unless we consider gravitational effects. These are related to an issue known as the 'cosmological constant problem'.

are excited quantum vibrational states of the underlying atoms. The creation operator $a^\dagger(k, \lambda)$ creates a photon of energy hf with (as shown in more advanced texts) momentum $\hbar\mathbf{k}$ and spin 1. The helicity λ characterizes the direction of the spin angular momentum relative to the direction of photon propagation. A massive spin 1 particle could have three values of angular momentum along the $\mathbf{k}$-direction, but a massless particle (for reasons connected to special relativity) can only have two values ± 1 (essentially the helicities) of spin angular momentum in the $\mathbf{k}$-direction.

There is still more: it is believed that *all* of the elementary particles are essentially excited states of quantized fields. Combine quantum mechanics with classical field theory, and particles—or, at least, excited states which carry energy and momentum and seem to behave in every way like point particles—are the inevitable result. In particular, combine quantum theory with Maxwell's equations, and out come photons. The beauty and the staggering implications of this result can hardly be overstated. It is the reason that modern elementary particle theory is essentially the theory of quantized fields.

Problems

1. While the potential V_{total} is minimized when the atoms are the same distance a apart, this does not determine the equilibrium positions x_{n0} even at the classical level. How does this fact appear in equation (27.14)? What is the physical interpretation of the momentum P_0 of the $k = 0$ mode?
2. Verify equations (27.53)–(27.58).
3. Canonical quantization of the electromagnetic field is complicated by the fact that, in Coulomb gauge and in the absence of sources, not all of the degrees of freedom are independent ($\nabla \cdot \mathbf{A} = \nabla \cdot \mathbf{E} = 0$). Defining

$$P_i(x) = \int \mathrm{d}^3k \sum_{\lambda=1}^{2} P(k, \lambda)\epsilon_i(k, \lambda)\mathrm{e}^{\mathrm{i}\mathbf{k}\cdot\mathbf{x}} \tag{27.64}$$

compute the commutator

$$\left[A_i(x), P_j(y)\right] \tag{27.65}$$

and show that it satisfies

$$\sum_{i=1}^{3} \frac{\partial_i}{\partial x_i}\left[A_i(x), P_j(y)\right] = \sum_{j=1}^{3} \frac{\partial_i}{\partial y_j}\left[A_i(x), P_j(y)\right] = 0 \tag{27.66}$$

4. Find the wave functional corresponding to a one photon state with wave vector $\mathbf{k}$ and polarization λ.

5. In chapter 9 it was shown how to compute harmonic oscillator expectation values such as $\langle\varphi_0|x^2|\varphi_0\rangle$ by operator methods. Use operator methods to compute the equal times correlation function

$$\left\langle \Psi_0 \middle| A_i(x) A_j(y) \middle| \Psi_0 \right\rangle$$

Your answer may be left in the form of a Fourier transform.

www.ingramcontent.com/pod-product-compliance
Ingram Content Group UK Ltd.
Pitfield, Milton Keynes, MK11 3LW, UK
UKHW051910100926
8556IPUK00005B/834

9 780750 3116